SOLUTIONS MANUAL
TO ACCOMPANY

PHYSICAL CHEMISTRY FOR THE LIFE SCIENCES

Second Edition

C. A. TRAPP
Professor of Chemistry
University of Louisville, Louisville, Kentucky, USA

M. P. CADY
Professor of Chemistry
Indiana University Southeast, New Albany, Indiana, USA

OXFORD
UNIVERSITY PRESS

OXFORD
UNIVERSITY PRESS
Great Clarendon Street, Oxford OX2 6DP
Oxford University Press is a department of the University of Oxford.
It furthers the University's objective of excellence in research, scholarship, and education by publishing worldwide in
Oxford New York
Auckland Cape Town Dar es Salaam Hong Kong Karachi
Kuala Lumpur Madrid Melbourne Mexico City Nairobi
New Delhi Shanghai Taipei Toronto
With offices in
Argentina Austria Brazil Chile Czech Republic France Greece
Guatemala Hungary Italy Japan Poland Portugal Singapore
South Korea Switzerland Thailand Turkey Ukraine Vietnam
Oxford is a registered trade mark of Oxford University Press
in the UK and in certain other countries
Published in the United States
by W. H. Freeman and Company

British Library Cataloguing in Publication Data
Data available
Library of Congress Cataloging in Publication Data
Data available
Typeset by Graphicraft Limited, Hong Kong
Printed in Great Britain on acid-free paper
by Ashford Colour Press Ltd, Gosport, Hampshire
ISBN 978–0–19–960032–8
11

Preface

This manual provides detailed solutions to all of the Discussion Questions, Exercises and Projects in the second edition of *Physical Chemistry for the Life Sciences* by Peter Atkins and Julio de Paula. We hope that these complete solutions will help in deepening your understanding of physical chemistry as it is applied to the Life Sciences.

The solutions to the Exercises in this edition rely somewhat more heavily on the mathematical and molecular modeling software that is now generally accessible to chemistry students, and this is particularly true for some of the Projects, which specifically request the use of such software for their solutions. But almost all of the Exercises can still be solved with a modern hand-held scientific calculator.

In general, we have adhered rigorously to the rules for significant figures in displaying the final answers. However, when intermediate answers are shown, they are often given with one more figure than would be justified by the data. These excess digits are usually indicated with an overline.

The solutions in this manual have been carefully cross-checked for errors not only by ourselves, but also very thoroughly by Valerie Walters, who made many helpful suggestions for improvement. We expect that most errors have been eliminated, but would be grateful to any readers who bring any remaining ones to our attention.

We warmly thank our publishers, especially Jonathan Crowe and Jessica Fiorilla, as well as David Quinn, for their patience in guiding this complex, detailed project to completion. We also thank Peter Atkins and Julio de Paula for the opportunity to participate in the development of this outstanding Physical Chemistry text.

C. A. T
M. P. C.

Contents

Fundamentals

Discussion questions

F.1 The loss of electrons by atoms results in **cations** (such as Na^+, Ca^{2+}, and Fe^{3+}), while the gain of electrons results in **anions** (such as H^-, Cl^-, and O^{2-}). The **ionic bond** arises from the attractive electrostatic force between ions of opposite electrical charge, a force that generally results in a solid, hard, brittle lattice of closely packed ions known as an **ionic compound** and, commonly, as a salt. Examples include sodium hydride (NaH), sodium chloride (NaCl), and calcium oxide (CaO). A **polyatomic ion** is a small group of covalently bonded atoms that has a net non-zero electrical charge. Examples include the ammonium cation (NH_4^+), the carbonate anion (CO_3^{2-}), and the phosphate anion (PO_4^{3-}) that appear, respectively, in ammonium chloride (NH_4Cl), sodium carbonate (Na_2CO_3), and sodium phosphate (Na_3PO_4). The lattice structure of sodium chloride is shown in Figure F.1; each ion is surrounded by six counter-ion nearest neighbors so six is the **coordination number** of each ion.

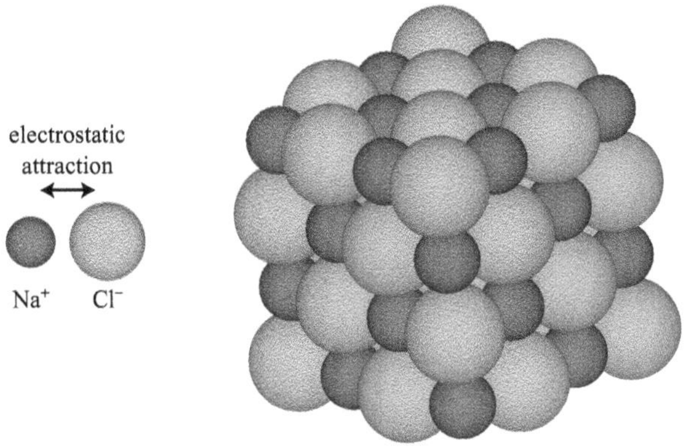

Figure F.1

A **molecule** is a distinctive group of atoms that has a net electrical charge of zero (i.e. each molecule is electrically **neutral**) while being held together by **covalent bonding** where a covalent bond is the sharing of an electron pair between adjacent atoms. Molecules of water (H_2O), methane (CH_4), and ethanol (CH_3CH_2OH, the alcohol molecule of beer and wine) are shown in Figure F.2 as space-filling models and in Figure F.3 as ball-and-stick models, and stick models. All are shown without the explicit display of **lone pairs**, pairs of valence electrons that are not shared and not directly involved in bonding.

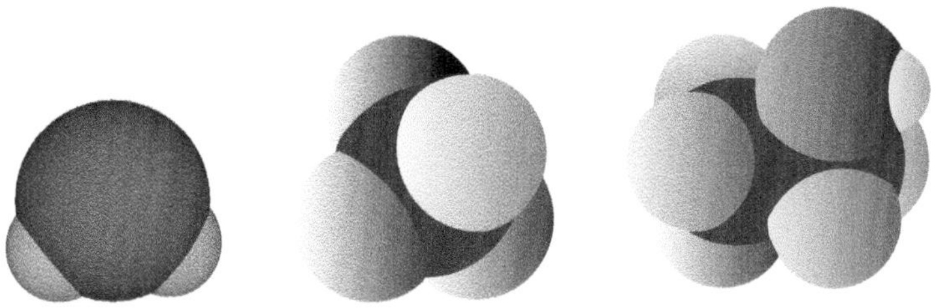

Figure F.2

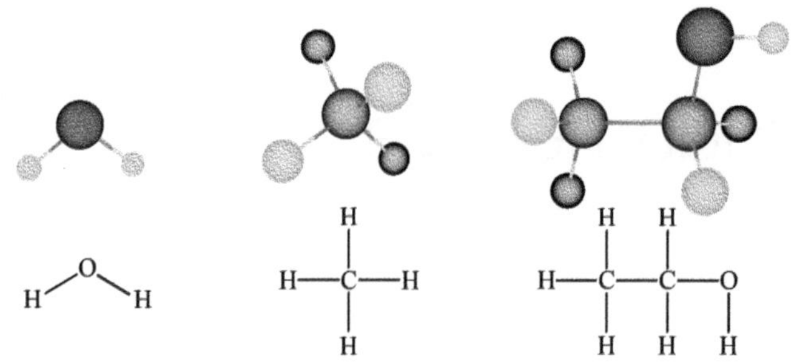

water, H_2O methane, CH_4 ethanol, CH_3CH_2OH

Figure F.3

Space-filling models are the most realistic way to view the three-dimensional shape of molecules. Stick molecular models do not necessarily show the actual three-dimensional molecular shape of molecules. However, stick models are easy to draw and they do show the connectivity of atomic bonding. In these models, a single stick between two atoms represents a shared pair of valence electrons (a single covalent bond) between the adjacent atoms. Two sticks between atoms represent the sharing of two electron pairs (a double bond) between the atoms. Three sticks between atoms represent the sharing of three electron pairs (a triple bond) between atoms. Biologically important molecules can be quite large. Figure F.4 displays stick and space-filling models of enflurane (a general anesthetic, upper molecule) and caffeine (a mild stimulant, lower molecule). In these molecules, hydrogen is observed to satisfy the **duet rule** as each hydrogen is surrounded by a total of 2 valence electrons for stability. Carbon, nitrogen, oxygen, and halogen atoms (F and Cl) satisfy the **octet rule** (i.e. each atom is surrounded by 8 valence electrons) with carbon having no lone pairs, while each nitrogen has 1 lone pair, each oxygen has 2 lone pairs, and each halogen has 3 lone pairs.

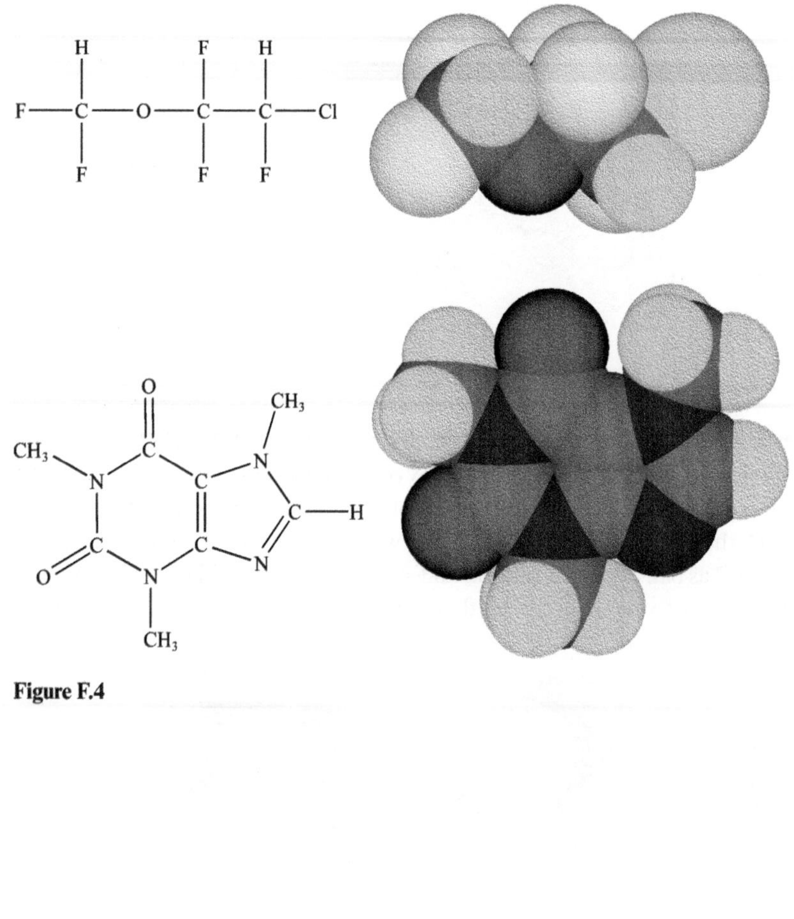

Figure F.4

Nitrogen, oxygen, and fluorine atoms have a greater ability to attract electrons than does the hydrogen atom. This ability is quantified as **electronegativity** and the atoms N, O, and F have successively larger electronegativities with fluorine having the highest electronegativity of all atoms. When these highly electronegative atoms are covalently bound to a hydrogen atom, the cloud of the shared electron pair is distorted toward the highly electronegative atom, thereby, giving them a partial (δ) negative charge and leaving hydrogen with a partial positive charge. These **polar bonds,** shown in Figure F.5 along with lone pairs, are characterized by significant separation of charge and at least one lone pair on the highly electronegative atom. These characteristics are the observed requirements for hydrogen bond formation between molecules.

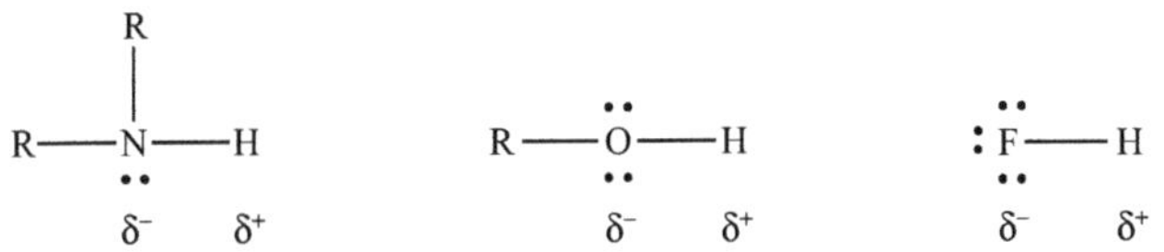

Figure F.5

A typical **hydrogen bond** is shown as a dashed line between the alcohol groups in Figure F.6. The polar bond of the alcohol group of one molecule aligns with the lone pair of the alcohol group of a second molecule. The hydrogen bond strength is about 10% of the covalent bond strength and consists of about 10% covalent character and 90% ionic character, which arises from the δ^+–δ^- attraction. The hydrogen bond is longer than a covalent bond. For example the O–H bond length of water is 101 pm, while the hydrogen bond length is about 175 pm.

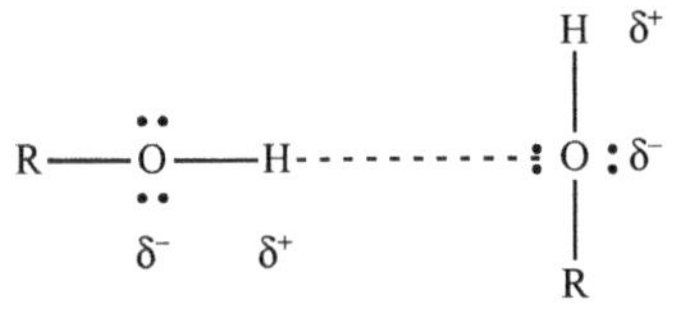

Figure F.6

van der Waals interactions are attractions between groups of atoms in different regions of a macromolecule or between different molecules. These interactions are weaker than hydrogen bonds and depend upon the distance r between the groups as $1/r^6$. They may be dipole–dipole interactions, dipole–induced-dipole interactions, or dispersion interactions between non-polar molecules or non-polar sections of a macromolecule. The **dispersion interaction** is the temporary, spontaneous polarity arising from an asymmetrical distribution of electrons.

F.2 α-Amino acids are the fundamental building blocks of protein. They are illustrated in the *Atlas of structures* in the *Resource section* of the textbook. The sequence of amino acids in a particular protein may be displayed as linked three-letter abbreviations of amino acid residues such as –Phe–Ala–Ser–His–. Each linkage is an amide bond between the α-carboxylic acid group of the left residue and the α-amine group of the right residue. The linkage is called a **peptide bond** and all proteins are polypeptides. Figure F.7 shows the formation of the Phe–Ala dipeptide as the net removal of water between carboxylic acid and amine groups.

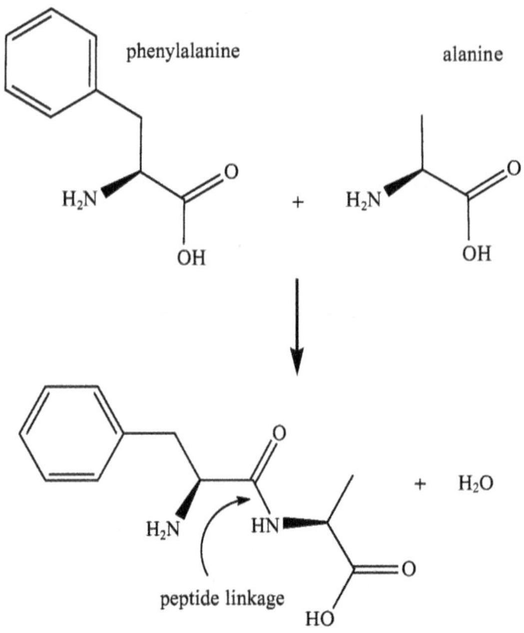

Figure F.7

The monomeric unit of a polynucleotide consists of a phosphate group and a nitrogen base attached, respectively, to the 5′ and 1′ carbons of a five-carbon sugar (ribose in RNA and 2′-deoxyribose in DNA). The nitrogen bases are adenine (A), guanine (G), cytosine (C), and uracil (U) in RNA; thymine (T) replaces uracil in DNA. The sequence of nucleotides in a particular polynucleotide is specified by a list of nucleotide bases such as ···ACU···, where each nucleotide is linked via its phosphate group to the 3′ carbon of the sugar within the nucleotide to its left, a 3′→5′ phosphodiester link. These characteristics are shown in the deoxyribose AC dinucleotide of Figure F.8.

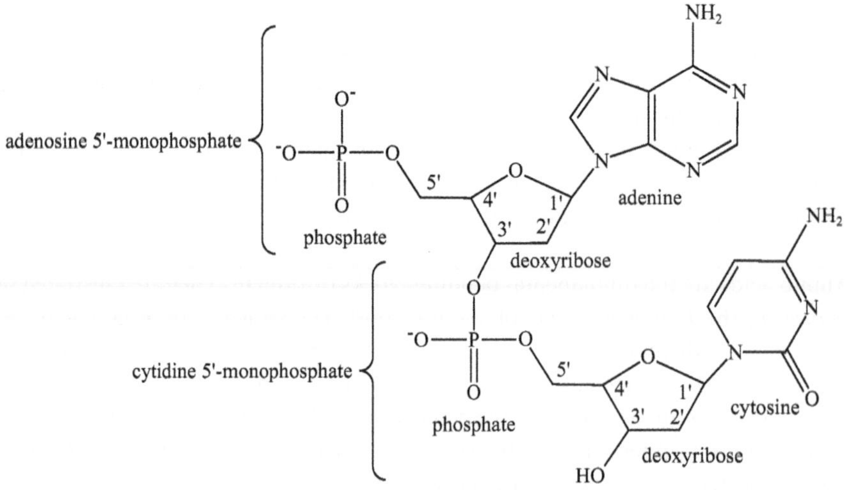

Figure F.8

The monomer of a polysaccharide is often a single simple sugar. In the storage macromolecules starch and glycogen (animal starch) the monomer is glucose (glu) alone. If two or more monosaccharides appear in the primary structure, the macromolecule is called a heteropolysaccharide. Amylopectin, a constituent of starch, exhibits $\alpha(1\rightarrow4)$ and $\alpha(1\rightarrow6)$ glycoside links as illustrated in Figure F.9 (most hydroxyl groups are omitted for simplicity). $\beta(1\rightarrow4)$ Glycosidic links join the glucose units of cellulose, the structural polysaccharide of plant cell walls, as shown in Figure F.10.

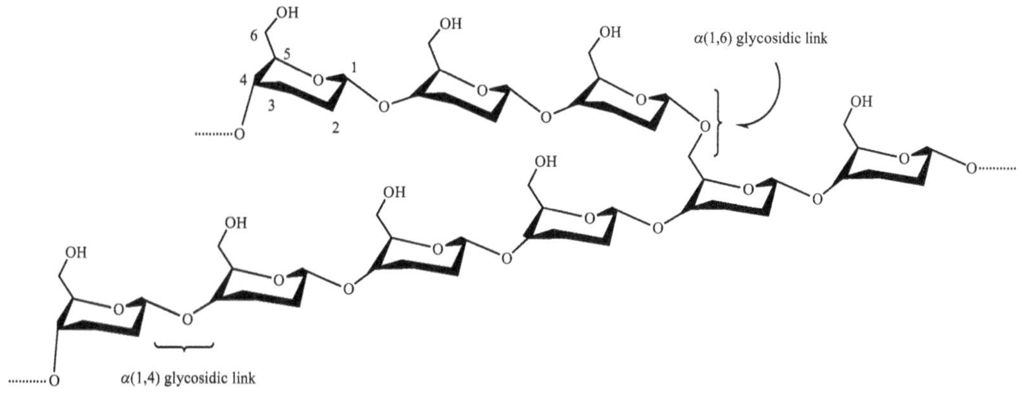

Figure F.9

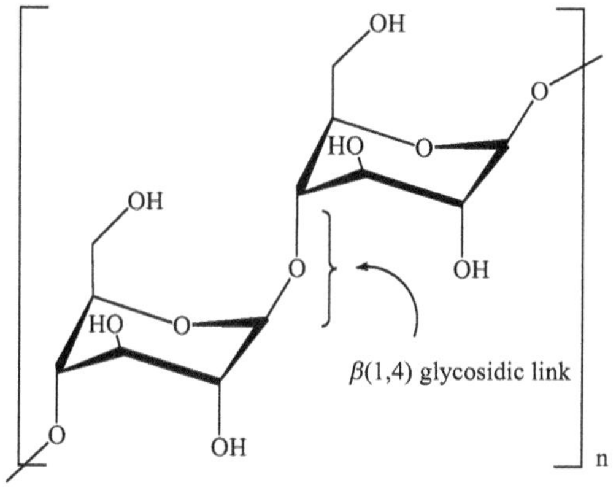

Figure F.10

F.3 Essential structural features of a lipid bilayer are illustrated in Figure F.11, which is a modification of text Figure F.1. The hydrophilic heads of lipid molecules line both sides of the bilayer where they are exposed to aqueous environments. Both hydrogen bonding and van der Waals interactions with water molecules stabilize the position of the heads. Hydrophobic, non-polar hydrocarbon lipid tails, which typically consist of 14–24 carbon atoms, form the interior of the bilayer because of the attraction caused by the dispersion interaction, a feature that excludes water molecules from the bilayer interior. The dispersion interaction is not strong enough, nor are the tails rigid enough, to hold each lipid molecule firm in a permanent lattice site so the lipid bilayer is more fluid with each molecule exhibiting a mobility of about 1 micrometer (a cell diameter) per minute.

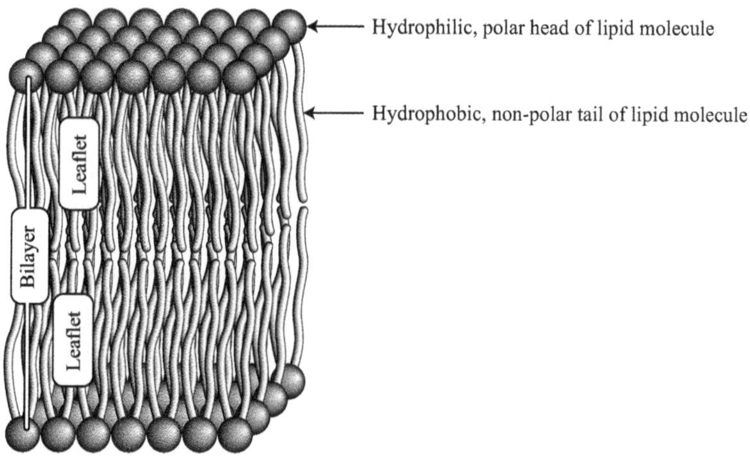

Figure F.11

F.4 See text Section F.1(c) along with text Figures F.2a–d.

F.5 The solid phase of matter has a shape that is independent of the container it occupies. It has a density compatible with the close proximity of either its constituent elemental atoms or its constituent molecules and, consequently, it has low compressibility. Constituent atoms, or molecules, are held firmly at specific lattice sites by relatively strong, net forces of attraction between neighboring constituents. Solids may be characterized by terms such as brittle, ductile, tensile strength, toughness, and hardness.

A liquid adopts the shape of the part of the container that it occupies; it can flow under the influence of gravitational attraction to occupy any shape. Like a solid, it has a density caused by the close proximity of either its constituent elemental atoms or its constituent molecules and it has low compressibility. Liquids can flow because the constituent atoms or molecules have enough average kinetic energy to overcome the attractive forces between neighboring constituents, thereby, making it possible for them to slip past immediate neighbors. This causes constituents to be placed randomly in contrast to the orderly array in crystals. Liquids are characterized by terms such as surface tension, viscosity, and capillary action. Liquids within a vertical, narrow tube exhibit a meniscus that is either concave-up or concave-down, depending upon the nature of the attractive or repulsive forces between the liquid and the material of the tube.

Gases have no fixed shape or volume. They expand to fill the container volume. The constituent molecules move freely and randomly. A perfect gas has a total molecular volume that is negligibly small compared to the container volume and, because of the relatively large average distance between molecules, intermolecular attractive forces are negligibly small. Gases are compressible.

F.6 In the classical physics of Newton the **force** $\boldsymbol{F}$ acting on a body so as to cause acceleration $\boldsymbol{a}$ equals the mass m of the object multiplied by its acceleration: $\boldsymbol{F} = m\boldsymbol{a}$ where the bold faced symbols represent vector quantities. Coulomb's law describes the particularly important electrostatic force between two point charges Q_1 and Q_2 separated by the distance r:

$$F = \frac{Q_1 Q_2}{4\pi\varepsilon_0 r^2} \text{ in a vacuum } (\varepsilon_0 \text{ is the vacuum permittivity})$$

and

$$F = \frac{Q_1 Q_2}{4\pi\varepsilon_r\varepsilon_0 r^2} \text{ in a medium that has the relative permittivity } \varepsilon_r \text{ (formerly, dielectric constant)}$$

Convention assigns a negative value to the Coulomb force when it is attractive and a positive value when it is repulsive. The SI unit of force is the newton (N) and $1\ \text{N} = 1\ \text{kg m s}^{-2}$.

The amount of **work**, w, done on a body when it experiences a displacement, d, is given by: $w = -Fd$ when F is a constant opposing force. The SI unit of work is the joule (J) and $1\ \text{J} = 1\ \text{N m} = 1\ \text{Pa m}^3 = 1\ \text{kg m}^2\ \text{s}^{-2}$.

Energy is the capacity to do work. It is a property and the SI unit of energy, like the unit of work, is the joule. The **law of conservation of energy** states that the total energy E of an isolated system is conserved; that is, the total energy of an isolated system is a constant. Even in an isolated system, however, energy can be transferred from one location to another and transformed from one form to another. The transfers and transformations involve heat, work, gravitational potential energy, Coulomb potential energy, and electromagnetic radiation.

All objects in motion have the ability to do work during the process of slowing. That is, they have energy, or, more precisely, the energy possessed by a body because of its motion is its **kinetic energy**, E_k. The law of conservation of energy tells us that the kinetic energy of an object equals the work done on the object in order to change its motion from an initial (i) state of $v_i = 0$ to a final (f) state of $v_f = v$. For an object of mass m travelling at a speed v,

$$E_k = \tfrac{1}{2}mv^2 \text{ [F.12]}$$

The **potential energy**, E_p, commonly V, of an object is the energy it possesses as a result of its position. For an object of mass m at an altitude h close to the surface of the Earth, the gravitational potential energy is

$$E_p = mgh \text{ [F.14]} \quad \text{where} \quad g = 9.81 \text{ m s}^{-2}$$

Eqn F.14 assigns the gravitational potential energy at the surface of the Earth a value equal to zero and g is called the **acceleration of free fall.**

The **Coulomb potential energy** describes the particularly important electrostatic interaction between two point charges Q_1 and Q_2 separated by the distance r:

$$E_p = \frac{Q_1Q_2}{4\pi\varepsilon_0 r} \text{ in a vacuum [F.13, } \varepsilon_0 \text{ is the vacuum permittivity]}$$

and

$$E_p = \frac{Q_1Q_2}{4\pi\varepsilon_r\varepsilon_0 r} \text{ in a medium that has the relative permittivity } \varepsilon_r \text{ (formerly, dielectric constant).}$$

Eqn F.13 assigns the Coulomb potential energy at infinite separation a value equal to zero. Convention assigns a negative value to the Coulomb potential energy when the interaction is attractive and a positive value when it is repulsive. The Coulomb potential energy and the force acting on the charges are related by the expression $F = -\mathrm{d}V/\mathrm{d}r$.

The energy provided by the disorderly motion of molecules is called the **energy of thermal motion.** Matter at a high temperature has the thermal energy to heat matter of a lower temperature. A useful rule of thumb is that the order of magnitude of the energy that a molecule possesses as a result of its thermal motion is kT where $k = 1.381 \times 10^{-23}$ J K^{-1} is the **Boltzmann constant** and the temperature T is in kelvin.

F.7 When the pressures on both sides of an object, for example a movable piston confined in a cylinder containing a gas, are equal, there is no net force acting upon the object and it will not move in either direction. The object is said to be in **mechanical equilibrium**. Two objects that exhibit no net flow of energy between them upon contact, are in **thermal equilibrium**. They have the same temperature.

F.8 The wave nature of electromagnetic radiation is characterized by its wavelength, λ (lambda), and its frequency, ν (nu). As shown in Figure F.6 of the text, the amplitude of the electric field varies sinusoidally while propagating at the speed c in a vacuum. **Frequency** in hertz (Hz) is the number of repetitive amplitude cycles occurring per second. **Wavelength** is the distance traveled during one amplitude cycle. The radiation also has a magnetic field component that is perpendicular to the electric field; both are perpendicular to the direction of propagation. Frequency and wavelength are related by

$$\lambda\nu = c \text{ [F.16]}$$

Electromagnetic radiation is delivered in discrete packets known as **photons**. The energy of a photon is related to frequency and wavelength by the expressions

$$E = h\nu \text{ [F.17]} = hc/\lambda$$

where h (= 6.626×10^{-34} J s) is **Planck's constant**.

Different ranges of the electromagnetic spectrum are able to excite one or another type of movement. These regions are summarized in text Figure F.7 and in the following table.

Radiation type	Wavelength range / nm	Excitation capability
X-rays and γ-rays	≤ 3	Core-electron; nuclear excitations
ultraviolet	3–400	Electronic; breaks chemical bonds
visible	400–700	Electronic; process of sight
infrared	700–10^6	Molecular vibration
microwaves	10^6–10^8	Molecular rotation
radio waves	10^8–10^{14}	Electrons in metal wires

F.9 The **Boltzmann distribution:** $\dfrac{N_2}{N_1} = e^{-(E_2-E_1)/kT}$ [F.19a, applicable at thermal equilibrium only]

The Boltzmann distribution indicates that temperature T is not a property of a single molecule. It is a characteristic of the state of a macroscopic system as a whole, appearing as a parameter characterizing the distribution of molecules over available energy states. Temperature is the single parameter we need in order to state the relative populations of energy levels. All molecules occupy the lowest energy level at absolute zero. As temperature rises, upper energy levels become occupied until all energy states are equally populated at extremely high temperature.

Exercises (All gases are perfect unless otherwise indicated.)

F.10 (a) Sulfite anion, SO_3^{2-}

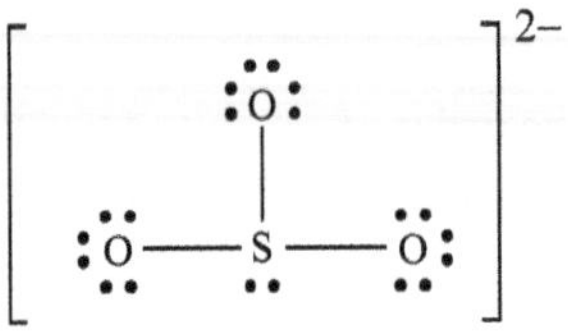

Alternatively, resonance structures may be drawn and, if desired, formal charges (shown in circles below) may be indicated.

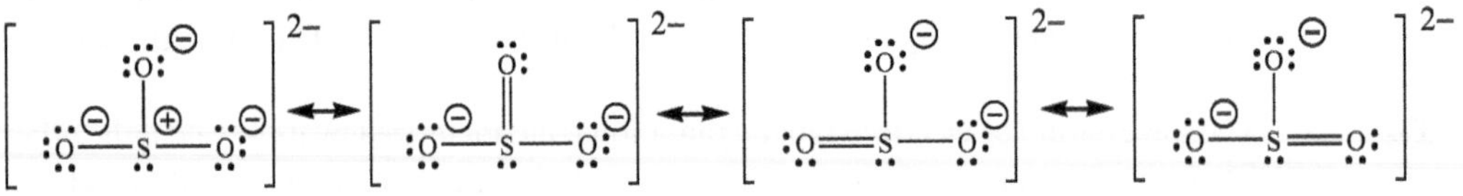

(b) Xenon tetrafluoride, XeF_4

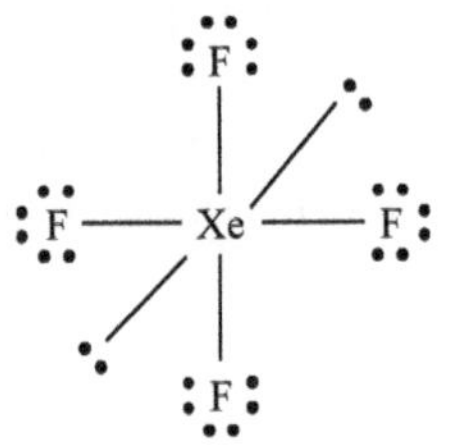

(c) White phosphorus, P_4

(d) Ozone, O_3. Formal charges (shown in circles) may be indicated.

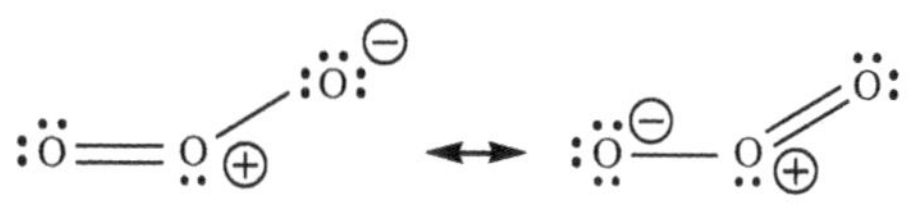

(e) ClF_3^+

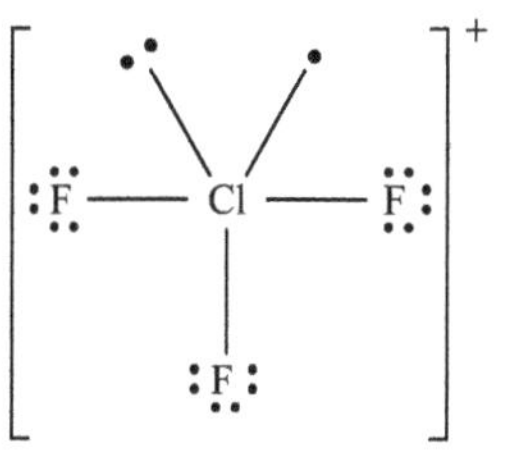

(f) azide anion, N_3^-

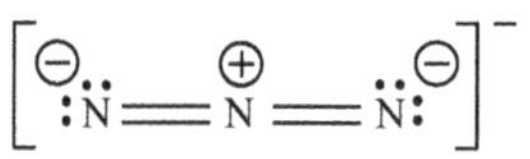

F.11 **Valence-shell electron-pair repulsion theory (VSEPR theory)** predicts molecular shape with the concept that regions of high electron density (as represented by single bonds, multiple bonds, and lone pairs) take up orientations around the central atom that maximize their separation. The resulting positions of attached atoms (not lone pairs) are used to classify the shape of the molecule. When the central atom has two or more lone pairs, the molecular geometry must minimize repulsion between the relatively diffuse orbitals of the lone pair. Furthermore, it is assumed that the repulsion between a lone pair and a bonding pair is stronger than the repulsion between two bonding pairs, thereby, making bond angles smaller than the idealized bond angles that appear in the absence of a lone pair.

Molecular and polyatomic ion shape are predicted by drawing the Lewis structure and applying the concepts of VSEPR theory.

(a) PCl_3
Lewis structure:

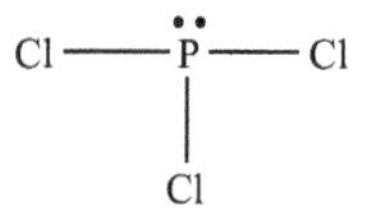

Orientations caused by repulsions between one lone pair and three bonding pairs:

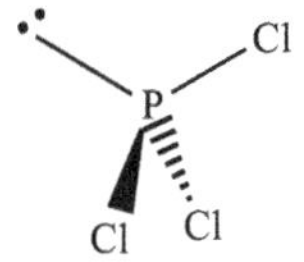

Molecular shape: trigonal pyramidal and bond angles somewhat smaller than 109.5°.

(b) PCl_5
Lewis structure:

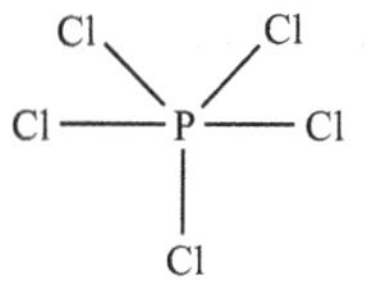

Orientations caused by repulsions between five bonding pairs (no lone pair):

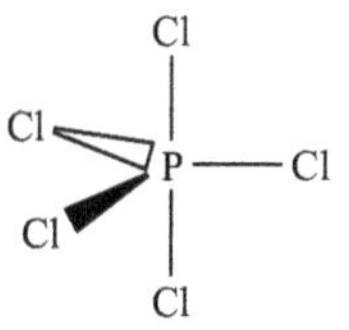

Molecular shape: trigonal bipyramidal with equatorial bond angles of 120° and axial bond angles of 90°.

(c) XeF_2
Lewis structure:

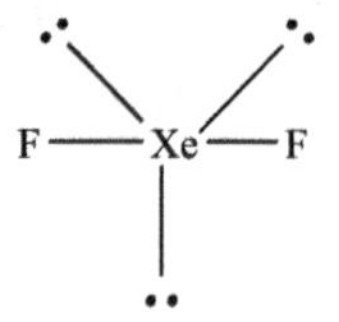

Orientations caused by repulsions between three lone pairs and two bonding pairs:

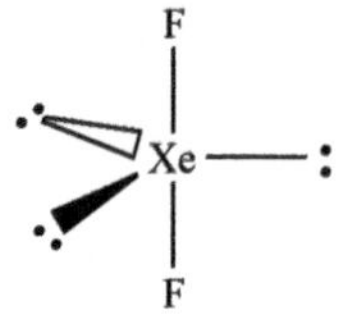

Molecular shape: linear with a 180° bond angle.

(d) XeF_4
Lewis structure:

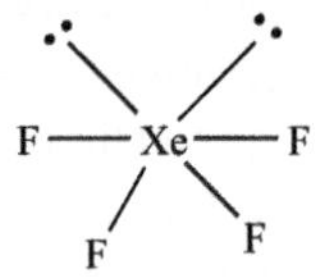

Orientations caused by repulsions between two lone pairs and four bonding pairs:

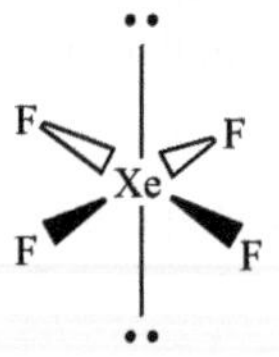

Molecular shape: square planar with a 90° bond angles.

(e) H_2O_2
Lewis structure:

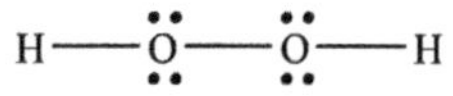

Orientations caused by repulsions between two lone pairs and two bonding pairs around each oxygen atom:

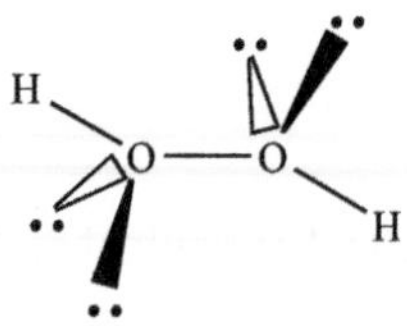

Molecular shape around each oxygen atom: bent (or angular) with bond angles somewhat smaller than 109.5°.

(f) FSO_3^-
Lewis structure:
(Formal charged is circled.)

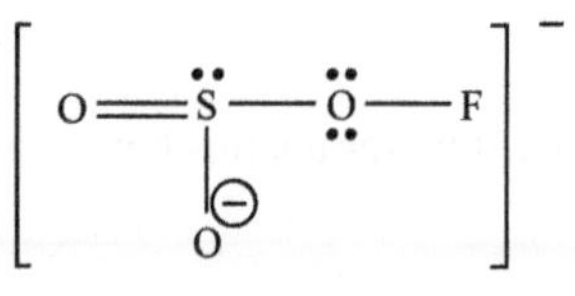

Orientations around the sulfur are caused by repulsions between one lone pair, one double bond, and two single bonds, while orientations around the oxygen to which fluorine is attached are caused by repulsions between two lone pairs and two single bonds:

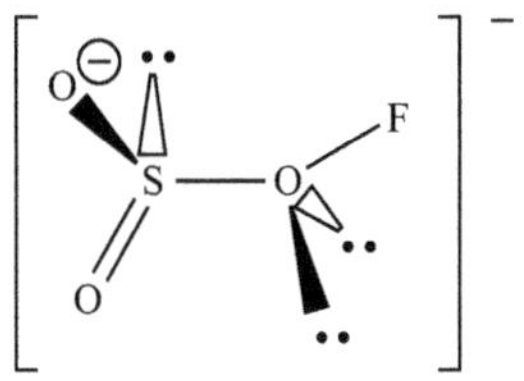

Molecular shape around the sulfur atom is trigonal pyramidal with bond angles somewhat smaller than 109.5°, while the shape around the oxygen to which fluorine is attached is bent (or angular) with a bond angle somewhat smaller than 109.5°.

(g) KrF_2
Lewis structure:

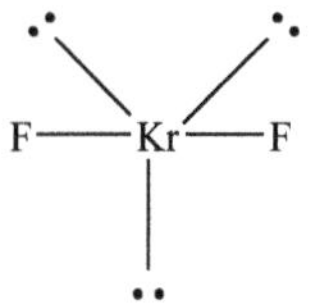

Orientations caused by repulsions between three lone pairs and two bonding pairs:

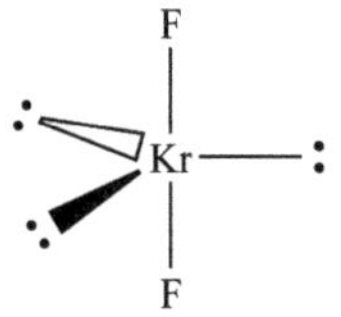

Molecular shape: linear with a 180° bond angle.

(h) PCl_4^+
Lewis structure:
(Formal charge is shown in a circle.)

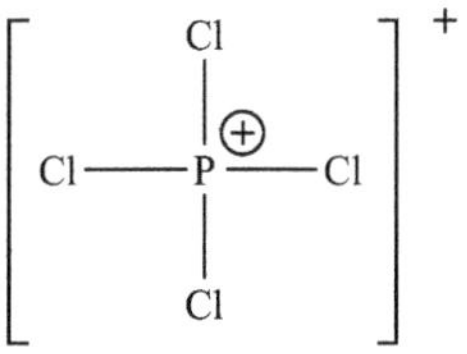

Orientations caused by repulsions between four bonding pairs (no lone pairs):

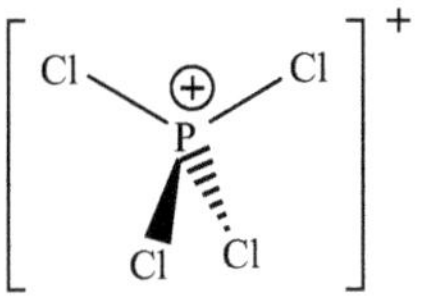

Molecular shape: tetrahedral and bond angles of 109.5°.

F.12 Refer to Table F.1 for pressure conversion factors.

(a) $p = 110\text{ kPa} \times \dfrac{760\text{ Torr}}{101.325\text{ kPa}} = \boxed{825\text{ Torr}}$.

(b) $p = 0.997\text{ bar} \times \dfrac{100\text{ kPa}}{1\text{ bar}} \times \dfrac{1\text{ atm}}{101.325\text{ kPa}} = \boxed{0.984\text{ atm}}$.

(c) $p = (21.5\text{ kPa}) \times \left(\dfrac{1\text{ atm}}{101.325\text{ kPa}}\right) = \boxed{0.212\text{ atm}}$.

(d) $p = 723\text{ Torr} \times \left(\dfrac{101.325 \times 10^3\text{ Pa}}{760\text{ Torr}}\right) = \boxed{9.64 \times 10^4\text{ Pa}}$.

F.13 $\theta_{\text{Celsius}}/°\text{C} = \frac{5}{9}(\theta_{\text{Fahrenheit}}/°\text{F} - 32)$

Solving for $\theta_{\text{Fahrenheit}}$ gives

$$\theta_{\text{Fahrenheit}}/°\text{F} = \tfrac{9}{5}\theta_{\text{Celsius}}/°\text{C} + 32$$

Since $\theta/°\text{C} = -273.15$ at $T = 0$, absolute zero on the Fahrenheit scale is

$$\theta_{\text{Fahrenheit}}/°\text{F} = \tfrac{9}{5}(-273.15) + 32$$
$$= -459.67 \quad \text{and} \quad \theta_{\text{Fahrenheit}} = \boxed{-459.67°\text{F}}$$

F.14 The mathematical equation for a straight line through two points $P_1(x_1, y_1)$ and $P_2(x_2, y_2)$ is

$$y = \frac{\Delta y}{\Delta x}(x - x_1) + y_1 \quad \text{where} \quad \frac{\Delta y}{\Delta x} = \frac{y_2 - y_1}{x_2 - x_1}.$$

Using $P_1(x_1, y_1) = (-209.9°\text{C}, 0°\text{P})$, $P_2(x_2, y_2) = (-195.8°\text{C}, 100°\text{P})$ as the two points and θ and θ_P to be temperature in degree Celsius and degree Plutonium, respectively, the function $\theta_\text{P}(\theta)$ is

$$\theta_\text{P} = \left(\frac{100°\text{P} - 0°\text{P}}{-195.8°\text{C} - (-209.9°\text{C})}\right) \times (\theta + 209.9°\text{C}) + 0$$

$$\theta_\text{P} = (7.092°\text{P}\ °\text{C}^{-1}) \times (\theta + 209.9°\text{C}) \quad \text{or} \quad \boxed{\theta_\text{P}/°\text{P} = 7.092 \times (\theta/°\text{C} + 209.9)}.$$

(a) Substitution of the definition $\theta/°\text{C} = T/\text{K} - 273.15$, where T is kelvin temperature into the above equation gives

$$\boxed{\theta_\text{P}/°\text{P} = 7.092 \times (T/\text{K} - 63.25)}.$$

(b) Substitution of the relationship $\theta/°\text{C} = \frac{5}{9}(\theta_\text{F}/°\text{F} - 32)$, where θ_F is fahrenheit temperature into the top equation gives

$$\boxed{\theta_\text{P}/°\text{P} = 3.940 \times (\theta_\text{F}/°\text{F} + 345.8°\text{F})}.$$

F.15 Solve the perfect gas law [F.6] for n and recognize the volume unit of the data needs to be converted to dm^3 so that units cancel conveniently in the calculation.

$$V = 250.0\ \text{cm}^3 = 0.2500\ \text{dm}^3 \quad \text{and} \quad T = (273.15 + 19.5)\ \text{K} = 292.65\ \text{K}.$$

$$n = \frac{pV}{RT} = \frac{(24.5\ \text{kPa}) \times (0.2500\ \text{dm}^3)}{(8.3145\ \text{dm}^3\ \text{kPa}\ \text{K}^{-1}\ \text{mol}^{-1}) \times (292.65\ \text{K})} = 2.52 \times 10^{-3}\ \text{mol} = \boxed{2.52\ \text{mmol}}$$

F.16 According to the perfect gas law, $p_1V_1 = p_2V_2$ when n and T are constant because $nRT = pV =$ a constant. (The subscripts 1 and 2 represent two different gaseous states.) Solving for p_2 gives

$$p_2 = \frac{V_1}{V_2} \times p_1 \text{ where}$$

$V_1 = 1.00\ \text{dm}^3 = 1.00 \times 10^3\ \text{cm}^3$, $p_1 = 1.00\ \text{atm}$, and $V_2 = 1.00 \times 10^2\ \text{cm}^3$. Therefore,

$$p_2 = \frac{V_1}{V_2} \times p_1 = \left(\frac{1.00 \times 10^3\ \text{cm}^3}{100\ \text{cm}^3}\right) \times (1.00\ \text{atm}) = \boxed{10.0\ \text{atm}}$$

F.17 The amount n and volume V are constant, hence by the perfect gas law: $nR/V = p/T$ = a constant. Thus,

$$\frac{p_1}{T_1} = \frac{p_2}{T_2} \text{ and, solving for } p_2 \text{ gives } p_2 = \frac{T_2 p_1}{T_1}.$$

$$p_2 = \frac{973\ \text{K} \times 125\ \text{kPa}}{291\ \text{K}} = 418\ \text{kPa} = \boxed{4.18\ \text{bar}}$$

F.18 According to the perfect gas law, $p_1V_1 = p_2V_2$ when n and T are constant because $nRT = pV$ = a constant. Thus, solving for p_2 gives

$$p_2 = \frac{p_1V_1}{V_2} = \frac{101\ \text{kPa} \times 7.20\ \text{dm}^3}{4.21\ \text{dm}^3} = \boxed{173\ \text{kPa}}$$

F.19 According to the perfect gas law, $\frac{T_1}{V_1} = \frac{T_2}{V_2}$ when n and p are constant because $p/nR = T/V$ = a constant.

$T_1 = 340$ K and the volume has increased by 14%, so $V_2 = 1.14\ V_1$. Thus, solving for T_2 gives

$$T_2 = \frac{V_2T_1}{V_1} = \frac{1.14\ V_1}{V_1} \times 340\ \text{K} = 1.14 \times 340\ \text{K} = \boxed{388\ \text{K}}$$

F.20 According to the perfect gas law, $\frac{p_1V_1}{T_1} = \frac{p_2V_2}{T_2}$ or $V_2 = \frac{p_1T_2}{p_2T_1} \times V_1$ when n is constant because $nR = pV/T$ = a constant.

(a) $$V_2 = \frac{p_1T_2}{p_2T_1} \times V_1 = \frac{(104\ \text{kPa}) \times (268.15\ \text{K})}{(52\ \text{kPa}) \times (294.25\ \text{K})} \times (2.0\ \text{m}^3) = \boxed{3.6\ \text{m}^3}$$

(b) $$V_2 = \frac{p_1T_2}{p_2T_1} \times V_1 = \frac{(104 \times 10^3\ \text{Pa}) \times (221.15\ \text{K})}{(880\ \text{Pa}) \times (294.25\ \text{K})} \times (2.0\ \text{m}^3) = \boxed{1.8 \times 10^2\ \text{m}^3}$$

F.21 Let the subscripts 1 and 2 represent the unsubmerged and submerged gaseous states, respectively. According to the perfect gas law, $p_1V_1 = p_2V_2$ when n and T are constant because $nRT = pV$ = a constant. The pressure in the submerged state will equal that of the unsubmerged state plus the hydrostatic pressure due to submersion. That is, at the depth d: $p_2 = p_1 + \rho_{\text{water}}gd$, where $p_1 = 1.00$ atm $= 1.01 \times 10^5$ Pa. Thus,

$$\begin{aligned} p_2 &= p_1 + \rho_{\text{water}}gd \\ &= (1.01 \times 10^5\ \text{Pa}) + (1.025 \times 10^3\ \text{kg m}^{-3}) \times (9.81\ \text{m s}^{-2}) \times (50\ \text{m}) \\ &= (1.01 \times 10^5\ \text{Pa}) + (5.0 \times 10^5\ \text{kg m}^{-1}\ \text{s}^{-2}) \\ &= 6.0 \times 10^5\ \text{Pa}\ [\text{Note: } 1\ \text{kg m}^{-1}\ \text{s}^{-2} = 1\ (\text{kg m s}^{-2})\text{m}^{-2} = 1\ \text{N m}^{-2} = 1\ \text{Pa}] \end{aligned}$$

$$V_2 = \frac{p_1}{p_2}V_1 = \frac{1.01 \times 10^5\ \text{Pa}}{6.0 \times 10^5\ \text{Pa}} \times 3.0\ \text{m}^3 = \boxed{0.50\ \text{m}^3}$$

F.22 $$\begin{aligned} \textit{Work}_{\text{lift}} &= (\textit{force}_{\text{gravity}}) \times (\textit{vertical displacement}) = mgd\ [\text{F.11 and associated } \textit{brief illustration}] \\ &= (65\ \text{kg}) \times (9.81\ \text{m s}^{-2}) \times (3.5\ \text{m}) = 2.2 \times 10^3\ \text{kg m}^2\ \text{s}^{-2} = \boxed{2.2\ \text{kJ}}\quad [1\ \text{kg m}^2\ \text{s}^{-2} = 1\ \text{J}] \end{aligned}$$

F.23 $$E_k = \tfrac{1}{2}mv^2\ [\text{F.12}] = \tfrac{1}{2}(58 \times 10^{-3}\ \text{kg}) \times (30\ \text{m s}^{-1})^2 = 26\ \text{kg m}^2\ \text{s}^{-2} = \boxed{26\ \text{J}}$$

F.24 $E_k = \frac{1}{2}mv^2$ [F.12]

$$= \tfrac{1}{2}(1.5\ \text{t}) \times (50\ \text{km h}^{-1})^2 \times \left(\frac{1000\ \text{kg}}{1\ \text{t}}\right) \times \left(\frac{1000\ \text{m}}{1\ \text{km}}\right)^2 \times \left(\frac{1\ \text{h}}{3600\ \text{s}}\right)^2$$

$$= 1.4 \times 10^5\ \text{J} = \boxed{1.4 \times 10^2\ \text{kJ}}.$$

F.25 We estimate the average kinetic energy of a molecule with $v = 400\ \text{m s}^{-1}$ and $m = M/N_A = (29\ \text{g mol}^{-1})/N_A$ in eqn F.12, $E_k = \frac{1}{2}mv^2$. The number of molecules, N, is given by $N = nN_A$ and the total energy stored as molecular kinetic energy is NE_k.

$$NE_k = (nN_A) \times (\tfrac{1}{2}mv^2) = (nN_A) \times \left\{\tfrac{1}{2}\left(\frac{M}{N_A}\right) \times v^2\right\} = \tfrac{1}{2}nMv^2$$

$$= \tfrac{1}{2}(1\ \text{mol}) \times (0.029\ \text{kg mol}^{-1}) \times (400\ \text{m s}^{-1})^2$$

$$= 2.3 \times 10^3\ \text{kg m}^2\ \text{s}^{-2} = 2.3 \times 10^3\ \text{J} = \boxed{2.3\ \text{kJ}}$$

F.26 The law of conservation of energy requires that the minimum kinetic energy [F.12] required to reach height h equals the increase in gravitational potential energy [F.14].

$$E_k = mgh$$

$$= (0.025\ \text{kg}) \times (9.81\ \text{m s}^{-2}) \times (50\ \text{m}) = 12\ \text{kg m}^2\ \text{s}^{-2} = \boxed{12\ \text{J}}$$

F.27 The Coulomb potential, ϕ, is

$$\phi = \frac{Q_2}{4\pi\varepsilon_0 r}$$ where r is the separation of point charge Q_1 and the nuclear charge Q_2

Q_1 interacts with two nuclei in this exercise and the interactions are additive.

$$\phi = \left(\frac{Q_2}{4\pi\varepsilon_0 r}\right)_{\text{Li nucleus}} + \left(\frac{Q_2}{4\pi\varepsilon_0 r}\right)_{\text{H nucleus}}$$

$$= \left(\frac{Ze}{4\pi\varepsilon_0 r}\right)_{\text{Li nucleus}} + \left(\frac{Ze}{4\pi\varepsilon_0 r}\right)_{\text{H nucleus}}$$

$$= \frac{e}{4\pi\varepsilon_0} \times \left\{\left(\frac{Z}{r}\right)_{\text{Li nucleus}} + \left(\frac{Z}{r}\right)_{\text{H nucleus}}\right\}$$

$$= \left(\frac{1.6022 \times 10^{-19}\ \text{C}}{1.113 \times 10^{-10}\ \text{J}^{-1}\ \text{C}^2\ \text{m}^{-1}}\right) \times \left\{\frac{3}{200 \times 10^{-12}\ \text{m}} + \frac{1}{150 \times 10^{-12}\ \text{m}}\right\}$$

$$= 31.2\ \text{J C}^{-1} = \boxed{31.2\ \text{V}}$$

F.28 The Coulomb potential, ϕ, is

$$\phi = \frac{Q_2}{4\pi\varepsilon_0 r},$$ where r is the separation of point charge Q_1 and the ion charge Q_2

Q_1 interacts with two ions, which are treated as point charges in this exercise, and the interactions are additive.

$$\phi = \left(\frac{Q_2}{4\pi\varepsilon_0 r}\right)_{Na^+} + \left(\frac{Q_2}{4\pi\varepsilon_0 r}\right)_{Cl^-}$$

$$= \left(\frac{e}{4\pi\varepsilon_0 r}\right)_{Na^+} + \left(\frac{-e}{4\pi\varepsilon_0 r}\right)_{Cl^-}$$

$$= \frac{e}{4\pi\varepsilon_0} \times \left\{\left(\frac{1}{r}\right)_{Na^+} - \left(\frac{1}{r}\right)_{Cl^-}\right\}$$

$$= \left(\frac{1.6022\times10^{-19}\text{ C}}{1.1127\times10^{-10}\text{ J}^{-1}\text{ C}^2\text{ m}^{-1}}\right) \times \left(\frac{1}{10^{-12}\text{ m}}\right) \times \left\{\left(\frac{1}{r/\text{pm}}\right)_{Na^+} - \left(\frac{1}{r/\text{pm}}\right)_{Cl^-}\right\}$$

$$= (1440\text{ V}) \times \left\{\left(\frac{1}{r/\text{pm}}\right)_{Na^+} - \left(\frac{1}{r/\text{pm}}\right)_{Cl^-}\right\} \quad [1\text{ J C}^{-1} = 1\text{ V}]$$

Figure F.12 shows the positions of the sodium and chloride ions as the charge Q_1 approaches the center point between the two ions along a straight line at the angle θ to the internuclear line. If we interpret the exercise as specifying that the approach be at the angle $\theta = 90°$, then $r_{Na^+} = r_{Cl^-}$ all along the approach and the above relation tells us that $\boxed{\phi_{\theta=90°} = 0}$ at all values of r (defined in Fig F.12). For angles other than $\theta = 90°$, the above equation for ϕ can be computed as a function of r at fixed θ. The law of cosines is used to calculate the requisite values r_{Na^+} and r_{Cl^-} at each value of r and θ.

$$r_{Na^+} = (r_c^2 + r^2 - 2r_c r\cos\theta)^{1/2} \quad \text{and} \quad r_{Cl^-} = (r_c^2 + r^2 - 2r_c r\cos(\pi - \theta))^{1/2}$$

Plots of ϕ against r at $\theta = 30°$, 45°, and 60° are presented in Figure F.13. It is apparent that as Q_1 approaches the center from infinity the Coulomb potential rises to a peak at about $\frac{1}{2}$ the internuclear distance because of the dominant interaction with the sodium cation. Upon closer approach to the center the influence of the chloride anion progressively increases, thereby, causing a decline in the Coulomb potential until the interactions with the two ions is exactly balanced when Q_1 is midway between the ions (i.e. at $r = 0$).

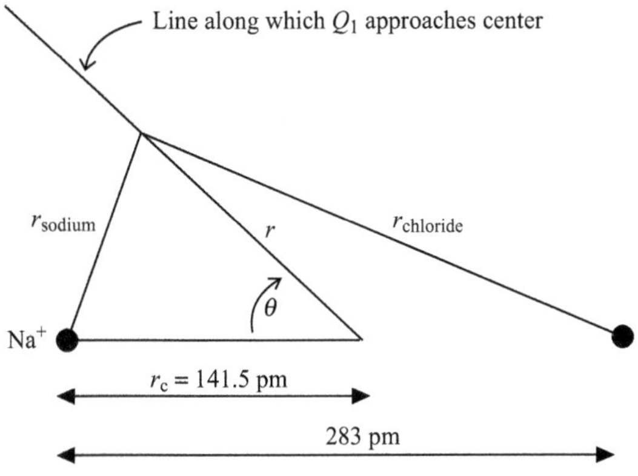

Figure F.12

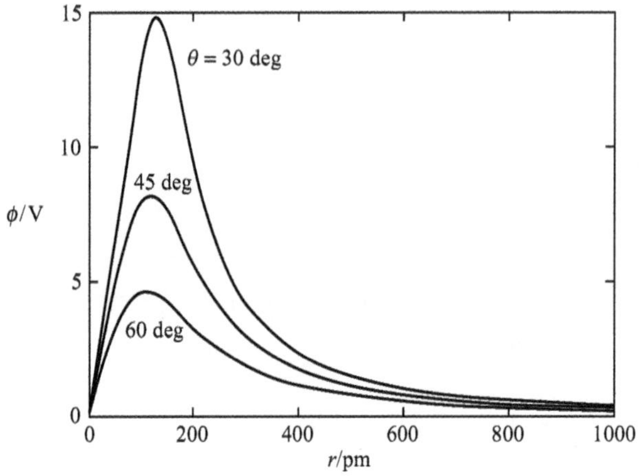

Figure F.13

F.29 $$\lambda = \frac{c}{\nu}\text{[F.16]} = \frac{2.998\times10^{8}\text{ m s}^{-1}}{92.0\times10^{6}\text{ s}^{-1}} = \boxed{3.26\text{ m}}$$

F.30 (a) $$E = h\nu = \frac{hc}{\lambda}\text{[F.17 and F.16]}$$

$$= \frac{(6.62608\times10^{-34}\text{ J s})\times(2.99792\times10^{8}\text{ m s}^{-1})}{670\times10^{-9}\text{ m}}$$

$$= 0.296\text{ aJ [atto, a} \equiv 10^{-18}] = \boxed{296\text{ zJ}}\text{ [zepto, z} \equiv 10^{-21}]$$

(b) $$E(\text{per mole}) = N_{\text{A}}E = \frac{N_{\text{A}}hc}{\lambda}$$

$$= (6.022\times10^{23}\text{ mol}^{-1})\times\frac{(6.62608\times10^{-34}\text{ J s})\times(2.99792\times10^{8}\text{ m s}^{-1})}{670\times10^{-9}\text{ m}}$$

$$= \boxed{179\text{ kJ mol}^{-1}}$$

F.31 Let the subscripts 1 and 2 represent the random coil and fully stretched macromolecules, respectively.

$$\frac{N_2}{N_1} = \mathrm{e}^{-(E_2-E_1)/RT}\text{ [F.19b]}$$

$$= \mathrm{e}^{-(2.4\times10^{3}\text{ J mol}^{-1})/\{(8.3145\text{ J K}^{-1}\text{ mol}^{-1})\times(293.15\text{ K})\}} = \boxed{0.37}$$

F.32 Let the subscripts 1 and 2 represent the lower and upper energies, respectively. Then, $E_1 = -\mu_{\text{B}}\mathcal{B}$ and $E_2 = \mu_{\text{B}}\mathcal{B}$.

$$\frac{N_2}{N_1} = \mathrm{e}^{-(E_2-E_1)/kT}\text{ [F.19a]} = \mathrm{e}^{-2\mu_{\text{B}}\mathcal{B}/kT}$$

$$= \mathrm{e}^{-2(9.274\times10^{-24}\text{ J T}^{-1})\times(1.0\text{ T})/\{(1.381\times10^{-23}\text{ J K}^{-1})\times T\}} = \mathrm{e}^{-1.3\overline{43}/(T/\text{K})}$$

(a) At 4.0 K: $$\frac{N_2}{N_1} = \mathrm{e}^{-1.3\overline{43}/(4.0)} = \boxed{0.71}$$

(b) At 298 K: $$\frac{N_2}{N_1} = \mathrm{e}^{-1.3\overline{43}/(298)} = \boxed{0.99\overline{6}}$$

These calculations demonstrate that a majority of molecules occupy the low energy state at low temperature while all energy states are equally occupied at extremely high temperature. The values of the "low temperature" and the "high temperature" depend upon the spacing of the energy levels.

F.33

$$\bar{c} = \left(\frac{8RT}{\pi M}\right)^{1/2} \quad [\text{F.20}]$$

Now we develop a convenient computational formula for various T and M.

$$\bar{c} = \left(\frac{8 \times (8.3145\ \text{J K}^{-1}\ \text{mol}^{-1}) \times T}{\pi M \times (10^{-3}\ \text{kg/g})}\right)^{1/2}$$

$$\bar{c}/(\text{m s}^{-1}) = 145.5 \times \left(\frac{T/\text{K}}{M/\text{g mol}^{-1}}\right)^{1/2}$$

(a) $M_{\text{He}} = 4.00\ \text{g mol}^{-1}$; $\bar{c}_{\text{He}}/(\text{m s}^{-1}) = 72.75 \times (T/\text{K})^{1/2}$

$$\bar{c}_{\text{He}}(77\ \text{K}) = \boxed{638\ \text{m s}^{-1}}$$

$$\bar{c}_{\text{He}}(298\ \text{K}) = \boxed{1.26\ \text{km s}^{-1}}$$

$$\bar{c}_{\text{He}}(1000\ \text{K}) = \boxed{2.30\ \text{km s}^{-1}}$$

(b) $M_{\text{CH}_4} = 16.04\ \text{g mol}^{-1}$; $\bar{c}_{\text{CH}_4}/(\text{m s}^{-1}) = 36.33 \times (T/\text{K})^{1/2}$

$$\bar{c}_{\text{CH}_4}(77\ \text{K}) = \boxed{319\ \text{m s}^{-1}}$$

$$\bar{c}_{\text{CH}_4}(298\ \text{K}) = \boxed{627\ \text{m s}^{-1}}$$

$$\bar{c}_{\text{CH}_4}(1000\ \text{K}) = \boxed{1.15\ \text{km s}^{-1}}$$

Projects

F.34

(a) $$E_{\text{gravitational potential}}\big|_{r=r_{\text{E}}+h} = -\frac{Gmm_{\text{E}}}{r_{\text{E}}+h} = -\left(\frac{Gmm_{\text{E}}}{r_{\text{E}}}\right) \times \left(\frac{1}{1+h/r_{\text{E}}}\right) = -\left(\frac{Gmm_{\text{E}}}{r_{\text{E}}}\right) \times \left(1+\frac{h}{r_{\text{E}}}\right)^{-1}$$

Since $h/r_{\text{E}} \ll 1$, the last factor may be expanded in a Taylor expansion series. Second- and higher-order terms may be discarded because powers of very small fractions produce yet smaller fractions. Using $x = h/r_{\text{E}}$, the Taylor expansion is: $(1+x)^{-1} = 1 - x + x^2 - x^3 + \cdots = 1 - x$.

Substitution gives:

$$E_{\text{gravitational potential}}\big|_{r=r_{\text{E}}+h} = -\left(\frac{Gmm_{\text{E}}}{r_{\text{E}}}\right) \times \left(1 - \frac{h}{r_{\text{E}}}\right) = -\frac{Gmm_{\text{E}}}{r_{\text{E}}} + \frac{Gmm_{\text{E}}h}{r_{\text{E}}^2} = -\frac{Gmm_{\text{E}}}{r_{\text{E}}} + m \times \left(\frac{Gm_{\text{E}}}{r_{\text{E}}^2}\right) \times h$$

$$= E_{\text{gravitational potential}}\big|_{r=r_{\text{E}}} + mgh, \quad \text{where} \quad g = Gm_{\text{E}}/r_{\text{E}}^2.$$

Thus, the difference between the gravitational potential at $r = r_{\text{E}} + h$ and the gravitation potential at $r = r_{\text{E}}$, when $h \ll r_{\text{E}}$, is mgh, where $\boxed{g = Gm_{\text{E}}/r_{\text{E}}^2}$. This difference is the gravitational potential above the surface and it is normally written as $E_p = mgh$ [F.14]. The values of m_E and r_E (at the equator) are found in the *CRC Handbook of Chemistry and Physics*, so it is possible to calculate the value of g at the equator.

$$g = \frac{(6.67259 \times 10^{-11}\ \text{N m}^2\ \text{kg}^{-2})(5.9763 \times 10^{24}\ \text{kg})}{(6378.077 \times 10^3\ \text{m})^2}\left(\frac{1\ \text{kg m s}^{-2}}{1\ \text{N}}\right) = 9.8027\ \text{m s}^{-2} \text{ at equator}$$

(b) The required fuel energy equals the difference between the gravitational potential at $r = \infty$ and the gravitational potential at $r = r_E$, where r_E is the radius of the Earth.

$$\text{Energy} = \left(-\frac{Gmm_E}{r}\right)_{\text{at } r=\infty} - \left(-\frac{Gmm_E}{r}\right)_{\text{at } r=r_E} = 0 + \frac{Gmm_E}{r_E} = \frac{Gmm_E}{r_E}$$

$$m_E = V_E\rho_E = \left(\frac{4\pi(6371 \times 10^3\ \text{m})^3}{3}\right) \times \left(\frac{5.5170\ \text{g}}{\text{cm}^3}\right) \times \left(\frac{1\ \text{cm}}{10^{-2}\ \text{m}}\right)^3 \times \left(\frac{1\ \text{kg}}{10^3\ \text{g}}\right) = 5.976 \times 10^{24}\ \text{kg}.$$

$$\text{Energy} = \frac{(6.673 \times 10^{-11}\ \text{N m}^2\ \text{kg}^{-2}) \times (185\ \text{kg}) \times (5.976 \times 10^{24}\ \text{kg})}{6371 \times 10^3\ \text{m}}$$

$$= 1.16 \times 10^{10}\ \text{J} = \boxed{11.6\ \text{GJ}}$$

(c) (i) An infinitesimally small amount of work is done upon an object of mass m when lifting it through the infinitesimal distance dr against the opposing gravitational attraction F. The work causes an infinitesimal increase dE_P in the potential energy of the object.

$$dE_P = F\,dr$$

$$F = \frac{dE_P}{dr} = \frac{d}{dr}\left(-\frac{Gmm_E}{r}\right) = -Gmm_E\frac{d}{dr}\left(\frac{1}{r}\right) = -Gmm_E\left(-\frac{1}{r^2}\right)$$

$$F = \frac{Gmm_E}{r^2}$$

(ii) It is shown in the solution to part (a) that $g = \frac{Gm_E}{r_E^2}$. Consequently, the gravitation attraction at the Earth's surface is given by $\boxed{F = \frac{Gmm_E}{r_E^2} = mg}$. For an 80.0 kg person the attraction is:

$$F = (80.0\ \text{kg}) \times (9.81\ \text{m s}^{-2}) = \boxed{78\ \text{N}}$$

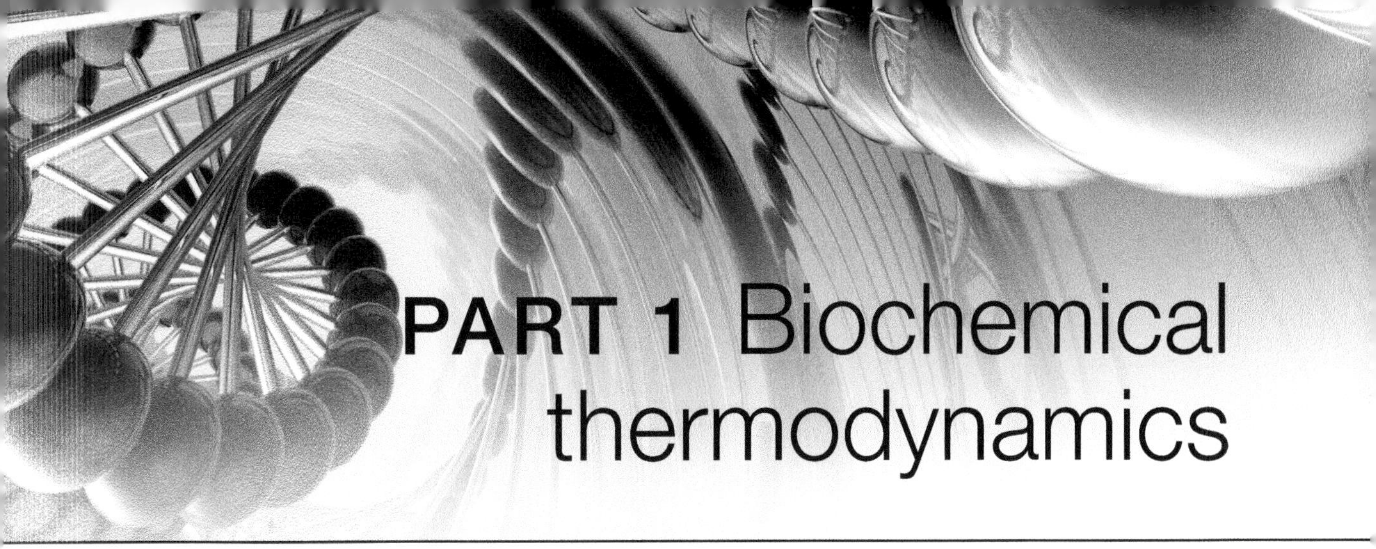

PART 1 Biochemical thermodynamics

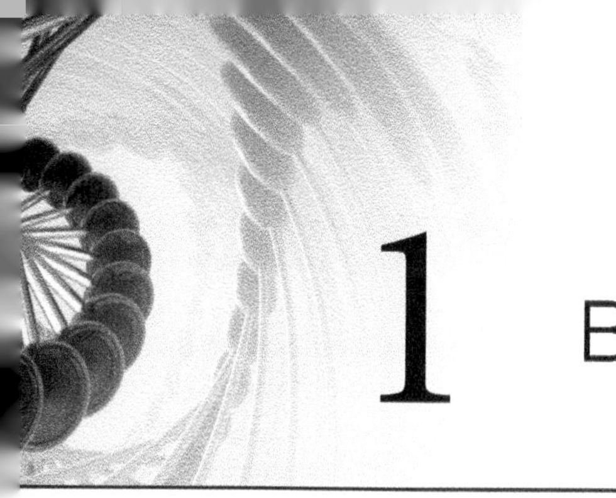

1 Bioenergetics: The First Law

Answers to discussion questions

D1.1 At the molecular level, **work** is a transfer of energy that results in orderly, uniform motion of the atoms and molecules in a system as illustrated in text Figure 1.5; the motion may even be that of an electrical current. In contrast, **heat** is the mode of transfer of energy that achieves or utilizes random motion of atoms, molecules, and electrons in the surroundings as illustrated in text Figure 1.6. Heat is energy in transit as a result of a temperature difference for which highly energetic molecules in a region of high temperature either directly carry energy into regions of low temperature during their random meanderings or transfer energy during collisions with lower-energy molecules in a region of lower temperature. Consequently, heat always flows from high temperature to low temperature.

Temperature, an intensive property possessed by local regions in thermal equilibrium, is the single parameter that tells us the relative molecular (and/or atomic) populations over the available energy levels of a system. This molecular interpretation is reflected in the Boltzmann distribution,

$$\frac{N_2}{N_1} = e^{-(E_2-E_1)/kT} \quad [1.1]\text{, where } k \text{ is the Boltzmann constant,}$$

which reveals that temperature T alone determines the distribution of particles over available energy levels.

The molecular interpretation of **heat capacity** C is based upon its definition, $C = q/\Delta T$ [1.5a] where q is a heat input and ΔT is the subsequent temperature change, and the effect that q has upon the Boltzmann distribution. When energy levels are very close together, the heat input will promote molecules to excited states that do not differ much from the states populated at the initial temperature. Thus, the Boltzmann distribution does not differ much at the final temperature and the final temperature differs only slightly from the initial temperature; $\Delta T = T_f - T_i$ is small. Placing a small value of ΔT into the heat capacity definition yields a large heat capacity. We conclude that large heat capacity reflects a multitude of molecular states that have closely spaced energies. For example, the vibrational energies of hydrogen bonds in liquid water are close together, and the heat capacity of water is larger than expected for a substance consisting of small molecules.

D1.2 Molecules can survive for long periods without undergoing chemical reaction at low temperatures when few molecules have the requisite speed and corresponding kinetic energy to promote excitation and bond breakage during collisions. This is shown by comparing the molar thermal energy at the conventional temperature with the mean energy of bond breakage (about 400 kJ mol^{-1}). The molar thermal energy is approximated as being

$$N_AkT = RT = (8.3145\ \mathrm{J\,K^{-1}\,mol^{-1}}) \times (298.15\ \mathrm{K}) = 2.48 \times 10^3\ \mathrm{J\,mol^{-1}} = 2.48\ \mathrm{kJ\,mol^{-1}},$$

a value that is much smaller than the mean bond energy.

D1.3 The general patterns of energy conversion in living organisms are discussed in the text section titled "Case study 1.1: Energy conversion in organisms". A flow diagram of the energy conversion is presented in text Figure 1.7. The primary source of energy that sustains the bulk of plant and animal life on Earth is the Sun. Solar radiation is ultimately stored during photosynthesis in the form of organic molecules, such as carbohydrates, fats, and proteins that are subsequently oxidized to meet the energy demands of organisms. Catabolism of the nutrients in highly controlled oxidation reactions utilize the energy of these biological fuels to synthesize energy transport molecules such as ATP, NADH, and NADPH and much of this energy is ultimately expended as useful work during the contraction of muscle, the establishment of ion gradients across cell membranes, the transport of molecules across membranes, and the anabolism of molecules like DNA.

A molecule like NADH is a reduced species and energy transfer can involve oxidized/reduced pairs like the NAD^+/NADH pair. Oxidation–reduction reactions ("redox reactions") transfer energy out of NADH and other reduced species, storing it in the mobile carrier ATP, and in ion gradients across membranes. The essence of ATP's action is the loss of its terminal phosphate group in an energy-releasing reaction.

Living organisms are not perfectly efficient machines, for not all the energy available from the Sun and oxidation of organic compounds is used to perform work as some is lost as heat. The dissipation of energy as heat is advantageous because it can be used to control the organism's temperature. Energy is eventually transferred as heat to the surroundings.

Key vocabulary terms and abbreviations that are relevant include:

Metabolism, the collection of chemical reactions that trap, store, and utilize energy in biological cells.

Anabolism, the biosynthesis of small and large molecules.

Catabolism, the collection of reactions associated with the oxidation of nutrients in the cell and may be regarded as highly controlled reactions, with the energy liberated as work rather than as heat.

ATP, adenosine triphosphate, a general energy-transport molecule that gives up its energy during the controlled conversion to **ADP** (adenosine diphosphate).

NADH, the reduced form of **NAD^+** (nicotinamide adenine dinucleotide); controlled, biological reactions transfer energy from NADH during oxidation to NAD^+.

NADPH, the phosphorylated derivative of NADH.

D1.4 The difference between ΔH and ΔU results from the enthalpy definition $H = U + pV$ [1.10]; hence, $\Delta H = \Delta U + \Delta(pV)$. As $\Delta(pV)$ is not usually zero, except for isothermal processes in a perfect gas, the difference between ΔH and ΔU is a non-zero quantity. As shown in Sections 1.5(a) and 1.6(b) of the text, ΔH can be interpreted as the heat associated with a process at constant pressure, and ΔU as the heat at constant volume.

D1.5 (a) $\Delta H = \Delta U + p\Delta V$ Limitation: constant pressure process. See the discussion of Eqn 1.11b.

(b) $\Delta_r H^\ominus(T') = \Delta_r H^\ominus(T) + (T' - T)\Delta_r C_p^\ominus$

Limitation: constant-pressure heat capacities of reactants and products are independent of temperature. See text Justification 1.4.

D1.6 The vaporization of the water is an endothermic transformation that cools the linen and its immediate environment: $H_2O(l) \rightarrow H_2O(g)$ $\Delta_{vap}H^\ominus = +44.01$ kJ mol^{-1}.

D1.7 A difference in the thermogram baseline at two different temperatures indicates that the sample has different heat capacities at the two temperatures. The sample may have experienced a phase transition from one crystalline, or non-crystalline, structure to another. It may have melted, evaporated, or decomposed. It may have reacted with the atmosphere or, if the sample is a mixture, components may have reacted. In all of these examples the sample has experienced either a physical or chemical change while being heated from one temperature to the other and, consequently, exhibits different heat capacities at those temperatures.

Solid polymers often exhibit a DSC baseline change for a "glass transition" at a well-defined temperature without the occurrence of reaction. A typical thermogram for a glass transition is shown in Figure 1.1. Upon reaching the temperature of the glass transition the polymer appears to change from a brittle glass-like state to a more rubber-like state that allows for increased rotational motion of the polymer chain and great randomness in the positions of molecules. It is a transition from a crystalline lattice to a glassy state. The glass transition of pure polystyrene, the plastic of cheap rulers and coffee spoons, occurs at about 90°C.

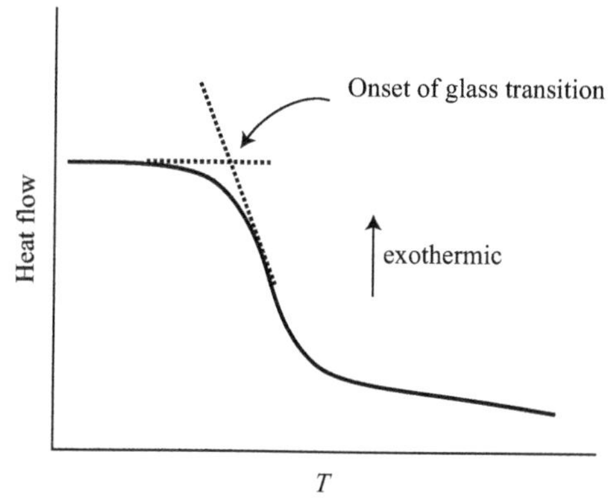

Figure 1.1

D1.8 Standard reaction enthalpies can be calculated from a knowledge of the standard enthalpies of formation of all the substances (reactants and products) participating in the reaction. This is an exact method that involves no approximations. The only disadvantage is that standard enthalpies of formation are not known for all substances.

Approximate values can be obtained from mean bond enthalpies. See Example 1.3 for an illustration of the method of calculation. This method is often quite inaccurate, though, because the average values of the bond enthalpies used may not be close to the actual values in the compounds of interest and the method ignores attractive forces between molecules, or is applicable to perfect gases only.

Computer-aided molecular modeling is now the method of choice for estimating standard reaction enthalpies, especially for large molecules with complex three-dimensional structures, but accurate numerical values are still difficult to obtain.

D1.9 The standard state of a substance at a specified temperature is its pure form at 1 bar. The reference state of a substance is its most stable state at the specified temperature and 1 bar. The distinction (the reference state must be the most stable state) is important because formation enthalpies are for formation of a substance from the elements in their reference states under prevailing conditions but other types of reactions may list a substance in a non-reference state.

Solutions to exercises

E1.10 $\Delta\varepsilon = E_{upper} - E_{lower} = 1\ \text{eV} = 1.6022 \times 10^{-19}\ \text{J}$

(a) $\dfrac{N_{upper}}{N_{lower}} = e^{-\Delta\varepsilon/kT}\,[1.1] = e^{-(1.6022\times10^{-19}\ \text{J})/\{(1.381\times10^{-23}\ \text{J K}^{-1})\times(300\ \text{K})\}} = \boxed{1.602\times10^{-17}}$

(b) $\dfrac{N_{upper}}{N_{lower}} = e^{-\Delta\varepsilon/kT} = e^{-(1.6022\times10^{-19}\ \text{J})/\{(1.381\times10^{-23}\ \text{J K}^{-1})\times(3000\ \text{K})\}} = \boxed{2.092\times10^{-2}}$

E1.11 The minimum metabolic energy expended by the bird equals the work needed to move to height h.

$$\text{Work} = mgh\,[1.2] = 0.200\ \text{kg}\times 9.81\ \text{m s}^{-2}\times 20\ \text{m} = \boxed{39\ \text{J}}$$

E1.12 Amount of glucose $= 1.0\ \text{g}\ C_6H_{12}O_6 \times \dfrac{1\ \text{mol}}{180.2\ \text{g}} = 5.5\overline{5}\times10^{-3}$ mol glucose

The balanced reaction equation for the complete combustion of glucose is written as

$$C_6H_{12}O_6(s) + 6\ O_2(g) \rightarrow 6\ CO_2(g) + 6\ H_2O$$

In Part (a) the net expansion work done on the chemical system, w, is computed for the case in which water is produced as a liquid. In Part (b) water is considered to form as a gas. The difference is important because gases have a large molar volume compared to the negligibly small molar volumes of solids and liquids. When a balanced reaction indicates a large net change in the number of moles of gas per reaction, $\Delta v_{gas} = v_{product\ gases} - v_{reactant\ gases}$, the magnitude of w is significantly larger. Should there be no change in Δv_{gas} (= 0), the extremely small changes in the volumes of reactant liquids and solids to product liquids and solids can be expressed as $\Delta V \simeq 0$ and, consequently, $w = -p_{ex}\Delta V\,[1.3] \simeq 0$.

(a) When the reaction water is written as liquid, gaseous carbon dioxide is produced and gaseous oxygen is consumed. The combustion of 1 mol glucose causes the gaseous change:

$$\Delta v_{gas} = v_{product\ gases} - v_{reactant\ gases} = 6 - 6 = 0$$

We conclude that $w = -p_{ex}\Delta V\,[1.3] \simeq \boxed{0}$.

(b) When the reaction water is written as gas, both gaseous carbon dioxide and gaseous water are produced, while gaseous oxygen is consumed. The combustion of 1 mol glucose causes the gaseous change: $\Delta v_{gas} = v_{product\ gases} - v_{reactant\ gases} = 6 + 6 - 6 = 6$. Expansion work is significant and it is caused by the appearance of 6 new moles of gas per reaction.

$$\Delta V \text{ per reaction} = \frac{\Delta v_{gas}\times RT}{p_{ex}} \quad \text{(Perfect gas law [F.6])}$$

$$w \text{ per reaction} = -p_{ex}\Delta V\,[1.3] = -p_{ex}\left(\frac{\Delta v_{gas}\times RT}{p_{ex}}\right) = -\Delta v_{gas}\times RT$$

$$w \text{ per mole glucose} = \frac{-\Delta v_{gas}\times RT}{1\ \text{mol glucose}}$$

$$w \text{ per amount glucose} = \left(\frac{-\Delta v_{gas}\times RT}{1\ \text{mol glucose}}\right)\times n_{glucose}$$

$$= -\left(\frac{(6\ \text{mol})(8.3145\ \text{J K}^{-1}\ \text{mol}^{-1})(293\ \text{K})}{1\ \text{mol glucose}}\right)\times(5.5\overline{5}\times10^{-3}\ \text{mol glucose})$$

$$= \boxed{-81\ \text{J}}$$

The reaction does 81 J of expansion work as the production of gas pushes on the surrounding atmospheric gases.

E1.13 $p_{ex} = 1.00 \times 10^5$ Pa, $\Delta V = 100\ \text{cm}^2 \times 10\ \text{cm} = 1.0 \times 10^3\ \text{cm}^3 = 1.0 \times 10^{-3}\ \text{m}^3$

$$w = -p_{ex}\Delta V\,[1.3]$$
$$= -1.00 \times 10^5\ \text{Pa} \times 1.0 \times 10^{-3}\ \text{m}^3 = -1.0 \times 10^2\ \text{Pa m}^3\ [1\ \text{Pa m}^3 = 1\ \text{J}]$$

$w = \boxed{-1.0 \times 10^2\ \text{J}}$ The expanding gas does 100 J of work while moving the piston.

E1.14 (a) For an isobaric expansion against p_{ex}:

$$w = -p_{ex}\Delta V\,[1.3] = -(30.0 \times 10^3\ \text{Pa}) \times (3.3\ \text{dm}^3) \times \left(\frac{1 \times 10^{-3}\ \text{m}^3}{1\ \text{dm}^3}\right) = \boxed{-99\ \text{J}}$$

(b) For an isothermal expansion:

$$n = \frac{4.50\ \text{g}}{16.04\ \text{g mol}^{-1}} = 0.280\bar{5}\ \text{mol},\ V_i = 12.7\ \text{dm}^3,\ V_f = (12.7 + 3.3)\ \text{dm}^3 = 16.0\ \text{dm}^3$$

For an infinitesimal change in volume the amount of work done will be infinitesimally small (i.e., dw). Thus, we write that in the case of mechanical equilibrium ($p_{ex} = p$, which is valid for a reversible expansion): $dw = -p_{ex}dV = -p\,dV$. For a perfect gas $p = nRT/V$ so the work expression becomes $dw = -(nRT/V)dV$. It follows that the total work of an expansion is given by the integrated form of our expression. At constant temperature, T moves outside the integration along with other constants giving:

$$w = -nRT\int_{V_i}^{V_f} \frac{1}{V}dV = -nRT \ln V\Big|_{V_i}^{V_f}$$
$$= -nRT\{\ln V_f - \ln V_i\}$$
$$= -nRT \ln \frac{V_f}{V_i} \quad \text{[reversible, isothermal expansion of perfect gas]}$$
$$= -(0.280\bar{5}\ \text{mol}) \times (8.3145\ \text{J K}^{-1}\ \text{mol}^{-1}) \times (310\ \text{K}) \times \ln\left(\frac{16.0\ \text{dm}^3}{12.7\ \text{dm}^3}\right) = \boxed{-167\ \text{J}}$$

E1.15 We take the room temperature and pressure to be $T = 298$ K and $p = 1.00$ atm. The volume is

$$V = (5.5\ \text{m}) \times (6.5\ \text{m}) \times (3.0\ \text{m}) = 1.0\bar{7} \times 10^5\ \text{dm}^3.$$

The amount of gas in the room is computed with the perfect gas law.

$$n = \frac{pV}{RT} = \frac{(1.00\ \text{atm}) \times (1.0\bar{7} \times 10^5\ \text{dm}^3)}{(8.206 \times 10^{-2}\ \text{dm}^3\ \text{atm K}^{-1}\ \text{mol}^{-1}) \times (298\ \text{K})} = 4.3\bar{8} \times 10^3\ \text{mol}$$

Thus, the energy required to raise the air temperature by 10°C (equivalent to a 10 K change) is

$$q = nC_{p,\text{m}}\Delta T$$
$$= (4.3\bar{8} \times 10^3\ \text{mol}) \times (21\ \text{J K}^{-1}\ \text{mol}^{-1}) \times (10\ \text{K}) = 9.2 \times 10^5\ \text{J} = \boxed{9.2\bar{0} \times 10^2\ \text{kJ}}.$$

Because $q = P \times t$, where P is the power of the heater and t is the time for which it operates,

$$t = \frac{q}{P} = \frac{9.2\bar{0} \times 10^5\ \text{J}}{1.5 \times 10^3\ \text{J s}^{-1}} = \boxed{6.1 \times 10^2\ \text{s}}.$$

In practice, the walls and furniture of a room are also heated.

E1.16 $q = nRT\ln\left(\frac{V_f}{V_i}\right)$ [See derivation of E1.14(b); reversible, isothermal expansion of perfect gas]

$$= (1.00\text{ mol}) \times (8.3145\text{ J K}^{-1}\text{ mol}^{-1}) \times (300\text{ K}) \times \ln\left(\frac{30.0}{22.0}\right) = \boxed{773\text{ J}}$$

E1.17 $w_{\text{lift}} = mgh\,[1.2] = (0.200\text{ kg}) \times (9.81\text{ m s}^{-2}) \times (1.55\text{ m}) = 3.04\text{ J}$

$w_{\text{animal}} = -w_{\text{lift}} = -3.04\text{ J}$

$\Delta U_{\text{animal}} = w_{\text{animal}} + q_{\text{animal}}\,[1.6] = (-3.04\text{ J}) + (-5.0\text{ J}) = \boxed{-8.04\text{ J}}$

E1.18 $q_p = \boxed{-1.2\text{ kJ}}$ (Heat leaves the sample.)

At constant pressure $\Delta H = q_p$ [1.13], hence $\Delta H = \boxed{-1.2\text{ kJ}}$.

$$C_p = \frac{\Delta H}{\Delta T}[1.14\text{a}] = \frac{-1.2\text{ kJ}}{-15\text{ K}} = \boxed{80\text{ J K}^{-1}}$$

E1.19 (a) The molar internal energy and enthalpy of a perfect gas are related by eqn 1.12b ($H_m = U_m + RT$), which we can write as $H_m - U_m = RT$. When the temperature increases by ΔT, the molar enthalpy increases by ΔH_m and the internal energy increases by ΔU_m, so $\Delta H_m - \Delta U_m = R\Delta T$. Now, divide both sides by ΔT, which gives

$$\frac{\Delta H_m}{\Delta T} - \frac{\Delta U_m}{\Delta T} = R$$

We recognize the first term on the left as the molar constant-pressure heat capacity, $C_{p,m}$, and the second term as the molar constant-volume heat capacity, $C_{V,m}$. Therefore,

$$C_{p,m} - C_{V,m} = R \quad [1.15, \text{perfect gas}]$$

(b) $C_p = \frac{q_p}{\Delta T}$ [1.13 and 1.14a] and division by n gives $C_{p,m} = \frac{q_p}{n\Delta T}$. Thus,

$$C_{p,m} = \frac{q_p}{n\Delta T} = \frac{229\text{ J}}{(3.00\text{ mol}) \times (2.06\text{ K})} = \boxed{37.1\text{ J K}^{-1}\text{ mol}^{-1}}$$

$$C_{V,m} = C_{p,m} - R\,[1.15] = (37.1 - 8.3145)\text{ J K}^{-1}\text{ mol}^{-1} = \boxed{28.8\text{ J K}^{-1}\text{ mol}^{-1}}$$

E1.20 (a) $\Delta H_m = C_{p,m}\Delta T$ [1.14a after division by n]

$$= (37.1\text{ J K}^{-1}\text{ mol}^{-1}) \times (37 - 15)\text{ K} = \boxed{81\bar{6}\text{ J mol}^{-1}}$$

(b) $\Delta U_m = \Delta H_m - \Delta(pV_m)$ [1.11a after division by n] $= \Delta H_m - \Delta(RT) = \Delta H_m - R\Delta T$

$$= 81\bar{6}\text{ J mol}^{-1} - (8.3145\text{ J K}^{-1}\text{ mol}^{-1}) \times (37 - 15\text{ K}) = \boxed{63\bar{3}\text{ J mol}^{-1}}$$

E1.21 $C_{V,m} = \frac{dU_m}{dT}$ [1.9b after division by n]

$$= \frac{d}{dT}(a + bT + cT^2) = \frac{da}{dT} + \frac{d(bT)}{dT} + \frac{d(cT^2)}{dT} = 0 + b + 2cT = \boxed{b + 2cT}$$

E1.22 $C_{p,\mathrm{m}} = \frac{\mathrm{d}H_\mathrm{m}}{\mathrm{d}T}$ [1.14b after division by n] so $\mathrm{d}H_\mathrm{m} = C_{p,\mathrm{m}}\mathrm{d}T$

ΔH_m is the sum of all infinitesimal changes, $\mathrm{d}H_\mathrm{m}$, between T_i and T_f, which is equivalent to the integration of $\mathrm{d}H_\mathrm{m}$ between T_i and T_f.

$$\Delta H_\mathrm{m} = \int_{T_\mathrm{i}}^{T_\mathrm{f}} \mathrm{d}H_\mathrm{m} = \int_{T_\mathrm{i}}^{T_\mathrm{f}} C_{p,\mathrm{m}}\mathrm{d}T$$

$$= \int_{T_\mathrm{i}}^{T_\mathrm{f}} \left(a + bT + \frac{c}{T^2}\right)\mathrm{d}T = \int_{T_\mathrm{i}}^{T_\mathrm{f}} a\mathrm{d}T + \int_{T_\mathrm{i}}^{T_\mathrm{f}} bT\mathrm{d}T + \int_{T_\mathrm{i}}^{T_\mathrm{f}} \frac{c}{T^2}\mathrm{d}T$$

$$= a\int_{T_\mathrm{i}}^{T_\mathrm{f}} \mathrm{d}T + b\int_{T_\mathrm{i}}^{T_\mathrm{f}} T\mathrm{d}T + c\int_{T_\mathrm{i}}^{T_\mathrm{f}} \frac{1}{T^2}\mathrm{d}T = \left[aT + \frac{bT^2}{2} - \frac{c}{T}\right]_{T_\mathrm{i}=288.15\ \mathrm{K}}^{T_\mathrm{f}=310.15\ \mathrm{K}}$$

$$H_\mathrm{m} = \left\{(44.22\ \mathrm{J\ K^{-1}\ mol^{-1}})(310.15\ \mathrm{K}) + \frac{(8.79\times10^{-3}\ \mathrm{J\ K^{-2}\ mol^{-1}})(310.15\ \mathrm{K})^2}{2} - \frac{(-8.62\times10^{5}\ \mathrm{J\ K\ mol^{-1}})}{310.15\ \mathrm{K}}\right\}$$

$$-\left\{(44.22\ \mathrm{J\ K^{-1}\ mol^{-1}})(288.15\ \mathrm{K}) + \frac{(8.79\times10^{-3}\ \mathrm{J\ K^{-2}\ mol^{-1}})(288.15\ \mathrm{K})^2}{2} - \frac{(-8.62\times10^{5}\ \mathrm{J\ K\ mol^{-1}})}{288.15\ \mathrm{K}}\right\}$$

$$\boxed{\Delta H_\mathrm{m} = 818\ \mathrm{J\ mol^{-1}}}$$

E1.23 (a) We arrive at an expression for ΔH_m as a function of T, $\Delta H_\mathrm{m}(T)$, by proceeding as in E1.22 with the difference that the upper integration limit is T. This gives

$$\Delta H_\mathrm{m}(T) = \left[aT + \frac{bT^2}{2} - \frac{c}{T}\right]_{T_\mathrm{i}=288.15\ \mathrm{K}}^{T=T}$$

$$= aT + \frac{bT^2}{2} - \frac{c}{T}$$

$$-\left\{(44.22\ \mathrm{J\ K^{-1}\ mol^{-1}})(288.15\ \mathrm{K}) + \frac{(8.79\times10^{-3}\ \mathrm{J\ K^{-2}\ mol^{-1}})(288.15\ \mathrm{K})^2}{2} - \frac{(-8.62\times10^{5}\ \mathrm{J\ K\ mol^{-1}})}{288.15\ \mathrm{K}}\right\}$$

$$\boxed{\Delta H_\mathrm{m}(T) = aT + \frac{bT^2}{2} - \frac{c}{T} - 16.1\ \mathrm{kJ\ mol^{-1}}}$$

(b) A plot of $\Delta H_\mathrm{m}(T)$ is presented in Figure 1.2.

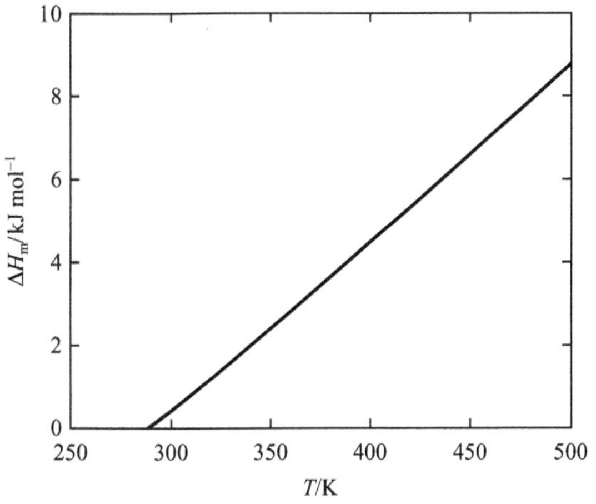

Figure 1.2

E1.24 A positive reaction enthalpy indicates an **endothermic reaction**, while a negative reaction enthalpy indicates an **exothermic reaction.**

(a) $\Delta_r H^\ominus < 0$, exothermic reaction
(b) $\Delta_r H^\ominus > 0$, endothermic reaction
(c) $\Delta_{vap} H^\ominus > 0$, endothermic reaction
(d) $\Delta_{fus} H^\ominus > 0$, endothermic reaction
(e) $\Delta_{sub} H^\ominus > 0$, endothermic reaction

E1.25 (a) $C(s, \text{graphite}) + O_2(g) \rightarrow CO_2(g)$ $\quad \Delta_{c1}H^\ominus = -393.5\ \text{kJ mol}^{-1}$
$C(s, \text{diamond}) + O_2(g) \rightarrow CO_2(g)$ $\quad \Delta_{c2}H^\ominus = -395.41\ \text{kJ mol}^{-1}$
$C(s, \text{graphite}) \rightarrow C(s, \text{diamond})$ $\quad \Delta_{trs}H^\ominus = \Delta_{c1}H^\ominus - \Delta_{c2}H^\ominus = \boxed{+1.9\ \text{kJ mol}^{-1}}$

(b) The reaction enthalpies of part (a) are those for $p^\ominus = 1$ bar = 100 kPa and we begin by asking how the transition enthalpy varies with pressure. If we assume that $\Delta_{trs}H(p)$ is a linear function of p, we can write the linear relation

$$\Delta_{trs}H(p) = \Delta_{trs}H^\ominus + \left\{\frac{\Delta(\Delta_{trs}H)}{\Delta p}\right\} \times (p - p^\ominus),$$

which suggests that knowledge of $\left(\frac{\Delta(\Delta_{trs}H)}{\Delta p}\right)$ facilitates the computation of $\Delta_{trs}H(p)$. Introduction of a volume change into this term, we have

$$\left(\frac{\Delta(\Delta_{trs}H)}{\Delta p}\right) = \left(\frac{\Delta(\Delta_{trs}H)}{\Delta(\Delta_{trs}V)}\right) \times \left(\frac{\Delta(\Delta_{trs}V)}{\Delta p}\right) \simeq 0$$

because the transition involves solids only and the volume of solids changes very little with a change in pressure so $\frac{\Delta(\Delta_{trs}V)}{\Delta p} \simeq 0$. Thus, we can make the reasonable estimate that $\Delta_{trs}H(p) = \Delta_{trs}H^\ominus =$ 1.9 kJ mol^{-1}.

$\Delta_{trs}U(p)$ and $\Delta_{trs}H(p)$ are related by the expression

$$\begin{aligned}
\Delta_{trs}U(p) &= \Delta_{trs}H(p) - p\Delta_{trs}V \quad [1.11b] \\
&= \Delta_{trs}H^\ominus - p\Delta_{trs}V = \Delta_{trs}H^\ominus - p\{V_{m,\text{diamond}} - V_{m,\text{graphite}}\} \\
&= \Delta_{trs}H^\ominus - p\left\{\left(\frac{M}{\rho}\right)_{\text{diamond}} - \left(\frac{M}{\rho}\right)_{\text{graphite}}\right\} \\
&= \Delta_{trs}H^\ominus - pM\left\{\left(\frac{1}{\rho}\right)_{\text{diamond}} - \left(\frac{1}{\rho}\right)_{\text{graphite}}\right\}
\end{aligned}$$

At $p = 150$ kbar $= 150 \times 10^8$ Pa:

$$\Delta_{trs}U = 1.9 \times 10^3\ \text{J mol}^{-1} - (150 \times 10^8\ \text{Pa}) \times \left(\frac{12.01\ \text{g}}{\text{mol}}\right) \times \left\{\frac{\text{cm}^3}{3.510\ \text{g}} - \frac{\text{cm}^3}{2.250\ \text{g}}\right\} \times \left\{\frac{\text{m}^3}{10^6\ \text{cm}^3}\right\} \quad [1\ \text{Pa m}^3 = 1\ \text{J}]$$

$$= +30.6 \times 10^3\ \text{J mol}^{-1} = \boxed{+30.6\ \text{kJ mol}^{-1}}$$

This calculation indicates that 94% of the internal energy change of the transformation is accounted for by the work of pushing the graphite crystal structure into the diamond structure.

E1.26 (a) $q = mC_{\text{s,water}}\Delta T$ [1.5b with $C = mC_{\text{s,water}}$ and $C_{\text{s,water}} = 4.18\ \text{J K}^{-1}\ \text{g}^{-1}$]

$$\Delta T = \frac{q}{mC_\text{s}} = \frac{10\times 10^6\ \text{J}}{(65\times 10^3\ \text{g})\times(4.18\ \text{J g}^{-1}\ \text{K}^{-1})} = 37\ \text{K} = \boxed{37°\text{C}}$$

(b) Let m_vap be the mass of water evaporated by q.

$$q = (m_\text{vap}/M)\Delta_\text{vap}H^\ominus$$

$$m_\text{vap} = \frac{Mq}{\Delta_\text{vap}H^\ominus} = \frac{(0.018016\ \text{kg mol}^{-1})\times(10\times 10^3\ \text{kJ})}{44\ \text{kJ mol}^{-1}} = \boxed{4.1\ \text{kg}}$$

This estimate ignores both the conduction of heat that occurs from high temperature to low temperature without evaporation and the small amount of heat used to bring ingested water to body temperature.

E1.27
$$n = \frac{m}{M} = \frac{100\ \text{g ice}}{18.0\ \text{g mol}^{-1}} = 5.55\ \text{mol}$$

The heat needed to melt 100 g of ice is

$$q_1 = n\times\Delta_\text{fus}H^\ominus = (5.55\ \text{mol})\times(6.01\ \text{kJ mol}^{-1}) = +33.4\ \text{kJ}.$$

The heat needed to raise the temperature of the water from 0°C to 100°C is

$$q_2 = mC_\text{s}(\text{l, water})\Delta T = (100\ \text{g})\times(4.18\ \text{J K}^{-1}\ \text{g}^{-1})\times(100\ \text{K}) = 4.18\times 10^4\ \text{J} = +41.8\ \text{kJ}.$$

The heat needed to vaporize the water is

$$q_3 = n\times\Delta_\text{vap}H^\ominus = (5.55\ \text{mol})\times(40.7\ \text{kJ mol}^{-1}) = +226\ \text{kJ}.$$

The total heat is $q = q_1 + q_2 + q_3 = 33.4\ \text{kJ} + 41.8\ \text{kJ} + 226\ \text{kJ} = \boxed{+301\ \text{kJ}}$.

The graph of temperature against time is sketched in Figure 1.3. Note that the length of the liquid + vapor, two-phase line is longer than the solid + liquid line in proportion to their $\Delta_\text{trs}H$ values.

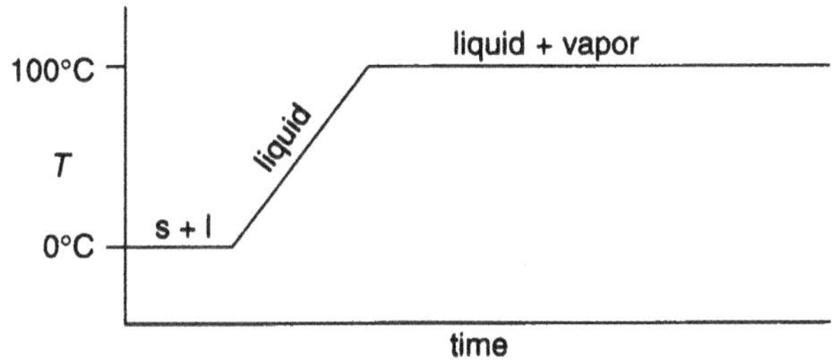

Figure 1.3

E1.28 Assuming that the calorimeter has constant volume,

$$\Delta U = w + q_\text{heater} = q_\text{heater} = IVt\ [1.17] = (0.02222\ \text{A})\times(11.8\ \text{V})\times(162\ \text{s}) = \boxed{+42.5\ \text{J}}$$

E1.29 The electric current provides the heat $q_{\text{heater}} = IVt$ [1.17], which vaporizes 0.798 g $H_2O(l)$ at 373.15 K and 1.0 atm pressure. Under these conditions:

$$\Delta_{\text{vap}}H = \frac{q_{\text{heater}}}{n_{\text{vaporized}}} = \frac{q_{\text{heater}}M}{m_{\text{vaporized}}} = \frac{IVtM}{m_{\text{vaporized}}}$$
$$= \frac{(0.50\text{ A})\times(12\text{ V})\times(300\text{ s})\times(18.015\text{ g mol}^{-1})}{0.798\text{ g}}$$
$$= \boxed{40.\bar{6}\text{ kJ mol}^{-1}}\quad [1\text{ A V s} = 1\text{ J}]$$

$$\Delta_{\text{vap}}U = \Delta_{\text{vap}}H - p\Delta_{\text{vap}}V_{\text{m}} \quad [1.11\text{b}]$$
$$= \Delta_{\text{vap}}H - p\{V_{\text{m}}(\text{water, g}) - V_{\text{m}}(\text{water, l})\} \quad [V_{\text{m}}(\text{water, l}) \ll V_{\text{m}}(\text{water, g})]$$
$$= \Delta_{\text{vap}}H - pV_{\text{m}}(\text{water, g})$$
$$= \Delta_{\text{vap}}H - RT \quad [\text{F.8}]$$
$$= 40.\bar{6}\times 10^3\text{ J mol}^{-1} - (8.3145\text{ J K}^{-1}\text{ mol}^{-1})\times(373.15\text{ K})$$
$$= \boxed{37.\bar{5}\text{ kJ mol}^{-1}}$$

E1.30 We begin by using the data in a computation of the calorimeter heat capacity C.

$$q = IVt\,[1.17] = (1.27\text{ A})\times(12.5\text{ V})\times(157\text{ s}) = 2.49\text{ kJ} \quad [1\text{ A s} = 1\text{ C}, 1\text{ C V} = 1\text{ J}]$$

$$C = \frac{q}{\Delta T}\,[1.16] = \frac{2.49\text{ kJ}}{3.88\text{ K}} = 0.642\text{ kJ K}^{-1}$$

With the use of an oxygen bomb calorimeter the combustion is at constant volume, giving

$$\Delta U_{\text{calorimeter}} = q_V = C\Delta T = (0.642\text{ kJ K}^{-1})\times(2.89\text{ K}) = \boxed{+1.86\text{ kJ}}$$

which is the energy released by the combustion reaction.

E1.31 We begin by using the benzoic acid data in a computation of the heat capacity C of a constant-volume combustion calorimeter. At constant volume $\Delta U = q_V$ [1.8] so

$$C = \frac{q_{V,\text{calorimeter}}}{\Delta T}\,[1.16] = \frac{-q_{V,\text{benzoic acid combustion}}}{\Delta T} = \frac{-(n\Delta_{\text{c}}U)_{\text{benzoic acid}}}{\Delta T} = \frac{-(m\Delta_{\text{c}}U/M)_{\text{benzoic acid}}}{\Delta T}$$
$$= \frac{-(0.917\text{ g})\times(-3226\text{ kJ mol}^{-1})}{(122.13\text{ g mol}^{-1})\times(1.940\text{ K})} = 12.4\overline{86}\text{ kJ K}^{-1}$$

The internal energy change of D-ribose combustion is

$$\Delta_{\text{c}}U = \left(\frac{q_V}{n}\right)_{\text{D-ribose}} = \frac{-q_{V,\text{calorimeter}}}{n_{\text{D-ribose}}} = -\frac{(C\Delta T)_{\text{calorimeter}}}{(m/M)_{\text{D-ribose}}}$$
$$= -\frac{(12.4\overline{86}\text{ kJ K}^{-1})\times(0.910\text{ K})\times(150.1\text{ g mol}^{-1})}{0.727\text{ g}}$$
$$= \boxed{-234\bar{6}\text{ kJ mol}^{-1}}$$

The combustion reaction for D-ribose is $C_5H_{10}O_5$(s, D-ribose) + 5 O_2(g) → 5 CO_2(g) + 5 H_2O(l). Since there is no change in the number of moles of gas, $\Delta_{\text{c}}H = \Delta_{\text{c}}U = -234\bar{6}$ kJ mol^{-1} [1.21]. By Hess's law the combustion enthalpy of D-ribose is related to standard formation enthalpies:

$$\Delta_{\text{c}}H^\ominus = 5\times\{\Delta_{\text{f}}H^\ominus(CO_2, \text{g}) + \Delta_{\text{f}}H^\ominus(H_2O, \text{l})\} - \{\Delta_{\text{f}}H^\ominus(\text{D-ribose, s}) + 5\times\Delta_{\text{f}}H^\ominus(O_2, \text{g})\} \quad [1.23]$$

Solving for $\Delta_f H$(s, D-ribose) gives:

$$\Delta_f H^{\ominus}(\text{D-ribose, s}) = 5 \times \{\Delta_f H^{\ominus}(\text{CO}_2, \text{g}) + \Delta_f H^{\ominus}(\text{H}_2\text{O, l})\} - 5 \times \Delta_f H^{\ominus}(\text{O}_2, \text{g}) - \Delta_c H^{\ominus}$$
$$= [5 \times \{(-393.51) + (-285.83)\} - 5 \times 0 - (-234\bar{6})] \text{ kJ mol}^{-1} \text{ [Resource data]}$$
$$= \boxed{-105\bar{1} \text{ kJ mol}^{-1}}$$

E1.32 Figure 1.4 is a modification of text Figure 1.27, the DSC thermogram of hen white lysozyme. It is apparent that the unfolding of the protein is endothermic with a so-called "melting temperature" of $T_m = 64.5°C$. The figure indicates the heat-capacity variation with temperature of both the native protein state and the denatured state. The baseline of the figure accounts for the heat-capacity variation of the native and denatured states during the period of denaturation. The additional contribution to the heat capacity, $C_{p,\text{ex}}$, due to the unfolding transition is given by the difference between the DSC curve and the baseline:

$$C_{p,\text{ex}} = C_{p,\text{m}} - C_{p,\text{baseline}}$$

The enthalpy of the transition is given by the thermogram area between the DSC curve and the baseline over the temperature range of 53°C to 79°C:

$$\Delta_{\text{trs}} H = \int_{326\text{ K}}^{352\text{ K}} C_{p,\text{ex}} \text{d}T \quad [1.18]$$

We can get a good estimate of the area represented by the above integral by dividing the temperature range into 2°C (i.e., 2 K) intervals with the middle of each interval aligned on the baseline and serving as the narrow base of a rectangle that extends vertically to a height at which the top of the rectangle intersects the DSC curve. We read both $C_{p,\text{m}}$ and $C_{p,\text{baseline}}$ for each rectangle and calculate $C_{p,\text{ex}}$. The thermogram area estimate for the integral of eqn 1.18 is the sum of all the rectangle areas:

$$\Delta_{\text{trs}} H = \sum_{\text{all rectangles}} (\text{rectangle area}) = (\text{rectangle width}) \sum_{\text{all rectangles}} (\text{rectangle height})$$
$$= (2\text{ K}) \sum_{\text{all rectangles}} (C_{p,\text{ex}})_{\text{rectangle}}$$

This procedure yields the following table of thermogram readings with the indicated temperature being the middle of each rectangle temperature interval. (The $C_{p,\text{ex}}$ values have been estimated to 0.1 kJ K^{-1} mol^{-1}. Since this estimate is very uncertain, at the end of the calculation we round-off to the next larger digit.)

θ/°C	$C_{p,\text{ex}}$/kJ K^{-1} mol^{-1}	θ/°C	$C_{p,\text{ex}}$/kJ K^{-1} mol^{-1}	θ/°C	$C_{p,\text{ex}}$/kJ K^{-1} mol^{-1}
54	0.6	64	59.4	74	3.8
56	5.6	66	51.9	76	1.3
58	15.0	68	30.0	78	0.3
60	30.0	70	15.0		
62	49.4	72	8.1		

Performing the requisite sum and calculating the transition enthalpy gives:

$$\sum_{\text{all rectangles}} (C_{p,\text{ex}})_{\text{rectangle}} = 270.\bar{4} \text{ kJ K}^{-1} \text{ mol}^{-1}$$

$$\Delta_{\text{trs}} H = (2\text{ K}) \sum_{\text{all rectangles}} (C_{p,\text{ex}})_{\text{rectangle}} = (2\text{ K}) \times (270.\bar{4} \text{ kJ K}^{-1} \text{ mol}^{-1}) = \boxed{541 \text{ kJ mol}^{-1}}$$

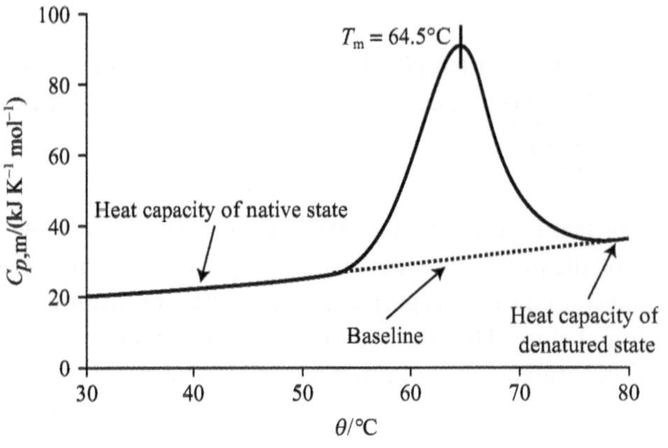

Figure 1.4

E1.33 The **specific enthalpy** is defined to be $-\Delta_c H^\circ/M$, so more positive values indicate more fuel-energy content per gram. When comparing a carbon-based series of compounds, all having similar molar masses, those that are least oxidized will have the most exothermic combustion enthalpies, $\Delta_c H^\circ$, and the higher specific enthalpies. This is illustrated by contrasting two important biological fuels: carbohydrates and fat. Carbon atoms in carbohydrates are bonded to oxygen atoms and are already partially oxidized, whereas most of the carbon atoms in fats are bonded to hydrogen and other carbon atoms and hence have lower oxidation numbers. Thus, glucose ($C_6H_{12}O_6$), having more highly oxidized carbon atoms than those of *n*-decanoic acid ($C_{10}H_{20}O_2$), has a lower specific enthalpy. The major origin of this difference between carbohydrates and fats of similar molar mass is that the combustion of the fat produces more of the very stable carbon dioxide and water molecules and, therefore, a net of more stable C=O and O−H bonds. To reinforce this point we use mean bond enthalpies of Tables 1.3 and 1.4 to estimate the gas-phase specific enthalpies of both glucose (180.16 g mol^{-1}, a carbohydrate) and capric acid (*n*-decanoic acid, 172.26 g mol^{-1}, a fatty acid). The bonding connections for these molecules are shown in Figure 1.5.

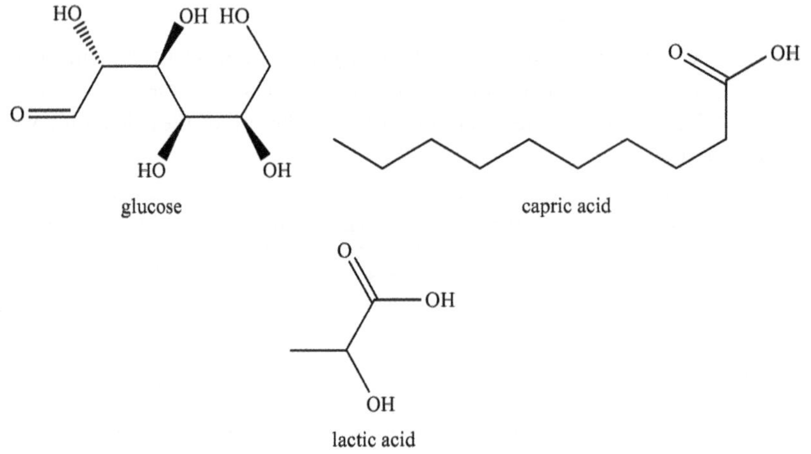

Figure 1.5

Glucose combustion (gas phase): $C_6H_{12}O_6(g) + 6\,O_2(g) \rightarrow 6\,CO_2(g) + 6\,H_2O(g)$

(See the note below for the justification that $\Delta H_B(C{=}O \text{ in } CO_2) = 804\ \text{kJ mol}^{-1}$.)

$$\begin{aligned}-\Delta_c H(\text{glucose, g}) &= \text{sum of product bond enthalpies} - \text{sum of reactant bond enthalpies}\\ &= 6\times[2\times\Delta H_B(C{=}O \text{ in } CO_2) + 2\times\Delta H_B(O{-}H)]\\ &\quad -[5\times\Delta H_B(C{-}C)+\Delta H_B(C{=}O)+7\times\Delta H_B(C{-}H)+5\times\Delta H_B(C{-}O)\\ &\quad +5\times\Delta H_B(O{-}H)+6\times\Delta H_B(O{=}O)]\end{aligned}$$

$$\begin{aligned}-\Delta_c H(\text{glucose, g})/\text{kJ mol}^{-1} &= 6\times[2\times(804)+2\times(463)]\\ &\quad -[5\times(348)+(743)+7\times(412)+5\times(360)+5\times(463)+6\times(497)]\\ &= 2740\end{aligned}$$

$$\begin{aligned}\text{Specific enthalpy of gas phase glucose} &= -\Delta_c H(\text{glucose, g})/M = (2740\ \text{kJ mol}^{-1})/(180.16\ \text{g mol}^{-1})\\ &= \boxed{15.2\ \text{kJ g}^{-1}}\end{aligned}$$

Capric acid combustion (gas phase): $C_{10}H_{20}O_2(g) + 14\,O_2(g) \rightarrow 10\,CO_2(g) + 10\,H_2O(g)$

$$\begin{aligned}-\Delta_c H(\text{capric acid, g}) &= \text{sum of product bond enthalpies} - \text{sum of reactant bond enthalpies}\\ &= 10\times[2\times\Delta H_B(C{=}O \text{ in } CO_2)+2\times\Delta H_B(O{-}H)]\\ &\quad -[9\times\Delta H_B(C{-}C)+\Delta H_B(C{=}O)+19\times\Delta H_B(C{-}H)\\ &\quad +\Delta H_B(C{-}O)+\Delta H_B(O{-}H)+14\times\Delta H_B(O{=}O)]\end{aligned}$$

$$\begin{aligned}-\Delta_c H(\text{capric acid, g})/\text{kJ mol}^{-1} &= 10\times[2\times(804)+2\times(463)]\\ &\quad -[9\times(348)+(743)+19\times(412)+(360)+(463)+14\times(497)]\\ &= 5856\end{aligned}$$

$$\begin{aligned}\text{Specific enthalpy of gas-phase capric acid} &= -\Delta_c H(\text{capric acid, g})/M\\ &= (5856\ \text{kJ mol}^{-1})/(172.26\ \text{g mol}^{-1}) = \boxed{34.0\ \text{kJ g}^{-1}}\end{aligned}$$

Glucose has less than half the specific enthalpy of capric acid even though these compounds have similar molar masses.

Note: The value of the mean bond energy $\Delta H_B(C{=}O \text{ in } CO_2)$ can be calculated with formation enthalpies from the *Resource data* appendix. Consider the atomization reaction: $CO_2(g) \rightarrow C(g) + 2\,O(g)$.

$$\Delta_r H^\circ = 2\times\Delta H_B(C{=}O \text{ in } CO_2) = \Delta_f H^\circ(C,\text{g}) + 2\times\Delta_f H^\circ(O,\text{g}) - \Delta_f H^\circ(CO_2,\text{g})$$

Therefore, $\Delta H_B(C{=}O \text{ in } CO_2) = \frac{1}{2}[716.68 + 2\times(249.17) - (-393.51)]\ \text{kJ mol}^{-1} = 804.27\ \text{kJ mol}^{-1}$

Alternatively, the mean bond energy can be calculated with the bond enthalpy, $H(A{-}B)$, values of Table 1.3 by consideration of the two-step atomization: $CO_2(g) \rightarrow O(g) + C{\equiv}O(g) \rightarrow C(g) + 2\,O(g)$.

$$\Delta_r H^\circ = 2\times\Delta H_B(C{=}O \text{ in } CO_2) = H(O{=}CO) + H(C{\equiv}O)$$

Therefore, $\Delta H_B(C{=}O \text{ in } CO_2) = \frac{1}{2}[531 + 1074]\ \text{kJ mol}^{-1} = 803\ \text{kJ mol}^{-1}$

E1.34 When using bond enthalpies (Tables 1.3 and 1.4) alone to estimate reaction enthalpies, it is not possible to include condensed phase interactions so the estimates are for gas-phase reactions.

(a) $C_6H_{12}O_6(g) \rightarrow 2\,CH_3CH(OH)COOH(g)$

The molecular formulas of these compounds are shown both below and in Figure 1.5.

7 C–H single bonds 5 O–H single bonds 5 C–O single bonds 1 C=O double bond 5 C–C single bond	HC=O | HC–OH | HO–CH | HC–OH | HC–OH | CH_2OH glucose

4 C–H single bonds × 2 2 O–H single bonds × 2 2 C–O single bonds × 2 1 C=O double bond × 2 2 C–C single bonds × 2	O // C–OH | HO–C–H | CH_3 lactic acid

$$\begin{aligned}\Delta_r H &= \text{sum of reactant bond enthalpies} - \text{sum of product bond enthalpies}\\ &= [5\times\Delta H_B(\text{C–C}) + \Delta H_B(\text{C=O}) + 7\times\Delta H_B(\text{C–H}) + 5\times\Delta H_B(\text{C–O}) + 5\times\Delta H_B(\text{O–H})]\\ &\quad - 2\times[2\times\Delta H_B(\text{C–C}) + \Delta H_B(\text{C=O}) + 4\times\Delta H_B(\text{C–H}) + 2\times\Delta H_B(\text{C–O}) + 2\times\Delta H_B(\text{O–H})]\\ &= \Delta H_B(\text{C–C}) - \Delta H_B(\text{C=O}) - \Delta H_B(\text{C–H}) + \Delta H_B(\text{C–O}) + \Delta H_B(\text{O–H})\\ &= [(348) - (743) - (412) + (360) + (463)]\ \text{kJ mol}^{-1}\\ &= \boxed{+16\ \text{kJ mol}^{-1}}\end{aligned}$$

(b) Glucose combustion (gas phase): $C_6H_{12}O_6(g) + 6\,O_2(g) \rightarrow 6\,CO_2(g) + 6\,H_2O(g)$

(See the note at the end of E1.33 for the justification that $\Delta H_B(\text{C=O in }CO_2) = 804\ \text{kJ mol}^{-1}$.)

$$\begin{aligned}\Delta_c H &= \text{sum of reactant bond enthalpies} - \text{sum of product bond enthalpies}\\ &= [5\times\Delta H_B(\text{C–C}) + \Delta H_B(\text{C=O}) + 7\times\Delta H_B(\text{C–H}) + 5\times\Delta H_B(\text{C–O}) + 5\times\Delta H_B(\text{O–H})\\ &\quad + 6\times\Delta H_B(\text{O=O})] - 6\times[2\times\Delta H_B(\text{C=O of }CO_2) + 2\times\Delta H_B(\text{O–H})]\\ &= \{[5\times(348) + (743) + 7\times(412) + 5\times(360) + 5\times(463) + 6\times(497)]\\ &\quad - 6\times[2\times(804) + 2\times(463)]\}\ \text{kJ mol}^{-1}\\ &= \boxed{-2740\ \text{kJ mol}^{-1}}\end{aligned}$$

E1.35 (a) $q = n\Delta_c H^{\ominus} = \dfrac{1.5\ \text{g}}{342.3\ \text{g mol}^{-1}} \times (-5645\ \text{kJ mol}^{-1})$ [Resource data] $= \boxed{-24.\bar{7}\ \text{kJ}}$

(b) The energy available for work is $\tfrac{1}{4}\times(+24.\bar{7}\ \text{kJ}) = +6.1\bar{8}\ \text{kJ}$

The energy expended as work during the climb is mgh [1.2]. Therefore, for a mass of 80.0 kg the energy available from the sugar lump will be expended after a climb of height given by

$$h = \frac{\text{energy available}}{mg} = \frac{+6.1\bar{8}\times 10^3\ \text{J}}{(80.0\ \text{kg})\times(9.81\ \text{m s}^{-2})} = \boxed{7.9\ \text{m}}$$

(c) Heat released $= -\Delta_r H = -n\Delta_c H^{\ominus} = -\left(\dfrac{2.5\ \text{g}}{180.16\ \text{g mol}^{-1}}\right)\times(-2808\ \text{kJ mol}^{-1}) = \boxed{+39.\bar{0}\ \text{kJ}}$

(d) With $\tfrac{1}{4}$ of the glucose energy available to do work and a mass of 80.0 kg:

$$h = \frac{\text{energy available}}{mg} = \frac{\tfrac{1}{4}\times(+39.\bar{0}\times 10^3\ \text{J})}{(80.0\ \text{kg})\times(9.81\ \text{m s}^{-2})} = \boxed{12.\bar{4}\ \text{m}}$$

E1.36 $C_3H_8(l) \rightarrow C_3H_8(g)$ $\Delta_{vap}H^{\ominus} = +15\ \text{kJ mol}^{-1}$

$C_3H_8(g) + 5\,O_2(g) \rightarrow 3\,CO_2(g) + 4\,H_2O(l)$ $\Delta_c H^{\ominus}(\text{propane, g}) = -2220\ \text{kJ mol}^{-1}$

(a) $\Delta_c H^{\ominus}(\text{propane, l}) = \Delta_{vap}H^{\ominus} + \Delta_c H^{\ominus}(\text{propane, g})$

$= +15\ \text{kJ mol} - 2220\ \text{kJ mol}^{-1} = \boxed{-2205\ \text{kJ mol}^{-1}}$

(b) $\Delta v_{gas} = -2[5\ O_2(g)$ replaced with $3\ CO_2(g)]$

$\Delta_c U^{\ominus}(\text{propane, l}) = \Delta_c H^{\ominus}(\text{propane, l}) - (-2)RT$ [1.21]

$= -2205\ \text{kJ mol}^{-1} + (2 \times 2.5\ \text{kJ mol}^{-1}) = \boxed{-2200\ \text{kJ mol}^{-1}}$

E1.37 $2\,C_2H_6(g) + 7\,O_2(g) \rightarrow 4\,CO_2(g) + 6\,H_2O(l)$ $\Delta_r H^{\ominus}(\text{ethane, g}) = -3120\ \text{kJ}$

(a) The standard enthalpy of combustion applies to the combustion of one mole, therefore

$$\Delta_c H^{\ominus}(\text{ethane, g}) = \frac{\Delta_r H^{\ominus}}{v_{ethane}} = \frac{-3120\ \text{kJ}}{2\ \text{mol}} = \boxed{-1560\ \text{kJ mol}^{-1}}$$

(b) The **specific enthalpy** is defined to be $-\Delta_c H^{\ominus}/M$, so more positive values indicate more fuel-energy content per gram and greater fuel efficiency.

Specific enthalpy of ethane $= -(\Delta_c H^{\ominus}/M)_{ethane}$

$= -(-1560\ \text{kJ mol}^{-1})/(30.07\ \text{g mol}^{-1}) = 51.88\ \text{kJ g}^{-1}$

Specific enthalpy of methane $= -(\Delta_c H^{\ominus}/M)_{methane}$

$= -(-890\ \text{kJ mol}^{-1})/(16.04\ \text{g mol}^{-1}) = 55.49\ \text{kJ g}^{-1}$

Ethane is a $\boxed{\text{slightly less efficient}}$ fuel than methane.

E1.38 $H_2(g) + \tfrac{1}{2}O_2(g) \rightarrow H_2O(l)$ $\Delta_f H(H_2O, l, p)$

$$\Delta_f H(H_2O, l, p) = \Delta_f H^{\ominus}(H_2O, l) + \left(\frac{\partial(\Delta_f H(H_2O, l))}{\partial p}\right)_T \times (p - p^{\ominus})$$

$$= \Delta_f H^{\ominus}(H_2O, l) + \left\{\left(\frac{\partial H_m(H_2O, l)}{\partial p}\right)_T - \left(\frac{\partial H_m(H_2, g)}{\partial p}\right)_T - \frac{1}{2}\left(\frac{\partial H_m(O_2, g)}{\partial p}\right)_T\right\} \times (p - p^{\ominus})$$

At the low pressures around 1 bar the gases are approximately perfect gases, which means that gas enthalpies do not change with pressure when T is held constant.

$$\Delta_f H(H_2O, l, p) = \Delta_f H^{\ominus}(H_2O, l) + \left(\frac{\partial H_m(H_2O, l)}{\partial p}\right)_T \times (p - p^{\ominus})$$

The partial derivative is simplified with the relationship

$$\left(\frac{\partial H_m}{\partial p}\right)_T = V_m - T\left(\frac{\partial V_m}{\partial T}\right)_p \simeq V_m \quad \text{for a condensed phase}$$

(See Problem 2.34 in P. Atkins and J. de Paula, *Physical Chemistry*, 8th edn, W.H. Freeman, 2006.)

$$\Delta_f H(H_2O, l, p) = \Delta_f H^{\ominus}(H_2O, l) + V_m \times (p - p^{\ominus})$$

$$\Delta_f H(H_2O, l, p) - \Delta_f H^{\ominus}(H_2O, l) = V_m \times (p - p^{\ominus}) \text{ where } V_m = \left(\frac{1.00\ \text{cm}^3}{1\ \text{g}}\right) \times \left(\frac{18.0\ \text{g}}{1\ \text{mol}}\right) = \frac{18.0 \times 10^{-6}\ \text{m}^3}{1\ \text{mol}}$$

Evaluating this difference at $p = 1.000$ atm $= 1.013$ bar yields

$$\Delta_f H(H_2O, l, 1\ atm) - \Delta_f H^{\circ}(H_2O, l) = \frac{18.0 \times 10^{-6}\ m^3}{1\ mol} \times (1.013 - 1)\ bar \times \left(\frac{10^5\ Pa}{1\ bar}\right)$$

$$= \boxed{0.0234\ J\ mol^{-1}}$$

E1.39 (a) glycylglycine(s) + H_2O(l) $\rightarrow$ 2 glycine(s)

$$\Delta_r H^{\circ} = 2\Delta_f H^{\circ}(glycine, s) - [\Delta_f H^{\circ}(dipeptide, s) + \Delta_f H^{\circ}(H_2O, l)]$$
$$= \{2 \times (-528.5) - [(-747.7) + (-285.83)]\}\ kJ\ mol^{-1}$$
$$= \boxed{-23.47\ kJ\ mol^{-1}}$$

(b) glycylglycine(s) + H_2O(l) $\rightarrow$ 2 glycine(aq)

$$\Delta_r H^{\circ} = 2\Delta_f H^{\circ}(glycine, aq) - [\Delta_f H^{\circ}(dipeptide, s) + \Delta_f H^{\circ}(H_2O, l)]$$
$$= \{2 \times (-469.8) - [(-747.7) + (-285.83)]\}\ kJ\ mol^{-1}$$
$$= \boxed{+93.9\ kJ\ mol^{-1}}$$

(c) $C_6H_{12}O_6$(β-D-fructose, s) + 6 O_2(g) $\rightarrow$ 6 CO_2(g) + 6 H_2O(l)

$$\Delta_c H^{\circ} = 6 \times \{\Delta_f H^{\circ}(CO_2, g) + \Delta_f H^{\circ}(H_2O, l)\} - \{\Delta_f H^{\circ}(\beta\text{-D-fructose}, s) + 6 \times \Delta_f H^{\circ}(O_2, g)\}\ [1.23]$$
$$= 6 \times \{(-393.51) + (-285.83)\} - \{(-1265.6) + 6 \times (0)\}$$
$$= \boxed{-2810.44\ kJ\ mol^{-1}}$$

(d) NO_2(g) $\rightarrow$ NO(g) + O(g)

$$\Delta_r H^{\circ} = \Delta_f H^{\circ}(NO, g) + \Delta_f H^{\circ}(O, g) - \Delta_f H^{\circ}(NO_2, g)$$
$$= [(+90.25) + (+249.17) - (+33.18)]\ kJ\ mol^{-1}$$
$$= \boxed{+306.24\ kJ\ mol^{-1}}$$

E1.40 $C_6H_{12}O_6$(α-glucose, s) + O_2(g) $\rightarrow$ 2 $CH_3COCOOH$(s) + 2 H_2O(l) $\quad \Delta_r H^{\circ} = -480.7\ kJ\ mol^{-1}$

$$\Delta_r H^{\circ} = 2\Delta_f H^{\circ}(pyruvic\ acid, s) + 2\Delta_f H^{\circ}(water, l) - \Delta_f H^{\circ}(\alpha\text{-glucose}, s) - \Delta_f H^{\circ}(O_2, g)$$

$$\Delta_f H^{\circ}(pyruvic\ acid, s) = \tfrac{1}{2}\{\Delta_r H^{\circ} - 2\Delta_f H^{\circ}(water, l) + \Delta_f H^{\circ}(\alpha\text{-glucose}, s) + \Delta_f H^{\circ}(O_2, g)\}$$
$$= \tfrac{1}{2}\{(-480.7) - 2(-285.83) + (-1273.3) + (0)\}\ kJ\ mol^{-1}$$
$$= \boxed{-591.2\ kJ\ mol^{-1}}$$

$CH_3COCOOH$(s) + $\tfrac{5}{2}O_2$(g) $\rightarrow$ 3 CO_2(g) + 2 H_2O(l)

$$\Delta_c H^{\circ} = 3\Delta_f H^{\circ}(CO_2, g) + 2\Delta_f H^{\circ}(H_2O, l) - \{\Delta_f H^{\circ}(pyruvic\ acid, s) + \tfrac{5}{2} \times \Delta_f H^{\circ}(O_2, g)\}\quad [1.23]$$
$$= [\{3 \times (-393.51) + 2 \times (-285.83)\} - \{(-591.2) + \tfrac{5}{2} \times (0)\}]\ kJ\ mol^{-1}$$
$$= \boxed{-1161.0\ kJ\ mol^{-1}}$$

E1.41 protein(native) $\rightarrow$ protein(denatured) $\quad \Delta_r H^{\circ}(298\ K) = +217.6\ kJ\ mol^{-1} \quad \Delta_r C_p^{\circ} = +6.3\ kJ\ K^{-1}\ mol^{-1}$

(a) $\Delta_r H^{\circ}(T) = \Delta_r H^{\circ}(298\ K) + \Delta_r C_p^{\circ} \times (T - 298\ K)$ [1.24, temperature-independent heat capacities]

(i) $\Delta_r H^{\circ}(351\ K) = +217.6\ kJ\ mol^{-1} + (+6.3\ kJ\ K^{-1}\ mol^{-1}) \times (351\ K - 298\ K)$

$$= \boxed{+55\bar{2}\ kJ\ mol^{-1}}$$

(ii) $\Delta_r H^{\ominus}(263\text{ K}) = +217.6\text{ kJ mol}^{-1} + (+6.3\text{ kJ K}^{-1}\text{ mol}^{-1}) \times (263\text{ K} - 298\text{ K})$

$$= \boxed{-2.9\text{ kJ mol}^{-1}}$$

(b) Answer (i) of part (a) is endothermic while answer (ii) is exothermic. Therefore, there is a temperature T_{zero} between 263 K and 351 K such that below T_{zero} the reaction is exothermic, above T_{zero} it is endothermic, and $\Delta_r H^{\ominus}(T_{zero}) = 0$. Solving the eqn of part (a) for T_{zero} gives

$$\begin{aligned} T_{zero} &= 298\text{ K} - \Delta_r H^{\ominus}(298\text{ K})/\Delta_r C_p^{\ominus} \\ &= 298\text{ K} - (+217.6\text{ kJ mol}^{-1})/(+6.3\text{ kJ K}^{-1}\text{ mol}^{-1}) \\ &= \boxed{263.\bar{5}\text{ K}} \quad \text{(slightly greater than 263 K)} \end{aligned}$$

E1.42 $\Delta_{vap} H^{\ominus}(T') = \Delta_{vap} H^{\ominus}(T) + \Delta_r C_p^{\ominus} \times (T' - T)$ [1.24]

$$\begin{aligned} \Delta_{vap} H^{\ominus}(373\text{ K}) &= 44.01\text{ kJ mol}^{-1} + (33.58\text{ J K}^{-1}\text{ mol}^{-1} - 75.29\text{ J K}^{-1}\text{ mol}^{-1}) \times (373\text{ K} - 298\text{ K}) \\ &= \boxed{40.88\text{ kJ mol}^{-1}} \end{aligned}$$

E1.43 $C_6H_{12}O_6(\alpha\text{-glucose, s}) + 6\,O_2(g) \rightarrow 6\,CO_2(g) + 6\,H_2O(l) \quad \Delta_c H^{\ominus} = -2808\text{ kJ mol}^{-1}$

$\Delta_c H^{\ominus}(T) = \Delta_c H^{\ominus}(298\text{ K}) + \Delta_c C_p^{\ominus} \times (T - 298\text{ K})$ [1.24, temperature-independent heat capacities]

The above application of Kirchhoff's law, which uses the 298 K reference temperature, indicates that, if $\Delta_c C_p^{\ominus} > 0$, the combustion enthalpy will be less exothermic at temperatures above 298 K. Using the *Resource data* to check the value of $\Delta_c C_p^{\ominus}$, we find

$$\begin{aligned} \Delta_c C_p^{\ominus} &= 6\{C_{p,m}^{\ominus}(CO_2, g) + C_{p,m}^{\ominus}(H_2O, l)\} - \{C_{p,m}^{\ominus}(\alpha\text{-glucose, s}) + 6C_p^{\ominus}(O_2, g)\} \\ &= 6 \times \{37.11 + 75.291\}\text{ J K}^{-1}\text{ mol}^{-1} - \{C_{p,m}^{\ominus}(\alpha\text{-glucose, s}) + 6 \times (29.355\text{ J K}^{-1}\text{ mol}^{-1})\} \\ &= 498.28\text{ J K}^{-1}\text{ mol}^{-1} - C_{p,m}^{\ominus}(\alpha\text{-glucose, s}) \end{aligned}$$

The CRC *Handbook of Chemistry and Physics* (49th edn, 1968–1969) reports that the heat capacity of glucose(s) is 219 J K^{-1} mol^{-1} so

$$\begin{aligned} \Delta_c C_p^{\ominus} &= 498.28\text{ J K}^{-1}\text{ mol}^{-1} - 219\text{ J K}^{-1}\text{ mol}^{-1} \\ &= \boxed{279\text{ J K}^{-1}\text{ mol}^{-1}} \end{aligned}$$

At the physiological temperature 37°C the combustion enthalpy is

$$\begin{aligned} \Delta_c H^{\ominus}(T) &= (-2808\text{ kJ mol}^{-1}) + (279 \times 10^{-3}\text{ kJ K}^{-1}\text{ mol}^{-1}) \times (310\text{ K} - 298\text{ K}) \\ &= \boxed{-2805\text{ kJ mol}^{-1}} \end{aligned}$$

and we concluded that the combustion enthalpy is slightly $\boxed{\text{less exothermic}}$ at blood temperature than at the conventional temperature (25°C).

E1.44 The enthalpy change due to cooling reactants from T to 298 K is given by

$$\Delta_{(a)} H^{\ominus} = \sum \nu C_{p,m}^{\ominus}(\text{reactants}) \times (298\text{ K} - T)$$

The reaction enthalpy at 298 K is simply $\Delta_{(b)} H^{\ominus}(298\text{ K}) = \Delta_r H^{\ominus}(298\text{ K})$.

The enthalpy change due to heating products from 298 K to T is given by

$$\Delta_{(c)} H^{\ominus} = \sum \nu C_{p,m}^{\ominus}(\text{products}) \times (T - 298\text{ K})$$

The reaction enthalpy at temperature T is the sum of the contributions of process (a), (b), and (c).

$$\begin{aligned}\Delta_r H^\circ(T) &= \Delta_{(a)}H^\circ + \Delta_{(b)}H^\circ + \Delta_{(c)}H^\circ \\ &= \sum \nu C^\circ_{p,m}(\text{reactants}) \times (298\text{ K} - T) + \Delta_r H^\circ(298\text{ K}) + \sum \nu C^\circ_{p,m}(\text{products}) \times (T - 298\text{ K}) \\ &= \Delta_r H^\circ(298\text{ K}) + \sum \nu C^\circ_{p,m}(\text{products}) \times (T - 298\text{ K}) - \sum \nu C^\circ_{p,m}(\text{reactants}) \times (T - 298\text{ K}) \\ &= \boxed{\Delta_r H^\circ(298\text{ K}) + \Delta_r C^\circ_p \times (T - 298\text{ K})}\end{aligned}$$

where $\Delta_r C^\circ_p = \sum \nu C^\circ_{p,m}(\text{products}) - \sum \nu C^\circ_{p,m}(\text{reactants})$

E1.45 The internal energy change due to cooling reactants from T to 298 K is given by

$$\Delta_{(a)}U^\circ = \sum \nu C^\circ_{V,m}(\text{reactants}) \times (298\text{ K} - T)$$

The reaction internal energy at 298 K is simply $\Delta_{(b)}U^\circ(298\text{ K}) = \Delta_r U^\circ(298\text{ K})$.

The internal energy change due to heating products from 298 K to T is given by

$$\Delta_{(c)}U^\circ = \sum \nu C^\circ_{V,m}(\text{products}) \times (T - 298\text{ K})$$

The reaction internal energy at temperature T is the sum of the contributions of process (a), (b), and (c).

$$\begin{aligned}\Delta_r U^\circ(T) &= \Delta_{(a)}U^\circ + \Delta_{(b)}U^\circ + \Delta_{(c)}U^\circ \\ &= \sum \nu C^\circ_{V,m}(\text{reactants}) \times (298\text{ K} - T) + \Delta_r U^\circ(298\text{ K}) + \sum \nu C^\circ_{V,m}(\text{products}) \times (T - 298\text{ K}) \\ &= \Delta_r U^\circ(298\text{ K}) + \sum \nu C^\circ_{V,m}(\text{products}) \times (T - 298\text{ K}) - \sum \nu C^\circ_{V,m}(\text{reactants}) \times (T - 298\text{ K}) \\ &= \boxed{\Delta_r U^\circ(298\text{ K}) + \Delta_r C^\circ_V \times (T - 298\text{ K})}\end{aligned}$$

where $\Delta_r C^\circ_V = \sum \nu C^\circ_{V,m}(\text{products}) - \sum \nu C^\circ_{V,m}(\text{reactants})$

E1.46 $\nu_A A + \nu_B B + \cdots \rightarrow \nu_P P + \nu_Q Q + \cdots \tfrac{1}{2}\Delta_r H(T)$

Let $dH_m(i)$ be the infinitesimal molar enthalpy change of the ith chemical species due to the infinitesimal temperature change dT. The infinitesimal reaction enthalpy change due to the temperature change is $d\Delta_r H(T)$, which equals the sum of $\nu_{\text{product}} dH_m(\text{product})$ minus the sum of $\nu_{\text{reactant}} dH_m(\text{reactant})$.

$$\begin{aligned}d\Delta_r H(T) &= \{\nu_P dH_m(P) + \nu_Q dH_m(Q) + \cdots\} - \{\nu_A dH_m(A) + \nu_B dH_m(B) + \cdots\} \\ &= \sum \nu dH_m(\text{products}) - \sum \nu dH_m(\text{reactants})\end{aligned}$$

Substitution of $dH_m(i) = C_{p,m}(i)\, dT$ [1.14b as molar quantities] gives

$$\begin{aligned}d\Delta_r H(T) &= \sum \nu C_{p,m}(\text{products})\, dT - \sum \nu C_{p,m}(\text{reactants})\, dT \\ &= \{\sum \nu C_{p,m}(\text{products}) - \sum \nu C_{p,m}(\text{reactants})\}\, dT\end{aligned}$$

or

$$d\Delta_r H^\circ(T) = \Delta_r C^\circ_p\, dT, \quad \text{where } \Delta_r C^\circ_p = \sum \nu C_{p,m}(\text{products}) - \sum \nu C_{p,m}(\text{reactants}).$$

Integration between T and T' gives

$$\int_T^{T'} d\Delta_r H^\circ(T) = \int_T^{T'} \Delta_r C^\circ_p\, dT$$

$$\Delta_r H^\circ(T') - \Delta_r H^\circ(T) = \int_T^{T'} \Delta_r C^\circ_p\, dT \quad \text{or} \quad \boxed{\Delta_r H^\circ(T') = \Delta_r H^\circ(T) + \int_T^{T'} \Delta_r C^\circ_p\, dT}.$$

If $\Delta_r C_p^\circ$ is either temperature independent or negligibly dependent upon temperature over the temperature range, the integral on the right simplifies to Kirchhoff's law.

$$\Delta_r H^\circ(T') = \Delta_r H^\circ(T) + \int_T^{T'} \Delta_r C_p^\circ \,\mathrm{d}T = \Delta_r H^\circ(T) + \Delta_r C_p^\circ \int_T^{T'} \mathrm{d}T = \Delta_r H^\circ(T) + \Delta_r C_p^\circ \times (T' - T)$$

If $\Delta_r C_p^\circ$ is temperature dependent, care must be taken with the integral on the right of the above boxed expression. For example, if $\Delta_r C_p^\circ = a + bT + c/T^2$ the integral is

$$\int_T^{T'} \Delta_r C_p^\circ \mathrm{d}T = \int_T^{T'} \left(a + bT + \frac{c}{T^2}\right)\mathrm{d}T = a\int_T^{T'} \mathrm{d}T + b\int_T^{T'} T\mathrm{d}T + c\int_T^{T'} \frac{1}{T^2}\mathrm{d}T = aT\Big|_T^{T'} + \frac{bT^2}{2}\Big|_T^{T'} - \frac{c}{T}\Big|_T^{T'}$$

$$= a(T' - T) + \frac{b}{2}(T'^2 - T^2) - c\left(\frac{1}{T'} - \frac{1}{T}\right)$$

and the reaction enthalpy is

$$\Delta_r H^\circ(T') = \Delta_r H^\circ(T) + \int_T^{T'} \Delta_r C_p^\circ \,\mathrm{d}T = \boxed{\Delta_r H^\circ(T) + a(T' - T) + \frac{b}{2}(T'^2 - T^2) - c\left(\frac{1}{T'} - \frac{1}{T}\right)}$$

Solutions to projects

E1.47 (a) $E_k = \frac{1}{2}mv_x^2 + \frac{1}{2}mv_y^2 + \frac{1}{2}mv_z^2$

There are three quadratic contributions to the kinetic energy of a particle free to move in three dimensions. The **equipartition theorem** states that, for a collection of particles at thermal equilibrium at temperature T, the mean value of each quadratic contribution to the energy is $\frac{1}{2}kT$. Thus, the mean kinetic energy of particles moving in three dimensions is $\boxed{\frac{3}{2}kT}$.

(b) The molar internal energy of a monatomic perfect gas is the sum of the mean kinetic energies of the atoms, which is $\frac{3}{2}N_A kT = \frac{3}{2}RT$, and the total internal electron energy of the atoms, which is given by the total internal energy of the atoms at absolute zero (i.e. $U_m(0)$). Being a perfect gas, the interaction energy between atoms is zero. Thus,

$$\boxed{U_m(T) = U_m(0) + \tfrac{3}{2}RT}$$

(c) The molecules of a perfect gas, which consists of linear molecules like N_2 and HCN and ethyne, have three quadratic contributions to the kinetic energy due to movement in three dimensions and two quadratic contributions to the rotational energy ($\frac{1}{2}I\omega_x^2$ and $\frac{1}{2}I\omega_y^2$) where the z-axis is along the molecular axis). According to the equipartition theorem, the five quadratic contributions provide a total mean molar kinetic energy of $\frac{5}{2}RT$. Quantization of the vibration motion, a quantum effect, separates the vibrational energies to the extent that only the lowest vibrational state is populated, thereby, making no thermal contribution to the total internal energy of the gas. The total internal electron energy of the molecules is given by the total internal energy of the molecules at absolute zero (i.e. $U_m(0)$). Being a perfect gas, the interaction energy between atoms is zero. The molar internal energy is the sum of these contributions. Thus,

$$\boxed{U_m(T) = U_m(0) + \tfrac{5}{2}RT}$$

(d) Non-linear polyatomic molecules of a perfect gas have three quadratic contributions to the kinetic energy due to movement in three dimensions and three quadratic contributions to the rotational energy ($\frac{1}{2}I\omega_x^2$, $\frac{1}{2}I\omega_y^2$, and $\frac{1}{2}I\omega_z^2$). According to the equipartition theorem, the six quadratic contributions provide a total mean molar kinetic energy of $\frac{6}{2}RT = 3RT$. Quantization of the vibration motion, a quantum effect, separates the vibrational energies to the extent that only the lowest vibrational state is populated, thereby, making no thermal contribution to the total internal energy of the gas. The total internal electron energy of the molecules is given by the total internal energy of the molecules at absolute zero (i.e. $U_m(0)$). Being a perfect gas, the interaction energy between atoms is zero. The molar internal energy is the sum of these contributions. Thus,

$$\boxed{U_m(T) = U_m(0) + 3RT}$$

(e) All possible perfect gas cases are covered in parts (a)–(d), which show that the internal energy of a perfect gas depends upon temperature only. Thus, the internal energy of a perfect gas does not change when the gas undergoes isothermal expansion because the temperature has not changed.

(f) (i) A gaseous argon atom has three translational degrees of freedom (the components of motion in the x, y, and z directions). Consequently, the equipartition theorem assigns a mean energy of $\frac{3}{2}kT$ to each atom. The molar internal energy is

$$U_m = \tfrac{3}{2}N_A kT = \tfrac{3}{2}RT = \tfrac{3}{2}(8.3145\ \text{J mol}^{-1}\ \text{K}^{-1})(293.15\ \text{K}) = 3.656\ \text{kJ mol}^{-1}$$

$$U = nU_m = mM^{-1}U_m = (10.0\ \text{g})\left(\frac{1\ \text{mol}}{39.95\ \text{g}}\right)\left(\frac{3.656\ \text{kJ}}{\text{mol}}\right) = \boxed{0.915\ \text{kJ}}$$

(ii) A gaseous, linear, carbon dioxide molecule has three quadratic translational degrees of freedom (the components of motion in the x, y, and z directions) but it has only two rotational quadratic degrees of freedom because there is no rotation energy along the internuclear line. There is a total of five quadratic degrees of freedom for the molecule. Consequently, the equipartition theorem assigns a mean energy of $\frac{5}{2}kT$ to each molecule. The molar internal energy is

$$U_m = \tfrac{5}{2}N_A kT = \tfrac{5}{2}RT = \tfrac{5}{2}(8.3145\ \text{J mol}^{-1}\ \text{K}^{-1})(293.15\ \text{K}) = 6.093\ \text{kJ mol}^{-1}$$

$$U = nU_m = mM^{-1}U_m = (10.0\ \text{g})\left(\frac{1\ \text{mol}}{44.01\ \text{g}}\right)\left(\frac{6.093\ \text{kJ}}{\text{mol}}\right) = \boxed{1.38\ \text{kJ}}$$

(iii) A gaseous, non-linear, methane molecule has three quadratic translational degrees of freedom (the components of motion in the x, y, and z directions) and three quadratic rotational degrees of freedom. Consequently, the equipartition theorem assigns a mean energy of $\frac{6}{2}kT$ to each molecule. The molar internal energy is

$$U_m = \tfrac{6}{2}N_A kT = 3RT = 3(8.3145\ \text{J mol}^{-1}\ \text{K}^{-1})(293.15\ \text{K}) = 7.312\ \text{kJ mol}^{-1}$$

$$U = nU_m = mM^{-1}U_m = (10.0\ \text{g})\left(\frac{1\ \text{mol}}{16.04\ \text{g}}\right)\left(\frac{7.312\ \text{kJ}}{\text{mol}}\right) = \boxed{4.56\ \text{kJ}}$$

(g) $\Delta H_m = \Delta U_m + \Delta(pV_m)$ [1.10] $= \Delta U_m + \Delta(RT)$ [perfect gas] $= \Delta U_m + R\Delta T$

For an isothermal expansion $\Delta U_m = 0$ and $\Delta T = 0$. Therefore, $\boxed{\Delta H_m = 0}$.

E1.48 (a) Hooke's law restoring force:

$F = -k_f x$, where x is the difference in the end-to-end distance from the equilibrium value

(i) This is a very limited one-dimensional model in which different sections of the polymer can lay at identical positions as it is assumed that segment volumes are zero and a continuum of x values is permitted. Bond angles, bond lengths, and persistence lengths are ignored. Nevertheless, if the polymer chain is large so that x values are much smaller than the polymer diameter, while simultaneously being much larger than the persistence length l, Hooke's law restoring force can be useful.

(ii) The work of expanding the end-to-end distance from $x = 0$ to $x = x$ is given by:

$$w = -\int_{x=0}^{x=x} F\mathrm{d}x = k_\mathrm{f}\int_{x=0}^{x=x} x\mathrm{d}x = \boxed{\tfrac{1}{2}k_\mathrm{f}x^2}$$

A sketch of w against x is shown in Figure 1.6. The minimum is $w = 0$ at $x = 0$.

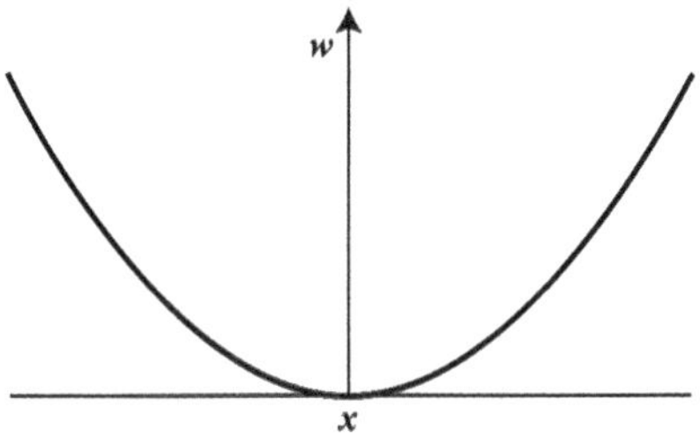

Figure 1.6

(b) (i) The freely jointed random coil model of a polymer chain of "units" or "residues" gives the simplest possibility for the conformation of the polymer that is not capable of forming hydrogen bonds or any other type of non-linkage bond. In this model, a bond that links adjacent units in the chain is free to make any angle with respect to the preceding one. We assume that the residues occupy zero volume, so different parts of the chain can occupy the same region of space. We also assume in the derivation of the expression for the probability of the ends of the chain being a distance nl apart, that the chain is compact in the sense that $n \ll N$. This model is obviously an oversimplification because a bond is actually constrained to a cone of angles around a direction defined by its neighbor. In a hypothetical one-dimensional freely jointed chain all the residues lie in a straight line, and the angle between neighbors is either 0° or 180°. The residues in a three-dimensional freely jointed chain are not restricted to lie in a line or a plane.

The random coil model ignores the role of the solvent: a poor solvent will tend to cause the coil to tighten; a good solvent does the opposite. Therefore, calculations based on this model are best regarded as lower bounds to the dimensions of a polymer in a good solvent and as an upper bound for a polymer in a poor solvent. The model is most reliable for a polymer in a bulk solid sample, where the coil is likely to have its natural dimensions.

(ii) $x = nl = 90$ nm so with $l = 45$ nm we see that $n = 2$ and $v = n/N = 2/200 = 0.01$. Using a negative sign to indicate a force that opposes changes from equilibrium:

$$F = -\frac{kT}{2l}\ln\left(\frac{1+v}{1-v}\right) = -\frac{(1.381\times10^{-23}\ \mathrm{J\ K^{-1}})\times(298.15\ \mathrm{K})}{2\times(45\times10^{-9}\ \mathrm{m})}\ln\left(\frac{1+v}{1-v}\right) = -(4.5\overline{7}\times10^{-14}\ \mathrm{N})\times\ln\left(\frac{1+v}{1-v}\right)$$

$$= -(4.5\overline{7}\times10^{-14}\ \mathrm{N})\times\ln\left(\frac{1+0.01}{1-0.01}\right)$$

$$= \boxed{-9.2\times10^{-16}\ \mathrm{N}}$$

(iii) A plot of F against v is shown in Figure 1.7. The figure uses the convention of a positive force for the compression of the polymer chain ends and a negative force for expansion. The tangent line at $v = 0$ is the Hooke's law force. It is apparent that the two models agree at small values of v and that the force described by the one-dimensional freely jointed chain is highly non-linear when $|v| > 0.3$. Hooke's force law is a linear law.

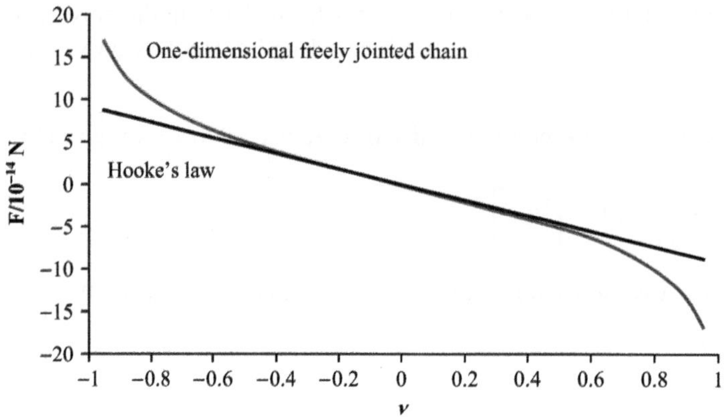

Figure 1.7

(iv) Since $x = nl = Nlv$, an infinitesimal change in x equals nl times an infinitesimal change in v: $dx = Nl\,dv$. The work of extending the ends of a polymer chain is related to v by the expression:

$$w = -\int_{x=0}^{x=x} F dx = -\int_{v=0}^{v=v} \left\{-\frac{kT}{2l}\ln\left(\frac{1+v}{1-v}\right)\right\} d(Nlv) = \frac{kNT}{2}\int_{v=0}^{v=v} \ln\left(\frac{1+v}{1-v}\right) dv$$

$$= \tfrac{1}{2}(1.381\times 10^{-23}\text{ J K}^{-1})\times(200)\times(298.15\text{ K})\times\int_{v=0}^{v=v}\ln\left(\frac{1+v}{1-v}\right)dv$$

$$= \boxed{(4.117\times 10^{-19}\text{ J})\times\int_{v=0}^{v=v}\ln\left(\frac{1+v}{1-v}\right)dv}$$

(v) For an extension to $v = 1$ the integral in the above expression can be performed numerically on a scientific calculator. The arguments of the typical integration function are shown in the following listing.

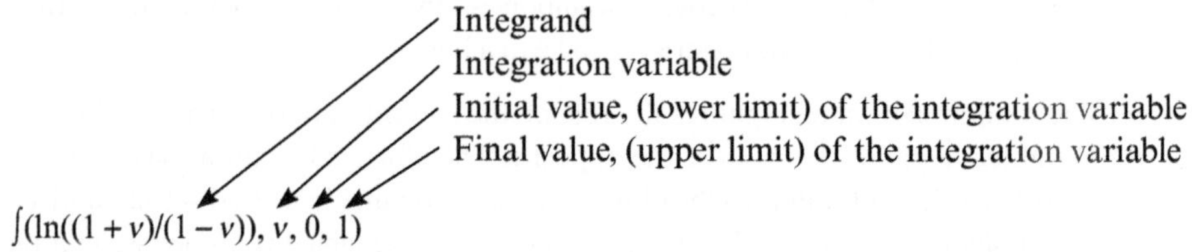

We find that $\int_{v=0}^{v=v}\ln\left(\frac{1+v}{1-v}\right)dv = 1.386$. Thus,

$$w = (4.117\times 10^{-19}\text{ J})\times\int_{v=0}^{v=v}\ln\left(\frac{1+v}{1-v}\right)dv = (4.117\times 10^{-19}\text{ J})\times(1.386) = \boxed{5.706\times 10^{-19}\text{ J}}$$

(c) Using a negative sign to indicate a force that opposes changes from equilibrium:

$F = -\frac{kT}{2l}\ln\left(\frac{1+v}{1-v}\right)$. Application of the logarithm property $\ln(x/y) = \ln x - \ln y$ gives

$F = -\frac{kT}{2l}\{\ln(1+v) - \ln(1-v)\}$. Using the Taylor series expansion $\ln(1+x) = x - \tfrac{1}{2}x^2 + \cdots$ yields

$$F = -\frac{kT}{2l}\{(v - \tfrac{1}{2}v^2 + \cdots) - (-v - \tfrac{1}{2}v^2 + \cdots)\}$$

$$= -\frac{kT}{2l}\{2v + \cdots\} \simeq -\frac{vkT}{l}$$ because higher powers of v are negligibly small when $v \ll 1$.

Since $v = n/N$, $\boxed{F \simeq -\frac{nkT}{Nl}}$

(d) Furthermore, $n = x/l$ so the above expression can be written in the form $F \simeq -\frac{kT}{Nl^2}x$.

Comparison with Hooke's law, $F = -k_f x$, reveals that the one-dimensional freely jointed chain model is identical with Hooke's law in the limit of very small values of v and that Hooke's law force constant is given by

$$\boxed{k_f = \frac{kT}{Nl^2}}$$

2 Bioenergetics: The Second Law

Answers to discussion questions

D2.1 $\Delta S_{\text{total}} = \Delta S_{\text{isolated system}} > 0$ (the universe is an isolated system) is a statement of the Second Law of thermodynamics. It is applicable to all macroscopic changes such as heat processes, chemical reactions, and phase transitions. It establishes criteria for spontaneity of changes and processes within an isolated system.

$dG \le 0$ is valid under conditions of constant temperature and pressure, with no additional work other than pressure–volume work. The statement originates with the Gibbs energy definition $G = H - TS$ and application of the Second Law. The inequality $dG < 0$ is the criterion for spontaneous change of constant T and constant p processes within non-isolated systems, including chemical reactions. The equality $dG = 0$ is the criterion for equilibrium with respect to constant T and constant p changes and processes within non-isolated systems.

D2.2 (a) The entropy of a sample increases as the temperature is raised from T_i to T_f, because the thermal disorder of the system is greater at the higher temperature due to the more vigorous molecular motion. Eqn 2.2 quantifies the entropy change when the heat capacity is constant over the range of temperatures:

$$\Delta S = C \ln \frac{T_f}{T_i} \quad [2.2]$$

where C is the heat capacity of the system; if the pressure is constant during heating, we use C_p, and if the volume is constant, we use C_V.

(b) The equation $\Delta G = \Delta H - T\Delta S$ [2.13] applies when a macroscopic finite change occurs in the system at constant temperature.

(c) The equation $\Delta G = w_{\text{max,non-exp}}$ [2.16] applies when a macroscopic finite change occurs in the system at constant temperature and pressure. See *Justification* 2.3.

D2.3 The entropy of unfolding of a protein can be determined in a manner similar to the determination of the enthalpy of unfolding of a protein using differential scanning calorimetry which is described in *In the Laboratory* 1.1. The entropy change for the process is given by

$$\Delta S = \int_{T_1}^{T_2} \frac{C_{p,\text{ex}}(T)}{T} dT$$

where T_1 is the temperature at which the unfolding begins and T_2 that where it ends. The data for $C_{p,\text{ex}}$ is obtained from the DSC and can be fitted to a polynomial as described in *In the Laboratory* 2.1.

$$C_{p,\text{ex}}(T) = a + bT + cT^2 + \cdots.$$

Then, after integration

$$\Delta S = a \ln\left(\frac{T_2}{T_1}\right) + b(T_2 - T_1) + \tfrac{1}{2}c(T_2^2 - T_1^2) + \cdots.$$

D2.4 Ludwig Boltzmann's fundamental equation $S = k \ln W$, where W is the number of ways that molecules of the system can be arranged yet correspond to the same total energy, provides the statistical definition of entropy that associates molecular motion and molecular quantum states with the thermodynamic entropy. We justify its identification with the thermodynamic entropy $\Delta S = q_{\text{rev}}/T$ [2.1] by the finding that the entropy computed with the Boltzmann equation matches the entropy value found by thermodynamically based experiments. This includes the entropy values at the absolute zero of temperature, residual entropies, the variation of entropy with temperature and volume, and the entropy changes of phase transitions.

D2.5 Residual entropy is due to the presence of some disorder in the system even at $T = 0$. It is observed in systems where there is very little energy difference—or none—between alternative arrangements of the molecules at very low temperatures. Consequently, the molecules cannot lock into a preferred orderly arrangement and some disorder persists.

D2.6 (a) $\boxed{\text{positive}}$, due to greater disorder in the product, though the difference may not be large;

(b) $\boxed{\text{negative}}$, less disorder (smaller number of moles of gas) in the product;

(c) $\boxed{\text{positive}}$, two new substances are formed, resulting in greater disorder on the product side.

D2.7 The increase in entropy of a solution when hydrophobic molecules or groups in molecules cluster together and reduce their structural demands on the solvent (water) is the origin of the hydrophobic interaction that tends to stabilize clustering of hydrophobic groups in solution. A manifestation of the hydrophobic interaction is the clustering together of hydrophobic groups in biological macromolecules. For example, the side chains of amino acids that are used to form the polypeptide chains of proteins are hydrophobic, and the hydrophobic interaction is a major contributor to the tertiary structure of polypeptides. At first thought, this clustering would seem to be a non-spontaneous process as the clustering of the solute results in a decrease in entropy of the solute. However, the clustering of the solute results in greater freedom of movement of the solvent molecules and an accompanying increase in disorder and entropy of the solvent. The total entropy of the system has increased and the process is spontaneous.

Solutions to exercises

E2.8 $$\Delta S_{\text{sur}} = \frac{q_{\text{sur}}}{T}\ [\text{Section 2.2(d)}] = \frac{120\ \text{J}}{293\ \text{K}} = \boxed{0.410\ \text{J K}^{-1}}$$

E2.9 $$\Delta_c H^\ominus = -2808\ \text{kJ mol}^{-1} = -q_{\text{body}}$$

$$\Delta S^\ominus_{\text{body}} = \frac{q_{\text{body}}}{T} = \frac{2808\ \text{kJ mol}^{-1} \times 100\ \text{g} \times \dfrac{1\ \text{mol}}{180\ \text{g}}}{273\ \text{K} + 37\ \text{K}} = \boxed{5.03\ \text{kJ K}^{-1}}$$

Note: The above calculation uses the value of $\Delta_c H^\ominus$ at 25°C. The value at 37°C should not be much different and can be calculated from knowledge of the heat capacities of all of the substances involved in the reaction.

E2.10 (a) We assume that the ice melts reversibly under the conditions described, therefore

$$\Delta S_{\text{ice}} = \frac{q_{\text{rev}}}{T} [2.1] = \frac{33\text{ kJ}}{273\text{ K}} = +0.12\text{ kJ K}^{-1}.$$

(b) $$\Delta S_{\text{sur}} = \frac{q_{\text{sur}}}{T} [\text{Section 2.2(d)}] = \frac{-33\text{ kJ}}{273\text{ K}} = \boxed{-0.12\text{ kJ K}^{-1}}$$

Note: Because this process is reversible, the total entropy change is zero.

E2.11 For the first step, melting 100 g ice:

$$\Delta_{\text{fus}}S = \frac{\Delta_{\text{fus}}H}{T_{\text{fus}}} = \frac{6.01\text{ kJ mol}^{-1}}{273\text{ K}} \times 100\text{ g} \times \frac{1\text{ mol}}{18.0\text{ g}} = \boxed{122\text{ J K}^{-1}}$$

For the second step, heating the water:

$$\Delta S = C_p \ln \frac{T_{\text{f}}}{T_{\text{i}}} = 4.18\text{ J K}^{-1}\text{ g}^{-1} \times 100\text{ g} \times \ln \frac{373}{273} = \boxed{130\text{ J K}^{-1}}$$

For the third step, vaporization:

$$\Delta_{\text{vap}}S = \frac{\Delta_{\text{vap}}H}{T_{\text{b}}} = \frac{40.7\text{ kJ mol}^{-1}}{373\text{ K}} \times 100\text{ g} \times \frac{1\text{ mol}}{18.0\text{ g}} = \boxed{606\text{ J K}^{-1}}$$

$$\Delta S_{\text{total}} = (122 + 130 + 606)\text{ J K}^{-1} = \boxed{858\text{ J K}^{-1}}$$

(a) A graph of temperature vs. time (Figure 2.1) shows a constant 273 K temperature until all the ice is melted. Temperature would increase until the boiling point, 373 K, is reached. Temperature again remains constant until all the liquid is vaporized.

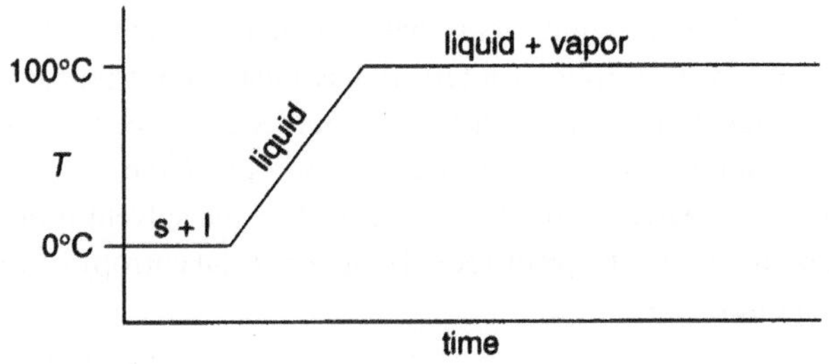

Figure 2.1

(b) A sketch of enthalpy as a function of time is shown in Figure 2.2. Note that absolute values of enthalpy are indeterminate.

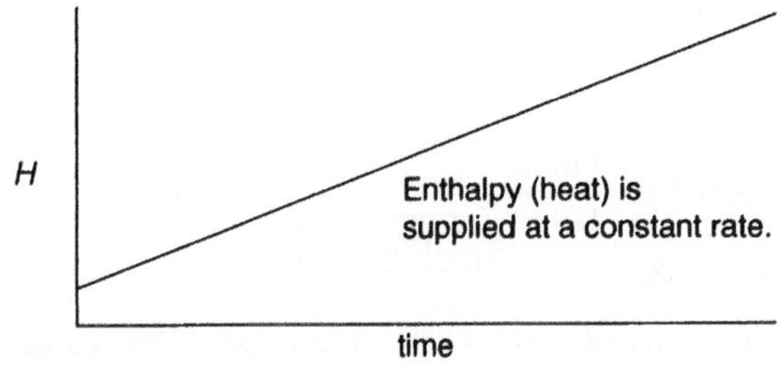

Figure 2.2

(c) A sketch of entropy as a function of time is shown in Figure 2.3.

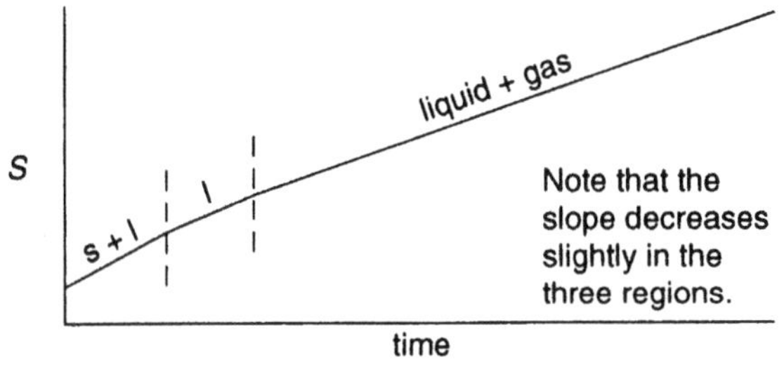

Figure 2.3

The graph of entropy against time does not look much different from that of enthalpy against time. The reason is that $\Delta t \propto \Delta H$. Therefore,

$$\frac{\Delta S}{\Delta t} \propto \frac{\Delta S}{\Delta H}.$$

In the three regions, we see that this ratio is roughly a constant.

$$\frac{\Delta S}{\Delta H} = \frac{122\ \text{J K}^{-1}}{33.3\ \text{kJ}} = 3.66\ \text{J K}^{-1}\ \text{kJ}^{-1}\ (\text{s} + \text{l region})$$

$$\frac{\Delta S}{\Delta H} = \frac{130\ \text{J K}^{-1}}{41.8\ \text{kJ}} = 3.11\ \text{J K}^{-1}\ \text{kJ}^{-1}\ (\text{liquid region})$$

$$\frac{\Delta S}{\Delta H} = \frac{606\ \text{J K}^{-1}}{226\ \text{kJ}} = 2.68\ \text{J K}^{-1}\ \text{kJ}^{-1}\ (\text{liquid} + \text{gas region})$$

E2.12

$$\Delta S = C_p \ln\frac{T_\text{f}}{T_\text{i}} = nC_{p,\text{m}} \ln\frac{T_\text{f}}{T_\text{i}} \quad [2.2]$$

$$= \frac{100\ \text{g}}{18.02\ \text{g mol}^{-1}} \times 75.5\ \text{J K}^{-1}\ \text{mol}^{-1} \times \ln\left(\frac{310\ \text{K}}{293\ \text{K}}\right) = \boxed{23.6\ \text{J K}^{-1}}$$

E2.13 We use eqn 2.7b.

$$S_\text{m}(T) - S_\text{m}(0) = \tfrac{1}{3}C_p(T)\ [2.7\text{b}] = \tfrac{1}{3} \times 1.2 \times 10^{-3}\ \text{J K}^{-1}\ \text{mol}^{-1} = 4.0 \times 10^{-4}\ \text{J K}^{-1}\ \text{mol}^{-1}$$

As $S_\text{m}(0)$ for the pure crystalline substance KCl is expected to be zero,

$$S_\text{m}(T) = \boxed{4.0 \times 10^{-4}\ \text{J K}^{-1}\ \text{mol}^{-1}}.$$

E2.14

$$\Delta S = \int_{T_\text{i}}^{T_\text{f}} \frac{C}{T}\,\text{d}T\ [2.3] = \int_{T_\text{i}}^{T_\text{f}} \left(\frac{a + bT + \dfrac{c}{T^2}}{T}\right)\text{d}T$$

$$= a\int_{T_\text{i}}^{T_\text{f}} \frac{1}{T}\,\text{d}T + b\int_{T_\text{i}}^{T_\text{f}} \text{d}T + c\int_{T_\text{i}}^{T_\text{f}} \frac{1}{T^3}\,\text{d}T = a\ln(T)\Big|_{T_\text{i}}^{T_\text{f}} + bT\Big|_{T_\text{i}}^{T_\text{f}} - \frac{c}{2T^2}\Big|_{T_\text{i}}^{T_\text{f}}$$

$$\Delta S = a\ln\left(\frac{T_\text{f}}{T_\text{i}}\right) + b(T_\text{f} - T_\text{i}) - \frac{c}{2}\left(\frac{1}{T_\text{f}^2} - \frac{1}{T_\text{i}^2}\right)$$

E2.15 First, find the common final temperature, T_f, by noting that the heat lost by the hot sample is gained by the cold sample.

$$q_{\text{cold sample 1}} = -q_{\text{warm sample 2}}$$
$$n_1 C_{p,m}(T_f - T_{i1}) = -n_2 C_{p,m}(T_f - T_{i2})$$

Solving for T_f

$$T_f = \frac{n_1 T_{i1} + n_2 T_{i2}}{n_1 + n_2}.$$

Because $n_1 = n_2 = \dfrac{100\text{ g}}{18.02\text{ g mol}^{-1}} = 5.55\text{ mol}$,

$$T_f = \tfrac{1}{2}(353\text{ K} + 283\text{ K}) = 318\text{ K}.$$

The total entropy change is therefore

$$\Delta S_{\text{total}} = \Delta S_1 + \Delta S_2 = n_1 C_{p,m} \ln\frac{T_f}{T_{i1}} + n_2 C_{p,m} \ln\frac{T_f}{T_{i2}} = n_1 C_{p,m} \ln\left\{\left(\frac{T_f}{T_{i1}}\right)\times\left(\frac{T_f}{T_{i2}}\right)\right\}$$
$$= (5.55\text{ mol}) \times (75.5\text{ J K}^{-1}\text{ mol}^{-1}) \times \ln\left\{\frac{318}{353}\times\frac{318}{283}\right\} = \boxed{5.11\text{ J K}^{-1}}.$$

E2.16 Taking the hint, we have

$$\Delta_{\text{trs}} S^\circ(25°\text{C}) = \Delta S_i + \Delta S_{ii} + \Delta S_{iii}.$$

We are not given the heat capacity of either the folded or unfolded protein, but if we let $C_{p,m}$ be the heat capacity of the folded protein, the heat capacity of the unfolded protein is $C_{p,m} + 6.28\text{ kJ K}^{-1}\text{ mol}^{-1}$. So for the heating and cooling steps, we have:

$$\Delta S_i = C_p \ln\left(\frac{T_f}{T_i}\right) = C_{p,m} \ln\left(\frac{348.7\text{ K}}{298.2\text{ K}}\right) \quad [2.2]$$

and $\Delta S_{iii} = (C_{p,m} + 6.28\text{ kJ K}^{-1}\text{ mol}^{-1}) \ln\left(\dfrac{298.2\text{ K}}{348.7\text{ K}}\right)$, so

$$\Delta S_i + \Delta S_{iii} = C_{p,m} \ln\left(\frac{348.7\text{ K}}{298.2\text{ K}}\right) + (C_{p,m} + 6.28\text{ kJ K}^{-1}\text{ mol}^{-1}) \ln\left(\frac{298.2\text{ K}}{348.7\text{ K}}\right)$$
$$= (6.28\text{ kJ K}^{-1}\text{ mol}^{-1}) \ln\left(\frac{298.2\text{ K}}{348.7\text{ K}}\right) = -0.983\text{ kJ K}^{-1}\text{ mol}^{-1}$$

For the transition itself, use the equation in the first Brief Illustration in Section 2.2(c):

$$\Delta S_{ii} = \frac{\Delta_{\text{trs}} H^\circ}{T_{\text{trs}}} = \frac{509\text{ kJ mol}^{-1}}{348.7\text{ K}} = 1.46\overline{0}\text{ kJ K}^{-1}\text{ mol}^{-1}$$

Hence, $\Delta_{\text{trs}} S^\circ = (1.46\overline{0} - 0.983)\text{ kJ K}^{-1}\text{ mol}^{-1} = 0.47\overline{7}\text{ kJ K}^{-1}\text{ mol}^{-1} = \boxed{47\overline{7}\text{ J K}^{-1}\text{ mol}^{-1}}$

E2.17 In a manner similar to eqns 2.4 and 2.5, we may write in general,

$$\Delta_{\text{trs}} S = \frac{\Delta_{\text{trs}} H}{T_{\text{trs}}};\ \text{therefore}\ \Delta_{\text{trs}} S = \frac{+1.9\text{ kJ mol}^{-1}}{2000\text{ K}} = \boxed{0.95\text{ J K}^{-1}\text{ mol}^{-1}}.$$

E2.18 (a) $\Delta_{\text{vap}}S = \dfrac{\Delta_{\text{vap}}H^{\ominus}}{T_{\text{b}}}\ [2.5] = \dfrac{35.27\times 10^{3}\ \text{J mol}^{-1}}{337.25\ \text{K}} = \boxed{+104.6\ \text{J K}^{-1}\ \text{mol}^{-1}}$

(b) Because the vaporization process can be accomplished reversibly, $\Delta S_{\text{total}} = 0$; hence

$\Delta S_{\text{sur}} = \boxed{-104.6\ \text{J K}^{-1}\ \text{mol}^{-1}}$.

E2.19 A liquid and its vapor are in equilibrium at the boiling point. Since the process is reversible under this condition, $\Delta S = q_{\text{rev}}/T$ [2.1] $= \Delta H/T$ [constant p] and the entropy of vaporization is given by $\Delta_{\text{vap}}S_{\text{m}} = \Delta_{\text{vap}}H_{\text{m}}/T_{\text{b}}$.

(a) **Trouton's rule** states that the ratio $\Delta_{\text{vap}}H/T_{\text{b}}$, and thus the vaporization entropy, is a constant. Explore the origin of the constancy by considering that the vaporization entropy has two components, only one of which depends upon liquid-phase properties. They are the molar entropy of the liquid and the molar entropy of the gas: $\Delta_{\text{vap}}S_{\text{m}} = S_{\text{m}}(\text{g}) - S_{\text{m}}(\text{l})$. Under ordinary conditions the value of $S_{\text{m}}(\text{g})$ is expected to be identical for all gases and relatively large with respect to $S_{\text{m}}(\text{l})$ because gases behave as perfect gases for which molecular volume and intermolecular forces are negligibly small. This allows completely random, very high entropy molecular motion, which is independent of molecular properties. In addition to the large $S_{\text{m}}(\text{g})$ value, should the value of $S_{\text{m}}(\text{l})$ be either negligibly small or a constant value for a series of compounds, $\Delta_{\text{vap}}S_{\text{m}}$, and the ratio $\Delta_{\text{vap}}H/T_{\text{b}}$, will be a constant. Trouton's rule is followed. Small, non-polar molecules provide examples that meet these conditions. The relative absence of molecular order in their liquid states gives relatively small and constant $S_{\text{m}}(\text{l})$ values. Thus, bromine, carbon tetrachloride, cyclohexane, and octane have approximate identical vaporization entropies (~85 J K^{-1} mol^{-1}).

It is also interesting to explore the origin of Trouton's rule with a careful analysis of the vaporization enthalpy (heat). Energy in the form of heat supplied to a liquid manifests itself as an increase in thermal motion. This is an increase in the kinetic energy of molecules. When the kinetic energy of the molecules is sufficient to overcome the attractive energy that holds them together the liquid vaporizes. The enthalpy of vaporization is the heat required to accomplish this at constant pressure. It seems reasonable that the greater the enthalpy of vaporization, the greater the kinetic energy required, and the greater the temperature needed to achieve this kinetic energy. Hence, we expect that $\Delta_{\text{vap}}H$ is proportional to T_{b}, which implies that their ratio is a constant.

(b) $\Delta_{\text{vap}}S = \dfrac{\Delta_{\text{vap}}H}{T_{\text{b}}} = +85\ \text{J K}^{-1}\ \text{mol}^{-1}$

$\Delta_{\text{vap}}H = \Delta_{\text{vap}}S \times T_{\text{b}} = +85\ \text{J K}^{-1}\text{mol}^{-1} \times 399\ \text{K} = \boxed{+34\ \text{kJ K}^{-1}\ \text{mol}^{-1}}$

(c) Exceptions to Trouton's rule include liquids in which the interactions between molecules result in the liquid being less disordered than the random jumble of molecules in something like carbon tetrachloride. This includes liquids in which hydrogen bonding creates local order, as in water and small alcohols. It also includes liquid metals in which the metallic bond creates atomic organization, as in mercury.

E2.20 $\Delta_{\text{fus}}C_p^{\ominus} = C_{p,\text{m}}^{\ominus}(\text{l}) - C_{p,\text{m}}^{\ominus}(\text{s})$ [similar to eqn 1.25] $= (28 - 19)\ \text{J K}^{-1}\ \text{mol}^{-1} = 9\ \text{J K}^{-1}\ \text{mol}^{-1}$

$$\Delta_{\text{fus}}S^{\ominus}(T') = \Delta_{\text{fus}}S^{\ominus}(T) + \int_T^{T'} \frac{\Delta_{\text{fus}}C_p^{\ominus}}{T}\,\text{d}T \quad \text{[phase transition analog of eqn 2.3]}$$

$$= \Delta_{\text{fus}}S^{\ominus}(T) + \Delta_{\text{fus}}C_p^{\ominus}\int_T^{T'} \frac{1}{T}\,\text{d}T = \Delta_{\text{fus}}S^{\ominus}(T) + \Delta_{\text{fus}}C_p^{\ominus}\ln(T)\Big|_T^{T'}$$

$$= \Delta_{\text{fus}}S^{\ominus}(T) + \Delta_{\text{fus}}C_p^{\ominus}\ln\left(\frac{T'}{T}\right)$$

At $T = T_{\text{fus}}$ the transition is reversible and $\Delta_{\text{fus}}S^{\ominus}(T_{\text{fus}}) = \dfrac{\Delta_{\text{fus}}H^{\ominus}(T_{\text{fus}})}{T_{\text{fus}}}$ [2.4].

$$\Delta_{\text{fus}}S^{\ominus}(T') = \frac{\Delta_{\text{fus}}H^{\ominus}(T_{\text{fus}})}{T_{\text{fus}}} + \Delta_{\text{fus}}C_p^{\ominus}\ln\left(\frac{T'}{T_{\text{fus}}}\right)$$

$$\Delta_{\text{fus}}S^{\ominus}(298\text{ K}) = \frac{32\text{ kJ mol}^{-1}}{419\text{ K}} + (9\text{ J K}^{-1}\text{ mol}^{-1})\times\ln\left(\frac{298\text{ K}}{419\text{ K}}\right)$$

$$= \boxed{73\text{ J K}^{-1}\text{ mol}^{-1}}$$

E2.21 Because the temperature is so low we can assume $C_p(T) = aT^3$ [2.7a]. Hence we can write

$$S(T) = S(0) + \int_0^T \frac{C_p(T)\mathrm{d}T}{T} = S(0) + \int_0^T \frac{aT^3\mathrm{d}T}{T} = S(0) + a\int_0^T T^2\mathrm{d}T = \tfrac{1}{3}aT^3$$

If we are dealing with a substance that has no residual entropy at $T = 0$ then $S(0) = 0$ and $S(T)$ becomes $S(T) = \frac{1}{3}C_p(T)$.

E2.22 In each case $S_{\text{m}} = R\ln s$, where s is the number of orientations of about equal energy that the molecule can adopt. Therefore,

(a) $S_{\text{m}} = R\ln 3 = 8.3145\text{ J K}^{-1}\text{ mol}^{-1}\times\ln 3 = \boxed{9.13\text{ J K}^{-1}\text{ mol}^{-1}}$

(b) $S_{\text{m}} = R\ln 5 = 8.3145\text{ J K}^{-1}\text{ mol}^{-1}\times\ln 5 = \boxed{13.4\text{ J K}^{-1}\text{ mol}^{-1}}$

(c) $S_{\text{m}} = R\ln 6 = 8.3145\text{ J K}^{-1}\text{ mol}^{-1}\times\ln 6 = \boxed{14.9\text{ J K}^{-1}\text{ mol}^{-1}}$

E2.23 The standard reaction entropy at 298 K can be calculated using eqn 2.9.

$$\Delta_{\text{r}}S^{\ominus} = \sum \nu S_{\text{m}}^{\ominus}(\text{products}) - \sum \nu S_{\text{m}}^{\ominus}(\text{reactants})$$
$$\Delta_{\text{r}}S^{\ominus} = (321.4\text{ J K}^{-1}\text{ mol}^{-1}) + (427.4\text{ J K}^{-1}\text{ mol}^{-1}) - (212\text{ J K}^{-1}\text{ mol}^{-1})$$
$$\Delta_{\text{r}}S^{\ominus} = \boxed{537\text{ J K}^{-1}\text{ mol}^{-1}}$$

E2.24 $\Delta S = nC_{p,\text{m}}\ln(T_{\text{f}}/T_{\text{i}})$ for each substance in each reaction, therefore

$\Delta_{\text{r}}S = \Delta_{\text{r}}C_p\ln(T_{\text{f}}/T_{\text{i}})$ for each reaction.

(a) $\Delta_{\text{r}}C_p = (2\text{ mol}\times 4R) - (3\text{ mol}\times\frac{7}{2}R) = -\frac{5}{2}R\text{ mol}$

$$\Delta_{\text{r}}S = -\tfrac{5}{2}R\text{ mol}\times\ln\left(\frac{283\text{ K}}{273\text{ K}}\right) = \boxed{-0.75\text{ J K}^{-1}}$$

(b) $\Delta_{\text{r}}C_p = (2\text{ mol}\times 4R) + (1\text{ mol}\times\frac{7}{2}R) - (1\text{ mol}\times 4R) - (2\text{ mol}\times\frac{7}{2}R) = +\frac{1}{2}R\text{ mol}$

$$\Delta_{\text{r}}S = +\tfrac{1}{2}R\text{ mol}\times\ln\left(\frac{283\text{ K}}{273\text{ K}}\right) = \boxed{+0.15\text{ J K}^{-1}}$$

E2.25 $N_2(g) + 3\,H_2(g) \rightarrow 2\,NH_3(g)$

$$\Delta_{\text{r}}S^{\ominus} = 2S_{\text{m}}^{\ominus}(NH_3,g) - S_{\text{m}}^{\ominus}(N_2,g) - 3S_{\text{m}}^{\ominus}(H_2,g) = (2\times 192.45 - 191.61 - 3\times 130.684)\text{ J K}^{-1}$$
$$= \boxed{-198.76\text{ J K}^{-1}}$$

$$\Delta_r H^\ominus = 2\Delta_f H^\ominus(\mathrm{NH_3,g}) = 2 \times (-46.11)\ \mathrm{kJ} = -92.22\ \mathrm{kJ}$$

$$\begin{aligned}\Delta_r G^\ominus &= \Delta_r H^\ominus - T\Delta_r S^\ominus\ [2.13]\\ &= -92.22\ \mathrm{kJ} - (298\ \mathrm{K}) \times (-0.19876\ \mathrm{kJ})\\ &= \boxed{-32.99\ \mathrm{kJ}}\end{aligned}$$

E2.26 (a) $\Delta G = \Delta H - T\Delta S$ [2.13]

$$= -125\ \mathrm{kJ\ mol^{-1}} - 310\ \mathrm{K} \times (-126\ \mathrm{J\ K^{-1}\ mol^{-1}}) = \boxed{-85.9\ \mathrm{kJ\ mol^{-1}}}$$

(b) Yes, ΔG is negative.

(c) $\Delta G = -T\Delta S_{\mathrm{total}}$ [2.14]

$$\Delta S_{\mathrm{total}} = -\frac{\Delta G}{T} = -\left(\frac{-85.9\ \mathrm{kJ\ mol^{-1}}}{310\ \mathrm{K}}\right) = \boxed{+0.277\ \mathrm{kJ\ K^{-1}\ mol^{-1}}}$$

E2.27 $\Delta G = w_{\mathrm{max,non\text{-}exp}} = -2808\ \mathrm{kJ\ mol^{-1}}$, so the maximum work that can be done is 2808 kJ mol^{-1}. We will assume that we will be able to extract the maximum work from the reaction.

$$\begin{aligned}w = mgh &= 65\ \mathrm{kg} \times 9.81\ \mathrm{m\ s^{-2}} \times 10\ \mathrm{m}\\ &= 6.4 \times 10^3\ \mathrm{J} = 6.4\ \mathrm{kJ}\end{aligned}$$

$$\mathrm{amount}(n) = \frac{6.4\ \mathrm{kJ}}{2808\ \mathrm{kJ\ mol^{-1}}} = 2.3 \times 10^{-3}\ \mathrm{mol}$$

$$\text{mass of glucose} = 2.3 \times 10^{-3}\ \mathrm{mol} \times 180\ \mathrm{g\ mol^{-1}} = \boxed{0.41\ \mathrm{g}}$$

E2.28 (a) Yes, coupling the two reactions can give a net ΔG that is negative, hence the overall process is spontaneous. For example, for one mole of glutamate and one mole of ATP,

$$\Delta G = (14.2 - 31)\ \mathrm{kJ\ mol^{-1}} = -17\ \mathrm{kJ\ mol^{-1}}.$$

(b) The minimum amount of ATP required is $\dfrac{1\ \mathrm{mol} \times (-14.2\ \mathrm{kJ\ mol^{-1}})}{-31\ \mathrm{kJ\ mol^{-1}}} = \boxed{0.46\ \mathrm{mol\ ATP}}$

E2.29 For the synthesis, $\Delta G = +42\ \mathrm{kJ\ mol^{-1}}$; hence, at least -42 kJ would need to be provided by the ATP in order to make ΔG overall negative.

$$\text{amount}(n)\ \text{of ATP} = \frac{-42\ \mathrm{kJ}}{-31\ \mathrm{kJ\ mol^{-1}}} = 1.35\ \mathrm{mol\ ATP}$$

$$1.35\ \mathrm{mol} \times 6.02 \times 10^{23}\ \mathrm{mol^{-1}} = \boxed{8.1 \times 10^{23}\ \text{molecules of ATP}}$$

E2.30 $$n(\mathrm{ATP}) = \frac{10^6}{6.02 \times 10^{23}\ \mathrm{mol^{-1}}} = 1.7 \times 10^{-18}\ \mathrm{mol}$$

$$\Delta G = 1.7 \times 10^{-18}\ \mathrm{mol\ s^{-1}} \times (-31\ \mathrm{kJ\ mol^{-1}}) = -5.3 \times 10^{-17}\ \mathrm{kJ\ s^{-1}} = -5.3 \times 10^{-14}\ \mathrm{J\ s^{-1}}$$

$$\text{Power density of cell} = \frac{\Delta G\ \text{of cell per second}}{\text{volume of cell}}$$

$$V_{\mathrm{cell}} = \tfrac{4}{3}\pi r^3 = \tfrac{4}{3}\pi(10 \times 10^{-6}\ \mathrm{m})^3 = 4.2 \times 10^{-15}\ \mathrm{m^3}$$

$$\text{Power density of cell} = \frac{5.3 \times 10^{-14}\ \mathrm{J\ s^{-1}}}{4.2 \times 10^{-15}\ \mathrm{m^3}} = \boxed{13\ \mathrm{W\ m^{-3}}}$$

$$\text{Power density of battery} = \frac{15\ \text{W}}{100\ \text{cm}^3 \times 10^{-6}\ \text{m}^3/\text{cm}^3} = \boxed{150\ \text{kW m}^{-3}}$$

The $\boxed{\text{battery}}$ has the greater power density.

Solutions to projects

P2.31 (a) The hydrocarbons in question form a homologous series. They are straight-chain alkanes of the formula C_nH_{2n+2}, or R–H where R = C_nH_{2n+1}. Draw up the following table:

n	1	2	3	4	5
π	0.5	1.0	1.5	2.0	2.5

The relationship here is evident by inspection: $\pi = n/2$, so we predict for the seven-carbon hydrocarbon in question:

$$\pi = 7/2 = \boxed{3.5}.$$

(b) The plot, shown below, is consistent with a linear relationship, for $R^2 = 0.997$ is close to unity. The best linear fit is:

$$\log K_I = -1.95 - 1.49\pi,$$

so $\boxed{\text{slope} = -1.49}$ and $\boxed{\text{intercept} = -1.95}$.

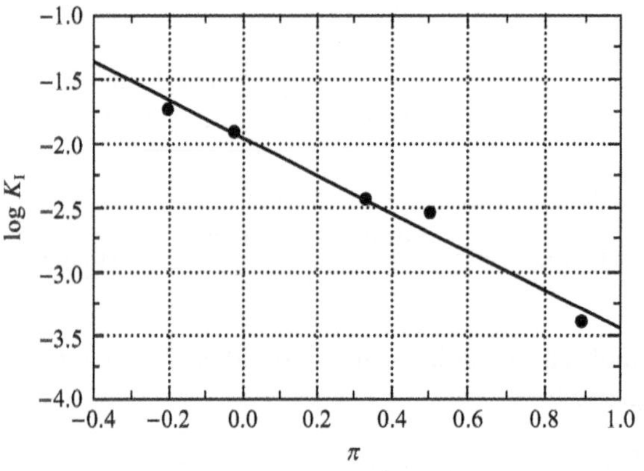

Figure 2.4

(c) If we know π for the substituent R = H, then we can use the linear SAR just derived. Our best estimate of π can be obtained by considering the zero-carbon "alkane" H_2, whose radical H ought to have a hydrophobicity constant $\pi = 0/2 = 0$. This value yields

$$\log K = -1.95 - 1.49(0) = -1.95, \quad \text{so} \quad K = 10^{-1.95} = \boxed{1.12 \times 10^{-2}}.$$

Note: the assumption that R = H is part of the homologous series of straight-chain alkanes is a reasonable but questionable one.

P2.32 (a) ATP hydrolysis at physiological pH, $\text{ATP(aq)} + H_2O(l) \rightarrow \text{ADP(aq)} + P_i^-(aq) + H_3O^+(aq)$, converts two reactant moles into three product moles. The increased number of chemical species present in solution increases the disorder of the system by increasing the number of molecular rotational, vibrational, and translational degrees of freedom. This is an effective increase in the number of available molecular states and an increase in entropy.

(b) At physiological pH the oxygen atoms of ATP are deprotonated, negatively charged, and the molecule is best represented as ATP^{4-}. The electrostatic repulsions between the highly charged oxygen atoms of ATP^{4-} are expected to give it an exergonic hydrolysis free energy by making the hydrolysis enthalpy negative.

The electrostatic repulsion between the highly charged oxygen atoms of ATP^{4-} is a hypothesis that is consistent with the observation that protonated ATP, H_4ATP, has an exergonic hydrolysis free energy of smaller magnitude because the negative repulsions of oxygen atoms are not present. Likewise for $MgATP^{2-}$, because the Mg^{2+} ion lies between negatively charged oxygen atoms, thereby, reducing repulsions and stabilizing the ATP molecule.

Adenosine Triphosphate, ATP^{4-}

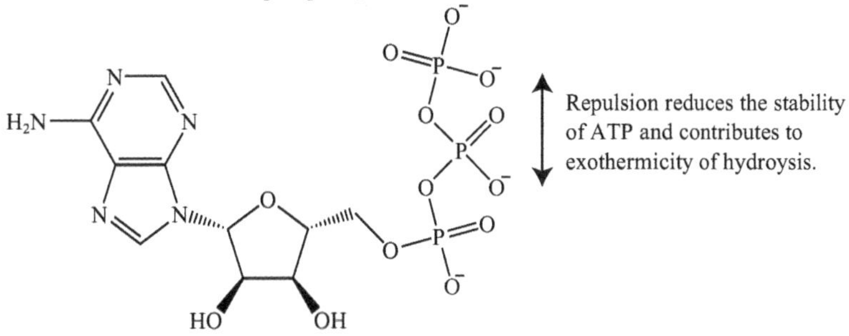

(c) The deprotonated phosphate species, $P_i(aq)$, produced in the hydrolysis ATP has more resonance structures than ATP^{4-}. Resonance lowers the energy of the dissociated phosphate making the hydrolysis enthalpy more negative and contributing to the exergonicity of the hydrolysis.

3 Bioenergetics: Phase equilibria

Answers to discussion questions

D3.1 For a one-component system the **chemical potential**, μ (mu), is equivalent to the molar Gibbs energy, G_m: $\mu = G_m = H_m - TS_m$. The chemical potential is the difference between the total stored energy and the energy stored randomly. This means that it is available energy for doing non-expansion work, or, in an equivalent view, it is the energy stored in the orderly motion and arrangement of the molecules of the system. It is a very important physical property because the phase of a one-component system that has the lowest chemical potential is the stable phase at a given temperature and pressure and the chemical potential decreases in a spontaneous change at constant temperature and pressure (see Section 2.6(b)). Furthermore, phase 1 and phase 2 of a pure substance are in equilibrium when $\mu_1 = \mu_2$.

(a) Eqn 3.2 tells us that $d\mu = V_m dp - S_m dT$ so that at constant p (i.e. $dp = 0$) we see that $d\mu = -S_m dT$. Our interpretation is that as temperature increases, the chemical potential varies as $-S_m$. The chemical potential varies with temperature because of the entropy of the system. Since the entropy is always positive, the chemical potential decreases with increasing temperature.

(b) At constant temperature (i.e. $dT = 0$) the expression $d\mu = V_m dp - S_m dT$ reduces to $d\mu = V_m dp$. Our interpretation is that as pressure increases, the chemical potential varies as V_m. The chemical potential varies with pressure because of the molar volume of the system. Since the molar volume is always positive, the chemical potential increases with increasing pressure.

D3.2 Consider two phases of a system, labeled α and β. The phase with the lower molar Gibbs energy under the given set of conditions is the more stable phase. First, consider the variation of the molar Gibbs energy of each phase with temperature at a fixed pressure by comparing the equation [3.4] expressions:

$$\frac{\Delta G_\alpha}{\Delta T} = -S_\alpha \quad \text{and} \quad \frac{\Delta G_\beta}{\Delta T} = -S_\beta \quad \text{at constant } p.$$

They clearly show that, if S_β is larger in magnitude than S_α, then ΔG_β decreases to a greater extent than ΔG_α as temperature increases. β phase becomes the more stable phase at higher temperature.

Secondly, consider the variation of the molar Gibbs energy of each phase with pressure at a fixed temperature by comparing the equation [3.1] expressions:

$$\frac{\Delta G_\alpha}{\Delta p} = V_\alpha \quad \text{and} \quad \frac{\Delta G_\beta}{\Delta p} = V_\beta \quad \text{at constant } T$$

These equations clearly show that, if V_β is larger in magnitude than V_α, then ΔG_β increases to a greater extent than ΔG_α as pressure increases. β phase becomes the unstable phase at higher pressure; α phase becomes the stable phase.

D3.3 At equilibrium, the chemical potentials of any component in both the liquid and vapor phases must be equal. This is justified with the equilibrium criterion that under constant temperature and pressure conditions, with no additional work, $\Delta G = 0$. Consider the relationships $dG_{m,J(\alpha)} = V_{m,J(\alpha)}dp - S_{m,J(\alpha)}dT + \mu_J(\alpha)dn_J$ and $dG_{m,J(\beta)} = V_{m,J(\beta)}dp - S_{m,J(\beta)}dT + \mu_J(\beta)dn_J$ for a chemical species "J" that is present in the two phases α and β (see eqn 3.2 and *Justification* 3.5). The terms with dp and dT on the right side of these expressions equal zero for constant p and constant T processes, and the near-equilibrium transformation $J(\beta) \rightleftharpoons J(\alpha)$ is such a process. Subtraction and application of the equilibrium criteria gives: $dG_J = dG_{m,J(\alpha)} - dG_{m,J(\beta)} = \{\mu_J(\alpha) - \mu_J(\beta)\}dn_J = 0$ or $\mu_J(\alpha) - \mu_J(\beta) = 0$ at equilibrium. The chemical potential of each chemical species must be equal in the liquid (α) and vapor (β) phases or in any two phases that are at equilibrium: $\boxed{\mu_J(\alpha) = \mu_J(\beta)}$.

D3.4 The Clapeyron equation is exact and applies rigorously to all first-order phase transitions. It shows how pressure and temperature vary with respect to each other (temperature or pressure) along the phase boundary line, and in that sense, it defines the phase boundary line.

The Clausius–Clapeyron equation serves the same purpose, but it is not exact; its derivation involves approximations, in particular the assumptions that the perfect gas law holds and that the volume of condensed phases can be neglected in comparison to the volume of the gaseous phase. It applies only to phase transitions between the gaseous state and condensed phases.

D3.5 Consult text Figure 3.16 and Example 3.1. The sharp step in the plot of fraction of unfolded protein against temperature is a result of the cooperativity of the denaturation process. Thus, we expect that the greater the degree of cooperativity, the sharper the step will be around the melting temperature.

D3.6 The activity of a solute is that property that determines how the chemical potential of the solute varies from its value in a specified standard state. This is seen from the general definition

$$\mu_J = \mu_J^{\ominus} + RT \ln a_J \quad [3.17]$$

where $\mu_J^{\ominus}$ is the value of the chemical potential of J in the standard state for which $a_J = 1$. The relation is true at all concentrations and for both the solvent and the solute. It is well worth remembering several useful activity forms.

Ideal solutions:	$a_J = x_J$
Ideal-dilute solutions:	$a_B = [B]/c^{\ominus}$, where $c^{\ominus} = 1$ mol dm^{-3}
Solvent A of a non-ideal solution:	$a_A = \gamma_A x_A$
Solute B of a non-ideal solution:	$a_B = \gamma_B[B]/c^{\ominus}$, where $c^{\ominus} = 1$ mol dm^{-3}

The dimensionless activity coefficients, γ_J, of non-ideal solutions must be deduced from experimental data. Also, the activity of a pure solid or a pure liquid at 1 bar always equals 1 as these are standard states.

D3.7 All the colligative properties (properties that depend only on the number of solute particles present, not their chemical identity) are a result of the lowering of the chemical potential of the solvent due to the presence of the solute. This reduction takes the form $\mu_A = \mu_A^* + RT \ln x_A$ or $\mu_A = \mu_A^* + RT \ln a_A$, depending on whether or not the solution can be considered ideal. The lowering of the chemical potential results in a freezing point depression and a boiling point elevation as illustrated in Figures 3.32 and 3.33 of the text. Both of these effects can be explained by the lowering of the vapor pressure of the solvent in solution due to the presence of the solute. The solute molecules get in the way of the solvent molecules, reducing their escaping tendency.

D3.8 The osmotic pressure, Π, method (see Example 3.4) for determination of the molar mass of a biological macromolecule involves measurement of Π for a series of successively more dilute mass concentrations c_{polymer}. The extrapolated intercept at $c_{\text{polymer}} = 0$ of a Π/c_{polymer} against c_{polymer} plot equals RT/M_{polymer}. Consequently, $M_{\text{polymer}} = RT/\text{intercept}$.

Solutions to exercises

E3.9 (a) We assume that the molar volume of water is approximately constant with respect to variation in pressure. Then,

$$V_m = \frac{18.02\ \text{g mol}^{-1}}{1.03\ \text{g cm}^{-3}} = 17.5\ \text{cm}^3\ \text{mol}^{-1}$$
$$= 1.75 \times 10^{-5}\ \text{m}^3\ \text{mol}^{-1}$$

$$\Delta p = g\rho h \quad [\Delta p = p_{\text{trench}} - p_{\text{surface}}]$$
$$= 9.81\ \text{m s}^{-2} \times 1.03\ \text{g cm}^{-3} \times \frac{1\ \text{kg}}{10^3\ \text{g}} \times \frac{10^6\ \text{cm}^3}{\text{m}^3} \times 11.5 \times 10^3\ \text{m}$$
$$= 1.16 \times 10^8\ \text{Pa} = 116\ \text{MPa}$$

$$\Delta G_m = V_m \Delta p = 1.75 \times 10^{-5}\ \text{m}^3\ \text{mol}^{-1} \times 1.16 \times 10^8\ \text{Pa}$$
$$= 2.03 \times 10^3\ \text{J mol}^{-1} = \boxed{+2.03\ \text{kJ mol}^{-1}}$$

(b) The pressure at the bottom of the mercury column is 1.000 atm = 1.013×10^5 Pa.

$$\Delta p = 1.013 \times 10^5\ \text{Pa} - 0.160\ \text{Pa} \approx 1.013 \times 10^5\ \text{Pa}$$

$$V_m = \frac{200.6\ \text{g mol}^{-1}}{13.6\ \text{g cm}^3} = 14.8\ \text{cm}^3\ \text{mol} = 1.48 \times 10^{-5}\ \text{m}^3\ \text{mol}^{-1}$$

$$\Delta G_m = V_m \Delta p = 1.48 \times 10^{-5}\ \text{m}^3\ \text{mol}^{-1} \times 1.013 \times 10^5\ \text{Pa}$$
$$= \boxed{+1.50\ \text{J mol}^{-1}}$$

E3.10 $$\Delta G_m = V_m \Delta p$$

$$V_m = \frac{891.51\ \text{g mol}^{-1}}{0.95\ \text{g cm}^{-3}} = 938\ \text{cm}^3\ \text{mol}^{-1} = 9.4 \times 10^{-4}\ \text{m}^3\ \text{mol}^{-1}$$

$$\Delta p = g\rho h = 9.81\ \text{m s}^{-2} \times 1.03 \times 10^3\ \text{kg m}^{-3} \times 2.0 \times 10^3\ \text{m} = 2.0 \times 10^7\ \text{Pa}$$

$$\Delta G_m = 9.4 \times 10^{-4}\ \text{m}^3\ \text{mol}^{-1} \times 2.0 \times 10^7\ \text{Pa} = +1.9 \times 10^4\ \text{J mol}^{-1} = \boxed{+19\ \text{kJ mol}^{-1}}$$

E3.11 $$\Delta G_m = RT \ln \frac{p_f}{p_i} \quad [3.3]$$

(a) $$\Delta G_m = 8.3145\ \text{J K}^{-1}\ \text{mol}^{-1} \times 293\ \text{K} \times \ln\left(\frac{2.0\ \text{bar}}{1.0\ \text{bar}}\right)$$
$$= 1.7 \times 10^3\ \text{J mol}^{-1} = \boxed{+1.7\ \text{kJ mol}^{-1}}$$

(b) $$\Delta G_m = 8.3145\ \text{J K}^{-1}\ \text{mol}^{-1} \times 293\ \text{K} \times \ln\left(\frac{0.00027\ \text{bar}}{1.0\ \text{bar}}\right)$$
$$= -2.0 \times 10^4\ \text{J mol}^{-1} = \boxed{-20\ \text{kJ mol}^{-1}}$$

E3.12 The slope of a graph of G_m against T is $-S_m$, that is $\frac{\Delta G_m}{\Delta T} = -S_m$ [3.4].

The slopes in all phases are negative, because S_m is always positive, but

$$\left|\frac{\Delta G_m}{\Delta T}(g)\right| > \left|\frac{\Delta G_m}{\Delta T}(l)\right| > \left|\frac{\Delta G_m}{\Delta T}(s)\right|$$

because $S_m(g) > S_m(l) > S_m(s)$.

Therefore, a graph of G_m against T appears as in Figure 3.1. Absolute values of G_m are not known, but ΔG_m in each phase could be calculated.

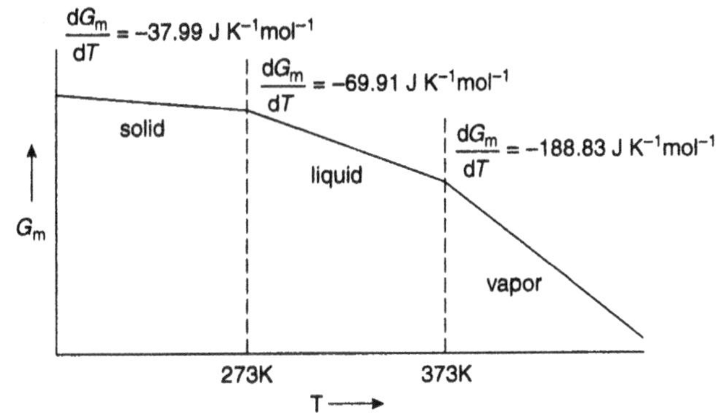

Figure 3.1

E3.13 Use the perfect gas law to calculate the amount and the mass.

$V = 6.0\ \text{m} \times 5.3\ \text{m} \times 3.2\ \text{m} = 102\ \text{m}^3 = 102 \times 10^3\ \text{dm}^3$

$$m = nM = \frac{pVM}{RT}$$

(a) $$m = \frac{2.3\ \text{kPa} \times (102 \times 10^3\ \text{dm}^3) \times (18.02\ \text{g mol}^{-1})}{(8.3145\ \text{dm}^3\ \text{kPa K}^{-1}\ \text{mol}^{-1}) \times (298.15\ \text{K})} = \boxed{1.7\ \text{kg}}$$

(b) $$m = \frac{10\ \text{kPa} \times (102 \times 10^3\ \text{dm}^3) \times (78.11\ \text{g mol}^{-1})}{(8.3145\ \text{dm}^3\ \text{kPa K}^{-1}\ \text{mol}^{-1}) \times (298.15\ \text{K})} = \boxed{32\ \text{kg}}$$

(c) $$m = \frac{(0.30 \times 10^{-3}\ \text{kPa}) \times (102 \times 10^3\ \text{dm}^3) \times (200.59\ \text{g mol}^{-1})}{(8.3145\ \text{dm}^3\ \text{kPa K}^{-1}\ \text{mol}^{-1}) \times (298.15\ \text{K})} = \boxed{2.5\ \text{g}}$$

E3.14 (a) The Clapeyron equation for the solid–liquid phase boundary is

$$\frac{dp}{dT} = \frac{\Delta_{fus}H}{T_{fus}\Delta_{fus}V}\ [3.5].$$

$$\Delta_{fus}V = V_m(l) - V_m(s) = M\left(\frac{1}{\rho_l} - \frac{1}{\rho_s}\right)$$

$$= 18.02\ \text{g mol}^{-1}\left(\frac{1}{0.99984\ \text{g cm}^{-3}} - \frac{1}{0.91671\ \text{g cm}^{-3}}\right)$$

$$= -1.634\ \text{cm}^3\ \text{mol}^{-1}$$

$$= -1.634 \times 10^{-6}\ \text{m}^3\ \text{mol}^{-1}$$

$$\frac{dp}{dT} = \frac{6.008 \times 10^3 \text{ J mol}^{-1}}{(273.15 \text{ K}) \times (-1.634 \times 10^{-6} \text{ m}^3 \text{ mol}^{-1})}$$
$$= -1.346 \times 10^7 \text{ Pa K}^{-1} = \boxed{-134.6 \text{ bar K}^{-1}}$$

The slope is very steep with an unusual negative slope that is caused by the decrease in volume that occurs when ice melts. The melting process destroys some of the hydrogen-bond scaffolding that holds ice in a larger molar volume.

(b) $\dfrac{\Delta p}{\Delta T} = -134.6 \text{ bar K}^{-1}$

For $\Delta T = -1$ K, $\Delta p = 134.6$ bar. Consequently, $p = p_i + \Delta p = 1.0 \text{ bar} + 134.6 \text{ bar} = \boxed{135.6 \text{ bar}}$.

E3.15 (a) To obtain the explicit expression for the vapor pressure at any temperature we rearrange the Clausius–Clapeyron equation into

$$\text{d}\ln p = \frac{\Delta_{\text{vap}}H}{RT^2}\text{d}T$$

and integrate both sides. If the vapor pressure is p at a temperature T and p' at a temperature T', this integration takes the form

$$\int_{\ln p}^{\ln p'} \text{d}\ln p = \int_T^{T'} \frac{\Delta_{\text{vap}}H}{RT^2}\text{d}T$$

The integral on the left evaluates to $\ln p' - \ln p$, which simplifies to $\ln(p'/p)$. To evaluate the integral on the right we assume that the enthalpy of vaporization is constant over the temperature range involved, so together with R it can be taken outside of the integral sign. Then,

$$\ln\frac{p'}{p} = \frac{\Delta_{\text{vap}}H}{R}\int_T^{T'} \frac{1}{T^2}\text{d}T = \frac{\Delta_{\text{vap}}H}{R}\left(\frac{1}{T} - \frac{1}{T'}\right)$$

or $\ln p' = \ln p + \dfrac{\Delta_{\text{vap}}H}{R}\left(\dfrac{1}{T} - \dfrac{1}{T'}\right)$

(b) We use the result from part (a).

$$\ln\frac{p'}{p} = \frac{\Delta_{\text{vap}}H}{R}\left(\frac{1}{T} - \frac{1}{T'}\right)$$

with $T = 293$ K, $p = 160$ mPa, and $T' = 313$ K; then solve for p'.

$$\ln\frac{p'}{p} = \frac{59.30 \times 10^3 \text{ J mol}^{-1}}{8.3145 \text{ J K}^{-1} \text{ mol}^{-1}}\left(\frac{1}{293 \text{ K}} - \frac{1}{313 \text{ K}}\right) = 1.56$$

$$\frac{p'}{p} = e^{1.56} = 4.74$$

$$p' = (4.74) \times (160 \text{ mPa}) = 758 \text{ mPa} = \boxed{0.758 \text{ Pa}}$$

E3.16 (a) We can rewrite the expression derived in Exercise 3.15 as

$$\ln p = \ln p' - \frac{\Delta_{\text{vap}}H}{R}\left(\frac{1}{T} - \frac{1}{T'}\right).$$

Then, because $\ln x = \ln 10 \times \log x$ this becomes

$$\log p = \log p' + \frac{\Delta_{\rm vap}H}{RT' \ln 10} - \frac{\Delta_{\rm vap}H}{RT \ln 10}$$

This expression has the form $\log p = A - \frac{B}{T}$, where

$$A = \log p' + \frac{\Delta_{\rm vap}H}{RT' \ln 10} \text{ and } B = \frac{\Delta_{\rm vap}H}{R \ln 10}.$$

(b) Because $B = 1785$ K, it follows from the preceding expression for B that

$$\Delta_{\rm vap}H = BR \ln 10 = (1785\ \text{K}) \times (8.3145\ \text{J K}^{-1}\ \text{mol}^{-1}) \times \ln 10 = \boxed{34.2\ \text{kJ mol}^{-1}}$$

E3.17 The vapor pressure of ice at -5°C is 3.9×10^{-3} atm, or 3.0 Torr (CRC *Handbook of Chemistry and Physics*). Since the partial pressure of water is lower (2 Torr), the frost will sublime. A partial pressure of 3.0 Torr or more will ensure that the frost remains.

E3.18 (a) The volume decreases as the vapor is cooled from 400 K, at constant pressure, in a manner described by the perfect gas equation $V = nRT/p$. That is, V is a linear function of T. This continues until 373 K is reached where the vapor condenses to a liquid and there is a large decrease in volume. As the temperature is lowered further to 273 K, liquid water freezes to ice. Only a small decrease in volume occurs in the liquid as the temperature is decreased, and a small (~9%) increase in volume occurs when the liquid freezes. Water remains as a solid at 260 K.

(b) The cooling curve appears roughly as sketched in Figure 3.2. The vapor and solid phases show a steeper rate of decline than for the liquid phase due to their smaller heat capacities. The temperature halt in the liquid plus vapor region is longer than for the liquid plus solid region due to its larger heat of transition.

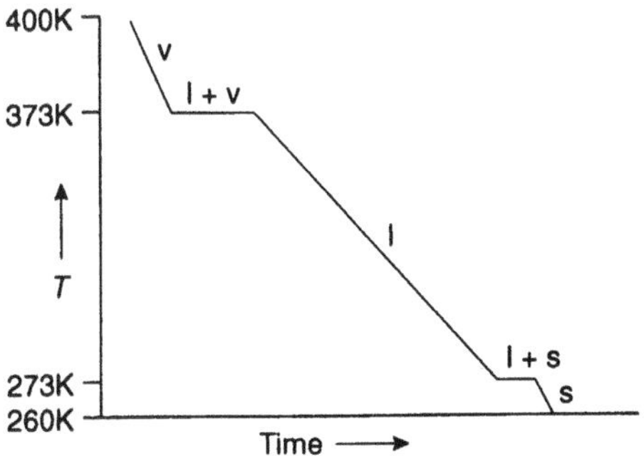

Figure 3.2

E3.19 Cooling from, say, 400 K at 0.006 bar (text Figure 3.13) will cause a decrease in volume of gaseous water until 273.16 K is reached, at which temperature liquid and solid water will appear. All three phases will remain in equilibrium until the constant cooling completely transforms the gas and liquid phases to ice. Then, cooling drops the temperature below 273.16 K.

E3.20 (a) $\Delta G_{\rm m} = (n-4)\Delta_{\rm hb}H_{\rm m} - (n-2)T_{\rm m}\Delta_{\rm hb}S_{\rm m}$ (3.1)

The enthalpy term is justified by $n - 4$ independent hydrogen bonds for which each requires $\Delta_{\rm hb}H_{\rm m}$ of heat to break during melting dissociation. The entropy term is justified by $n - 2$ highly ordered, but independent, structures for which each experiences an entropy increase of $\Delta_{\rm hb}S_{\rm m}$ during the melting process. According to [2.13], the enthalpy and entropy terms give a Gibbs energy change of $\Delta G = \Delta H - T\Delta S$ for a constant temperature process. Eqn (3.1) above has this necessary form.

(b) $\Delta_{trs}S = \dfrac{\Delta_{trs}H}{T_{trs}}$ [analogous to eqns 2.4 and 2.5] yields $T_{trs} = \dfrac{\Delta_{trs}H}{\Delta_{trs}S}$, which here becomes

$$T_m = \frac{(n-4)\Delta_{hb}H_m}{(n-2)\Delta_{hb}S_m}$$

(c) See Figure 3.3

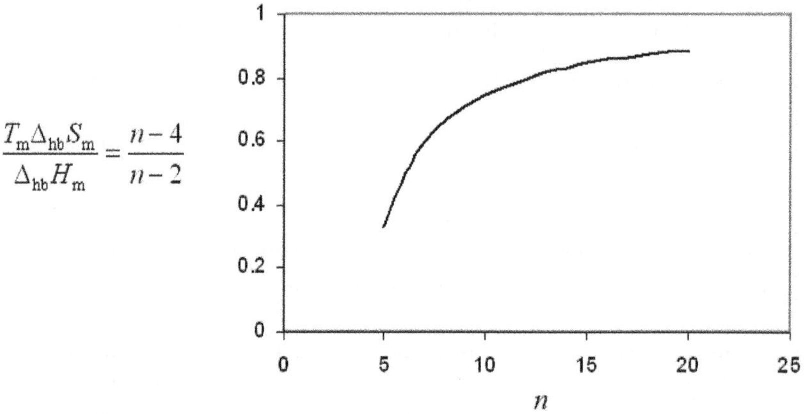

Figure 3.3

$$\text{Consider } \frac{1}{T_m}\frac{dT_m}{dn} = \frac{\Delta_{hb}H_m}{T_m\Delta_{hb}S_m}\frac{d(T_m\Delta_{hb}S_m/\Delta_{hb}H_m)}{dn} = \left(\frac{n-2}{n-4}\right)\frac{d}{dn}\left(\frac{n-4}{n-2}\right)$$

$$= \left(\frac{n-2}{n-4}\right)\left(\frac{2}{(n-2)^2}\right) = \frac{2}{(n-4)(n-2)}$$

This expression will be less than 1% when $\dfrac{2}{(n-4)(n-2)} \lesssim 0.01$ or when n equals, or is larger than the value given by $n^2 - 6n + 8 = 200$. The positive root of this quadratic is $n \cong \boxed{17}$. T_m changes by about 1% or less upon addition of another amino acid residue when the polypeptide consists of 17 or more residues.

E3.21 (a) $\Delta_{init}G^{\ominus}$ is positive, because by bringing together the two strands to form DNA, $\Delta_{init}S^{\ominus}$ should be negative, resulting in a positive value for $\Delta_{init}G^{\ominus}$. $\Delta_{seq}G^{\ominus}$ is negative, because of hydrogen bonding and π-stacking interactions between base pairs. The hydrogen bonding is greater for G–C base pairs because of three hydrogen bonds.

(b) The standard Gibbs function change can be estimated using the $\Delta_{init}G^{\ominus}$ and $\Delta_{seq}G^{\ominus}$ provided,

$$\Delta_{DNA}G^{\ominus} = \Delta_{init}G^{\ominus} + \sum_{seq}\Delta_{seq}G^{\ominus}$$

$$\Delta_{DNA}G^{\ominus} = \Delta_{init}G^{\ominus} + [2\times\Delta_{seq}G^{\ominus}(AG/TC) + \Delta_{seq}G^{\ominus}(GC/CG) + \Delta_{seq}G^{\ominus}(TG/AC)$$

$$\Delta_{DNA}G^{\ominus} = +14.2\text{ kJ mol}^{-1} + (-10.8 - 10.5 - 6.7)\text{ kJ mol}^{-1} = \boxed{-13.8\text{ kJ mol}^{-1}}$$

In a similar manner, the standard enthalpy and entropy changes are calculated below:

$$\Delta_{DNA}H^{\ominus} = \Delta_{init}H^{\ominus} + \sum_{seq}\Delta_{seq}H^{\ominus}$$

$$\Delta_{DNA}H^{\ominus} = \Delta_{init}H^{\ominus} + [2\times\Delta_{seq}H^{\ominus}(AG/TC) + \Delta_{seq}H^{\ominus}(GC/CG) + \Delta_{seq}H^{\ominus}(TG/AC)$$

$$\Delta_{DNA}H^{\ominus} = +2.5\text{ kJ mol}^{-1} + (-51.0 - 46.4 - 31.0)\text{ kJ mol}^{-1} = \boxed{-126\text{ kJ mol}^{-1}}$$

$$\Delta_{DNA}S^\ominus = \Delta_{init}S^\ominus + \sum_{seq} \Delta_{seq}S^\ominus$$

$$\Delta_{DNA}S^\ominus = \Delta_{init}S^\ominus + [2\times\Delta_{seq}S^\ominus(AG/TC) + \Delta_{seq}S^\ominus(GC/CG) + \Delta_{seq}S^\ominus(TG/AC)$$

$$\Delta_{DNA}S^\ominus = -37.7\ J\ K^{-1}\ mol^{-1} + (-134.8 - 118.8 - 80.8)\ J\ K^{-1}\ mol^{-1} = \boxed{-372\ J\ K^{-1}\ mol^{-1}}$$

(c) Text Figure 3.17 shows that the melting temperature changes linearly with the fraction of G–C base pairs: $T_m/K = 325 + 39.7f$. The fraction of G–C base pairs in this example is 3/5. Thus an estimate of the melting temperature of the piece of DNA shown in part (b) is:

$$T_m/K = 325 + 39.7 \times 0.6 = \boxed{348\ K}$$

E3.22 The total pressure p is given by **Dalton's law**: $p = p_A + p_B + \cdots$. Therefore,

$$p = p_{dry\ air} + p_{water\ vapor}.$$

$$p_{dry\ air} = p - p_{water\ vapor} = (760 - 47)\ Torr = \boxed{713\ Torr}$$

E3.23 (a) There is only one volume, so using the amount of nitrogen and partial pressure of nitrogen we can calculate the volume.

$$n_{N_2} = \frac{0.225\ g}{28.02\ g\ mol^{-1}} = 8.03\times10^{-3}\ mol, \quad p_{N_2} = 15.2\ kPa, \quad T = 300\ K$$

$$V = \frac{n_{N_2}RT}{p_{N_2}}\ \text{[Each component of the mixture satisfies the perfect gas law.]}$$

$$= \frac{(8.03\times10^{-3}\ mol)\times(8.3145\ dm^3\ kPa\ K^{-1}\ mol^{-1})\times(300\ K)}{15.2\ kPa} = \boxed{1.32\ dm^3}$$

(b) $n_{CH_4} = \dfrac{0.320\ g}{16.04\ g\ mol^{-1}} = 2.00\times10^{-2}\ mol$

$$n_{Ar} = \frac{0.175\ g}{39.95\ g\ mol^{-1}} = 4.38\times10^{-3}\ mol$$

$$n = n_{CH_4} + n_{Ar} + n_{N_2} = (2.00 + 0.438 + 0.803)\times10^{-2}\ mol = 3.24\times10^{-2}\ mol$$

Solving the perfect gas law for the total pressure p of n moles of gas, we find

$$p = \frac{nRT}{V} = \frac{(3.24\times10^{-2}\ mol)\times(8.3145\ dm^3\ kPa\ K^{-1}\ mol^{-1})\times(300\ K)}{1.32\ dm^3} = \boxed{61.2\ kPa}.$$

E3.24 (a) $\text{mass} = 250.0\ cm^3 \times \dfrac{1\ dm^3}{10^3\ cm^3} \times \dfrac{0.112\ mol}{1\ dm^3} \times \dfrac{180.16\ g}{mol} = \boxed{5.04\ g}$

(b) $\text{mass} = 250.0\ g \times \dfrac{1\ kg}{10^3\ g} \times \dfrac{0.112\ mol}{1\ kg} \times \dfrac{180.16\ g}{mol} = \boxed{5.04\ g}$

E3.25 $0.134\ m = \dfrac{0.134\ mol}{1\ kg\ water}$

$$n_{H_2O} = \frac{1000\ g}{18.02\ g\ mol^{-1}} = 55.5\ mol$$

$$x_{alanine} = \frac{0.134\ mol}{0.134\ mol + 55.5\ mol} = \boxed{2.41\times10^{-3}}$$

E3.26

$$x_{\text{sucrose}} = 0.124 = \frac{n_{\text{sucrose}}}{n_{\text{sucrose}} + n_{H_2O}} = \frac{m/M_{\text{sucrose}}}{m/M_{\text{sucrose}} + \dfrac{100\text{ g}}{18.02\text{ g mol}^{-1}}}$$

Solve the above equation for mass (m).

$$0.124\left(\frac{m}{M_{\text{sucrose}}} + 5.55\text{ mol}\right) = \frac{m}{M_{\text{sucrose}}}$$

$$0.124 \times 5.55\text{ mol} = \frac{m}{M_{\text{sucrose}}}(1 - 0.124)$$

$$m = \frac{0.124 \times 5.55\text{ mol} \times M_{\text{sucrose}}}{0.876} \quad \left[M_{\text{sucrose}} = \frac{342.30\text{ g}}{\text{mol}}\right]$$

$$\text{mass} = \boxed{269\text{ g sucrose}}$$

E3.27

(a) $\Delta G_m = RT(x_A \ln x_A + x_B \ln x_b)$ [3.20], where $A = N_2(g)$ and $B = O_2(g)$

$$\Delta G_m = 2.479\text{ kJ mol}^{-1}\{0.78\ln(0.78) + 0.22\ln(0.22)\}$$
$$= \boxed{-1.31\text{ kJ mol}^{-1}}$$

Because ΔG_m is negative, the mixing is $\boxed{\text{spontaneous}}$.

(b) $\Delta S_m = -R(x_A \ln x_A + x_B \ln x_B)$ [3.21b]

$$= -(8.3145\text{ J K}^{-1}\text{ mol}^{-1}) \times \{0.78\ln(0.78) + 0.22\ln(0.22)\}$$
$$= \boxed{+4.38\text{ J K}^{-1}\text{ mol}^{-1}}$$

E3.28

$\Delta G_m = RT(x_A \ln x_A + x_B \ln x_B + x_C \ln x_C)$ [3.20], where $A = N_2(g)$, $B = O_2(g)$, and $C = Ar(g)$

$$\Delta G_m = 2.4790\text{ kJ mol}^{-1}\ \{0.780\ln(0.780) + 0.210\ln(0.210) + 0.0096\ln(0.0096)\}$$
$$= \boxed{-1.40\text{ kJ mol}^{-1}}$$

Because the change in ΔG_m is negative upon the addition of argon, the mixing is $\boxed{\text{spontaneous}}$.

$$\Delta S_m = -R(x_A \ln x_A + x_B \ln x_B + x_C \ln x_C)$$
$$= \boxed{+4.71\text{ J K}^{-1}\text{ mol}^{-1}}$$

By adding to the mixture of Exercise 3.27 a third gas at about 1% of the whole, the Gibbs energy is lowered by about 10% and the entropy of mixing is increased by about 10%.

E3.29

For the sake of convenience we make computations for 1.000 dm³ of seawater. Taking this liberty is valid because it quickly gives us the mole fraction (an intensive property that is independent of the volume of solution used in the calculation) of water in the solution. Furthermore, being a dilute solution, we assume that 1.000 dm³ of seawater contains roughly 1000 g of water.

$$n_{\text{water}} = (m/M)_{\text{water}} = (1000\text{ g})/(18.02\text{ g mol}^{-1}) = 55.5\text{ mol}$$

$$n_{\text{ions}} = 2Vc_{\text{NaCl}} = 2 \times (1.000\text{ dm}^3)(0.50\text{ mol dm}^{-3}) = 1.0\text{ mol}$$

$$x_{\text{water}} = \frac{n_{\text{water}}}{n_{\text{water}} + n_{\text{ions}}} = \frac{55.5}{56.5} = 0.982$$

$$p_{\text{water}} = x_{\text{water}}p^*_{\text{water}} = 0.982 \times 2.338\text{ kPa} = \boxed{2.30\text{ kPa}}$$

E3.30 The 97 per cent saturated haemoglobin (Hb) in the lungs releases oxygen in the capillary until the haemoglobin is 75 per cent saturated.

100 cm^3 (= 0.100 dm^3) of blood in the lung containing 150 g dm^{-3} of Hb at 97 per cent saturated with O_2 binds

$$0.97 \times \left(\frac{1.34\ \text{cm}^3\ \text{O}_2}{\text{g Hb}}\right) \times \left(\frac{150\ \text{g Hb}}{\text{dm}^3}\right) \times (0.100\ \text{dm}^3) = 20\ \text{cm}^3\ \text{O}_2.$$

The same 100 cm^3 of blood in the arteries would contain

$$20\ \text{cm}^3\ \text{O}_2 \times \frac{75\%}{97\%} = 15\ \text{cm}^3\ \text{O}_2.$$

Therefore, about (20 – 15) cm^3 or $\boxed{5\ \text{cm}^3}$ of O_2 is given up in the capillaries to body tissue.

E3.31 (a) In this case, we write the Henry's law expression as

mass of $N_2 = p_{N_2} \times$ mass of $H_2O \times K_{N_2}$.

At $p_{N_2} = 0.78 \times 4.0\ \text{atm} = 3.1\ \text{atm}$,

mass of $N_2 = 3.1\ \text{atm} \times 100\ \text{g H}_2\text{O} \times 1.8 \times 10^{-4}\ \text{mg N}_2/(\text{g H}_2\text{O atm}) = \boxed{0.056\ \text{mg N}_2}$

At $p_{N_2} = 0.78$ atm, mass of $N_2 = \boxed{0.014\ \text{mg N}_2}$

(b) In fatty tissue the increase in N_2 concentration from 1 atm to 4 atm is

$4 \times (0.056 - 0.014)\ \text{mg N}_2 = \boxed{0.17\ \text{mg N}_2}$

E3.32 $K_{CO_2/\text{lipid}} = (8.6 \times 10^4\ \text{Torr}) \times (101.325\ \text{kPa}/760\ \text{Torr}) = 1.1\bar{5} \times 10^4\ \text{kPa}$

$x_{CO_2} = p_{CO_2}/K_{CO_2/\text{lipid}}$ [3.13] $= (55\ \text{kPa})/(1.1\bar{5} \times 10^4\ \text{kPa}) = \boxed{4.8 \times 10^{-3}}$

E3.33 $[CO_2] = p_{CO_2} \times K_{CO_2}$ [3.14], $K_{CO_2} = 0.339\ \text{kPa mol m}^{-3}$

(a) $[CO_2] = 4.0\ \text{kPa} \times 0.339\ \text{kPa}^{-1}\ \text{mol m}^{-3} = 1.36\ \text{mol m}^{-3} = \boxed{1.36\ \text{mmol dm}^{-3}}$

(b) $[CO_2] = 100\ \text{kPa} \times 0.339\ \text{kPa}^{-1}\ \text{mol m}^{-3} = 33.9\ \text{mol m}^{-3} = \boxed{33.9\ \text{mmol dm}^{-3}}$

E3.34 $[J] = p_J \times K_H(J)$ [3.14] $= x_J(\text{gas}) \times p \times K_H(J)$

We assume that $p = p^{\ominus} = 1.00\ \text{bar} = 100\ \text{kPa}$.

$[N_2] = 0.78 \times (100\ \text{kPa}) \times 6.48 \times 10^{-3}\ \text{kPa}^{-1}\ \text{mol m}^{-3}$ [Table 3.2] $= 0.51\ \text{mol m}^{-3} = 0.51\ \text{mmol dm}^{-3}$

$[O_2] = 0.21 \times (100\ \text{kPa}) \times 1.30 \times 10^{-2}\ \text{kPa}^{-1}\ \text{mol m}^{-3}$ [Table 3.2] $= 0.27\ \text{mol m}^{-3} = 0.27\ \text{mmol dm}^{-3}$

The magnitudes of molarity and molality concentrations are equal in very dilute solutions such as these. Consequently, $\boxed{b_{N_2} = 0.51\ \text{mmol kg}^{-1}}$ and $b_{O_2} = \boxed{0.27\ \text{mmol kg}^{-1}}$.

E3.35 Assume 150 cm^3 of water has a mass of 0.150 kg.

$$\Delta T_f = K_f b_B\ [3.22] = 1.86\ \text{K kg mol}^{-1} \times \frac{7.5\ \text{g}}{342.3\ \text{g mol}^{-1} \times 0.150\ \text{kg}} = 0.27\ \text{K}$$

The freezing point will be approximately $\boxed{-0.27°\text{C}}$

E3.36 $K = \dfrac{[A_2]}{[A]^2}$ and let n denote the initial number of moles A.

At equilibrium, $n_{A2} = f\,n$, $n_A = (1-2f)n$, and the total amount of solute is $(1-f)n$.

Therefore, if the volume is V,

$$K = \frac{fnV}{(1-2f)^2n^2} = \frac{f}{(1-2f)^2c}, \text{ where } c = n/V.$$

Vapor pressure, p is $p = x_{\text{solvent}}p^*$.

$$p = x_{\text{solvent}}p^* = \frac{n_{\text{solvent}}\,p^*}{n_A + n_{A_2} + n_{\text{solvent}}} = \frac{n_{\text{solvent}}\,p^*}{(1-f)n + n_{\text{solvent}}}$$

$n_{\text{solvent}} = Vr$ with $r = \rho/M$ and ρ = density of solvent.

$$p = \frac{rp^*}{(1-f)c + r}, \text{ rearranging } f = 1 - \frac{r(p^*-p)}{cp} \text{ and, finally}$$

$$\boxed{K = \frac{1 - \dfrac{r(p^*-p)}{cp}}{c\left(1 - \dfrac{2r(p^*-p)}{cp}\right)^2}}.$$

E3.37 $\Pi V = n_B RT$ [3.23a]

$$\frac{n_B}{V} = M_B \text{ (molarity)} \approx b\rho \quad [\rho = \text{density}]$$

with $\rho = 10^3$ kg m^{-3} for dilute aqueous solutions

Then,

$$b \approx \frac{n_B}{V\rho} = \frac{\Pi}{RT\rho}$$

$$\Delta T_f = K_f b_B \approx K_f \times \frac{\Pi}{RT\rho}$$

Therefore, with $K_f = 1.86$ K kg mol^{-1} (Table 3.4)

$$\Delta T_f = \frac{(1.86 \text{ K kg mol}^{-1}) \times (120 \times 10^3 \text{ Pa})}{(8.3145 \text{ J K}^{-1}\text{ mol}^{-1}) \times (300 \text{ K}) \times (1.00 \times 10^3 \text{ kg m}^{-3})} = 0.089 \text{ K}$$

Therefore, the solution will freeze at about $\boxed{-0.09°\text{C}}$.

E3.38 Our strategy is to avoid assuming that these solutions behave as ideal-dilute solutions and to analyze the data as illustrated in Example 3.4. The method of analysis is suggested by the equation:

$$\frac{\Pi}{c} = \frac{RT}{M} + \left(\frac{RTB}{M^2}\right)c, \quad \text{where } c = m/V.$$

$\Pi = \rho gh$ [hydrostatic pressure] so

$$\frac{h}{c} = \frac{\Pi}{\rho gc} = \left(\frac{RT}{\rho gM}\right) + \left(\frac{RTB}{\rho gM^2}\right)c.$$

This says that a plot of h/c against c has an intercept equal to $RT/\rho gM$ and a slope equal to $RTB/\rho gM^2$, where B is the osmotic virial coefficient. We draw up a table to calculate h/c values

$c/(\text{mg cm}^{-3})$	3.221	4.618	5.112	6.722
h/cm	5.746	8.238	9.119	11.990
$h/c/(\text{mg}^{-1}\text{ cm}^4)$	1.784	1.783	1.784	1.784

Inspection of the h/c values reveals that they are a constant for this experimental set. This implies that the virial coefficient equals zero and that these enzyme solutions are behaving as ideal-dilute solutions. The last term in the above equation vanishes giving the van't Hoff equation [3.23b]. Solving for M (assuming a density of 1.000 g cm^{-3}):

$$M = \frac{RT}{\rho g \times (h/c)}$$

$$= \frac{(8.3145\text{ J K}^{-1}\text{ mol}^{-1}) \times (293.15\text{ K})}{(1.000\text{ g cm}^{-3}) \times (9.807\text{ m s}^{-2}) \times (1.784 \times 10^3\text{ g}^{-1}\text{ cm}^4)} \times \left(\frac{1\text{ cm}}{10^{-2}\text{ m}}\right)$$

$$= \boxed{13.9\bar{3}\text{ kg mol}^{-1}}$$

Solutions to projects

P3.39 (a) The number of degrees of freedom in a multi-component system is given by the Gibbs phase rule, $F = C - P + 2$, where C is the number of components and P is the number of phases. The number 2 in this formula stands for T (temperature) and p (pressure). But if pressure is fixed in a temperature–composition diagram, then we write $F' = C - P + 1$ for the number of degrees of freedom.

(1) $P = 2$, $C = 2$, then $F' = 2 - 2 + 1 = 1$. One degree of freedom defines a line.

(2) $P = 3$, $C = 2$, then $F' = 2 - 3 + 1 = 0$. No degree of freedom defines a point.

(b) (i) Below a denaturant concentration of 0.1 only the native and unfolded forms are stable.

(ii) At a denaturant concentration of 0.15 only the native form is stable below a temperature of about 0.70. At a temperature of 0.70 the native and molten-globule forms are at equilibrium. Heating above 0.70 causes all native forms to become molten-globules. At a temperature of 0.90, equilibrium between molten-globule and unfolded protein is observed and above this temperature only the unfolded form is stable.

(c) Above about 33°C the membrane has the highly mobile liquid crystal form. At 33°C the membrane consists of liquid crystal in equilibrium with a relatively small amount of the gel form. Cooling from 33°C to about 20°C, the equilibrium persists but shifts to a greater relative abundance of the gel form. Below 20°C the gel form alone is stable.

P3.40 (a) $K = \dfrac{[\text{MA}]}{[\text{M}]_{\text{free}}[\text{A}]_{\text{free}}} = \dfrac{[\text{A}]_{\text{bound}}}{([\text{M}] - [\text{A}]_{\text{bound}})[\text{A}]_{\text{free}}}$, where we have used

$[\text{MA}] = [\text{A}]_{\text{bound}}$ and $[\text{M}]_{\text{free}} = [\text{M}] - [\text{MA}] = [\text{M}] - [\text{A}]_{\text{bound}}$. On division by [M], and replacement of $[\text{A}]_{\text{free}}$ by $[\text{A}]_{\text{out}}$ the last expression becomes

$$K = \frac{\nu}{(1 - \nu)[\text{A}]_{\text{out}}}, \text{ as was to be shown.}$$

(b) Substituting v/N for v, this expression becomes

$$K = \frac{v/N}{(1 - v/N)[\mathrm{A}]_{\mathrm{out}}}.$$ It then follows that

$$\frac{v}{[\mathrm{A}]_{\mathrm{out}}} = KN - Kv,$$ as was to be shown.

(c) $v = \dfrac{[\mathrm{EB}]_{\mathrm{bound}}}{[\mathrm{M}]}$ and $[\mathrm{EB}]_{\mathrm{bound}} = [\mathrm{EB}]_{\mathrm{in}} - [\mathrm{EB}]_{\mathrm{out}}$

Draw up the following table:

$[\mathrm{EB}]_{\mathrm{out}}/(\mu\mathrm{mol\ dm^{-3}})$	0.042	0.092	0.204	0.526	1.150
$[\mathrm{EB}]_{\mathrm{bound}}/(\mu\mathrm{mol\ dm^{-3}})$	0.250	0.498	1.000	2.005	3.000
v	0.250	0.498	1.000	2.005	3.000
$\dfrac{v/[\mathrm{EB}]_{\mathrm{out}}}{\mu\mathrm{mol^{-1}\ dm^{3}}}$	5.95	5.41	4.90	3.81	2.61

A plot of $v/[\mathrm{EB}]_{\mathrm{out}}$ is shown in Figure 3.4 below.

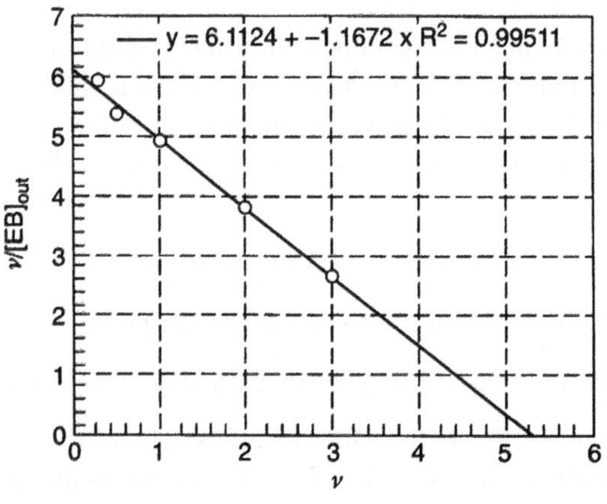

Figure 3.4

The slope is $-1.167\ \mathrm{dm^3\ \mu mol^{-1}}$, hence $K = \boxed{1.167\ \mathrm{dm^3\ \mu mol^{-1}}}$. The intercept at $v = 0$ is $\boxed{N = 5.24}$ and this is the average number of binding sites per oligonucleotide. The close fit of the data to a straight line indicates that identical and independent sites $\boxed{\text{is applicable}}$.

(d) (i) $i = 1$ only, $N_1 = 4$, $K_1 = 1.0 \times 10^7\ \mathrm{dm^3\ mol^{-1}}$

$$\frac{v}{[\mathrm{A}]} = \frac{4 \times 10\ \mathrm{dm^3\ \mu mol^{-1}}}{1 + 10\ \mathrm{dm^3\ \mu mol^{-1}} \times [\mathrm{A}]}$$

The plot is shown in Figure 3.5(a)

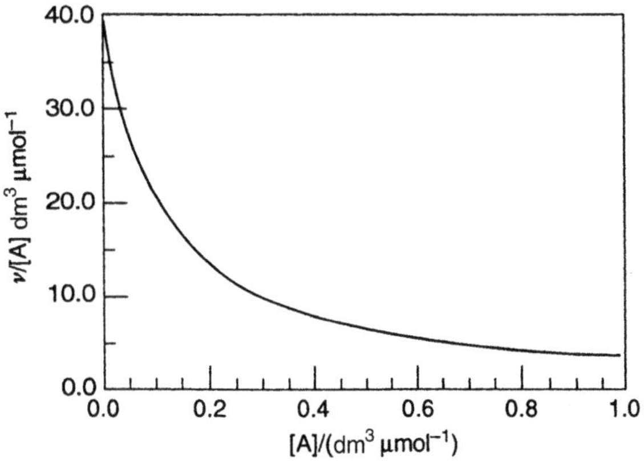

Figure 3.5(a)

(ii) $i = 2$; $N_1 = 4$, $N_2 = 2$; $K_1 = 1.0 \times 10^5$ dm^3 mol^{-1} = 0.10 dm^3 µmol^{-1}, $K_2 = 2.0 \times 10^6$ dm^3 mol^{-1} = 2.0 dm^3 µmol^{-1}

$$\frac{v}{[A]} = \frac{4 \times 0.10\ \text{dm}^3\ \mu\text{mol}^{-1}}{1 + 0.10\ \text{dm}^3\ \mu\text{mol}^{-1} \times [A]} + \frac{2 \times 2.0\ \text{dm}^3\ \mu\text{mol}^{-1}}{1 + 2.0\ \text{dm}^3\ \mu\text{mol}^{-1} \times [A]}$$

The plot is shown in Figure 3.5(b).

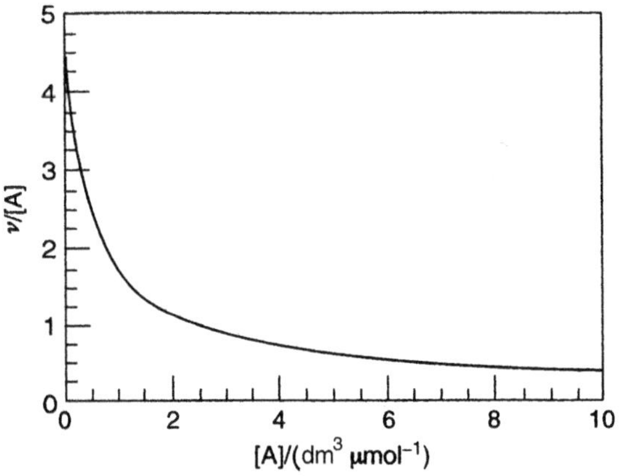

Figure 3.5(b)

4 Chemical equilibrium

Answers to discussion questions

D4.1 The position of equilibrium is always determined by the condition that the reaction Gibbs energy equals zero at equilibrium:

$\Delta_r G = 0$ at equilibrium and, therefore, $\Delta_r G^\circ = -RT \ln Q_{eq} = -RT \ln K$ [4.7, 4.10a].

If the mixing of reactants and products gives $\Delta_r G < 0$, reactant activities will spontaneously diminish to increase product activities until $\Delta_r G = 0$. If the mixing gives $\Delta_r G > 0$, product activities will spontaneously diminish to increase reactant activities until $\Delta_r G = 0$. For the general reaction

$a\,\mathrm{A} + b\,\mathrm{B} \rightleftharpoons c\,\mathrm{C} + d\,\mathrm{D} \quad Q = a_C^c a_D^d / a_A^a a_B^b$ [4.6]

we say that, when the mixing gives $\Delta_r G < 0$, the reaction proceeds spontaneously to the right (*forward* reaction) until equilibrium is achieved but, when $\Delta_r G > 0$, the reaction proceeds spontaneously to the left (*reverse* direction). If reactant or product is added to an equilibrium mixture, the reaction spontaneously shifts in the direction that lowers the Gibbs energy of the reaction mixture (see text Figures 4.1–4.3, remember that $\Delta_r G$ is the *slope* of G plotted against composition). The reaction spontaneously shifts to the right upon addition of reactant to an equilibrium mixture; left upon addition of product. We must also remember that thermodynamics says nothing about the rate at which the reaction occurs or shifts. A spontaneous reaction may occur very rapidly, infinitely slowly, or at any intermediate speed.

D4.2 A non-spontaneous, endergonic reaction ($\Delta_r G > 0$) may be driven forward by a spontaneous, exergonic reaction ($\Delta_r G < 0$) that can supply the requisite reaction Gibbs energy. The total Gibbs energy change for this reaction coupling must be exergonic and the coupling is accomplished in many enzyme-catalyzed biochemical reactions where the enzyme serves as the transfer agent for the Gibbs energy. The exergonic reaction gives up a portion of its Gibbs energy not as heat but to the conversion of a low-potential biochemical intermediate to a high-potential one. The high-potential species carries the energy to the endergonic reaction (on the enzyme surface for example) and in the process of releasing its chemical energy to the endergonic reaction it returns to its low-potential form. The energy-carrying intermediate effectively "couples" the two reactions. Adenosine diphosphate (ADP) and adenosine triphosphate (ATP) is an example set of coupling intermediates:

$\mathrm{ATP(aq)} + \mathrm{H_2O(l)} \rightarrow \mathrm{ADP(aq)} + \mathrm{P_i^-(aq)} + \mathrm{H^+(aq)} \qquad \Delta_r G^\oplus = -31\ \mathrm{kJ\ mol^{-1}}$.

D4.3 For lactate production: $\Delta_r H^\oplus = -120\ \mathrm{kJ\ mol^{-1}}$ and $\Delta_r G^\oplus = -218\ \mathrm{kJ\ mol^{-1}}$ at blood temperature (310 K).

A reaction is more exergonic than it is exothermic when the reaction entropy is positive, a condition that indicates an increase in disorder. From the definition of the Gibbs energy, $G = H - TS$, the reaction Gibbs energy is given by $\Delta_r G^\oplus = \Delta_r H^\oplus - T\Delta_r S^\oplus$. Solving for $\Delta_r S^\oplus$ for lactate production, we find that it is indeed positive:

$$\Delta_r S^\ominus = \frac{\Delta_r H^\ominus - \Delta_r G^\ominus}{T}$$

$$= \frac{(-120 \times 10^3\ \text{J mol}^{-1}) - (-218 \times 10^3\ \text{J mol}^{-1})}{310\ \text{K}}$$

$$= +316\ \text{J K}^{-1}\ \text{mol}^{-1}$$

D4.4 **Le Chatelier's principle**, an empirical rule: When a system at equilibrium is subjected to a disturbance, the composition of the system adjusts so as to tend to minimize the effect of the disturbance.

Thermodynamics provides both understanding of the rule's origin and relations that quantify catalytic, temperature change, and compression (pressure change) effects.

(1) Response to the presence of a catalyst. Neither the quantity $\Delta_r G^\ominus$ nor the equilibrium constant K is affected by a catalyst so the presence of a catalyst does not illicit a reaction response within a chemical mixture at equilibrium. The catalyst increases both the reaction rate to the left and the reaction rate to the right but these rates remain equal at dynamic equilibrium.

(2) Response to change in temperature. The **van't Hoff equation** shows that K decreases with increasing temperature when the reaction is exothermic ($\Delta_r H^\ominus < 0$); thus the reaction shifts to the left, the opposite occurs in endothermic reactions ($\Delta_r H^\ominus > 0$).

$$\ln K_2 = \ln K_1 + \frac{\Delta_r H^\ominus}{R}\left(\frac{1}{T_1} - \frac{1}{T_2}\right) \quad [4.18]$$

The reaction in which reactants and products exhibit identical standard enthalpies ($\Delta_r H^\ominus \simeq 0$) has an equilibrium constant that is independent of temperature ($K' = K$). A simple example is provided by mixing to form an ideal solution.

(3) Response to change in pressure. The relation $K = e^{-\Delta_r G^\ominus/RT}$ [4.10a] indicates that the equilibrium constant for a gas-phase reaction has no dependence upon pressure because the right side of the relation depends upon T alone and, therefore, the left side must depend upon T alone. Nonetheless, individual partial pressures and mole fractions can change as the total pressure changes. This will happen when there is a difference, $\Delta\nu_{\text{gas}}$, between the sums of the number of moles of gases on the product and reactant sides of the chemical equation. The requirement of an unchanged equilibrium constant implies that the side with the smaller number of moles of gas be favored as pressure increases.

These statements are based upon the thermodynamic analysis of the general gas-phase reaction equation:

$$a\,\text{A} + b\,\text{B} \rightleftharpoons c\,\text{C} + d\,\text{D} \qquad Q = a_\text{C}^c a_\text{D}^d / a_\text{A}^a a_\text{B}^b \quad [4.6]$$

Using the equilibrium (eq) definition $K_c = [\text{C}]_{\text{eq}}^c[\text{D}]_{\text{eq}}^d(c^\ominus)^{-\Delta\nu_{\text{gas}}}/[\text{A}]_{\text{eq}}^a[\text{B}]_{\text{eq}}^b$, where $c^\ominus = 1$ mol dm^{-3}, the derivation of the relevant thermodynamic relation for a gas-phase reaction is

$$K = \frac{(p_\text{C}/p^\ominus)^\text{c}(p_\text{D}/p^\ominus)^\text{d}}{(p_\text{A}/p^\ominus)^\text{a}(p_\text{B}/p^\ominus)^\text{b}} \quad [\text{for a perfect gas } p_\text{J} = n_\text{J}RT/V = [\text{J}]RT]$$

$$= \frac{([\text{C}]RT)^\text{c}([\text{D}]RT)^\text{d}}{([\text{A}]RT)^\text{a}([\text{B}]RT)^\text{b}}\left(\frac{1}{p^\ominus}\right)^{(\text{c+d})-(\text{a+b})}$$

$$= K_c \times \left(\frac{c^\ominus RT}{p^\ominus}\right)^{\Delta\nu_{\text{gas}}} = K_c \times \left(\frac{T}{12.027\ \text{K}}\right)^{\Delta\nu_{\text{gas}}}, \quad \text{where } \Delta\nu_{\text{gas}} = \sum_{\text{J=product gases}} \nu_\text{J} - \sum_{\text{J=reactant gases}} \nu_\text{J}.$$

This equation clearly shows that, when a larger pressure is isothermally applied to an equilibrium mixture, the value of K_c remains unchanged because neither K nor the factor $T^{\Delta\nu_{gas}}/12.027$ K change with changes in p. To find the Le Chatelier response to an increase in pressure, we now take a careful look at K_c in terms of mole fractions. For perfect gas component J we substitute the relation $[J]_{eq} = x_J p/RT$ into the expression for K_c where x_J is the mole fraction of J in the mixture. This gives

$$K_c = K_x \times (c^{\ominus}RT/p)^{-\Delta\nu_{gas}}, \quad \text{where} \quad K_x = x_C^c x_D^d / x_A^a x_B^b.$$

For an isothermal compression this simplifies to

$$K_x \propto 1/p^{\Delta\nu_{gas}},$$

which says that the mole fraction reaction quotient depends upon pressure! If $\Delta\nu_{gas} > 0$, K_x is diminished by an increase in p. This is a decrease in the mole fractions of reaction products, while the mole fractions of reactants increase – a shift to the left, the side with fewer moles of gas in the balanced reaction equation. For the same reason, the shift is to the right when $\Delta\nu_{gas} < 0$. If $\Delta\nu_{gas} = 0$, there will be no reaction shift when the pressure is changed.

Failure of Le Chatelier's principle is very unusual. The expected response to pressure may prove wrong should the gas components be non-perfect, real gases. Should the gases of the reaction side that is favored by Le Chatelier's principle strongly repel while the gas molecules of the other side strongly attract, a compression may shift the reaction in the direction opposite to that expected for perfect gases.

D4.5 See text Section 4.3(c).

D4.6 An acid buffer is a solution of approximately equal concentrations of a weak acid and its salt. This solution maintains some constancy in the acidic pH range through its ability to neutralize both a small amount of strong base and a small amount of strong acid. A base buffer, a solution of equal concentrations of a weak base and its salt, does much the same thing but in the base pH range.

D4.7 (a) The pH of the solution of an amphiprotic species is estimated with the relation $\text{pH} = \frac{1}{2}(\text{p}K_{a1} + \text{p}K_{a2})$ [4.34]. This relation is valid when the **formal concentration** S (the concentration as prepared) of the salt MHA, where HA^- is an amphiprotic anion, satisfies the condition $S/c^{\ominus} \gg K_w/K_{a2}$ and $S/c^{\ominus} \gg K_{a1}$. Under these conditions eqn 4.34 indicates that the formal concentration does not determine the pH because S does not appear in the equation (i.e. the pH is constant over a considerable range of formal concentration). These conditions also cause $[H_2A] \approx [A^-]$.

(b) The Henderson–Hasselbalch equation, $\text{pH} = \text{p}K_a - \log([\text{acid}]/[\text{base}])$ [4.35], is limited to the condition that the weak acid solution has a **formal concentration** S (the concentration as prepared) that is dilute to the extent that it is valid to replace activities with concentrations in equilibrium expressions. Other conditions become more important when the equation is used in the form $\text{pH} \approx \text{p}K_a - \log(c_{HA}/c_{MA})$, where c_{HA} and c_{MA} are the formal concentrations of a solution prepared from a weak acid HA and a salt of its conjugate base A^-. The reversible acid dissociation reaction causes some dynamic adjustment in the concentration of both the acid and the conjugate base so we draw an equilibrium table to examine the changes closely.

	HA	$\rightleftharpoons$	A^-	H_3O^+
Initial molar concentration/mol dm^{-3}	c_{HA}		c_{MA}	0
Change to reach equilibrium/mol dm^{-3}	$-x$		$+x$	$+x$
Equilibrium concentration/mol dm^{-3}	$c_{HA} - x$		$c_{MA} + x$	x

$$K_a = \frac{a_{H_3O^+}a_{A^-}}{a_{HA}} \simeq \frac{[H_3O^+][A^-]}{[HA]} \simeq \frac{[H_3O^+](c_{MA}+x)}{(c_{HA}-x)}$$

$$K_a \approx [H_3O^+]\left(\frac{c_{MA}}{c_{HA}}\right) \quad \text{and} \quad pH \approx pK_a - \log\left(\frac{c_{HA}}{c_{MA}}\right) \quad \text{provided that } |x| \ll c_{HA}, c_{MA}.$$

The condition $|x| \ll c_{HA}, c_{MA}$ is somewhat satisfied in practice. For example, the common 0.025 molar phosphate buffer is prepared with $c_{KH_2PO_4} = 25.0$ mmol dm^{-3} and $c_{Na_2HPO_4} = 25.0$ mmol dm^{-3} and has pH = 6.86 at 25°C but the equation $pH \approx pK_a - \log(c_{HA}/c_{MA})$ predicts that $pH \approx pK_{a,H_2PO_4^-} = 7.21$. The discrepancy between the actual pH and the prediction of the Henderson–Hasselbalch eqn has two origins: the activity coefficients do not equal 1 exactly and the condition $|x| \ll c_{HA}, c_{MA}$ is not entirely satisfied. To see the latter, solve the above eqn for x; if $|x|/c_{HA} < 0.01$, the condition $|x| \ll c_{HA}, c_{MA}$ is satisfied.

$$x = \frac{c_{HA}K_a - c_{MA}[H_3O^+]}{K_a + [H^+]} = \frac{(25\text{ mmol dm}^{-3}) \times (10^{-7.21} - 10^{-6.86})}{(10^{-7.21} + 10^{-6.86})} = -9.6\text{ mmol dm}^{-3}$$

$$\frac{|x|}{c_{HA}} = \frac{9.6}{25} = 0.38 \text{ or } 38\%$$

We see that although $|x| < c_{HA}$, since the condition $|x|/c_{HA} < 0.01$ is not satisfied, the condition $|x| \ll c_{HA}, c_{MA}$ is not satisfied. Nevertheless, the Henderson–Hasselbalch equation in the form $pH \approx pK_a - \log(c_{HA}/c_{MA})$ provides a very useful approximation when working with a solution of a weak acid and its conjugate base.

(c) The van't Hoff equation, written as $\ln K_2 = \ln K_1 + \frac{\Delta_r H^\ominus}{R}\left(\frac{1}{T_1} - \frac{1}{T_2}\right)$ [4.18], is valid over small temperature ranges in which neither $\Delta_r H^\ominus$ nor $\Delta_r S^\ominus$ vary much with temperature. Kirchhoff's law [1.24] indicates that the former criterion is often satisfied because $\Delta_r C_p^\ominus$ is often small. The latter criterion is justified by the relation $\Delta_r S^\ominus = \Delta_r H^\ominus/T$ for a constant T and p process; the numerator does not change much with T and the large magnitude of T means the $1/T$ does not change much either so there is little variation of $\Delta_r S^\ominus$ over a small temperature range.

Solutions to exercises

E4.8 The general reaction equation and corresponding reaction quotient are:

$$a\,A + b\,B \rightleftharpoons c\,C + d\,D \quad K = Q_{\text{equilibrium}} = (a_C^c a_D^d / a_A^a a_B^b)_{\text{equilibrium}} \text{ [4.9].}$$

The activities of pure solids and liquids are equal to 1. Substitute $a_{\text{solute}} = \gamma_{\text{solute}}[\text{solute}]/c^\ominus \approx [\text{solute}]/c^\ominus$ for solute activities; assume perfect gas behavior with the substitution $a_{\text{gas}} = p_{\text{gas}}/p^\ominus$. $c^\ominus$ and $p^\ominus$ (1 mol dm^{-3} and 1 bar, respectively) are omitted for convenience and must be replaced during computations. The concentrations and partial pressures of the reaction quotient change in time as the reaction occurs until equilibrium is achieved.

(a) $G6P(aq) + H_2O(l) \rightarrow G(aq) + P_i(aq)$ $\quad K = \frac{[G][P_i]}{[G6P]}$

(b) $Gly(aq) + Ala(aq) \rightarrow Gly\text{–}Ala(aq) + H_2O(l)$ $\quad K = \frac{[Gly\text{–}Ala]}{[Gly][Ala]}$

(c) $Mg^{2+}(aq) + ATP^{4-}(aq) \rightarrow MgATP^{2-}(aq)$ $\quad K = \frac{[MgATP^{2-}]}{[Mg^{2+}][ATP^{4-}]}$

(d) $2\,CH_3COCOOH(aq) + 5\,O_2(g) \rightarrow 6\,CO_2(g) + 4\,H_2O(l)$ $\quad K = \frac{p_{CO_2}^6}{[CH_3COCOOH]^2 p_{O_2}^5}$

E4.9 $A + B \rightleftharpoons 2\,C \qquad K = \dfrac{[C]^2}{[A][B]} = 3.4 \times 10^4$

(a) $2\,C \rightleftharpoons A + B \qquad K = \left(\dfrac{[C]^2}{[A][B]}\right)^{-1} = (3.4 \times 10^4)^{-1} = \boxed{2.9 \times 10^{-5}}$

(b) $2\,A + 2\,B \rightleftharpoons 4\,C \qquad K = \left(\dfrac{[C]^2}{[A][B]}\right)^{2} = (3.4 \times 10^4)^{2} = \boxed{1.2 \times 10^{9}}$

(c) $\frac{1}{2}A + \frac{1}{2}B \rightleftharpoons C \qquad K = \left(\dfrac{[C]^2}{[A][B]}\right)^{1/2} = (3.4 \times 10^4)^{1/2} = \boxed{1.8 \times 10^{2}}$

E4.10 Gly–Ala(aq) + H_2O(l) $\rightarrow$ Gly(aq) + Ala(aq) $\quad K = 8.1 \times 10^2$ at 310 K

$$\begin{aligned}\Delta_r G^{\circ} &= -RT \ln K \quad [4.10a] \\ &= -(8.3145\ \text{J K}^{-1}\ \text{mol}^{-1}) \times (310\ \text{K}) \times \ln(8.1 \times 10^2) \\ &= \boxed{-17\ \text{kJ mol}^{-1}}\end{aligned}$$

E4.11 $K_1 = 10 \times K_2$

$$\begin{aligned}\Delta_{r2} G^{\circ} &= -RT \ln(K_1/10) \quad [4.10a] = -RT \ln K_1 + RT \ln(10) \\ &= \Delta_{r1} G^{\circ} + RT \ln(10) \quad [\text{4.10a; assume a temperature of 298.15 K}] \\ &= -300\ \text{kJ mol}^{-1} + (8.3145 \times 10^{-3}\ \text{kJ K}^{-1}\ \text{mol}^{-1}) \times (298.15\ \text{K}) \times \ln(10) \\ &= \boxed{-294\ \text{kJ mol}^{-1}}\end{aligned}$$

E4.12 $\Delta_r G^{\circ} = -RT \ln K \quad [4.10a] = 0$, so $\ln K = 0$ and $\boxed{K = 1}$.

E4.13 Let glucose-1-phosphate = G1P, glucose-6-phosphate = G6P, and glucose-3-phosphate = G3P.

$\ln K = -\Delta_r G^{\oplus}/RT \quad [4.10b] \quad$ or $\quad K = e^{-\Delta_r G^{\oplus}/RT}$ (The biological standard state, $\oplus$, has pH = 7.)

At 37°C, $RT = (8.3145\ \text{J K}^{-1}\ \text{mol}^{-1}) \times (310\ \text{K}) = 2.57\overline{7}\ \text{kJ mol}^{-1}$.

$$K_{G1P} = \exp\left(\frac{-[-21\ \text{kJ mol}^{-1}]}{2.57\overline{7}\ \text{kJ mol}^{-1}}\right) = \boxed{3.5 \times 10^3}$$

$$K_{G6P} = \exp\left(\frac{-[-14\ \text{kJ mol}^{-1}]}{2.57\overline{7}\ \text{kJ mol}^{-1}}\right) = \boxed{2.3 \times 10^2}$$

$$K_{G3P} = \exp\left(\frac{-[-9.2\ \text{kJ mol}^{-1}]}{2.57\overline{7}\ \text{kJ mol}^{-1}}\right) = \boxed{36}$$

E4.14 ATP(aq) + H_2O(l) $\rightarrow$ ADP(aq) + P_i^-(aq) + H^+(aq) $\quad \Delta_r G^{\oplus} = -31\ \text{kJ mol}^{-1}$

The Gibbs energy quoted applies to the biological standard state where $a_{H^+} = 10^{-7}$ (pH = 7) and all other activities are 1. The chemical standard state is defined with $a_J = 1$ for all reactants and products including $a_{H^+} = 1$ (pH = 0). The relation between the two standard states is provided by noting that:

$$\Delta_r G = \Delta_r G^{\circ} + RT \ln Q\ [4.7] \quad \text{and} \quad \Delta_r G^{\oplus} = \Delta_r G^{\circ} + RT \ln Q^{\oplus}.$$

Thus, when using the biological standard state, we may replace eqn 4.7 with the relation:

$$\Delta_r G = \Delta_r G^{\oplus} - RT \ln Q^{\oplus} + RT \ln Q.$$

$$\boxed{\Delta_r G = \Delta_r G^{\oplus} + RT\ln(Q/Q^{\oplus})}$$

When discussing a solution with the hydrogen ion activity fixed at pH = 7, the reaction quotient ratio $Q/Q^{\oplus}$ effectively cancels out a_{H^+} factors. If a_{H^+} does not appear in Q (because $H^+(aq)$ does not appear in the reaction equation), $Q^{\oplus} = 1$ and $\Delta_r G^{\oplus} = \Delta_r G^{\circ}$.

For the above reaction

$$\Delta_r G = \Delta_r G^{\oplus} + RT\ln\left(\frac{a_{ADP}a_{P_i^-}}{a_{ATP}}\right) = \Delta_r G^{\oplus} + RT\ln\left(\frac{[ADP]\times[P_i^-]}{[ATP]c^{\circ}}\right).$$

At 37°C, $RT = 8.3145\ \mathrm{J\ K^{-1}\ mol^{-1}} \times 310\ \mathrm{K} = 2.57\overline{7}\ \mathrm{kJ\ mol^{-1}}$.

(a) $\Delta_r G = -31\ \mathrm{kJ\ mol^{-1}} + 2.57\overline{7}\ \mathrm{kJ\ mol^{-1}} \times \ln(1.0\times10^{-3}) = \boxed{-49\ \mathrm{kJ\ mol^{-1}}}$

(b) $\Delta_r G = -31\ \mathrm{kJ\ mol^{-1}} + 2.577\ \mathrm{kJ\ mol^{-1}} \times \ln(1.0\times10^{-6}) = \boxed{-67\ \mathrm{kJ\ mol^{-1}}}$

E4.15 $\mu_J = \mu_J^{\circ} + RT\ln a_J$ [4.3]

Assume the ideal-dilute activity: $a_{Na^+} = [Na^+]$. Then,

$$\Delta G = \mu_{Na^+,outside} - \mu_{Na^+,inside} = RT\ln([Na^+]_{outside}) - RT\ln([Na^+]_{inside})$$
$$= RT\ln\left(\frac{[Na^+]_{outside}}{[Na^+]_{inside}}\right) \quad [\ln x - \ln y = \ln(x/y)]$$
$$= (8.3145\ \mathrm{J\ K^{-1}\ mol^{-1}}) \times (310\ \mathrm{K}) \times \ln(140/10)$$
$$= 6.8\times10^3\ \mathrm{J\ mol^{-1}} = \boxed{6.8\ \mathrm{kJ\ mol^{-1}}}.$$

E4.16 $ATP(aq) + H_2O(l) \rightarrow ADP(aq) + P_i^-(aq) + H^+(aq)$

$\Delta_r H^{\oplus} = -20\ \mathrm{kJ\ mol^{-1}}$ and $\Delta_r S^{\oplus} = +34\ \mathrm{J\ K^{-1}\ mol^{-1}}$

Since $\Delta_r G^{\oplus} = \Delta_r H^{\oplus} - T\Delta_r S^{\oplus}$ at pH = 7, we see that $\Delta_r G^{\oplus} < 0$ and, therefore, $K > 1$ at $\boxed{\text{all temperatures}}$ because the reaction is exothermic and $\Delta_r S^{\oplus} > 0$.

E4.17 $\text{Pyruvate}^{-1}(aq) + NADH(aq) + H^+(aq) \rightarrow \text{lactate}^{-1}(aq) + NAD^+(aq)$

$\Delta_r G^{\circ} = -66.6\ \mathrm{kJ\ mol^{-1}}$ at 37°C

Since $\Delta_r G = \Delta_r G^{\circ} + RT\ln Q$ [4.7], we immediately have $\Delta_r G^{\oplus} = \Delta_r G^{\circ} + RT\ln Q^{\oplus}$, where $Q^{\oplus}$ is the reaction quotient for the biological standard state defined with $a_{H^+} = 10^{-7}$ (pH = 7) and all other activities are 1.

$$\Delta_r G^{\oplus} = \Delta_r G^{\circ} + RT\ln\left(\frac{1}{a_{H^+}}\right) = (-66.6\ \mathrm{kJ\ mol^{-1}}) + (8.3145\ \mathrm{J\ K^{-1}\ mol^{-1}}) \times (310\ \mathrm{K}) \times \ln\left(\frac{1}{10^{-7}}\right)$$
$$= \boxed{-25.1\ \mathrm{kJ\ mol^{-1}}}$$

E4.18 The reaction is $AMP(aq) + H_2O(l) \rightarrow A(aq) + P_i^-(aq) + H^+(aq)$ $\quad \Delta_r G^{\oplus} = -14\ \mathrm{kJ\ mol^{-1}}$ at 298 K.

Since $\Delta_r G = \Delta_r G^{\circ} + RT\ln Q$ [4.7], we immediately have $\Delta_r G^{\oplus} = \Delta_r G^{\circ} + RT\ln Q^{\oplus}$, where $Q^{\oplus}$ is the reaction quotient for the biological standard state defined with $a_{H^+} = 10^{-7}$ (pH = 7) and all other activities are 1. Therefore,

$$\Delta_r G^{\circ} = \Delta_r G^{\oplus} - RT\ln Q^{\oplus} = \Delta_r G^{\oplus} - RT\ln(a_{H^+})$$
$$= (-14\ \mathrm{kJ\ mol^{-1}}) - (8.3145\ \mathrm{J\ K^{-1}\ mol^{-1}}) \times (298\ \mathrm{K}) \times \ln(10^{-7})$$
$$= \boxed{26\ \mathrm{kJ\ mol^{-1}}}$$

E4.19 The phosphate transfer potentials of Table 4.3 are used to calculate $\Delta_r G^\circ$ values at 298.15 K after which we calculate $\Delta_r G^\oplus$ with eqn 4.8: $\Delta_r G^\oplus = \Delta_r G^\circ + 7\nu RT \ln 10$, where ν is the reaction coefficient for $H^+(aq)$, which is taken to be positive when H^+ is a reactant and negative when H^+ is a product. A balanced reaction equation is needed to find the value of ν so we write balanced half-reactions for the phosphate transfer and add them to get a balanced net reaction.

(a) $GTP^{4-}(aq) + H_2O(l) \rightarrow GDP^{3-}(aq) + HPO_4^{2-}(aq) + H_3O^+(aq) \quad \Delta_r G^\circ = -31\ kJ\ mol^{-1}$

$ADP^{3-}(aq) + HPO_4^{2-}(aq) + H_3O^+(aq) \rightarrow ATP^{4-}(aq) + H_2O(l) \quad \Delta_r G^\circ = +31\ kJ\ mol^{-1}$

$GTP^{4-}(aq) + ADP^{3-}(aq) \rightarrow GDP^{3-}(aq) + ATP^{4-}(aq)$

$\nu = 0$ and $\Delta_r G^\circ = (-31 + 31)\ kJ\ mol^{-1} = 0$

$\Delta_r G^\oplus = \Delta_r G^\circ + 7\nu RT \ln 10 = \Delta_r G^\circ = \boxed{0}$

(b) $\text{glycerol}(aq) + HPO_4^{2-}(aq) \rightarrow \text{glycerol-1-phosphate}^{2-}(aq) + H_2O(l) \quad \Delta_r G^\circ = +10\ kJ\ mol^{-1}$

$ATP^{4-}(aq) + H_2O(l) \rightarrow ADP^{3-}(aq) + HPO_4^{2-}(aq) + H_3O^+(aq) \quad \Delta_r G^\circ = -31\ kJ\ mol^{-1}$

$\text{glycerol}(aq) + ATP^{4-}(aq) \rightarrow \text{glycerol-1-phosphate}^{2-}(aq) + ADP^{3-}(aq) + H_3O^+(aq)$

$\nu = -1$ and $\Delta_r G^\circ = (+10 - 31)\ kJ\ mol^{-1} = -21\ kJ\ mol^{-1}$

$$\begin{aligned}\Delta_r G^\oplus &= \Delta_r G^\circ + 7\nu RT \ln 10 \\ &= (-21\ kJ\ mol^{-1}) + 7 \times (-1) \times (8.3145 \times 10^{-3}\ kJ\ K^{-1}\ mol^{-1}) \times (298.15\ K) \times \ln 10 \\ &= \boxed{-61\ kJ\ mol^{-1}}\end{aligned}$$

(c) $\text{3-phosphoglycerate}^{3-}(aq) + HPO_4^{2-}(aq) + H_3O^+(aq)$

$\rightarrow \text{1,3-bis(phospho)glycerate}^{4-}(aq) + H_2O(l) \quad \Delta_r G^\circ = +49\ kJ\ mol^{-1}$

$ATP^{4-}(aq) + H_2O(l) \rightarrow ADP^{3-}(aq) + HPO_4^{2-}(aq) + H_3O^+(aq) \quad \Delta_r G^\circ = -31\ kJ\ mol^{-1}$

$\text{3-phosphoglycerate}^{3-}(aq) + ATP^{4-}(aq) \rightarrow \text{1,3-bis(phospho)glycerate}^{4-}(aq) + ADP^{3-}(aq)$

$\nu = 0$ and $\Delta_r G^\circ = (+49 - 31)\ kJ\ mol^{-1} = +18\ kJ\ mol^{-1}$

$\Delta_r G^\oplus = \Delta_r G^\circ + 7\nu RT \ln 10 = \Delta_r G^\circ = \boxed{+18\ kJ\ mol^{-1}}$

E4.20 (a) Electrostatic, solvent, base-stacking, and base-pairing interactions all contribute to the DNA double-helix structure. The latter two, which originate with π electron sharing with a hydrophobic interaction between aromatic rings and hydrogen bonding, respectively, are probably the most important. However, hydrogen bonding between base pairs is probably the major contributor to the stability of double-helix formation from two single strands. In going from the formation of a double helix from two A_nU_n single strands to the formation of a double helix from two $A_{n+1}U_{n+1}$ single strands there is an addition of two A–U, hydrogen-bonded base pairs. Thus, $\Delta_r G^\circ$ for the formation of $A_{n+1}U_{n+1}$ is more negative, which yields a greater equilibrium constant.

(b) Since there is an increase of two A–U, hydrogen-bonded base pairs in going from the A_5U_5 double helix to the A_6U_6 double helix, the stabilization contribution of a single A–U base pair to the Gibbs energy of formation equals the difference between respective formation Gibbs energies with the result divided by two.

$2\ ssA_5U_5 \rightarrow dsA_5U_5$

$\Delta_r G^\circ = -RT \ln K \quad [4.10a] = -(8.3145\ J\ K^{-1}\ mol^{-1}) \times (298.15\ K) \times \ln(5.0 \times 10^3) = -21\ kJ\ mol^{-1}$

$2\ ssA_6U_6 \rightarrow dsA_6U_6$

$\Delta_r G^\circ = -RT \ln K \quad [4.10a] = -(8.3145\ J\ K^{-1}\ mol^{-1}) \times (298.15\ K) \times \ln(2.0 \times 10^5) = -30\ kJ\ mol^{-1}$

Contribution of a single A–U base pair $= \{(-30\ kJ\ mol^{-1}) - (-21\ kJ\ mol^{-1})\}/2 \approx \boxed{-5\ kJ\ mol^{-1}}$

E4.21 The text section titled Case Study 4.3 reports that the biological hydrolysis of ATP releases sufficient energy to do a maximum of 31 kJ mol⁻¹ of non-expansion work at 310 K:

$ATP(aq) + H_2O(l) \rightarrow ADP(aq) + P_i^-(aq) + H_3O^+(aq) \quad \Delta_r G^\oplus_{310\,K} = -31\ kJ\ mol^{-1}$

The *Resource section 3*: *Data* reports that for the combustion of glucose at 298 K:

$$C_6H_{12}O_6(\text{glucose, s}) + 6\,O_2(g) \rightarrow 6\,CO_2(g) + 6\,H_2O(l) \quad \Delta_c H^{\ominus}_{298\,K} = -2808\ \text{kJ mol}^{-1}$$

$$\begin{aligned}\Delta_c S^{\ominus}_{298\,K}(\text{glucose, s}) &= \{6 \times S^{\ominus}_m(CO_2, g) + 6 \times S^{\ominus}_m(H_2O, l) - S^{\ominus}_m(\text{sucrose, s}) - 6 \times S^{\ominus}_m(O_2, g)\}\ \text{J K}^{-1}\,\text{mol}^{-1}\\ &= \{6 \times (213.74) + 6 \times (69.91) - (212.1) - 6 \times (205.138)\}\ \text{J K}^{-1}\,\text{mol}^{-1}\\ &= 259.0\ \text{J K}^{-1}\,\text{mol}^{-1}\end{aligned}$$

Therefore,

$$\begin{aligned}\Delta_c G^{\ominus}_{298\,K}(\text{glucose, s}) &= \Delta_c H^{\ominus}_{298\,K}(\text{glucose, s}) - T\Delta_c S^{\ominus}_{298\,K}(\text{glucose, s})\\ &= -2808\ \text{kJ mol}^{-1} - (298.15\ \text{K}) \times (0.2590\ \text{kJ K}^{-1}\,\text{mol}^{-1})\\ &= -2885\ \text{kJ mol}^{-1}\end{aligned}$$

The combustion Gibbs energy for glucose at 310 K is calculated with the van't Hoff equation after making the substitution $\ln K = -\Delta_r G^{\ominus}/RT$ [4.10a].

$$\ln K_2 = \ln K_1 + \frac{\Delta_r H^{\ominus}}{R}\left(\frac{1}{T_1} - \frac{1}{T_2}\right) \quad [4.18]$$

$$-\Delta_r G^{\ominus}_2/RT_2 = -\Delta_r G^{\ominus}_1/RT_1 + \frac{\Delta_r H^{\ominus}}{R}\left(\frac{1}{T_1} - \frac{1}{T_2}\right)$$

$$\Delta_r G^{\ominus}_2/T_2 = \Delta_r G^{\ominus}_1/T_1 - \Delta_r H^{\ominus}\left(\frac{1}{T_1} - \frac{1}{T_2}\right)$$

$$\Delta_r G^{\ominus}_2 = T_2\left\{\frac{\Delta_r G^{\ominus}_1}{T_1} - \Delta_r H^{\ominus}\left(\frac{1}{T_1} - \frac{1}{T_2}\right)\right\}$$

$$\begin{aligned}\Delta_c G^{\ominus}_{310\,K} &= (310\ \text{K})\left\{\frac{(-2885\ \text{kJ mol}^{-1})}{298.15\ \text{K}} - (-2808\ \text{kJ mol}^{-1})\left(\frac{1}{298.15\ \text{K}} - \frac{1}{310\ \text{K}}\right)\right\}\\ &= -2888\ \text{kJ mol}^{-1}\end{aligned}$$

Thus, the combustion of glucose at 310 K releases energy capable of doing a maximum of 2888 kJ mol^{-1} of non-expansion work and we recognize that the computation indicates that $\Delta_c G^{\ominus}$ is almost constant between 298 K and 310 K. Consequently, $\Delta_c G^{\ominus}_{298\,K}$ could be used in the computations below without significant disadvantage. Furthermore, we recognize that, because the combustion reaction contains no solvated protons, $\Delta_r G^{\oplus} = \Delta_r G^{\ominus}$ at all temperatures.

(a) If we assume that each mole of ATP formed during the aerobic breakdown of glucose produces about -31 kJ mol^{-1} and 38 moles of ATP are produced per mole of glucose consumed, then

$$\text{percentage efficiency} = \frac{38 \times (-31\ \text{kJ mol}^{-1})}{-2888\ \text{kJ mol}^{-1}} \times 100\% = \boxed{41\%}.$$

(b) For the oxidation of glucose under the biological conditions of $T = 310$ K, $p_{CO_2} = 0.053$ bar, $p_{O_2} = 0.132$ bar, and [glucose] $= 0.056$ mol dm^{-3} we have

$$\begin{aligned}\Delta_r G &= \Delta_r G^{\ominus} + RT\ln Q = \Delta_r G^{\ominus} + RT\ln\left(\frac{(p_{CO_2}/p^{\ominus})^6 \times c^{\ominus}}{[\text{glucose}] \times (p_{O_2}/p^{\ominus})^6}\right)\\ &= -2888\ \text{kJ mol}^{-1} + (8.3145 \times 10^{-3}\ \text{kJ K}^{-1}\,\text{mol}^{-1}) \times (310\ \text{K})\ln\left(\frac{(0.053)^6}{(0.056) \times (0.132)^6}\right)\\ &= -2895\ \text{kJ mol}^{-1}\end{aligned}$$

This is not much different from the standard value of -2888 kJ mol^{-1}.

For the ATP hydrolysis under the given conditions of $T = 310$ K, $p_{CO_2} = 0.053$ bar, $p_{O_2} = 0.132$ bar, [glucose] = 0.056 mol dm^{-3}, [ATP] = [ADP] = [P$_i$] = 1.0 × 10^{-4} mol dm^{-3}, and pH = 7.4 we have

$$\Delta_r G = \Delta_r G^{\oplus} + RT\ln(Q/Q^{\oplus}) = \Delta_r G^{\oplus} + RT\ln\left(\frac{[\text{ADP}]\times[\text{P}_i^-]}{[\text{ATP}]\times(1\ \text{mol dm}^{-3})}\times\frac{[\text{H}^+]}{10^{-7}\ \text{mol dm}^{-3}}\right)$$

[i.e., $[\text{H}^+]^{\oplus} = 10^{-7}$ M, $c^{\oplus} = 1$ mol dm^{-3}]

$$= -31\ \text{kJ mol}^{-1} + (8.3145\ \text{J K}^{-1}\ \text{mol}^{-1})\times(310\ \text{K})\ln(1.0\times10^{-4}\times10^{-7.4}/10^{-7})$$
$$= -57\ \text{kJ mol}^{-1}$$

With this value for $\Delta_r G$ the efficiency becomes

$$\text{efficiency} = \frac{38\times(-57\ \text{kJ mol}^{-1})}{-2895\ \text{kJ mol}^{-1}} = \boxed{75\%}$$

E4.22 G6P → F6P $\quad Q = [\text{F6P}]/[\text{G6P}] \quad \Delta_r G^{\circ} = +1.7$ kJ mol^{-1} [Example 4.3]

$$f = \frac{[\text{F6P}]}{[\text{F6P}]+[\text{G6P}]}\ [\text{Example 4.3}] = \frac{1}{1+\dfrac{[\text{G6P}]}{[\text{F6P}]}} = \frac{1}{1+\dfrac{1}{Q}}$$

Solving for Q gives $Q = \dfrac{f}{1-f}$.

$$\Delta_r G = \Delta_r G^{\circ} + RT\ln Q = \Delta_r G^{\circ} + RT\ln\left(\frac{f}{1-f}\right) \quad \text{where} \quad T = 298\ \text{K}$$

$$= 1.7\ \text{kJ mol}^{-1} + (2.479\ \text{kJ mol}^{-1})\times\ln\left(\frac{f}{1-f}\right)$$

Figure 4.1 gives a plot of $\Delta_r G$ against f. When $\Delta_r G < 0$, the reaction proceeds spontaneously to the right until $\Delta_r G = 0$ at the equilibrium value of f, i.e. $f_{eq} = 0.33$ (Example 4.3). When $\Delta_r G > 0$, the reaction proceeds spontaneously to the left until $\Delta_r G = 0$ at the equilibrium value of f.

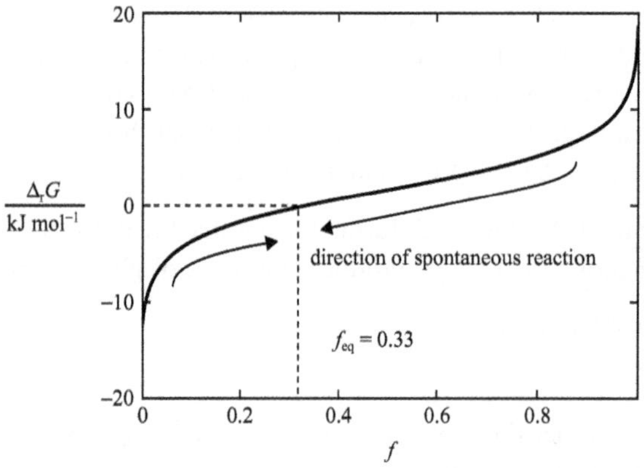

Figure 4.1

E4.23 (a) Hill equation: $\log\left(\dfrac{s}{1-s}\right) = \nu\log p - \nu\log K = \log\left(\dfrac{p}{K}\right)^{\nu}$ where s is the saturation at the partial pressure p of oxygen. K and ν are Hill coefficients. Solving the Hill equation for K yields $K = \left(\dfrac{1-s}{s}\right)^{1/\nu} p$. Thus, at 50% saturation ($s = 0.50$) we see that $K = p_{50\%\ \text{saturation}}$ irrespective of the value of ν. Reading the value of $p_{50\%\ \text{saturation}}$ for both myoglobin (Mb) and hemoglobin (Hb) off text Figure 4.9, we find that

$\boxed{K_{Mb} = 2.33 \text{ torr} = 0.311 \text{ kPa}}$

and

$\boxed{K_{Hb} = 34.7 \text{ torr} = 4.62 \text{ kPa}}$.

Solving the Hill equation for s we obtain $s = \dfrac{1}{(K/p)^{\nu} + 1}$. This equation is used to compute the saturation of both Mb and Hb over a range of p values as summarized in the following table. We have used $\nu_{Mb} = 1$ and $\nu_{Hb} = 2.8$.

p/kPa	1.0	1.5	2.5	4.0	8.0
s_{Mb}	0.76	0.83	0.89	0.93	0.96
s_{Hb}	0.01	0.04	0.15	0.40	0.82

(b) The computed values of s_{Hb} using the maximum value $\nu_{Hb} = 4$ are summarized in the following table.

p/kPa	1.0	1.5	2.5	4.0	8.0
s_{Hb}	0.003	0.01	0.8	0.36	0.90

By comparing the values of s_{Hb} calculated with $\nu = 2.8$ in part (a) and $\nu = 4$ in part (b) with the values reported in text Fig. 4.9 we see that the values calculated with $\nu = 2.8$ are in better agreement with experimental values. For example, at the partial oxygen pressure of the lungs (100 Torr = 13.3 kPa) the experimental saturation is given as 0.94 in Fig. 4.9 while the calculated values are $s_{Hb,\nu=2.8} = 0.95$ and $s_{Hb,\nu=4} = 0.99$. See Case Study 4.1 for a discussion of the cooperative binding of oxygen by hemoglobin.

E4.24 We look up $\Delta_f G^{\ominus}$ for each compound (*Resource section 3*: *Data*) and note the sign.

(a) –, exergonic (b) –, exergonic (c) +, endergonic (d) –, exergonic

E4.25 Sucrose combustion with liquid water product: $C_{12}H_{22}O_{11}(s) + 12\,O_2(g) \rightarrow 12\,CO_2(g) + 11\,H_2O(l)$.

(a) Using *Resource section 3*: *Data*, we have $\Delta_c H^{\ominus}(\text{sucrose, s}) = -5645 \text{ kJ mol}^{-1}$ and

$$\begin{aligned}
\Delta_c S^{\ominus} &= \sum \nu S^{\ominus}_m(\text{products}) - \sum \nu S^{\ominus}_m(\text{reactants}) \\
&= \{12 \times S^{\ominus}_m(CO_2, g) + 11 \times S^{\ominus}_m(H_2O, l) - S^{\ominus}_m(\text{sucrose, s}) - 12 \times S^{\ominus}_m(O_2, g)\} \text{ J K}^{-1} \text{ mol}^{-1} \\
&= \{12 \times (213.74) + 11 \times (69.91) - (360.2) - 12 \times (205.138)\} \text{ J K}^{-1} \text{ mol}^{-1} \\
&= 512.0 \text{ J K}^{-1} \text{ mol}^{-1}
\end{aligned}$$

Therefore,

$$\begin{aligned}
\Delta_c G^{\ominus}(\text{sucrose, s}) &= \Delta_c H^{\ominus}(\text{sucrose, s}) - T\Delta_c S^{\ominus}(\text{sucrose, s}) \\
&= -5645 \text{ kJ mol}^{-1} - (298.15 \text{ K}) \times (0.5120 \text{ kJ K}^{-1} \text{ mol}^{-1}) \\
&= \boxed{-5798 \text{ kJ mol}^{-1}}
\end{aligned}$$

Alternatively, we could use the standard formation Gibbs energies to calculate $\Delta_c G^{\ominus}(\text{sucrose, s})$.

$$\begin{aligned}
\Delta_c G^{\ominus}(\text{sucrose, s}) &= \sum \nu \Delta_f G^{\ominus}(\text{products}) - \sum \nu \Delta_f G^{\ominus}(\text{reactants}) \\
&= 12 \times \Delta_f G^{\ominus}(CO_2, g) + 11 \times \Delta_f G^{\ominus}(H_2O, l) - \Delta_f G^{\ominus}(\text{sucrose}) - 12 \times \Delta_f G^{\ominus}(O_2, g) \\
&= \{12 \times (-394.36) + 11 \times (-237.13) - (-1543) - 12 \times (0)\} \text{ kJ mol}^{-1} \\
&= -5798 \text{ kJ mol}^{-1}
\end{aligned}$$

(b) (i) At constant pressure the heat given off by the combustion of 1.0 kg sucrose is:

$$q = n_{\text{sucrose}}\Delta_c H^{\ominus} = (m/M)_{\text{sucrose}}\Delta_c H^{\ominus}$$
$$= \{(1.0\text{ kg})/(0.34230\text{ kg mol}^{-1})\} \times (-5645\text{ kJ mol}^{-1})$$
$$= \boxed{-16.\bar{5}\text{ MJ}}$$ The negative sign indicates that the heat is given off.

(ii) The maximum non-expansion work that can be performed by the combustion of 1.0 kg of sucrose is:

$$w_{\text{non-expansion}} = n_{\text{sucrose}}\Delta_c G^{\ominus} = (m/M)_{\text{sucrose}}\Delta_c G^{\ominus}$$
$$= \{(1.0\text{ kg})/(0.34230\text{ kg mol}^{-1})\} \times (-5798\text{ kJ mol}^{-1})$$
$$= \boxed{-16.\bar{9}\text{ MJ}}$$ The negative sign indicates that the combustion system does the work.

E4.26 We find the combustion enthalpies for glucose and sucrose at 298 K in the *Resource section 3: Data.*

$C_{12}H_{22}O_{11}(\text{sucrose, s}) + 12\,O_2(g) \rightarrow 12\,CO_2(g) + 11\,H_2O(l)$

$\Delta_c H^{\ominus} = -5645\text{ kJ mol}^{-1}$ or -16.49 kJ g^{-1}

$C_6H_{12}O_6(\text{glucose, s}) + 6\,O_2(g) \rightarrow 6\,CO_2(g) + 6\,H_2O(l)$ $\Delta_c H^{\ominus} = -2808\text{ kJ mol}^{-1}$ or -15.59 kJ g^{-1}

Since the specific enthalpy of sucrose (16.45 kJ g^{-1}) is slightly greater than the specific enthalpy of glucose (15.59 kJ g^{-1}), $\boxed{\text{sucrose is a slightly more effective fuel}}$.

The standard Gibbs energy for the combustion of glucose at 298 K is calculated with the standard formation Gibbs energies found in the *Resource section 3: Data.*

$$\Delta_c G^{\ominus}(\text{glucose, s}) = \sum \nu\Delta_f G^{\ominus}(\text{products}) - \sum \nu\Delta_f G^{\ominus}(\text{reactants})$$
$$= 6 \times \Delta_f G^{\ominus}(CO_2, g) + 6 \times \Delta_f G^{\ominus}(H_2O, l) - \Delta_f G^{\ominus}(\text{glucose}) - 6 \times \Delta_f G^{\ominus}(O_2, g)$$
$$= \{6 \times (-394.36) + 6 \times (-237.13) - (-917.2) - 6 \times (0)\}\text{ kJ mol}^{-1}$$
$$= -2871.7\text{ kJ mol}^{-1}$$

At constant pressure the heat given off by the combustion of 1.0 kg glucose is:

$$q = n_{\text{glucose}}\Delta_c H^{\ominus} = (m/M)_{\text{glucose}}\Delta_c H^{\ominus}$$
$$= \{(1.0\text{ kg})/(0.18016\text{ kg mol}^{-1})\} \times (-2808\text{ kJ mol}^{-1})$$
$$= \boxed{-15.\bar{6}\text{ MJ}}$$ The negative sign indicates that the heat is given off.

The maximum non-expansion work that can be performed by the combustion of 1.0 kg of glucose is:

$$w_{\text{non-expansion}} = n_{\text{glucose}}\Delta_c G^{\ominus} = (m/M)_{\text{glucose}}\Delta_c G^{\ominus}$$
$$= \{(1.0\text{ kg})/(0.18016\text{ kg mol}^{-1})\} \times (-2871.7\text{ kJ mol}^{-1})$$
$$= \boxed{-15.\bar{9}\text{ MJ}}$$ The negative sign indicates that the combustion system does the work.

Comparing $w_{\text{non-expansion}}$ (glucose, 1 kg) with $w_{\text{non-expansion}}$ (sucrose, 1 kg), which is calculated in E4.25(b), shows that the combustion of sucrose provides a greater capacity to do non-expansion work than does glucose.

The expansion work for the combustion of 1.0 kg of glucose is given by the constant p_{ex} expression:

$$w_{\text{expansion}} = -p_{\text{ex}}\Delta V_{\text{m}}(m/M)_{\text{glucose}} = -p^{\ominus}\Delta V_{\text{m}}(m/M)_{\text{glucose}}$$

where $\Delta V_{\text{m}} = V_{\text{products}} - V_{\text{reactants}} = V_{\text{gaseous products}} - V_{\text{gaseous reactants}}$
$= \Delta\nu_{\text{gas}}RT/p^{\ominus}$ [volumes of solids and liquids are negligibly small].

Thus,

$$w_{\text{expansion}} = -\Delta v_{\text{gas}} RT(m/M)_{\text{glucose}} = -(6-6)\times RT(m/M)_{\text{glucose}} = \boxed{0}$$

$$w_{\text{total}} = w_{\text{expansion}} + w_{\text{non-expansion}} = (0 - 15.\bar{9})\text{ MJ} = \boxed{-15.\bar{9}\text{ MJ}}$$

E4.27 $CH_3COCO_2^-(aq) + H^+(aq) \rightarrow CH_3CHO(g\text{ or }aq) + CO_2(g)$

The catalytic activity of pyruvate decarboxylase affects the kinetics of the reaction only. Like all enzymes, pyruvate decarboxylase does not affect the thermodynamic properties of the reaction and plays no role in the thermodynamic computations.

(a) $CH_3COCO_2^-(aq) + H^+(aq) \rightarrow CH_3CHO(g) + CO_2(g)$

We begin by calculating $\Delta_r G^\ominus$ at 298.15 K using *Resource section 3: Data*. Since the reaction equation contains hydrogen ions with $v = 1$, $\Delta_r G^\oplus \neq \Delta_r G^\ominus$ and eqn 4.8 must be used to calculate $\Delta_r G^\oplus$ at 298.15 K.

$$\begin{aligned}\Delta_r G^\ominus &= \sum v\Delta_f G^\ominus\text{ (products)} - \sum v\Delta_f G^\ominus\text{ (reactants)}\\ &= \{(-133) + (-394.36) - (-474) - (0)\}\text{ kJ mol}^{-1}\\ &= -53.\bar{4}\text{ kJ mol}^{-1}\end{aligned}$$

$$\begin{aligned}\Delta_r G^\oplus &= \Delta_r G^\ominus + 7vRT\ln 10\\ &= -53.\bar{4}\text{ kJ mol}^{-1} + 7\times 1\times(8.3145\times 10^{-3}\text{ kJ K}^{-1}\text{ mol}^{-1})\times(298.15\text{ K})\times\ln 10\\ &= \boxed{-13\text{ kJ mol}^{-1}}\end{aligned}$$

(b) $CH_3COCO_2^-(aq) + H^+(aq) \rightarrow CH_3CHO(aq) + CO_2(g)$

The complete miscibility of ethanal in water suggestions that solvation of gaseous ethanal is exergonic. This makes both $\Delta_r G^\ominus$ and $\Delta_r G^\oplus$ $\boxed{\text{more exergonic}}$ than the values calculated in part (a).

E4.28 $$\ln K = -\frac{\Delta_r G^\ominus}{RT}[4.10\text{a}] = -\frac{\Delta_r H^\ominus - T\Delta_r S^\ominus}{RT}[4.15] = \left(-\frac{\Delta_r H^\ominus}{R}\right)\times\frac{1}{T} + \frac{\Delta_r S^\ominus}{R}$$

For many reactions both $\Delta_r H^\ominus$ and $\Delta_r S^\ominus$ are weakly dependent upon temperature over a small temperature range and they may be treated as constant. In such cases, the factors $-\Delta_r H^\ominus/R$ and $\Delta_r S^\ominus/R$ are constants, thereby, making the above expression of ln K linear in the variable $1/T$. A plot of ln K against $1/T$ will be linear with a slope equal to $-\Delta_r H^\ominus/R$ and intercept equal to $\Delta_r S^\ominus/R$.

$$\ln K = slope\times\frac{1}{T} + intercept,\quad\text{where}\quad slope = -\Delta_r H^\ominus/R\quad\text{and}\quad intercept = \Delta_r S^\ominus/R$$

Consequently, $\Delta_r H^\ominus$ and $\Delta_r S^\ominus$ may often be determined by (i) measuring K over a range of temperatures, (ii) checking the linearity of the ln K against $1/T$ plot, and (iii) performing a linear regression fit of the plot to acquire the slope and intercept values. The thermodynamic properties are determined with the expressions

$$\Delta_r H^\ominus = -R\times slope\quad\text{and}\quad \Delta_r S^\ominus = R\times intercept.$$

E4.29 $\text{Fumarate}^{2-}(aq) + H_2O(l) \rightarrow \text{malate}^-(aq)$

As discussed in E4.28 we need a data plot of ln K against $1/T$. Toward that end we draw up the following data table and prepare the plot of Figure 4.2. The plot appears to be straight so a linear regression fit, shown in the figure, is appropriate. The square of the regression coefficient R is satisfactorily close to 1 and the desired slope and intercept are: *slope* = $1.80\overline{87}\times 10^3$ K and *intercept* = $-4.67\overline{96}$. Thus,

$$\Delta_r H^\ominus = -R \times slope$$
$$= -(8.3145 \text{ J K}^{-1} \text{ mol}^{-1}) \times (1.80\overline{87} \times 10^3 \text{ K})$$
$$= \boxed{-15.0 \text{ kJ mol}^{-1}}$$

$$\Delta_r S^\ominus = R \times intercept$$
$$= (8.3145 \text{ J K}^{-1} \text{ mol}^{-1}) \times (-4.67\overline{96})$$
$$= \boxed{-38.9 \text{ J K}^{-1} \text{ mol}^{-1}}$$

The reaction is exothermic and lowers entropy.

θ/°C	15	20	25	30	35	40	45	50
K	4.786	4.467	4.074	3.631	3.311	3.090	2.754	2.399
1000 K/T	3.4704	3.4112	3.3540	3.2987	3.2452	3.1934	3.1432	3.0945
ln K	1.5657	1.4967	1.4046	1.2895	1.1973	1.1282	1.0131	0.8751

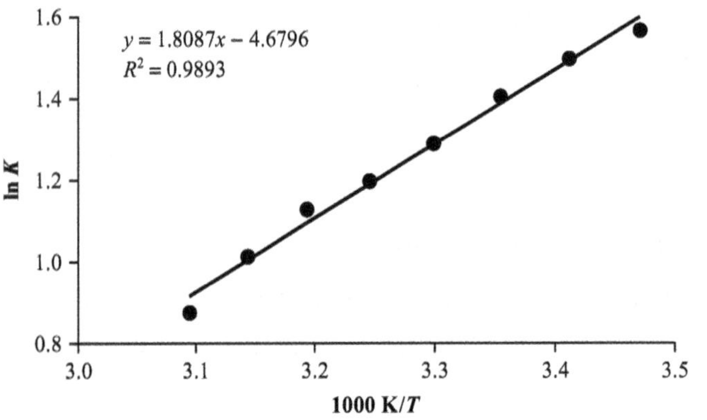

Figure 4.2

E4.30 $$\ln\left(\frac{K_2}{K_1}\right) = \frac{\Delta_r H^\ominus}{R}\left(\frac{1}{T_1} - \frac{1}{T_2}\right) \text{ [4.18] and, therefore, } \Delta_r H^\ominus = R\left(\frac{1}{T_1} - \frac{1}{T_2}\right)^{-1} \ln\left(\frac{K_2}{K_1}\right)$$

(a) $K_{308\text{ K}} = 2\, K_{298\text{ K}}$

$$\Delta_r H^\ominus = R\left(\frac{1}{298\text{ K}} - \frac{1}{308\text{ K}}\right)^{-1} \ln\left(\frac{2K_1}{K_1}\right) = (6.362 \times 10^3 \text{ K}) \times R = \boxed{52.9 \text{ kJ mol}^{-1}}$$

(b) $K_{308\text{ K}} = \frac{1}{2} K_{298\text{ K}}$

$$\Delta_r H^\ominus = R\left(\frac{1}{298\text{ K}} - \frac{1}{308\text{ K}}\right)^{-1} \ln\left(\frac{\frac{1}{2}K_1}{K_1}\right) = (-6.362 \times 10^3 \text{ K}) \times R = \boxed{-52.9 \text{ kJ mol}^{-1}}$$

E4.31 (a) $H_2PO_4^- + H_2O \rightleftharpoons HPO_4^{2-} + H_3O^+$

(b) $CH_3CH(OH)COOH + H_2O \rightleftharpoons CH_3CH(OH)CO_2^- + H_3O^+$

(c) $HOOC(CH_2)_2CH(NH_2)COOH + H_2O \rightleftharpoons HOOC(CH_2)_2CH(NH_2)CO_2^- + H_3O^+$
$HOOC(CH_2)_2CH(NH_2)CO_2^- + H_2O \rightleftharpoons {}^-O_2C(CH_2)_2CH(NH_2)CO_2^- + H_3O^+$

These are the carboxylic acid dissociation reactions. The amine group (R–NH_2) is a base so proton-transfer reactions can also be written that show the proton transfer from water to the amine for each of the above four species. Example:

$HOOC(CH_2)_2CH(NH_2)COOH + H_2O \rightleftharpoons HOOC(CH_2)_2CH(NH_3^+)COOH + OH^-$

(d) $NH_2CH_2COOH + H_2O \rightleftharpoons NH_2CH_2CO_2^- + H_3O^+$
$NH_2CH_2COOH + H_2O \rightleftharpoons {}^+NH_3CH_2COOH + OH^-$
$NH_2CH_2CO_2^- + H_2O \rightleftharpoons {}^+NH_3CH_2CO_2^- + OH^-$

(e) $HOOCCOOH + H_2O \rightleftharpoons HOOCCO_2^- + H_3O^+$
$HOOCCO_2^- + H_2O \rightleftharpoons {}^-O_2CCO_2^- + H_3O^+$

E4.32 (a) $2\,H_2O \rightleftharpoons H_3O^+ + OH^-$ and $[OH^-] = [H_3O^+]$ in pure water.
$K_w = 2.5 \times 10^{-14} = [H_3O^+][OH^-] = [H_3O^+]^2$
$[H_3O^+] = K_w^{1/2} = \sqrt{2.5 \times 10^{-14}} = \boxed{1.6 \times 10^{-7}\ \text{mol dm}^{-3}}$
$\text{pH} = -\log[H_3O^+] = \boxed{6.80}$

(b) $[OH^-] = [H_3O^+] = \boxed{1.6 \times 10^{-7}\ \text{mol dm}^{-3}}$

$\text{pOH} = -\log[OH^-] = \boxed{6.80}$

E4.33 (a) $2\,D_2O \rightleftharpoons D_3O^+ + OD^-$ and $[OD^-] = [D_3O^+]$ in pure dideuterium oxide.

(b) $K_{w,D} = [D_3O^+][OD^-] = 1.35 \times 10^{-15}$ and $pK_{w,D} = -\log K_{w,D} = \boxed{14.870}$

(c) $K_{w,D} = 1.35 \times 10^{-15} = [D_3O^+][OD^-] = [D_3O^+]^2$

$[D_3O^+] = K_{w,D}{}^{1/2} = \sqrt{1.35 \times 10^{-15}} = \boxed{3.67 \times 10^{-8}\ \text{mol dm}^{-3}}$

$[OD^-] = [D_3O^+] = \boxed{3.67 \times 10^{-8}\ \text{mol dm}^{-3}}$

(d) $\text{pD} = -\log(3.67 \times 10^{-8}) = \boxed{7.435} = \text{pOD}$

(e) $\text{pD} + \text{pOD} = pK_{w,D} = \boxed{14.870}$

E4.34 $\text{pH} = -\log a_{H_3O^+}$ [4.21] $\approx -\log([H_3O^+])$ and $\text{pOH} = pK_w - \text{pH}$ [4.27] $= 14.00 - \text{pH}$

	$[H_3O^+]/\text{mol dm}^{-3}$	pH	pOH
(a)	1.5×10^{-5}	4.82	9.18
(b)	1.5×10^{-3}	2.82	11.18
(c)	5.1×10^{-14}	13.29	0.71
(d)	5.01×10^{-5}	4.30	9.70

E4.35 These are strong acid/base neutralization reactions.

(a) amount (moles) $H_3O^+ = (0.0250\ \text{dm}^3) \times (0.144\ \text{mol dm}^{-3}) = 3.60 \times 10^{-3}$ mol
amount (moles) $OH^- = (0.0250\ \text{dm}^3) \times (0.125\ \text{mol dm}^{-3}) = 3.12 \times 10^{-3}$ mol
excess $H_3O^+ = (3.60 \times 10^{-3} - 3.12 \times 10^{-3}) = 0.48 \times 10^{-3}$ mol H_3O^+

$$[H_3O^+] = \frac{4.8 \times 10^{-4}\ \text{mol}}{0.0500\ \text{dm}^3} = \boxed{9.60 \times 10^{-3}\ \text{mol dm}^{-3}}$$

$\text{pH} = -\log(9.60 \times 10^{-3}) = \boxed{2.02}$

(b) amount of $H_3O^+ = (0.0250\ \text{dm}^3) \times (0.15\ \text{mol dm}^{-3}) = 3.75 \times 10^{-3}$ mol H_3O^+
amount of $OH^- = (0.0350\ \text{dm}^3) \times (0.15\ \text{mol dm}^{-3}) = 5.25 \times 10^{-3}$ mol OH^-
excess $OH^- = (5.25 \times 10^{-3} - 3.75 \times 10^{-3}) = 1.50 \times 10^{-3}$ mol OH^-

$$[OH^-] = \frac{1.50 \times 10^{-3}\ \text{mol}}{0.0600\ \text{dm}^3} = \boxed{0.025\ \text{mol dm}^{-3}}$$

$pOH = -\log(0.025) = 1.60$

$pH = 14.00 - 1.60 = \boxed{12.40}$

(c) amount of $H_3O^+ = (0.0212\ dm^3) \times (0.22\ mol\ dm^{-3}) = 4.7 \times 10^{-3}\ mol\ H_3O^+$
amount of $OH^- = (0.0100\ dm^3) \times (0.30\ mol\ dm^{-3}) = 3.0 \times 10^{-3}\ mol\ OH^-$

$$\text{concentration of excess } H_3O^+ = \frac{1.7 \times 10^{-3}\ mol}{0.0312\ dm^3} = 5.4 \times 10^{-2}\ M$$

$pH = \boxed{1.26}$

E4.36 General acidity/basicity rules predict that an aqueous solution containing a/an

(1) salt of a strong acid and strong base is neutral;
(2) acid or a conjugate acid of a weak base is acidic;
(3) base or a conjugate base of a weak acid is basic;
(4) small, highly charged metal cation is acidic. Example: The solution $FeCl_3(aq)$ is acidic because Fe^{3+} acts as a Lewis acid in water: $[Fe(H_2O)_6]^{3+}(aq) + H_2O(l) \rightleftharpoons [Fe(H_2O)_5OH]^{2+}(aq) + H_3O^+(aq)$.

(a) acidic; $NH_4^+(aq) + H_2O(l) \rightleftharpoons H_3O^+(aq) + NH_3(aq)$
(b) basic; $H_2O(l) + CO_3^{2-}(aq) \rightleftharpoons HCO_3^-(aq) + OH^-(aq)$
(c) basic; $H_2O(l) + F^-(aq) \rightleftharpoons HF(aq) + OH^-(aq)$
(d) neutral; $KBr(s) \xrightleftharpoons{\text{water}} K^+(aq) + Br^-(aq)$

Here are two additional applications of rule (4):

acidic; $[Al(H_2O)_6]^{3+}(aq) + H_2O(l) \rightleftharpoons [Al(H_2O)_5OH]^{2+}(aq) + H_3O^+(aq)$
acidic; $[Co(H_2O)_6]^{2+}(aq) + H_2O(l) \rightleftharpoons [Co(H_2O)_5OH]^+(aq) + H_3O^+(aq)$

E4.37 (a) $KCH_3CO_2(s) \xrightarrow{\text{water}} K^+(aq) + CH_3CO_2^-(aq)$
and $CH_3CO_2^-(aq) + H_2O(l) \rightleftharpoons CH_3COOH(aq) + OH^-(aq)$
$c_{KCH_3CO_2} = n/V = m/M/V = (8.4\ g)/(98.14\ g\ mol^{-1})/(0.250\ dm^3) = 0.34\bar{2}\ mol\ dm^{-3}$

$pK_b = 9.25$ [Table 4.4]	$CH_3CO_2^-(aq)$	+ $H_2O(l)$	$\rightleftharpoons$	$CH_3COOH(aq)$	+	$OH^-(aq)$
Formal conc./mol dm^{-3}	$c_{CH_3CO_2^-} = 0.34\bar{2}$					
Change/mol dm^{-3}	$-x$			$+x$		$+x$
Equil. conc./mol dm^{-3}	$c_{CH_3CO_2^-} - x \cong c_{CH_3CO_2^-}$			x		x

$$K_b = \left(\frac{[CH_3COOH][OH^-]}{[CH_3CO_2^-]}\right)_{eq} = \frac{x^2}{c_{CH_3CO_2^-}}$$

$$x = \sqrt{c_{CH_3CO_2^-} K_b} = \sqrt{(0.34\bar{2}) \times (10^{-9.25})}\ mol\ dm^{-3} = 1.3\bar{9} \times 10^{-5}\ mol\ dm^{-3}$$

Note that we can now justify the approximation $c_{CH_3CO_2^-} - x \cong c_{CH_3CO_2^-}$ with the observation that the calculated value of x is much, much smaller than the value of $c_{CH_3CO_2^-}$.

$pOH = -\log[OH^-] = -\log(1.3\bar{9} \times 10^{-5}) = 4.9$
$pH = 14.0 - pOH = 14.0 - 4.9$
$= \boxed{9.1}$

(b) $NH_4Br(s) \xrightarrow{\text{water}} NH_4^+(aq) + Br^-(aq)$ and $NH_4^+(aq) + H_2O(l) \rightleftharpoons NH_3(aq) + H_3O^+(aq)$
$c_{NH_4Br} = n/V = m/M/V = (3.75\ g)/(97.94\ g\ mol^{-1})/(0.100\ dm^3) = 0.383\ mol\ dm^{-3}$

$pK_a = 9.25$ [Table 4.4]	$NH_4^+(aq)$	+	$H_2O(l)$	$\rightleftharpoons$	$NH_3(aq)$	+	$H_3O^+(aq)$
Formal conc./mol dm^{-3}	$c_{NH_4^+} = 0.383$						
Change/mol dm^{-3}	$-x$				$+x$		$+x$
Equil. conc./mol dm^{-3}	$c_{NH_4^+} - x \cong c_{NH_4^+}$				x		x

$$K_a = \left(\frac{[NH_3][H_3O^+]}{[NH_4^+]}\right)_{eq} = \frac{x^2}{c_{NH_4^+}}$$

$$x = \sqrt{c_{NH_4^+}K_a} = \sqrt{(0.383)\times(10^{-9.25})}\ \text{mol dm}^{-3} = 1.47\times10^{-5}\ \text{mol dm}^{-3}$$

Note that we can now justify the approximation $c_{NH_4^+} - x \cong c_{NH_4^+}$ with the observation that the calculated value of x is much, much smaller than the value of $c_{NH_4^+}$.

$$pH = -\log[H_3O^+] = -\log(1.47\times10^{-5}) = \boxed{4.83}$$

(c) KBr is the salt of the strong acid HBr. Therefore, $\boxed{\text{none of the Br}^- \text{ is protonated}}$.

E4.38 (a) Lactic acid(aq) + $H_2O(l) \rightleftharpoons$ lactate(aq) + $H_3O^+(aq)$.
$[H_3O^+] = 10^{-pH} = 10^{-3.08}$ mol dm^{-3} = 8.32×10^{-4} mol dm^{-3}
Since [lactic acid] = [lactate], their ratio equals 1 in the equilibrium constant expression.

$$K_a = \frac{[H_3O^+][\text{lactate}]}{[\text{lactic acid}]} = [H_3O^+] = \boxed{8.32\times10^{-4}}$$. Additionally, $pK_a = pH = 3.08$.

(b) [Lactic acid] = 2 [lactate]

$$K_a = \frac{[H_3O^+][\text{lactate}]}{[\text{lactic acid}]} = \frac{[H_3O^+][\text{lactate}]}{2\,[\text{lactate}]} = \frac{[H_3O^+]}{2}$$

$[H_3O^+] = 2K_a = 2\times(8.32\times10^{-4})$ mol dm^{-3} = 1.66×10^{-3} mol dm^{-3}

$pH = -\log[H_3O^+] = -\log(1.66\times10^{-3}) = \boxed{2.78}$

E4.39 See E4.37 for detailed information on setting up the equilibrium calculations for weak acids. The general criterion (see E4.41) for application of the weak-acid approximation, $c_{HA} - [H^+] \simeq c_{HA}$, is that $4c_{HA} \gg K_a$. The general criterion for application of the weak-base approximation, $c_B - [OH^-] \simeq c_B$, is that $4c_B \gg K_b$.

(a) Lactic acid(aq) $\rightleftharpoons H^+(aq)$ + lactate(aq)
$K_a = 8.4\times10^{-4}$ [Table 4.4] and $c_{\text{lactic acid}} = 0.120$ mol dm^{-3}, which satisfies the criterion for the weak-acid approximation. Let $x = [H^+]$ = [lactate].

$$K_a = \frac{[H^+][\text{lactate}]}{[\text{lactic acid}]} = \frac{x^2}{c_{\text{lactic acid}} - x} \simeq \frac{x^2}{c_{\text{lactic acid}}}$$ [Weak-acid approximation.]

$$x = \sqrt{K_a c_{\text{lactic acid}}} = \sqrt{(8.4\times10^{-4})\times(0.120)} = 0.010$$

$pH = -\log([H^+]) = -\log(0.010) = \boxed{2.0}$

$pOH = 14.00 - 2.0 = \boxed{12.0}$

$$\text{Fraction deprotonated} = \frac{0.010}{0.120} = \boxed{0.083}$$

(b) Lactic acid(aq) $\rightleftharpoons$ H^+(aq) + lactate(aq)
$K_a = 8.4 \times 10^{-4}$ [Table 4.4] and $c_{\text{lactic acid}} = 1.4 \times 10^{-4}$ mol dm^{-3}, which does not satisfy the criterion for the weak-acid approximation. Let $x = [H^+] = [\text{lactate}]$.

$$K_a = \frac{[H^+][\text{lactate}]}{[\text{lactic acid}]} = \frac{x^2}{c_{\text{lactic acid}} - x}$$

$x^2 + K_a x - K_a c_{\text{lactic acid}} = 0$ [Need positive solution of this quadratic equation in x.]

$$x = \frac{-K_a + \sqrt{K_a^2 - 4 \times (-K_a c_{\text{lactic acid}})}}{2}$$

$$= \frac{-(8.4 \times 10^{-4}) + \sqrt{(8.4 \times 10^{-4})^2 + 4 \times (8.4 \times 10^{-4}) \times (1.4 \times 10^{-4})}}{2}$$

$$= 1.2\bar{2} \times 10^{-4}$$

$$\text{pH} = -\log([H^+]) = -\log(1.2\bar{2} \times 10^{-4}) = \boxed{3.9}$$

$$\text{pOH} = 14.00 - 3.9 = \boxed{10.1}$$

$$\text{Fraction deprotonated} = \frac{1.2\bar{2} \times 10^{-4}}{1.4 \times 10^{-4}} = \boxed{0.87}$$

(c) NH_4^+(aq) $\rightleftharpoons$ H^+(aq) + NH_3(aq)
$K_a = 5.6 \times 10^{-10}$ [Table 4.4] and $c_{\text{ammonium}} = 0.15$ mol dm^{-3}, which satisfies the criterion for the weak-acid approximation. Let $x = [H^+] = [NH_3]$.

$$K_a = \frac{[H^+][NH_3]}{[NH_4^+]} = \frac{x^2}{c_{\text{ammonium}} - x} \simeq \frac{x^2}{c_{\text{ammonium}}} \quad \text{[Weak-acid approximation.]}$$

$$x = \sqrt{K_a c_{\text{ammonium}}} = \sqrt{(5.6 \times 10^{-10}) \times (0.15)} = 9.1\bar{7} \times 10^{-6}$$

$$\text{pH} = -\log([H^+]) = -\log(9.1\bar{7} \times 10^{-6}) = \boxed{5.0}$$

$$\text{pOH} = 14.00 - 5.0 = \boxed{9.0}$$

$$\text{Fraction deprotonated} = \frac{9.1\bar{7} \times 10^{-6}}{0.15} = \boxed{6.1 \times 10^{-5}}$$

(d) $CH_3CO_2^-$(aq) + H_2O(l) $\rightleftharpoons$ CH_3COOH(aq) + OH^-(aq)
$K_b = 5.6 \times 10^{-10}$ [Table 4.4] and $c_{\text{acetate}} = 0.15$ mol dm^{-3}, which satisfies the criterion for the weak-base approximation. Let $x = [OH^-] = [\text{acetic acid}]$.

$$K_b = \frac{[OH^-][CH_3COOH]}{[CH_3COO^-]} = \frac{x^2}{c_{\text{acetate}} - x} \simeq \frac{x^2}{c_{\text{acetate}}} \quad \text{[Weak-base approximation.]}$$

$$x = \sqrt{K_b c_{\text{acetate}}} = \sqrt{(5.6 \times 10^{-10}) \times (0.15)} = 9.1\bar{7} \times 10^{-6}$$

$$\text{pOH} = -\log([OH^-]) = -\log(9.1\bar{7} \times 10^{-6}) = \boxed{5.0}$$

$$\text{pH} = 14.00 - 5.0 = \boxed{9.0}$$

$$\text{Fraction deprotonated} = \frac{9.1\bar{7} \times 10^{-6}}{0.15} = \boxed{6.1 \times 10^{-5}}$$

(e) $(CH_3)_3N$(aq) + H_2O(l) $\rightleftharpoons$ $(CH_3)_3NH^+$(aq) + OH^-(aq)
$K_b = 6.5 \times 10^{-5}$ [Table 4.4] and $c_{\text{amine}} = 0.112$ mol dm^{-3}, which satisfies the criterion for the weak-base approximation. Let $x = [OH^-] = [(CH_3)_3NH^+]$.

$$K_b = \frac{[OH^-][(CH_3)_3NH^+]}{[(CH_3)_3N]} = \frac{x^2}{c_{amine} - x} \approx \frac{x^2}{c_{amine}} \text{ [Weak-base approximation.]}$$

$$x = \sqrt{K_b c_{amine}} = \sqrt{(6.5 \times 10^{-5}) \times (0.112)} = 2.7\bar{0} \times 10^{-3}$$

$$pOH = -\log([OH^-]) = -\log(2.7\bar{0} \times 10^{-3}) = \boxed{2.6}$$

$$pH = 14.00 - 2.6 = \boxed{11.4}$$

$$\text{Fraction protonated} = \frac{2.7\bar{0} \times 10^{-3}}{0.112} = \boxed{0.024}$$

E4.40 (a) The glycine (Gly) deprotonation reactions are:

$$H_2Gly^+(aq) + H_2O(l) \rightleftharpoons H_3O^+(aq) + HGly(aq) \quad pK_{a1} = 2.34 \text{ [Table 4.6]}$$
$$HGly + H_2O(l) \rightleftharpoons H_3O^+(aq) + Gly^- \quad pK_{a2} = 9.58$$

The fraction of a species at any chosen pH is computed with equations that are analogous to eqns 4.31 and 4.32 for a triprotic acid. Let D be the function: $D = [H_3O^+]^2 + [H_3O^+]K_{a1} + K_{a1}K_{a2}$.

$f_1 = f(H_2Gly^+) = [H_3O^+]^2/D$,
$f_2 = f(HGly) = [H_3O^+]K_{a1}/D$, and finally
$f_3 = f(Gly^-) = K_{a1}K_{a2}/D$.

These fractions are plotted against pH in Figure 4.3. The molarity of a species at a specified pH is calculated with the relation: [species] = $c_{glycine} \times f$(species). For example, if pH = 9 and $c_{glycine}$ = 10 mmol dm^{-3}, then we find from the plot that $f(HGly) = 0.80$ and $f(Gly^-) = 0.20$; therefore,

$$[HGly] = (10 \text{ mmol dm}^{-3}) \times (0.80) = 8.0 \text{ mmol dm}^{-3}$$

and

$$[Gly^-] = (10 \text{ mmol dm}^{-3}) \times (0.20) = 2.0 \text{ mmol dm}^{-3}.$$

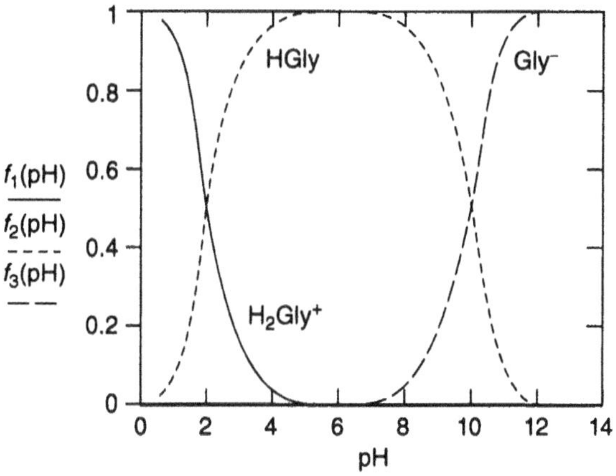

Figure 4.3

(b) The tyrosine (Tyr) deprotonation reactions are:

$$H_3Tyr^+(aq) + H_2O(l) \rightleftharpoons H_3O^+(aq) + H_2Tyr(aq) \quad pK_{a1} = 2.24 \text{ [Table 4.6]}$$
$$H_2Tyr(aq) + H_2O(l) \rightleftharpoons H_3O^+(aq) + HTyr^-(aq) \quad pK_{a2} = 9.04$$
$$HTyr^- + H_2O(l) \rightleftharpoons H_3O^+(aq) + Tyr^{2-} \quad pK_{a3} = 10.10$$

Equations that are analogous to eqns 4.32 and 4.33 are used to calculate the fractional composition of each deprotonated species as a function of pH. Let D be the function:

$D = [H_3O^+]^3 + [H_3O^+]^2K_{a1} + [H_3O^+]K_{a1}K_{a2} + K_{a1}K_{a2}K_{a3}$. Then, by generalizing the symmetry of the equations in eqns 4.32 and 4.33 we find that

$f_1 = f(H_3Tyr^+) = [H_3O^+]^3/D,$
$f_2 = f(H_2Tyr) = [H_3O^+]^2K_{a1}/D,$
$f_3 = f(HTyr^-) = [H_3O^+]K_{a1}K_{a2}/D$, and finally
$f_4 = f(Tyr^{2-}) = K_{a1}K_{a2}K_{a3}/D.$

These fractions are plotted against pH in Figure 4.4. The molarity of a species at a specified pH is calculated with the relation: [species] = $c_{\text{tyrosine}} \times f$(species). For example, if pH = 9 and c_{tyrosine} = 10 mmol dm^{-3}, then we find from the plot that $f(H_2Tyr) = 0.55$, $f(HTyr^-) = 0.40$, and $f(Tyr^{2-}) = 0.05$; therefore,

$[H_2Tyr] = (10 \text{ mmol dm}^{-3}) \times (0.55) = 5.5 \text{ mmol dm}^{-3},$
$[HTyr^-] = (10 \text{ mmol dm}^{-3}) \times (0.40) = 4.0 \text{ mmol dm}^{-3},$

and

$[Tyr^{2-}] = (10 \text{ mmol dm}^{-3}) \times (0.05) = 0.05 \text{ mmol dm}^{-3}.$

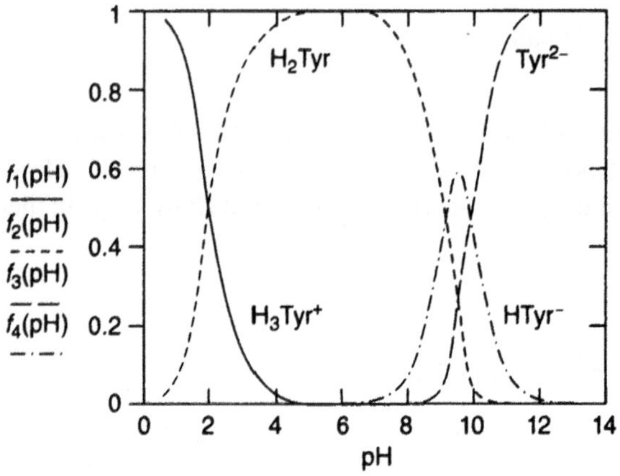

Figure 4.4

E4.41 Before justifying the various approximations that will be analyzed in this exercise, let us find the criterion that justifies the approximation [acid] = $c - x \simeq c$ where c is the formal, as prepared, molarity of a weak acid that has acid constant K and x is the conjugate base molarity. (Or, if c is the formal molarity of a weak base that has base constant K and x is the conjugate acid molarity, the approximation is [base] = $c - x \simeq c$.) The equilibrium constant has the general form

$$K = \frac{x^2}{c-x} \quad \text{or, writing the expression as a polynomial,} \quad x^2 + Kx - Kc = 0.$$

The quadratic equation solution for positive x (concentrations are always positive) is

$$x = \tfrac{1}{2}\left\{-K + \sqrt{K^2 + 4Kc}\right\}.$$

This expression can be used to calculate the value of x with given values of c and K. However, in many instances $4c \gg K$, a case that causes the expression to simplify into a very attractive computational equation: $x = \sqrt{Kc}$. In fact this is the equation we get by simply using the approximation [acid] = $c - x \simeq c$ in our equilibrium tables and expressions for equilibrium constant. The bottom line is: compare the value of $4c$ to K; if $4c \gg K$, immediately apply the approximation [acid] = $c - x \simeq c$.

(a) Boric acid solution with $c_{B(OH)_3} = 1.0 \times 10^{-4}$ mol dm^{-3} and $K_a = 7.2 \times 10^{-10}$ [Table 4.4]. Since $4c_{B(OH)_3} \gg K_a$, we immediately apply the approximation $[B(OH)_3] = c_{B(OH)_3} - x \simeq c_{B(OH)_3}$.

$K_a = 7.2 \times 10^{-10}$	$B(OH)_3(aq)$	$+\ 2\,H_2O(l)$	$\rightleftharpoons$	$B(OH)_4^-(aq)$	$+$	$H_3O^+(aq)$
Formal conc./mol dm^{-3}	$c_{B(OH)_3} = 1.0 \times 10^{-4}$					
Change/mol dm^{-3}	$-x$			$+x$		$+x$
Equil. conc./mol dm^{-3}	$c_{B(OH)_3} - x \simeq c_{B(OH)_3}$			x		x

$$K_a = \frac{[H_3O^+][B(OH)_4^-]}{[B(OH)_3]} = \frac{x^2}{c_{B(OH)_3}}$$

$$x \approx \sqrt{K_a c_{B(OH)_3}} = \sqrt{(7.2 \times 10^{-10}) \times (1.0 \times 10^{-4})}\ \text{mol dm}^{-3} = 2.6\overline{8} \times 10^{-7}\ \text{mol dm}^{-3}$$

$$[H_3O^+] = 2.6\overline{8} \times 10^{-7}\ \text{mol dm}^{-3}$$

$$\text{pH} = -\log[H_3O^+] = -\log(2.6\overline{8} \times 10^{-7}) = 6.6$$

This value of $[H_3O^+]$ is not much different from the value for pure water, 1.0×10^{-7} mol dm^{-3}; hence, it is at the lower limit of safely ignoring the contribution to $[H_3O^+]$ from the autoprotolysis of water. Autoprotolysis contributes to the hydronium ion concentration so that $[H_3O^+]$ is not equal to $[B(OH)_4^-]$, nor is it equal to $[OH^-]$, as in pure water. However, the principle of electroneutrality specifies that all solutions must be electrically neutral so the sum of anion charge concentrations must equal the sum of cation charge concentrations. For our single-charge ions: $[B(OH)_4^-] + [OH^-] = [H_3O^+]$. Substituting $[OH^-] = K_w/[H_3O^+]$, letting $[H_3O^+] = x$, and solving for $[B(OH)_4^-]$ gives $[B(OH)_4^-] = x - K_w/x$. This is now substituted into the equilibrium constant expression to give an equation for x, the hydronium ion molarity.

$$K_a = \frac{[H_3O^+][B(OH)_4^-]}{[B(OH)_3]} = \frac{x \times (x - K_w/x)}{c_{B(OH)_3}} = \frac{x^2 - K_w}{c_{B(OH)_3}} \quad \text{or} \quad x = \sqrt{K_w + K_a c_{B(OH)_3}}$$

$$x = \sqrt{1.00 \times 10^{-14} + (7.2 \times 10^{-10}) \times (1.0 \times 10^{-4})} = 2.9 \times 10^{-7}\ \text{mol dm}^{-3}$$

$$\text{pH} = -\log[H_3O^+] = -\log(2.9 \times 10^{-7}) = \boxed{6.5}$$

This value is slightly, but measurably, different from the value 6.6 obtained by ignoring the contribution to $[H_3O^+]$ from water. This last equation reduces to the approximate equation used above, $x \simeq \sqrt{K_a c_{B(OH)_3}}$, when $K_a c_{B(OH)_3} \gg K_w$. This condition is not entirely satisfied in this exercise because $K_a c_{B(OH)_3} = 7.2 \times 10^{-14}$ is only 7.2 times K_w, and it must be, say, no less than 10 or more times K_w before the condition is satisfied.

(b) Phosphoric acid solution with $c_{H_3PO_4} = 0.015$ mol dm^{-3}; $K_{a1} = 7.6 \times 10^{-3}$, $K_{a2} = 6.2 \times 10^{-8}$, and $K_{a3} = 2.1 \times 10^{-13}$. Since $4c_{H_3PO_4}$ is not much, much greater than K_{a1}, we should not immediately apply the approximation $[H_3PO_4] = c_{H_3PO_4} - x \simeq c_{H_3PO_4}$. Furthermore, as the first acid constant is much, much larger than that of boric acid (see part (a) of this exercise) we expect that the autoprotolysis of water contributes negligibly to the hydronium ion concentration and we ignore autoprotolysis. We can check that the criterion, $K_{a1}c_{H_3PO_4} \gg K_w$, derived in part (a) quantifies this expectation. It does because $K_{a1}c_{H_3PO_4} = 1.1 \times 10^{-4} \gg K_w$ so we ignore autoprotolysis.

We also suspect that, since $K_{a2} \ll K_{a1}$, the pH contribution of the second acid constant is negligibly small and can be ignored. But how do we quantify this suspicion? The principle of electroneutrality (see part (a)) in an H_3A solution provides a useful relation after making substitutions that replace $[H_2A^-]$, $[HA^-]$ and $[A^{3-}]$ with equilibrium constants and $[H_3A]$, which must be approximately equal to c_{H_3A} for a weak acid.

$$\begin{aligned}[H^+] &= [H_2A^-] + 2[HA^{2-}] + 3[A^{3-}] + [OH^-] \text{ (The last term is due to the autoprotolysis of water.)}\\ &= [H_2A^-] + 2[HA^{2-}] + 3[A^{3-}] \text{ ([OH}^-\text{] is negligible in all but the most weakly acid solutions)}\\ &= \frac{K_{a1}[H_3A]}{[H^+]} + \frac{2K_{a2}[H_2A^-]}{[H^+]} + \frac{3K_{a3}[HA^{2-}]}{[H^+]}\end{aligned}$$

$$\begin{aligned}[H^+]^2 &= K_{a1}[H_3A] + 2K_{a2}[H_2A^-] + 3K_{a3}[HA^{2-}]\\ &= K_{a1}[H_3A] + \frac{2K_{a1}K_{a2}[H_3A]}{[H^+]} + \frac{3K_{a2}K_{a3}[H_2A^-]}{[H^+]}\\ &= K_{a1}[H_3A] + \frac{2K_{a1}K_{a2}[H_3A]}{[H^+]} + \frac{3K_{a1}K_{a2}K_{a3}[H_3A]}{[H^+]^2}\\ &= K_{a1}[H_3A] \times \left\{1 + \frac{2K_{a2}}{[H^+]} + \frac{3K_{a2}K_{a3}}{[H^+]^2}\right\}\\ &= K_{a1}[H_3A] \times \left\{1 + \frac{2K_{a2}}{[H^+]}\left(1 + \frac{3K_{a3}}{2[H^+]^2}\right)\right\}\end{aligned}$$

Substitute the weak acid estimate that $[H_3A] \simeq c_{H_3A}$ and apply the approximation that $2K_{a2}/[H^+] \ll 1$ to get the usual approximation that $[H^+] \simeq \sqrt{K_{a1}c_{H_3A}}$. Substitute the approximations $[H_3A] \simeq c_{H_3A}$ and $[H^+] \simeq \sqrt{K_{a1}c_{H_3A}}$ into the above working equation to find the improved approximation that $[H^+]^2 \simeq K_{a1}c_{H_3A}\left(1 + 2K_{a2}/\sqrt{K_{a1}c_{H_3A}}\right)$. This tells us that, if $2K_{a2}/\sqrt{K_{a1}c_{H_3A}} \ll 1$, the second acid constant contributes negligibly to the pH. In this exercise $2K_{a2}/\sqrt{K_{a1}c_{H_3A}} = 1.2 \times 10^{-5}$ so the criterion for neglect of K_{a2} is satisfied. We have justified neglect of both autoprotolysis and K_{a2} and proceed with these common approximations.

$K_{a1} = 7.6 \times 10^{-3}$	$H_3PO_4(aq)$	+	$H_2O(l)$	$\rightleftharpoons$	$H_2PO_4^-(aq)$	+	$H_3O^+(aq)$
Formal conc./mol dm^{-3}	$c_{H_3PO_4} = 0.0150$						
Change/mol dm^{-3}	$-x$				$+x$		$+x$
Equil. conc./mol dm^{-3}	$c_{H_3PO_4} - x$				x		x

$$K_{a1} = \frac{x^2}{c_{H_3PO_4} - x}$$

$$x^2 + K_{a1}x - K_{a1}c_{H_3PO_4} = 0$$

$$\begin{aligned}x &= \tfrac{1}{2}\left\{-K_{a1} + \sqrt{K_{a1}^2 + 4K_{a1}c_{H_3PO_4}}\right\}\\ &= \tfrac{1}{2}\left\{-(7.6 \times 10^{-3}) + \sqrt{(7.6 \times 10^{-3})^2 + 4(7.6 \times 10^{-3}) \times (0.015)}\right\}\\ &= 7.5 \times 10^{-3}\ \text{mol dm}^{-3}\end{aligned}$$

$$\text{pH} = -\log[H_3O^+] = -\log(7.5 \times 10^{-3}) = \boxed{2.1}$$

(c) Sulfurous acid solution with $c_{H_2SO_3} = 0.10$ mol dm^{-3}; $K_{a1} = 1.5 \times 10^{-2}$ and $K_{a2} = 1.2 \times 10^{-7}$. Since $4c_{H_2SO_3}$ is not much, much greater than K_{a1}, we will not apply the approximation $[H_2SO_3] = c_{H_2SO_3} - x \simeq c_{H_2SO_3}$. Furthermore, as the first acid constant is much, much larger than that of boric acid (see part (a) of this exercise) we expect that the autoprotolysis of water contributes negligibly to the hydronium ion concentration and we ignore autoprotolysis. We can check that the criterion, $K_{a1}c_{H_2SO_3} \gg K_w$, derived in part (a) to quantify this expectation. Since $K_{a1}c_{H_2SO_3} = 1.5 \times 10^{-3} \gg K_w$, we can ignore autoprotolysis. We also suspect that, since $K_{a2} \ll K_{a1}$, the pH contribution of the second acid constant is negligibly small and can be ignored. We check this suspicion with the criterion $2K_{a2}/\sqrt{K_{a1}c_{H_2A}} \ll 1$ (see part (b)). In this exercise $2K_{a2}/\sqrt{K_{a1}c_{H_2A}} = 6.2 \times 10^{-6}$ so the criterion for neglect of K_{a2} is satisfied. We have justified neglect of both autoprotolysis and K_{a2}.

$K_{a1} = 1.5 \times 10^{-2}$	$H_2SO_3(aq)$	+	$H_2O(l)$	$\rightleftharpoons$	$HSO_3^-(aq)$	+	$H_3O^+(aq)$
Formal conc./mol dm^{-3}	$c_{H_2SO_3} = 0.10$						
Change/mol dm^{-3}	$-x$				$+x$		$+x$
Equil. conc./mol dm^{-3}	$c_{H_2SO_3} - x$				x		x

$$K_{a1} = \frac{x^2}{c_{H_2SO_3} - x}$$

$$x^2 + K_{a1}x - K_{a1}c_{H_2SO_3} = 0$$

$$x = \tfrac{1}{2}\left\{-K_{a1} + \sqrt{K_{a1}^2 + 4K_{a1}c_{H_2SO_3}}\right\}$$

$$= \tfrac{1}{2}\left\{-(1.5 \times 10^{-2}) + \sqrt{(1.5 \times 10^{-2})^2 + 4(1.5 \times 10^{-2}) \times (0.10)}\right\}$$

$$= 3.2 \times 10^{-2} \text{ mol dm}^{-3}$$

$$\text{pH} = -\log[H_3O^+] = -\log(3.2 \times 10^{-2}) = \boxed{1.5}$$

E4.42 Use the Henderson–Hasselbalch equation: $\text{pH} = \text{p}K_a - \log\frac{[\text{acid}]}{[\text{base}]}$ [4.35] or $\frac{[\text{acid}]}{[\text{base}]} = 10^{\text{p}K_a - \text{pH}}$

(a) $\frac{[\text{acid}]}{[\text{base}]} = 10^{2.20-7.00} = \boxed{1.58 \times 10^{-5}}$

(b) $\frac{[\text{acid}]}{[\text{base}]} = 10^{2.20-2.20} = \boxed{1}$

(c) $\frac{[\text{acid}]}{[\text{base}]} = 10^{2.20-1.50} = \boxed{5.01}$

E4.43 (a) Oxalic acid solution with $c_{(COOH)_2} = 0.15$ mol dm^{-3}; $K_{a1} = 5.9 \times 10^{-2}$ and $K_{a2} = 6.5 \times 10^{-5}$. Since $4c_{(COOH)_2}$ is not much, much greater than K_{a1}, we will not apply the approximation $[(COOH)_2] = c_{(COOH)_2} - x \approx c_{(COOH)_2}$. (See the first paragraph of Exercise 4.41 for discussion of this justification.) Furthermore, neither autoprotolysis nor the second acid constant provide a significant pH contribution (an assertion that can be checked with criteria summarized in Exercise 4.41 (c)) so common pH approximations are appropriate.

$K_{a1} = 5.9 \times 10^{-2}$	$(COOH)_2(aq)$	+	$H_2O(l)$	$\rightleftharpoons$	$HOOCCO_2^-(aq)$	+	$H_3O^+(aq)$
Formal conc./mol dm^{-3}	$c_{(COOH)_2} = 0.15$						
Change/mol dm^{-3}	$-x$				$+x$		$+x$
Equil. conc./mol dm^{-3}	$c_{(COOH)_2} - x$				x		x

$$K_{a1} = \frac{x^2}{c_{(COOH)_2} - x}$$

$$x^2 + K_{a1}x - K_{a1}c_{(COOH)_2} = 0$$

$$x = \tfrac{1}{2}\left\{-K_{a1} + \sqrt{K_{a1}^2 + 4K_{a1}c_{(COOH)_2}}\right\}$$

$$= \tfrac{1}{2}\left\{-(5.9 \times 10^{-2}) + \sqrt{(5.9 \times 10^{-2})^2 + 4(5.9 \times 10^{-2}) \times (0.15)}\right\} = 6.9 \times 10^{-2} \text{ mol dm}^{-3}$$

$$[H_3O^+] = \boxed{6.9 \times 10^{-2} \text{ mol dm}^{-3}} \quad \text{and} \quad [OH^-] = K_w/[H_3O^+] = \boxed{1.4 \times 10^{-13} \text{ mol dm}^{-3}}$$

$$[(COOH)_2] = c_{(COOH)_2} - x = \boxed{8.1 \times 10^{-2} \text{ mol dm}^{-3}}$$

The $HOOCCO_2^-$(aq) molarity is very slightly diminished below $x = 0.069$ mol dm^{-3} by the second acid constant. However, this does not contribute significantly to pH. It is the value of y in the following equilibrium table that we must find.

$K_{a2} = 6.5 \times 10^{-5}$	$HOOCCO_2^-(aq)$	+	$H_2O(l)$	$\rightleftharpoons$	$(CO_2)^-_2(aq)$	+	$H_3O^+(aq)$
First est./mol dm^{-3}	$x = 0.069$				0		x
Change/mol dm^{-3}	$-y$				$+y$		0
Equil. conc./mol dm^{-3}	$x - y$				y		x

$$K_{a2} = \frac{xy}{x-y} \quad \text{or} \quad y = \frac{K_{a2}x}{x+K_{a2}}$$

But $x \gg K_{a2}$, so the above equation reduces to $y = \dfrac{K_{a2}x}{x} = K_{a2}$.

$$[^-O_2CCO_2^-] = K_{a2} = \boxed{6.5 \times 10^{-5}\text{ mol dm}^{-3}}$$

$$[HOOCCO_2^-] = x - K_{a2} = \boxed{0.069\text{ mol dm}^{-3}}$$

(b) When the solution of an amphiprotic anion satisfies the criteria that $K_{a2} \ll K_{a1}$, $c_{\text{anion}} \gg K_W/K_{a2}$, and $c_{\text{anion}} \gg K_{a1}$, the solution pH is calculated with eqn 4.34: $\text{pH} = \frac{1}{2}(\text{p}K_{a1} + \text{p}K_{a2})$. A solution of the amphiprotic hydrogenoxalate anion ($K_{a1} = 0.0589$, $K_{a2} = 6.46 \times 10^{-5}$, Table 4.5) for which the preparation concentration of potassium hydrogenoxalate is 0.15 mol dm^{-3} satisfies the first two criteria but the solution does not completely satisfy the last criterion. Nevertheless, we use eqn 4.34 to estimate the pH.

$$\text{pH} \approx \tfrac{1}{2}(1.23 + 4.19) \approx 2.71 \text{ and } H_3O^+(aq) \approx 1.95 \times 10^{-3}\text{ mol dm}^{-3}$$

To explore features of a more accurate computation, we draw up the following table and solve for x and y, which are defined in the table. Then, the hydrogen ion concentration is give by $y - x$.

Species	$(COOH)_2(aq)$	$HOOCCO_2^-(aq)$	$(CO_2)^-_2(aq)$	$H_3O^+(aq)$
Initial molar conc./mol dm^{-3}	0	$c = 0.15$	0	0
Change to reach equilibrium/mol dm^{-3}	$+x$	$-(x+y)$	$+y$	$+(y-x)$
Equilibrium conc./mol dm^{-3}	x	$c - x - y$	y	$y - x$

Both x and y are likely to be much smaller than c (an approximation to be checked later) and the two acidity constants are

$$K_{a1} = \frac{[H_3O^+][HOOCCO_2^-]}{[(COOH)_2]} = \frac{[H_3O^+](c-x-y)}{x} \simeq \frac{[H_3O^+]c}{x} \quad \text{or} \quad x = \frac{[H_3O^+]c}{K_{a1}}$$

$$K_{a2} = \frac{[H_3O^+][(CO_2)_2^{2-}]}{[HOOCCO_2^-]} = \frac{[H_3O^+]y}{c-x-y} \simeq \frac{[H_3O^+]y}{c} \quad \text{or} \quad y = \frac{K_{a2}c}{[H_3O^+]}$$

Subtraction of these gives $[H_3O^+]$.

$$[H_3O^+] = \frac{K_{a2}c}{[H_3O^+]} - \frac{[H_3O^+]c}{K_{a1}}$$

$$\left(1 + \frac{c}{K_{a1}}\right)[H_3O^+]^2 = K_{a2}c$$

$$[H_3O^+] = \left\{ K_{a2}c\left(1 + \frac{c}{K_{a1}}\right)^{-1} \right\}^{1/2}$$

$$= \left\{ (6.46 \times 10^{-5}) \times (0.15) \times \left(1 + \frac{0.15}{0.0589}\right)^{-1} \right\}^{1/2}$$

$$= \boxed{1.65 \times 10^{-3}\ \text{mol dm}^{-3}}$$

$$\text{pH} = -\log[H_3O^+] = -\log(1.65 \times 10^{-3}) = \boxed{2.78}$$

Insertion of the $[H_3O^+]$ value into the above expressions for x and y gives $x/c = 0.028$ and $y/c = 0.039$, thereby justifying the approximation that x and y are much smaller than c.

E4.44 Hydrosulfuric acid solution with $c_{H_2S} = 0.065$ mol dm^{-3}; $K_{a1} = 1.32 \times 10^{-7}$ and $K_{a2} = 7.08 \times 10^{-15}$. Since $4c_{H_2S}$ is much, much greater than K_{a1}, we will apply the approximation $[H_2S] = c_{H_2S} - x \simeq c_{H_2S}$. (See the first paragraph of Exercise 4.41 for discussion of this justification.) Neither autoprotolysis nor the second acid constant provide a significant pH contribution (an assertion that can be checked with criteria summarized in Exercise 4.41(c)) so common pH approximations are appropriate.

$K_{a1} = 1.32 \times 10^{-7}$	$H_2S(aq)$	$+\ H_2O(l)\ \rightleftharpoons$	$HS^-(aq)$	$+$	$H_3O^+(aq)$
Formal conc./mol dm^{-3}	$c_{H_2S} = 0.065$				
Change/mol dm^{-3}	$-x$		$+x$		$+x$
Equil. conc./mol dm^{-3}	$c_{H_2S} - x \simeq c_{H_2S}$		x		x

$$K_{a1} = \frac{x^2}{c_{H_2S}}$$

$$x = \sqrt{K_{a1}c_{H_2S}}$$

$$= \sqrt{(1.32 \times 10^{-7}) \times (0.065)} = 9.26 \times 10^{-5}\ \text{mol dm}^{-3}$$

$$[H_3O^+] = \boxed{9.26 \times 10^{-5}\ \text{mol dm}^{-3}} \quad \text{and} \quad [OH^-] = K_w/[H_3O^+] = \boxed{1.08 \times 10^{-10}\ \text{mol dm}^{-3}}$$

$$[H_2S] = c_{H_2S} = \boxed{0.065\ \text{mol dm}^{-3}}$$

The $HS^-(aq)$ molarity is very slightly diminished below $x = 9.26 \times 10^{-5}$ mol dm^{-3} by the second acid constant. However, this does not contribute significantly to pH. It is the value of y in the following equilibrium table that we must find. Since $4x \gg K_{a2}$, we can use the approximation $x - y \simeq x$.

$K_{a2} = 7.08 \times 10^{-15}$	$HS^-(aq)$	$+\ H_2O(l)\ \rightleftharpoons$	$S^{2-}(aq)$	$+$	$H_3O^+(aq)$
First est./mol dm^{-3}	$x = 9.26 \times 10^{-5}$		0		x
Change/mol dm^{-3}	$-y$		$+y$		0
Equil. conc./mol dm^{-3}	$x - y \simeq x$		y		x

$$K_{a2} = \frac{xy}{x} \quad \text{or} \quad y = K_{a2}$$

$$[S^{2-}] = K_{a2} = \boxed{7.08 \times 10^{-15}\ \text{mol dm}^{-3}}$$

$$[HS^-] = x = \boxed{9.26 \times 10^{-5}\ \text{mol dm}^{-3}}$$

E4.45 (a) The deprotonation equilibrium of an amino acid for which the R-group on the α-carbon is neither acidic nor basic can be written in the form:

$$NH_3^+ - CHR - COOH \xrightleftharpoons{-H^+,\ K_{a1}} NH_3^+ - CHR - COO^- \xrightleftharpoons{-H^+,\ K_{a2}} NH_2 - CHR - COO^-$$

The zwitterion ion form of the amino acid is $NH_3^+ - CHR - COO^-$. The **isoelectric point, pI,** is the pH at which the zwitterion form dominates and $[NH_3^+ - CHR - COOH] = [NH_2 - CHR - COO^-]$. This implies equality of the species fractions on either side of the zwitterion in the deprotonation equilibrium sequence. These fractions are most easily acquired by making an analogy between these equilibria and those of eqn 4.33 of the text. They are

$$f(NH_3^+ - CHR - COOH) = \frac{[H_3O^+]^2}{H}, \text{ where } H = [H_3O^+]^2 + [H_3O^+]K_{a1} + K_{a1}K_{a2}$$

$$f(NH_2 - CHR - COO^-) = \frac{K_{a1}K_{a2}}{H}$$

Since these fractions are equal at the isoelectric point,

$$[H_3O^+]^2 = K_{a1}K_{a2}$$
$$2\log[H_3O^+] = \log(K_{a1}K_{a2}) = \log(K_{a1}) + \log(K_{a2})$$
$$-\log[H_3O^+] = \tfrac{1}{2}(-\log(K_{a1}) - \log(K_{a2}))$$

$$\boxed{pI = pH = \tfrac{1}{2}(pK_{a1} + pK_{a2})}$$

(b) Let RH signify an acidic R-group. The deprotonation equilibria are

$$NH_3^+ - CH(RH) - COOH \xrightleftharpoons{-H^+,\ K_{a1}} NH_3^+ - CH(RH) - COO^-$$
$$NH_3^+ - CH(RH) - COO^- \xrightleftharpoons{-H^+,\ K_{a2}} NH_3^+ - CH(R^-) - COO^-$$
$$NH_3^+ - CH(R^-) - COO^- \xrightleftharpoons{-H^+,\ K_{a3}} NH_2 - CH(R^-) - COO^-$$

The zwitterion ion form of the amino acid is $NH_3^+ - CH(RH) - COO^-$ and the species fractions on either side of the zwitterion in the deprotonation sequence are

$$f(NH_3^+ - CH(RH) - COOH) = \frac{[H_3O^+]^3}{H}, \text{ where } H = [H_3O^+]^3 + [H_3O^+]^2K_{a1} + [H_3O^+]K_{a1}K_{a2} + K_{a1}K_{a2}K_{a3}$$

$$f(NH_2 - CH(R^-) - COO^-) = \frac{[H_3O^+]K_{a1}K_{a2}}{H}$$

The equality of these fraction at the isoelectric point gives

$$[H_3O^+]^3 = [H_3O^+]K_{a1}K_{a2}$$
$$[H_3O^+]^2 = K_{a1}K_{a2}$$

Taking the log of each side and performing the manipulations detailed in part (a) gives

$$\boxed{pI = pH = \tfrac{1}{2}(pK_{a1} + pK_{a2})}$$

(c) Let RH^+ signify a basic R-group. The deprotonation equilibria are

$$NH_3^+ - CH(RH^+) - COOH \xrightleftharpoons{-H^+,\ K_{a1}} NH_3^+ - CH(RH^+) - COO^-$$
$$NH_3^+ - CH(RH^+) - COO^- \xrightleftharpoons{-H^+,\ K_{a2}} NH_2 - CH(RH^+) - COO^-$$
$$NH_2 - CH(RH^+) - COO^- \xrightleftharpoons{-H^+,\ K_{a3}} NH_2 - CH(R) - COO^-$$

The zwitterion ion form of the amino acid is $NH_2 - CH(RH^+) - COO^-$ and the species fractions on either side of the zwitterion in the deprotonation sequence are

$$f(NH_3^+-CH(RH^+)-COO^-)=\frac{[H_3O^+]^2K_{a1}}{H}$$

where $H=[H_3O^+]^3+[H_3O^+]^2K_{a1}+[H_3O^+]K_{a1}K_{a2}+K_{a1}K_{a2}K_{a3}$

$$f(NH_2-CH(R)-COO^-)=\frac{K_{a1}K_{a2}K_{a3}}{H}$$

The equality of these fraction at the isoelectric point gives

$$K_{a1}[H_3O^+]^2=K_{a1}K_{a2}K_{a3}$$
$$[H_3O^+]^2=K_{a2}K_{a3}$$

Taking the log of each side and performing the manipulations detailed in part (a) gives

$$\boxed{pI=pH=\tfrac{1}{2}(pK_{a2}+pK_{a3})}$$

E4.46 The rule of thumb we use is that the effective range of a buffer is roughly within plus or minus one pH unit of the pK_a of the acid.

Buffer system	pK_a	Effective pH buffer range
Lactate/lactic acid	3.08	2–4
Benzoate/benzoic acid	4.19	3–5
HPO_4^{2-}/PO_4^{3-}	12.67	12–13
$HPO_4^{2-}/H_2PO_4^-$	7.21	6–8
$NH_2OH/^+NH_3OH$	6.03	5–7

E4.47 Choose a buffer system in which the conjugate acid has a pK_a close to the desired pH. Therefore,

(a) H_3PO_4 and NaH_2PO_4
(b) NaH_2PO_4 and Na_2HPO_4, or $NaHSO_3$ and Na_2SO_3

E4.48 Let [acid] represent the concentration of the conjugate acid of tris, which is a base of concentration [tris].

(a) Since [acid] = [tris], the Henderson–Hasselbalch equation gives the pH as

$$pH=pK_a-\log([acid]/[tris])\ [4.35]=pK_a-\log(1)=pK_a=\boxed{8.3}$$

This is the pH at which the buffering action is best, as solution components provide the capacity to neutralize small amounts of either base or acid that may be added without drastic effect on solution pH.

(b) The base concentration that is added to the buffer without change in volume is

$[OH^-]_{addition}=0.0033\ mol/0.100\ dm^3=0.033\ mol\ dm^{-3}$.

The addition reacts with the conjugate acid of tris, giving

$[acid]=[acid]_{buffer}-0.033\ mol\ dm^{-3}$ and $[tris]=[tris]_{buffer}+0.033\ mol\ dm^{-3}$.

By the Henderson–Hasselbalch equation

$$pH=pK_a-\log([acid]/[tris])\ [4.35]=pK_a-\log\left(\frac{[acid]_{buffer}-0.033\ mol\ dm^{-3}}{[tris]_{buffer}+0.033\ mol\ dm^{-3}}\right)$$

We now suppose that the buffer system has the values $[acid]_{buffer}=[tris]_{buffer}=0.15\ mol\ dm^{-3}$ and use the Henderson–Hasselbalch equation for the pH computation.

$$\text{pH} = 8.3 - \log\left(\frac{0.15 - 0.033}{0.15 + 0.033}\right) = \boxed{8.5}$$

(c) The acid concentration that is added to the buffer without change in volume is

$[\text{OH}^-]_{\text{addition}} = 0.0060 \text{ mol}/0.100 \text{ dm}^3 = 0.060 \text{ mol dm}^{-3}$.

The addition reacts with tris, giving

$[\text{tris}] = [\text{tris}]_{\text{buffer}} - 0.060 \text{ mol dm}^{-3}$ and $[\text{acid}] = [\text{acid}]_{\text{buffer}} + 0.060 \text{ mol dm}^{-3}$ and.

By the Henderson–Hasselbalch equation

$$\text{pH} = \text{p}K_\text{a} - \log([\text{acid}]/[\text{tris}])\ [4.35] = \text{p}K_\text{a} - \log\left(\frac{[\text{acid}]_{\text{buffer}} + 0.060 \text{ mol dm}^{-3}}{[\text{tris}]_{\text{buffer}} - 0.060 \text{ mol dm}^{-3}}\right)$$

We now suppose that the buffer system has the values $[\text{acid}]_{\text{buffer}} = [\text{tris}]_{\text{buffer}} = 0.15 \text{ mol dm}^{-3}$ and use the Henderson–Hasselbalch equation for the pH computation.

$$\text{pH} = 8.3 - \log\left(\frac{0.15 + 0.060}{0.15 - 0.060}\right) = \boxed{7.9}$$

Solutions to projects

P4.49 native $\rightleftharpoons$ denatured $K_\text{d} = [\text{denatured}]/[\text{native}]$

(a) The fraction of native macromolecules, θ_{native}, is given by

$$\theta_{\text{native}} = \frac{[\text{native}]}{[\text{native}] + [\text{denatured}]} = \frac{1}{1 + \dfrac{[\text{denatured}]}{[\text{native}]}} = \frac{1}{1 + K_\text{d}}$$

The fraction denatured is

$$\theta_{\text{denatured}} = 1 - \theta_{\text{native}} = 1 - \frac{1}{1 + K_\text{d}} = \boxed{\frac{K_\text{d}}{1 + K_\text{d}}}$$

(b) $$\ln K_\text{d} = -\frac{\Delta_\text{d}G^\ominus}{RT}\ [4.10\text{a}] = -\frac{\Delta_\text{d}H^\ominus - T\Delta_\text{d}S^\ominus}{RT}\ [4.15] = \left(-\frac{\Delta_\text{d}H^\ominus}{R}\right) \times \frac{1}{T} + \frac{\Delta_\text{d}S^\ominus}{R}$$

Thus, $\boxed{K_\text{d} = \exp\{-(\Delta_\text{d}H^\ominus - T\Delta_\text{d}S^\ominus)/RT\}}$

For many reactions both $\Delta_\text{d}H^\ominus$ and $\Delta_\text{d}S^\ominus$ are weakly dependent upon temperature over a small temperature range and they may be treated as constant. In such cases, the factors $-\Delta_\text{d}H^\ominus/R$ and $\Delta_\text{d}S^\ominus/R$ are constants, thereby, making the above expression of ln K linear in the variable $1/T$. A plot of ln K against $1/T$ will be linear with a slope equal to $-\Delta_\text{d}H^\ominus/R$ and intercept equal to $\Delta_\text{d}S^\ominus/R$.

(c) $\Delta_\text{d}H^\ominus = +418 \text{ kJ mol}^{-1}$ and $\Delta_\text{d}S^\ominus = 1.32 \text{ kJ K}^{-1} \text{ mol}^{-1}$ for chymotrypsin denaturation.

Toward the goal of creating a plot of $\theta_{\text{denatured}}$ against T we prepare the following table assuming the standard enthalpy and entropy of chymotrypsin denaturation are constants over the temperature range. The plot, shown in Figure 4.5, has the characteristic sigmoidal, S-shape seen previously in Fig. 3.16. Below about 305 K the protein is solely in the native state; above about 325 K it is solely in the denatured state.

T/K	K_d	$\theta_{\text{denatured}}$
302	0.00045	0.0004
304	0.00134	0.0013
306	0.00395	0.0039
308	0.01148	0.0113
310	0.03290	0.0319
312	0.09305	0.0851
314	0.2597	0.2062
316	0.7154	0.4170
318	1.946	0.6605
320	5.226	0.8394
322	13.87	0.9327
324	36.35	0.9732
326	94.18	0.9895

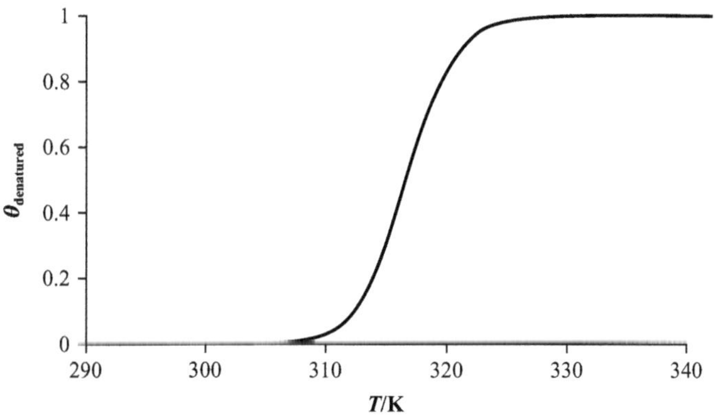

Figure 4.5

(d) The plot of Figure 4.5 indicates that the melting temperature is 317 K. The value is confirmed with an eqn 4.11 computation.

$$T = \frac{\Delta_d H^\ominus}{\Delta_d S^\ominus}\,[4.11] = \frac{+418\ \text{kJ mol}^{-1}}{+1.32\ \text{kJ K}^{-1}\ \text{mol}^{-1}} = \boxed{317\ \text{K}}$$

(e) $\Delta_d G^\ominus = \Delta_d H^\ominus - T\Delta_d S^\ominus$

$$= (+418\ \text{kJ mol}^{-1}) - (310\ \text{K}) \times (+1.32\ \text{kJ K}^{-1}\ \text{mol}^{-1})$$

$$= \boxed{9\ \text{kJ mol}^{-1}}$$

$$K_d = e^{-\frac{\Delta_d G^\ominus}{RT}} \quad [4.10a]$$

$$= e^{-\frac{9\times 10^3\ \text{J mol}^{-1}}{(8.3145\ \text{J K}^{-1}\ \text{mol}^{1})\times(310\ \text{K})}}$$

$$= 0.03$$

Since denaturation is endergonic at the physiological temperature, we conclude that the native state is stable at this temperature.

P4.50 $\text{native} \xrightleftharpoons{[\text{GuHCl}]\ \text{at}\ 300\ \text{K}} \text{denatured} \qquad K_d = [\text{denatured}]/[\text{native}]$

$\Delta_d G^\ominus = \Delta_d G^\ominus(\text{water}) - m[\text{D}]$, where [D] is the concentration of denaturant

(a) Given a value for the fraction θ_d of denatured macromolecules at 300 K, the eqn derived in E4.49 is used to calculate the denaturation equilibrium constant K_d at this temperature.

$$\theta_{\text{denatured}} = \frac{K_d}{1+K_d} \quad \text{or} \quad K_d = \frac{\theta_{\text{denatured}}}{1-\theta_{\text{denatured}}}$$

Having calculated K_d, the value of $\Delta_d G^{\ominus}$ is calculated with the expression:

$$\Delta_d G^{\ominus} = -RT \ln K_d \quad [4.10a]$$

This is done for each of the provided data points after which we prepare a plot of $\Delta_d G^{\ominus}$ against [D], shown in Figure 4.6. The plot is seen to be straight so it is valid to perform the linear regression fit of the plot, which is shown in the figure. According to the relation $\Delta_d G^{\ominus} = \Delta_d G^{\ominus}(\text{water}) - m[\text{D}]$, where [D] is the concentration of denaturant (GuHCl in this example), the regression slope equals $-m$ and the intercept equals $\Delta_d G^{\ominus}$(water). As a computational check, a table of the transformed data is included below. We find that

$$\boxed{m = 9.47 \text{ kJ mol}^{-1}/c^{\ominus}_{\text{GuHCl}} \quad \text{and} \quad \Delta_d G^{\ominus}(\text{water}) = +16.8 \text{ kJ mol}^{-1}}$$

The plot indicates that the native form of the macromolecule is most stable when [GuHCl] < 1.78 mol dm^{-3} and the denatured form is most stable when [GuHCl] > 1.78 mol dm^{-3}.

θ_{native}	$\theta_{\text{denatured}}$	[GuHCl]/mol dm^{-3}	K_d	$\Delta_d G^{\ominus}$/kJ mol^{-1}
1	0	0		
0.99	0.01	0.75	0.01	11.46
0.78	0.22	1.35	0.28	3.16
0.44	0.56	1.70	1.27	−0.60
0.23	0.77	2.00	3.35	−3.01
0.08	0.92	2.35	11.50	−6.09
0.06	0.94	2.70	15.67	−6.86
0.01	0.99	3.00	99.00	−11.46

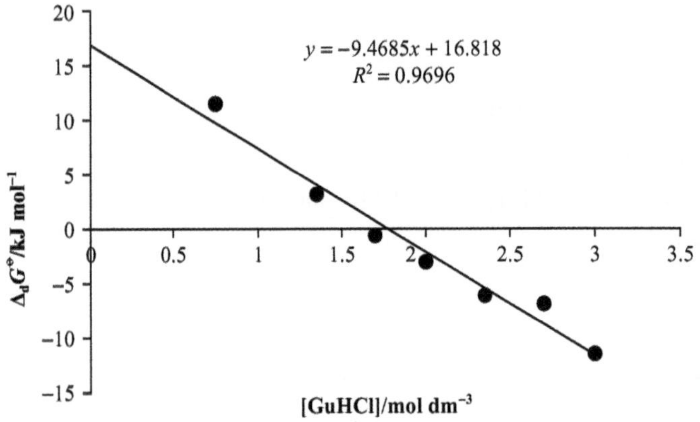

Figure 4.6

(b) The plot of $\theta_{\text{denatured}}$ against [GuHCl] is shown in Figure 4.7. As in Figure 4.5 for denaturation via temperature, denaturation via a denaturant has a sigmoidal shape.

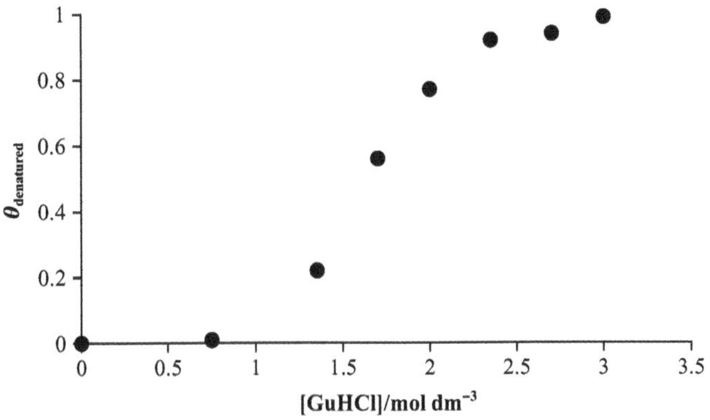

Figure 4.7

(c) When $\theta_{denatured} = \frac{1}{2}$, $K_d = 1$ and $\Delta_d G^\circ = 0$. Consequently,

$$\Delta_d G^\circ = \Delta_d G^\circ(\text{water}) - m[\text{D}]$$
$$0 = \Delta_d G^\circ(\text{water}) - m[\text{D}]_{1/2}$$
$$\boxed{\Delta_d G^\circ(\text{water}) = m[\text{D}]_{1/2}} \quad \text{and therefore} \quad \Delta_d G^\circ = ([\text{D}]_{1/2} - [\text{D}]) \times m$$

For the GuHCl denaturant of part (a) the value of $[\text{D}]_{1/2}$ is found to be

$$[\text{GuHCl}]_{1/2} = \frac{\Delta_d G^\circ(\text{water})}{m} = \frac{+16.8\ \text{kJ mol}^{-1}}{9.47\ \text{kJ mol}^{-1}/c^\circ_{\text{GuHCl}}} = \boxed{1.77\ \text{mol dm}^{-3}}$$

and examination of Figure 4.7 confirms this value.

The following is a derivation of an equation relating $\theta_{denatured}$ to [D], $[\text{D}]_{1/2}$, m, and T. Calculations with this equation provide the plot of Figure 4.8, which indicates that it provides a very good "smoothed" sigmoidal representation of the experimental data.

$$\Delta_d G^\circ = ([\text{D}]_{1/2} - [\text{D}]) \times m$$
$$-RT \ln K_d = ([\text{D}]_{1/2} - [\text{D}]) \times m \quad \text{or} \quad \boxed{K_d = e^{-([\text{D}]_{1/2} - [\text{D}]) \times m/RT}}$$

$$\theta_{denatured} = \frac{K_d}{1 + K_d} = \boxed{\frac{e^{-([\text{D}]_{1/2} - [\text{D}]) \times m/RT}}{1 + e^{-([\text{D}]_{1/2} - [\text{D}]) \times m/RT}}}$$

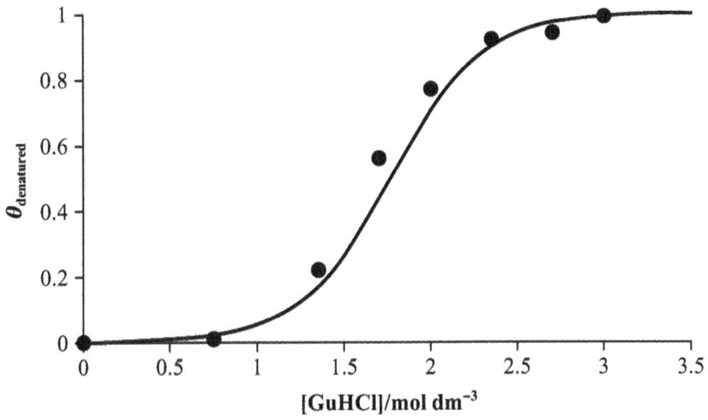

Figure 4.8

P4.51 With the simplified model that hemoglobin has a single acidic proton and HbH and Hb^- represent the protonated and deprotonated form, the equilibria are

$$HbH \rightleftharpoons Hb^- + H^+ \qquad pK_{a,HbH} = 8.18$$
$$HbHO_2 \rightleftharpoons HbO_2^- + H^+ \qquad pK_{a,HbHO_2} = 6.62$$

(a) (i) The fraction of deprotonated, deoxygenated hemoglobin, $f_{deoxy}(Hb^-)$, at pH = 7.4 is given by

$$f_{deoxy}(Hb^-) = \frac{[Hb^-]}{[HbH]+[Hb^-]} = \frac{[Hb^-]/[HbH]}{1+[Hb^-]/[HbH]}$$
$$= \frac{K_{a,HbH}/[H^+]}{1+K_{a,HbH}/[H^+]} = \frac{K_{a,HbH}}{[H^+]+K_{a,HbH}} = \frac{10^{-pK_{a,HbH}}}{10^{-pH}+10^{-pK_{a,HbH}}}$$
$$= \frac{10^{-8.18}}{10^{-7.4}+10^{-8.18}}$$
$$= \boxed{0.142}$$

(ii) The fraction of deprotonated, oxygenated hemoglobin, $f_{oxy}(HbO_2^-)$, at pH = 7.4 is given by

$$f_{oxy}(HbO_2^-) = \frac{[HbO_2^-]}{[HbHO_2]+[HbO_2^-]} = \frac{[HbO_2^-]/[HbHO_2]}{1+[HbO_2^-]/[HbHO_2]}$$
$$= \frac{K_{a,HbHO_2}/[H^+]}{1+K_{a,HbHO_2}/[H^+]} = \frac{K_{a,HbHO_2}}{[H^+]+K_{a,HbHO_2}} = \frac{10^{-pK_{a,HbHO_2}}}{10^{-pH}+10^{-pK_{a,HbHO_2}}}$$
$$= \frac{10^{-6.62}}{10^{-7.4}+10^{-6.62}}$$
$$= \boxed{0.858}$$

(iii) Using the computation of (i) we find that $f_{deoxy}(HbH) = 1 - f_{deoxy}(Hb^-) = 0.858$, which equals the value of $f_{oxy}(HbO_2^-)$ found in the computation of (ii). This suggests that in an oxygen-deficient environment hemoglobin has the protonated HbH form but converts to the HbO_2^- form in an oxygen-rich environment. The high, identical fractions implies a direct conversion between the two accounts for most of the hemoglobin with only small fractions of alternative structures. Thus, the oxygenation of hemoglobin is accompanied by the release of protons in a reversible process. In the following net reaction equation, the left-to-right process (oxygenation) occurs in the oxygen-rich environment of the alveoli of the lungs and the right-to-left process (deoxygenation) occurs in an oxygen-deficient environment such as in actively metabolizing tissue (e.g., muscle).

$$HbH + O_2 \rightleftharpoons HbO_2^- + H^+$$

(b) (i) The amount of hydronium ion bound per mole of oxygenated hemoglobin molecules at pH = 7.4 is $f_{oxy}(HbHO_2)$. $f_{oxy}(HbHO_2) = 1 - f_{oxy}(HbO_2^-) = 1 - 0.858 = \boxed{0.142}$

(ii) The amount of hydronium ion bound per mole of deoxygenated hemoglobin molecules at pH = 7.4 is $f_{deoxy}(HbH)$. $f_{deoxy}(HbH) = 1 - f_{deoxy}(Hb^-) = 1 - 0.142 = \boxed{0.858}$

(iii) The amount of hydronium ion that can be bound per mole of hemoglobin molecules as a result of the release of O_2 by the fully oxygenated protein is $0.858 - 0.142 = \boxed{0.716}$.

(iv) $CO_2 + H_2O \xrightleftharpoons{\text{carbonic anhydrase}} H_2CO_3 \rightleftharpoons HCO_3^- + H^+$

Table 4.5 reports that $pK_{a1} = 6.37$ for carbonic acid at 25°. At the physiological temperature (37°C) and the ionic strength of blood the value is $pK_{a1} = 6.1$.

The hydronium ion released by each mole of CO_2 equals the amount of HCO_3^- released and the ratio of HCO_3^- to H_2CO_3 is related to the pH by the Henderson–Hasselbalch equation.

$$pH = pK_a - \log([H_2CO_3]/[HCO_3^-]) \quad [4.35]$$

and, therefore,

$$[H_2CO_3]/[HCO_3^-] = 10^{pK_a - pH} = 10^{6.1-7.4} = 0.050$$

which means that 95% of the carbonic acid is dissociated at pH = 7.4. Multiplication by the amount of hydronium ion that can be bound per mole of hemoglobin gives the amount of CO_2 that can be released into the blood per mole of hemoglobin: $(0.95) \times (0.716) = \boxed{0.68}$. Although these model calculations simplify the actual behavior of hemoglobin, they do show that hemoglobin is a very significant pH buffer in the blood. The other significant buffering system is provided by the H_2CO_3/HCO_3^- acid/base buffer. The $H_2PO_4^-$/HPO_4^{2-} buffer does not contribute significantly because there is little unbound, "free" phosphate available for this buffer system.

5 Bioenergetics: Thermodynamics of ion and electron transport

Answers to discussion questions

D5.1 The **Debye–Hückel theory** of very dilute ionic solutions yields a method for calculating the **mean activity coefficient, $\gamma_{\pm}$**, of an electrolyte. The theory emphasizes the long-range Coulombic (electrostatic) interaction between ions that brings anions and cations into energetically favorable proximity. Averaged over time, **counter ions** (ions of opposite charge) are more likely to be found near any given ion. This time-averaged, spherical haze around the central ion, in which counter ions outnumber ions of the same charge as the central ion, has a net charge equal in magnitude but opposite in sign to that on the central ion, and is called the **ionic atmosphere**. The energy, and therefore the chemical potential, of any given central ion is lowered below its ideal value as a result of its net electrostatic attraction with its ionic atmosphere. The ideal value of both an activity coefficient and a mean activity coefficient is equal to 1 but the attraction of the ionic atmosphere causes $\gamma_{\pm}$ to be less than 1. The Debye–Hückel theory is correct in the limit of infinitely dilute electrolyte molality, but a rule-of-thumb accepts its application when the total ion concentration is less than about 10^{-3} mol kg^{-1}. When interested in property trends alone, we often accept, within limits, the estimate that activity coefficients are approximately equal to 1 and activities equal concentration.

The Debye–Hückel limiting theory of activity coefficients fails in more concentrated electrolyte solutions because ionic size becomes important and the effective hydration spheres of ions is reduced by competitive attraction by ions for available water molecules. These affect the ionic atmosphere stabilization effect of the Debye–Hückel limiting theory. First, direct association of oppositely charged ions yields ion pairs of zero net charge. The ion pairs have a dipole but a reduced attraction to the ionic atmosphere. Secondly, strong repulsions between either cations or anions diminish the validity of the hazy ionic atmosphere model. The charge imbalance of the ionic atmosphere may decrease or ions may begin to align in very small, localized patterns that resemble an ionic crystal lattice when time averaged. Consequently, as concentration approaches moderate levels the activity coefficient stops the decline of the Debye–Hückel limiting theory and begins to grow toward a value of 1 as the ionic atmosphere stabilization effect is lost. At higher concentrations the activity coefficient may even become larger than 1.

D5.2 The migration of aqueous protons is mechanistically very different from the migration of other ions. In a simplified view a proton on one water molecule (H_3O^+) migrates to a neighbor water molecule, a proton on that water molecule then migrates to its neighbor, and so on along a chain of water molecules. The motion of protons is therefore an *effective* motion of a proton, not the actual motion of a single proton. This causes proton migration to be far more rapid than the migration of other ions, which must move as a single, individual unit from one position to another.

According to the **Grotthuss mechanism**, there is an effective motion of a proton that involves the rearrangement of bonds in a group of water molecules. However, the actual mechanism is still highly contentious. Attention now focuses on the $H_9O_4^+$ unit in which the nearly trigonal planar

H_3O^+ ion is linked to three strongly solvating H_2O molecules. This cluster of atoms is itself hydrated, but the hydrogen bonds in the secondary sphere are weaker than in the primary sphere. It is envisaged that the rate-determining step is the cleavage of one of the weaker hydrogen bonds of this secondary sphere (Figure 5.1a). After this bond cleavage has taken place, and the released molecule has rotated through a few degrees (a process that takes about 1 ps), there is a rapid adjustment of bond lengths and angles in the remaining cluster, to form a $H_5O_2^+$ cation of structure $H_2O\cdots H^+\cdots OH_2$ (Figure 5.1b). Shortly after this reorganization has occurred, a new $H_9O_4^+$ cluster forms as other molecules rotate into a position where they can become members of a secondary hydration sphere, but now the positive charge is located one molecule to the right of its initial location (Figure 5.1c). According to this model, there is no coordinated motion of a proton along a chain of molecules, simply a very rapid hopping between neighboring sites, with a low activation energy. The model is consistent with the observation that the molar conductivity of protons increases as the pressure is raised, for increasing pressure ruptures the hydrogen bonds in water.

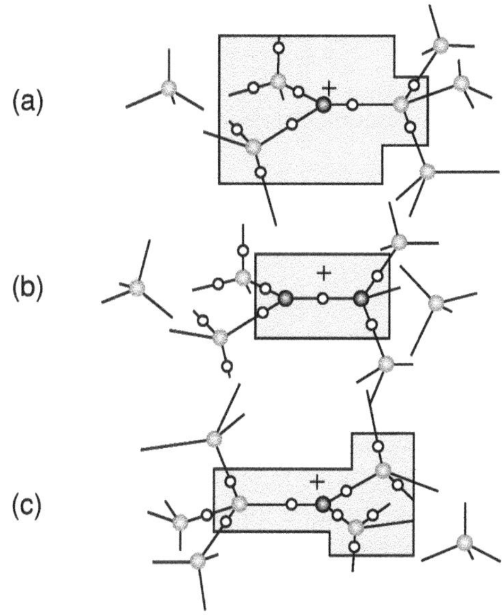

Figure 5.1

D5.3 A **galvanic cell** uses a spontaneous chemical reaction to generate a potential difference and deliver an electric current to an external device. An **electrolytic cell** uses an external potential difference to drive a chemical reaction in the cell that is by itself non-spontaneous. In their essential features, these two kinds of cells can be considered opposites of each other, in the sense that an electrolytic cell can be thought of as a galvanic cell operating in the reverse direction. For some electrochemical cells, this is easy to accomplish. We say they are rechargeable. The most common example is the lead-acid battery used in automobiles. For many other cells, however, this kind of reversibility cannot be achieved. A **fuel cell**, like the galvanic cell, uses a spontaneous chemical reaction to generate a potential difference and deliver an electric current to an external device. Unlike the galvanic cell, the fuel cell must receive reactants from an external storage tank.

D5.4 On a very basic level we observe that a concentration gradient establishes a chemical potential gradient and it is the potential gradient that can generate an electric current. The extreme example is provided by the **electrolyte concentration cell**, which is by definition a galvanic cell consisting of two electrodes of the same metal in different concentrations of the same salt of that metal. The concentration cell having two $M^+(aq)/M$ electrodes is $M(s)|M^+(aq,L)||M^+(aq,R)|M(s)$ and the net reaction is $M^+(aq,R) \rightarrow M^+(aq,L)$. For such a cell $E^{\ominus}_{cell} = 0$ and the Nernst equation gives the cell potential:

$$E_{cell} = E^{\ominus}_{cell} - \frac{RT}{\nu F}\ln Q \quad [5.13]$$

$$= -\frac{RT}{F}\ln\frac{a_{M^+(L)}}{a_{M^+(R)}}$$

With $[M^+(R)] = 10 \times [M^+(L)]$ and an assumption that $\gamma_{M^+(L)} \approx \gamma_{M^+(R)}$:

$$E_{cell} = -(25.693\ \text{mV}) \times \left(\ln\frac{1}{10}\right) = 59.16\ \text{mV}$$

The cell potential is half as large for the $M(s)|M^{2+}(aq,L)||M^{2+}(aq,R)|M(s)$ cell.

D5.5 Construct a cell using a standard hydrogen electrode and an electrode designed around the redox couple of interest. The cell potential E is measured with a high-impedance voltmeter under zero-current conditions. When using SHE as a reference electrode, E_{cell} is the desired half-reaction potential. Should the redox couple have one or more electroactive species (i) that are solvated with concentration b_i, E_{cell} must be measured over a range of b_i values.

The Nernst equation [5.13], with Q being the cell reaction quotient, is the starting point for analysis of the $E_{cell}(b_i)$ data.

$$E_{cell} = E^{\ominus}_{cell} - \frac{RT}{\nu F}\ln Q \quad [5.13]$$

It would seem that substitution of E_{cell} and Q values would allow the computation of the standard redox potential $E^{\ominus}_{cell}$ for the couple. However, a problem arises because the calculation of Q requires not only knowledge of the concentrations of the species involved in the cell reaction but also of their activity coefficients. These coefficients are not usually available, so the calculation cannot be directly completed. However, at very low concentrations, the Debye–Hückel limiting law for the coefficients holds. The procedure then is to substitute the Debye–Hückel law for the activity coefficients into the specific form of the Nernst equation for the cell under investigation and carefully examine the equation to determine what kind of plot to make of the $E_{cell}(b_i)$ data so that extrapolation of the plot to zero concentration, where the Debye−Hückel law is valid, gives a plot intercept that equals $E^{\ominus}_{cell}$. For example, when applied to the "Harned cell",

$$Pt(s)|H_2(g,1\ \text{bar})|HCl(aq,b)|AgCl(s)|Ag(s) \qquad \tfrac{1}{2}H_2(g) + AgCl(s) \rightarrow HCl(aq,b) + Ag(s),$$

the procedure indicates that a plot of $E_{cell} + (2RT/F)\ln b$ against $b^{1/2}$ is linear with an intercept (at $b = 0$) that equals the standard cell potential.

D5.6 Calorimetric methods for the experimental determination of reaction enthalpies and third-law entropies are discussed in Chapters 1 and 2, respectively. These are used to formulate tables of standard enthalpies of formation and standard entropies that are extremely useful when computing the standard Gibbs energy of a constant-temperature and constant-pressure process like a phase transition or chemical reaction. Simply use the tables to find $\Delta_r H^{\ominus}$ and $\Delta_r S^{\ominus}$ for the balanced reaction equation using

$$\Delta_r H^{\ominus} = \sum \nu \Delta_f H^{\ominus}(\text{products}) - \sum \nu \Delta_f H^{\ominus}(\text{reactants}) \quad [1.23]$$

and

$$\Delta_r S^{\ominus} = \sum \nu S^{\ominus}(\text{products}) - \sum \nu S^{\ominus}(\text{reactants}) \quad [2.9]$$

The standard Gibbs energy for the process is given by

$$\Delta_r G^{\ominus} = \Delta_r H^{\ominus} - T\Delta_r S^{\ominus} \quad [2.11]$$

The Gibbs energy is therefore determined indirectly through the arduous measurement of thermodynamic enthalpies and entropies.

Electrochemistry provides a second method for the determination of the Gibbs energy of a redox reaction. An electrochemical cell, based upon the redox reaction, is designed and the standard cell potential is measured as described in D5.5. The Gibbs reaction energy is given by

$$\Delta_r G^\ominus = -\nu F E^\ominus_{cell} \quad [5.14]$$

For a hydrogen electrode half-reaction, $\frac{1}{2}H_2(g,1\text{ bar}) + e^- \rightarrow H^+(aq)$, the standard potential is defined to be $E^\ominus = 0$, a definition that makes it possible to use $E^\ominus_{cell}$ measurements to form tables of standard potentials for reduction half-reactions. When available, the standard potential table is used to find $E^\ominus_{cell}$ using the relation

$$E^\ominus_{cell} = E^\ominus_R - E^\ominus_L \quad [5.17a]$$

after which eqn 5.14 is used in the calculation of $\Delta_r G^\ominus$ if it is needed. In eqn 5.17a, $E^\ominus_R$ is the standard potential of the right-hand electrode (the cathode) and $E^\ominus_L$ is that of the left-hand electrode (the anode).

D5.7 The net glucose catabolic decomposition and reaction properties at 298 K are:

$$C_6H_{12}O_6(\text{glucose, s}) + 6\,O_2(g) \rightarrow 6\,CO_2(g) + 6\,H_2O(l)$$

$$\Delta_c H^\ominus = -2808\text{ kJ mol}^{-1},\ \Delta_c S^\ominus = 259.0\text{ J K}^{-1}\text{ mol}^{-1},\text{ and } \Delta_c G^\ominus = -2885\text{ kJ mol}^{-1}$$

The catabolism of glucose releases 2885 kJ mol^{-1} of energy that can be used to perform non-expansion work. Some of this Gibbs energy is lost as heat but much of it is stored in the energy-transport molecules ATP, NADH, and $FADH_2$ during enzyme-controlled mechanistic reaction steps. These high-potential species carry the energy to endergonic reactions on the surfaces of specific enzymes and in the process of releasing the stored chemical energy to an endergonic reaction, the species returns to its low potential form ADP, NAD^+, or FAD, respectively. For example, the hydrolysis of ATP to ADP releases 31 kJ mol^{-1}:

$$ATP^{4-}(aq) + H_2O(l) \rightarrow ADP^{3-}(aq) + HPO_4^{2-}(aq) + H_3O^+(aq) \qquad \Delta_r G^\oplus = -31\text{ kJ mol}^{-1}.$$

NADH and $FADH_2$ produced in the catabolism of glucose can transfer their energy to a proton gradient across the inner membrane of cell mitochondria with the aid of **respiratory chain** protein. In the process they are oxidized to NAD^+ and FAD and the released electrons reduce O_2 to water. **Chemiosmotic theory** explains how H^+-ATPases use the energy stored in the transmembrane proton gradient to synthesize ATP from ADP. Experiments show that 11 molecules of ATP are made for every three molecules of NADH and one molecule of $FADH_2$. Since the catabolism of 1 mole of glucose produces the equivalence of 4 moles of ATP, 10 moles of NADH, and 2 moles of $FADH_2$, a maximum of 38 moles of ATP per mole of glucose can be produced by the actions of glycolysis, the citric acid cycle, respiratory chain reactions, and activity of H^+-ATPase. Each mole of ATP molecules extracts 31 kJ from the 2880 kJ supplied by 1 mole of glucose, so 1178 kJ (41%) is stored for later use.

The breakdown of glucose in the cell begins with **glycolysis**, a partial oxidation of glucose by nicotinamide adenine dinucleotide (NAD^+) to pyruvate ion, $CH_3COCO_2^-$. Metabolism continues in the form of the citric acid cycle, in which pyruvate ions are oxidized to CO_2, and ends with oxidative phosphorylation, in which O_2 is reduced to H_2O. Glycolysis is the main source of energy during **anaerobic metabolism**, a form of metabolism in which inhaled O_2 does not play a role. The citric acid cycle and oxidative phosphorylation are the main mechanisms for the extraction of energy from carbohydrates during **aerobic metabolism**, a form of metabolism in which inhaled O_2 does play a role.

Glycolysis occurs in the **cytosol**, the aqueous material encapsulated by the cell membrane, and consists of 10 enzyme-catalyzed reactions. The process needs to be initiated by consumption of two molecules of ATP per molecule of glucose. The first ATP molecule is used to drive the phosphorylation of glucose to glucose-6-phosphate (G6P):

$$\text{glucose(aq)} + \text{ATP(aq)} \rightarrow \text{G6P(aq)} + \text{ADP(aq)} + \text{H}^+\text{(aq)} \qquad \Delta_r G^{\oplus} = -17\ \text{kJ mol}^{-1}$$

The next step is the isomerization of G6P to fructose-6-phosphate (F6P). The second ATP molecule consumed during glycolysis drives the phosphorylation of F6P to fructose-1,6-diphosphate (FDP):

$$\text{F6P(aq)} + \text{ATP(aq)} \rightarrow \text{FDP(aq)} + \text{ADP(aq)} + \text{H}^+\text{(aq)} \qquad \Delta_r G^{\oplus} = -14\ \text{kJ mol}^{-1}$$

In the next step, FDP is broken into two three-carbon units, dihydroxyacetone phosphate (1,3-dihydroxypropanone phosphate, $CH_2OHCOCH_2OPO_3^{2-}$, and glyceraldehyde-3-phosphate, which exist in mutual equilibrium. Only the glyceraldehyde-3-phosphate is oxidized by NAD^+ to pyruvate ion, with formation of two ATP molecules. As glycolysis proceeds, all the dihydroxyacetone phosphate is converted to glyceraldehyde-3-phosphate, so the result is the consumption of two NAD^+ molecules and the formation of four ATP molecules per molecule of glucose.

The oxidation of glucose by NAD^+ to pyruvate ions has $\Delta_r G^{\oplus} = -147\ \text{kJ mol}^{-1}$ at blood temperature. In glycolysis, the oxidation of one glucose molecule is coupled to the net conversion of two ADP molecules to two ATP molecules, so the net reaction of glycolysis is

$$\begin{aligned}&\text{glucose(aq)} + 2\ \text{NAD}^+\text{(aq)} + 2\ \text{ADP(aq)} + 2\ \text{P}_i\text{(aq)} + 2\ \text{H}_2\text{O(l)} \rightarrow \\ &\qquad 2\ \text{CH}_3\text{COCO}_2^-\text{(aq)} + 2\ \text{NADH(aq)} + 2\ \text{ATP(aq)} + 2\ \text{H}_3\text{O}^+\text{(aq)}\end{aligned}$$

The biological standard reaction Gibbs energy is $(-147) - 2(-31)\ \text{kJ mol}^{-1} = -85\ \text{kJ mol}^{-1}$. The reaction is exergonic and therefore spontaneous under biological standard conditions: the oxidation of glucose is used to "recharge" the ATP.

In the presence of O_2, pyruvate is oxidized further during the citric acid cycle and oxidative phosphorylation, which occur in the mitochondria of cells. The further oxidation of carbon derived from glucose begins with a reaction between pyruvate ion, NAD^+, and coenzyme A (CoA) to give acetyl CoA, NADH, and CO_2:

$$\text{pyruvate} + \text{CoA} + \text{NAD}^+ \xrightarrow{\text{pyruvate dehydrogenase}} \text{acetyl-CoA} + \text{CO}_2 + \text{NADH}$$

Acetyl CoA is then oxidized by NAD^+ and flavin adenine dinucleotide (FAD) in the citric acid cycle, which requires eight enzymes and results in the synthesis of GTP from GDP or ATP from ADP:

$$\begin{aligned}&\text{Acetyl CoA(aq)} + 3\ \text{NAD}^+\text{(aq)} + \text{FAD(aq)} + \text{GDP(aq)} + \text{P}_i\text{(aq)} + 2\ \text{H}_2\text{O(l)} \rightarrow \\ &\qquad 2\ \text{CO}_2\text{(g)} + 3\ \text{NADH(aq)} + 2\ \text{H}_3\text{O}^+\text{(aq)} + \text{FADH}_2\text{(aq)} + \text{GTP(aq)} + \text{CoA(aq)}\end{aligned}$$

$$\Delta_r G^{\oplus} = -57\ \text{kJ mol}^{-1}$$

In cells that produce GTP, the enzyme nucleoside diphosphate kinase catalyzes the transfer of a phosphate group to ADP to form ATP: $\text{GTP(aq)} + \text{ADP(aq)} \rightarrow \text{GDP(aq)} + \text{ATP(aq)}$. For this reaction, $\Delta_r G^{\oplus} = 0$ because the phosphate group transfer potentials for GTP and ATP are essentially identical. Overall, we write the oxidation of glucose as a result of glycolysis and the citric acid cycle as

$$\begin{aligned}&\text{glucose(aq)} + 10\ \text{NAD}^+\text{(aq)} + 2\ \text{FAD(aq)} + 4\ \text{ADP(aq)} + 4\ \text{P}_i\text{(aq)} + 2\ \text{H}_2\text{O(l)} \rightarrow \\ &\qquad 6\ \text{CO}_2\text{(g)} + 10\ \text{NADH(aq)} + 6\ \text{H}_3\text{O}^+\text{(aq)} + 2\ \text{FADH}_2\text{(aq)} + 4\ \text{ATP(aq)}\end{aligned}$$

The NADH and $FADH_2$ go on to reduce O_2 during oxidative phosphorylation, which also produces ATP.

We end this summary of ATP production with an explanation of why the hydrolysis of ATP is exergonic. At physiological pH the oxygen atoms of ATP are deprotonated, negatively charged, and the molecule is best represented as ATP^{4-}. The electrostatic repulsions between the highly charged oxygen atoms of ATP^{4-} is expected to give it an exergonic hydrolysis free energy by making the hydrolysis enthalpy negative. Also, the deprotonated phosphate species, $P_i(aq)$, produced in the hydrolysis of ATP has more resonance structures than ATP^{4-}. Resonance lowers the energy of the dissociated phosphate making the hydrolysis enthalpy more negative and contributing to the exergonicity of the hydrolysis.

The electrostatic repulsion between the highly charged oxygen atoms of ATP^{4-} is a hypothesis that is consistent with the observation that protonated ATP, H_4ATP, has an exergonic hydrolysis free energy of smaller magnitude because the negative repulsions of oxygen atoms are not present. Likewise for $MgATP^{2-}$ because the Mg^{2+} ion lies between negatively charged oxygen atoms, thereby, reducing repulsions and stabilizing the ATP molecule.

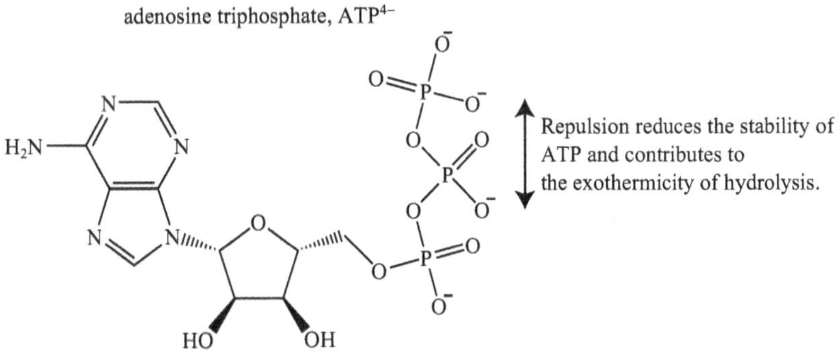
adenosine triphosphate, ATP4−
H2N
N
HO
OH
Repulsion reduces the stability of ATP and contributes to the exothermicity of hydrolysis.

Solutions to exercises

For notational simplicity, we have used both the molality concentration expression $a_J = \gamma_J b_J / b^{\circ}$ where $b^{\circ} = 1\ \text{mol kg}^{-1}$ [5.1a] and $a_J = \gamma_J b_J$ [5.1b] where b_J is the unitless magnitude of molality. The convention of eqn 5.1b is most often used in calculations of ionic strength while the convention of eqn 5.1a appears in Nernst equation computations.

E5.8 For a solution that contains a single cation and anion:

$$I = \tfrac{1}{2}\sum_i z_i^2 b_i / b^{\circ}\ [5.5\text{b}] = \tfrac{1}{2}\{z_+^2 b_+ + z_-^2 b_-\}/b^{\circ},\ \text{where } b^{\circ} = 1\ \text{mol kg}^{-1}$$

(a) $KCl(s) \xrightarrow{\text{water}} K^+(aq) + Cl^-(aq)$ Therefore, $z_+ = 1$, $z_- = -1$, and $b_+ = b_- = b$.

$$I = \tfrac{1}{2}\{1^2 b + 1^2 b\}/b^{\circ} = \boxed{b/b^{\circ}}$$

(b) $FeCl_3(s) \xrightarrow{\text{water}} Fe^{3+}(aq) + 3\,Cl^-(aq)$ Therefore, $z_+ = 3$, $z_- = -1$, and $b_+ = b$, $b_- = 3b$.

$$I = \tfrac{1}{2}\{3^2 b + 1^2(3b)\}/b^{\circ} = \boxed{6b/b^{\circ}}$$

(c) $CuSO_4(s) \xrightarrow{\text{water}} Cu^{2+}(aq) + SO_4^{2-}(aq)$ Therefore, $z_+ = 2$, $z_- = -2$, and $b_+ = b_- = b$.

$$I = \tfrac{1}{2}\{2^2 b + 2^2 b\}/b^{\circ} = \boxed{4b/b^{\circ}}$$

E5.9 $$I = I_{KCl} + I_{CuSO_4} = \tfrac{1}{2}(z_+^2 b_+ + z_-^2 b_-)_{KCl} + \tfrac{1}{2}(z_+^2 b_+ + z_-^2 b_-)_{CuSO_4}\quad [5.5\text{b}]$$

Let the preparation molalities, the formal concentrations, be b_{KCl} and b_{CuSO_4}. Determination of solution ionic strength requires the deduction of z_+^2, b_+, z_-^2, and b_- for each ionic compound. These values are substituted into the above equation.

$KCl(s) \xrightarrow{\text{water}} K^+(aq) + Cl^-(aq)$ Therefore, $z_+ = 1$, $z_- = -1$ and $b_+ = b_- = b_{KCl}$.

$CuSO_4(s) \xrightarrow{\text{water}} Cu^{2+}(aq) + SO_4^{2-}(aq)$ Therefore, $z_+ = 2$, $z_- = -2$, and $b_+ = b_- = b_{CuSO_4}$.

$$I = \tfrac{1}{2}\{1^2 b_{KCl} + 1^2 b_{KCl}\}/b^\ominus + \tfrac{1}{2}\{2^2 b_{CuSO_4} + 2^2 b_{CuSO_4}\}/b^\ominus = (b_{KCl} + 4b_{CuSO_4})/b^\ominus$$
$$= 0.10 + 4 \times (0.20)$$
$$= \boxed{0.90}$$

COMMENT: Note that the ionic strength of a solution of more than one electrolyte may be calculated by summing the ionic strengths of each electrolyte considered as a separate solution, as in the solution to this exercise, or by summing the product $\frac{1}{2}b_J z_J^2$ for each individual ion, as in the definition of I [5.5b].

E5.10 $I_{KNO_3} = b_{KNO_3} = 0.150$

Therefore, the ionic strengths of the added salts must be 0.100 to result in a total of 0.250.

(a) $Ca(NO_3)_2(s) \xrightarrow{\text{water}} Ca^{2+}(aq) + 2\,NO_3^-(aq)$ Therefore, $z_+ = 2$, $z_- = -1$, and $b_+ = b$, $b_- = 2b$.

$$I_{Ca(NO_3)_2} = \tfrac{1}{2}\sum_i z_i^2 b_i/b^\ominus\ [5.5b] = \tfrac{1}{2}\{z_+^2 b_+ + z_-^2 b_-\}/b^\ominus, \text{ where } b^\ominus = 1\ \text{mol kg}^{-1}$$

$$0.100 = \tfrac{1}{2}\{2^2 + 1^2 \times 2\}b/b^\ominus$$

$$b = 0.0333\ \text{mol kg}^{-1}$$

$$m_{Ca(NO_3)_2} = bM_{Ca(NO_3)_2}m_{\text{solvent}}$$
$$= (0.0333\ \text{mol kg}^{-1}) \times (164.1\ \text{g mol}^{-1}) \times (0.500\ \text{kg})$$
$$= \boxed{2.74\ \text{g}}$$

(b) $NaCl(s) \xrightarrow{\text{water}} Na^+(aq) + Cl^-(aq)$ Therefore, $z_+ = 1$, $z_- = -1$, and $b_+ = b_- = b$.

$$I_{NaCl} = \tfrac{1}{2}\sum_i z_i^2 b_i/b^\ominus\ [5.5b] = \tfrac{1}{2}\{z_+^2 b_+ + z_-^2 b_-\}/b^\ominus, \text{ where } b^\ominus = 1\ \text{mol kg}^{-1}$$

$$0.100 = \tfrac{1}{2}\{1^2 + 1^2\}b/b^\ominus$$

$$b = 0.100\ \text{mol kg}^{-1}$$

$$m_{NaCl} = bM_{NaCl}m_{\text{solvent}}$$
$$= (0.100\ \text{mol kg}^{-1}) \times (58.44\ \text{g mol}^{-1}) \times (0.500\ \text{kg})$$
$$= \boxed{2.92\ \text{g}}$$

E5.11 For the salt M_pX_q: $\gamma_\pm = (\gamma_+^p \gamma_-^q)^{1/s}$ where $s = p + q$ [5.3b]

For $CaCl_2$: $p = 1$, $q = 2$, $s = 3$, $\boxed{\gamma_\pm = (\gamma_+ \gamma_-^2)^{1/3}}$.

E5.12 The concentrations $b_{CaCl_2}/b^\ominus = 0.010$ and $b_{NaF}/b^\ominus = 0.030$ are sufficiently dilute for the Debye–Hückel limiting law to give a reasonable estimate of the mean ionic activity coefficients.

$$I = \tfrac{1}{2}\sum_i z_i^2 b_i\ [5.5b] = I_{CaCl_2} + I_{NaF}$$
$$= \tfrac{1}{2}\{(4 \times 0.010) + (1 \times 0.020)\} + \tfrac{1}{2}\{(1 \times 0.030) + (1 \times 0.030)\}$$
$$= \boxed{0.060}$$

For $CaCl_2$(aq):

$$\log(\gamma_\pm)_{CaCl_2} = -A\,|z_+z_-|\,I^{1/2}\ [5.4] = -0.509 \times |2\times(-1)| \times (0.060)^{1/2} = -0.24\overline{94}$$

$$(\gamma_\pm)_{CaCl_2} = 10^{-0.24\overline{94}} = \boxed{0.56\overline{3}}$$

$$a_{Ca^{2+}} = (\gamma_\pm)_{CaCl_2} b_{Ca^{2+}} = (0.56\overline{3}) \times (.010) = \boxed{0.0056}$$

$$a_{Cl^-} = (\gamma_\pm)_{CaCl_2} b_{Cl^-} = (0.56\overline{3}) \times (.020) = \boxed{0.011}$$

For NaF(aq):

$$\log(\gamma_\pm)_{NaF} = -A\,|z_+z_-|\,I^{1/2}\ [5.4] = -0.509 \times |1\times(-1)| \times (0.060)^{1/2} = -0.125$$

$$(\gamma_\pm)_{NaF} = 10^{-0.11\overline{4}} = \boxed{0.750}$$

$$a_{Na^+} = (\gamma_\pm)_{NaF} b_{Na^+} = (0.750) \times (.030) = \boxed{0.0225}$$

$$a_{Cl^-} = (\gamma_\pm)_{NaF} b_{F^-} = (0.750) \times (.030) = \boxed{0.0225}$$

E5.13 $$\log\gamma_\pm = -\frac{A\,|z_+z_-|\,I^{1/2}}{1+BI^{1/2}} \quad [5.6 \text{ with } C=0]$$

Solving for B,

$$B = -\left(\frac{1}{I^{1/2}} + \frac{A\,|z_+z_-|}{\log\gamma_\pm}\right).$$

Recognizing that for HBr we know that $|z_+z_-| = 1$ and $I = \frac{1}{2}(b_{H^+} + b_{Br^{-1}}) = b_{HBr} = b$, the equation for B simplifies to

$$B = -\left(\frac{1}{b^{1/2}} + \frac{0.509}{\log\gamma_\pm}\right).$$

We do a table calculation of the right side of this equation for each solution and average the results.

$b/b^\ominus$	0.0050	0.0100	0.0200
$\gamma_\pm$	0.930	0.907	0.879
B	2.01	2.01	2.02

The constancy of B indicates that the mean activity coefficient of HBr obeys the extended Debye–Hückel law very well with $\boxed{B = 2.01}$.

E5.14 $PX_\nu(s) \rightleftharpoons P^{\nu+}(aq) + \nu\,X^-(aq)$, where $P^{\nu+}$ is a polycationic protein.

The solubility of a polycationic protein is sensitive to ionic strength because of the large protein surface area and the hydration requirements of its many acidic sites. At very low ionic strengths of a salt that contains an anion of a very strong acid and large electric charge (like the SO_4^{2-} anion in $(NH_4)_2SO_4$ and Na_2SO_4), the large negative charge displaces many of the X^- anions from the immediate hydration sphere around the protein. Having a smaller precipitation tendency than X^-, the displacement by sulfate anion pulls the solubility equilibrium to the right in the above reaction equation. This is the **salting-in effect** of increasing protein solubility with increasing ionic strength.

At some point an increase in ionic strength begins to decrease the number of water molecules that are available to hydrate the polycationic protein because ions of salt additions require their own hydration spheres. There are fewer "insulating" water molecules between $P^{\nu+}$ ions and the precipitating X^- ions and the above equilibrium shifts to the left. This is the **salting-out effect** of decreasing protein solubility with increasing ionic strength.

Thus, solubility depends upon ionic strength as well as temperature, pH, and solvent. At very low ionic strengths solubility of a protein may increase with increasing ionic strength, reach a peak at an intermediate ionic strength, and decrease at ever higher ionic strengths.

The **extended Debye–Hückel law**, $\log \gamma_\pm = -\dfrac{A\,|z_+z_-|\,I^{1/2}}{1+BI^2} + CI$ with B and C empirically determined constants (text Fig. 5.2), is used to relate solubility and ionic strength. The equilibrium constant for the solubility reaction is

$$K = a_{P^{\nu+}} a^{\nu}_{X^-} = \gamma_\pm^2 [P^{\nu+}][X^-]^\nu$$

Taking the base 10 logarithm of both sides, solving for the logarithm of the solubility product $[P^{\nu+}][X^-]^\nu$, and substitution of the $\log \gamma_\pm$ term with the extended Debye–Hückel law gives

$$\begin{aligned}
\log([P^{\nu+}][X^-]^\nu) &= \log K - 2\log\gamma_\pm \\
&= \log K - 2\left(-\frac{A\,|z_+z_-|\,I^{1/2}}{1+BI^{1/2}} + CI\right) \\
&= \log K + \frac{2A\,|z_+z_-|\,I^{1/2}}{1+BI^{1/2}} - 2CI
\end{aligned}$$

$\log([P^{\nu+}][X^-]^\nu)$ is a measure of the protein solubility. Low values indicate low solubility; high values correspond to a large solubility. The empirical constants B and C are small enough that terms containing them are negligibly small at very low ionic strengths. Thus, at very low ionic strengths:

$$\begin{aligned}
\log([P^{\nu+}][X^-]^\nu) &= \log K + 2A\,|z_+z_-|\,I^{1/2} \quad \text{where } A = 0.509 \text{ at } 25°\text{C} \\
&= \text{constant} + (\text{another, positive constant}) \times I^{1/2}
\end{aligned}$$

and we see that a plot of $\log([P^{\nu+}][X^-]^\nu)$ against $I^{1/2}$ is linear with a positive slope at very low ionic strengths. This is the salting-in effect.

At high ionic strengths, $BI^{1/2} \gg 1$ and our working equation becomes

$$\begin{aligned}
\log([P^{\nu+}][X^-]^\nu) &= \log K + \frac{2A\,|z_+z_-|\,I^{1/2}}{BI^{1/2}} - 2CI = \log K + \frac{2A\,|z_+z_-|}{B} - 2CI \\
&= \text{constant} - 2CI
\end{aligned}$$

and we see that a plot of $\log([P^{\nu+}][X^-]^\nu)$ against I is linear with a negative slope at high ionic strengths when the empirical constant C is positive. This is the salting-out effect.

E5.15 $3\,Na^+(\text{aq,inside}) + 2\,K^+(\text{aq,outside}) + \text{ATP}$

$$\xrightarrow{Na^+/K^+\ \text{pump}} \text{ADP} + P_i + 3\,Na^+(\text{aq,outside}) + 2\,K^+(\text{aq,inside}) \qquad \Delta_r G = ?$$

The above reaction equation is the Na^+/K^+ pump coupling of ATP hydrolysis with the transport of the sodium and potassium ions across the cell membrane at pH = 7 and physiologically reasonable concentrations. $\Delta_r G$ is the sum of the ATP hydrolysis and the Gibbs energy for the ion transport, $\Delta_r G_{\text{transport}}$. If $\Delta_r G < 0$, the ATP-coupled transport will be spontaneous. We begin with the calculation of the ATP hydrolysis Gibbs energy $\Delta_r G_{ATP}$ when $[P_i] = 1.0$ mol dm^{-3} and [ATP]/[ADP] = 100. The balanced hydrolysis reaction is

$$ATP^{4-}(aq) + H_2O(l) \rightarrow ADP^{3-}(aq) + HPO_4^{2-}(aq) + H_3O^+(aq) \qquad \Delta_r G^{\oplus} = -31.3 \text{ kJ mol}^{-1}$$

$$\Delta_r G_{ATP} = \Delta_r G^{\oplus}_{ATP} + RT\ln Q\,[4.7] = \Delta_r G^{\oplus}_{ATP} + RT\ln\left(\frac{[ADP^{3-}][HPO_4^{2-}]a_{H_3O^+}}{[ATP^{4-}]}\right)$$

Since $a_{H_3O^+} = 1$ when using the biological standard state at pH = 7, we find

$$\Delta_r G_{ATP} = -31.3 \text{ kJ mol}^{-1} + (0.0083145 \text{ kJ K}^{-1}\text{ mol}^{-1}) \times (310 \text{ K}) \ln\left(\frac{1}{100}\right) = -43.2 \text{ kJ mol}^{-1}$$

To find the work performed by the Na^+/K^+ pump when pumping 3 moles of sodium cation out of the cell while simultaneously pumping 2 moles of potassium cation into the cell, we calculate the work per mole of each ion and add them using the stoichiometric weights 3 and 2. Computations are based upon eqn 5.8 with the physiologically reasonable values (see Case Study 5.1 *Action potentials*) of $[Na^+]_{outside}/[Na^+]_{inside} = 10$, $[K^+]_{inside}/[K^+]_{outside} = 20$ with a cell membrane **resting potential** of $\Delta\phi = \phi_{outside} - \phi_{inside} = +62$ mV for the sodium cation transport (the potential inside the membrane is lower than the potential outside), and $\Delta\phi = \phi_{inside} - \phi_{outside} = -62$ mV for the potassium cation transport.

$$\begin{aligned}
\Delta_r G_{transport} &= 3\Delta G_m(Na^+) + 2\Delta G_m(K^+) \\
&= 3\left\{RT\ln\frac{[Na^+]_{outside}}{[Na^+]_{inside}} + zF\Delta\phi(Na^+)\right\} + 2\left\{RT\ln\frac{[K^+]_{inside}}{[K^+]_{outside}} + zF\Delta\phi(K^+)\right\} \quad [5.8] \\
&= RT\ln\left\{\left(\frac{[Na^+]_{outside}}{[Na^+]_{inside}}\right)^3\left(\frac{[K^+]_{inside}}{[K^+]_{outside}}\right)^2\right\} + zF\{3\Delta\phi(Na^+) + 2\Delta\phi(K^+)\} \\
&= (0.0083145 \text{ kJ K}^{-1}\text{ mol}^{-1}) \times (310 \text{ K}) \ln\{10^3 \times 20^2\} + 1 \times (96.4853 \text{ kC mol}^{-1}) \\
&\quad \times (3-2) \times (62 \times 10^{-3} \text{ V}) \\
&= 39.2 \text{ kJ mol}^{-1} \quad [1 \text{ kC V} = 1 \text{ kJ}]
\end{aligned}$$

Thus,

$$\begin{aligned}
\Delta_r G &= \Delta_r G_{ATP} + \Delta_r G_{transport} = -43.2 \text{ kJ mol}^{-1} + 39.2 \text{ kJ mol}^{-1} \\
&= -4.0 \text{ kJ mol}^{-1}
\end{aligned}$$

Since $\Delta_r G < 0$, the hydrolysis of one mole of ATP does supply sufficient Gibbs energy to transport 3 moles of sodium cation and 2 moles of potassium cation across the cell membrane in the direction of each ion's concentration gradient (random diffusion would move the ions in the opposite directions). We have used $[P_i] = 1.0$ mol dm^{-3} in our computation of the ATP hydrolysis Gibbs energy. At lower concentrations of the hydrogen phosphate anion the value of $\Delta_r G_{ATP}$ is even more exergonic and, therefore, provides an even greater capacity to do the work of 3:2 ion transport.

E5.16 We solve the Goldman equation for the relative membrane permeability of the Na^+ cation.

$$\Delta\phi = \frac{RT}{F}\ln\left(\frac{\sum_i P_i[M_i^+]_{out} + \sum_j P_j[X_j^-]_{in}}{\sum_i P_i[M_i^+]_{in} + \sum_j P_j[X_j^-]_{out}}\right) \qquad \text{[Goldman equation, 5.10)}$$

$$\left\{\sum_i P_i[M_i^+]_{in} + \sum_j P_j[X_j^-]_{out}\right\}e^{F\Delta\phi/RT} = \sum_i P_i[M_i^+]_{out} + \sum_j P_j[X_j^-]_{in}$$

$$\{P_{Na^+}[Na^+]_{in}+P_{K^+}[K^+]_{in}+P_{Cl^-}[Cl^-]_{out}\}e^{F\Delta\phi/RT}=P_{Na^+}[Na^+]_{out}+P_{K^+}[K^+]_{out}+P_{Cl^-}[Cl^-]_{in}$$

$$\begin{aligned}\{[Na^+]_{out}-[Na^+]_{in}e^{F\Delta\phi/RT}\}P_{Na^+}&=\{P_{K^+}[K^+]_{in}+P_{Cl^-}[Cl^-]_{out}\}e^{F\Delta\phi/RT}-\{P_{K^+}[K^+]_{out}+P_{Cl^-}[Cl^-]_{in}\}\\&=\{[Cl^-]_{out}e^{F\Delta\phi/RT}-[Cl^-]_{in}\}P_{Cl^-}+\{[K^+]_{in}e^{F\Delta\phi/RT}-[K^+]_{out}\}P_{K^+}\end{aligned}$$

$$P_{Na^+}=\frac{\{[Cl^-]_{out}e^{F\Delta\phi/RT}-[Cl^-]_{in}\}P_{Cl^-}+\{[K^+]_{in}e^{F\Delta\phi/RT}-[K^+]_{out}\}P_{K^+}}{[Na^+]_{out}-[Na^+]_{in}e^{F\Delta\phi/RT}}$$

The value of the exponential term is $e^{F\Delta\phi/RT}=e^{-0.030\text{ V}/0.02671\text{ V}}=0.32\overline{52}$ at 310 K. Thus,

$$P_{Na^+}=\frac{\{100(0.32\overline{52})-10\}0.45+\{100(0.32\overline{52})-5\}1.0}{140-10(0.32\overline{52})}=\boxed{0.28}$$

E5.17 $CH_3COCO_2^-(aq)+NADH(aq)+H^+(aq)\rightarrow CH_3CH(OH)CO_2^-(aq)+NAD^+(aq)\quad \nu=2$

Yes, this a redox reaction. NADH is oxidized as it loses H^-, which is equivalent to the loss of electrons. Pyruvate is reduced as it has a reduction in the number of carbon-to-oxygen bonds during the conversion of the carbonyl group to an alcohol group, a conversion that is equivalent to the acquisition of electrons.

E5.18 The reduction half-reactions of E5.17 are:

R: $CH_3COCO_2^-(aq)+2\,H^+(aq)+2\,e^-\rightarrow CH_3CH(OH)CO_2^-(aq)$

L: $NAD^+(aq)+H^+(aq)+2\,e^-\rightarrow NADH(aq)$ [$H^+(aq)+2\,e^-$ is equivalent to $H^-(aq)$.]

The overall reaction is the difference between the first half-reaction and the second (R–L).

E5.19 Overall reaction: $CH_3CH_2OH(aq)+NAD^+(aq)\rightarrow CH_3CHO(aq)+NADH(aq)+H^+(aq)\quad \nu=2$

The reduction half-reactions are:

R: $NAD^++2\,e^-+H^+\rightarrow NADH$

L: $CH_3CHO+2\,e^-+2\,H^+\rightarrow CH_3CH_2OH$

The difference between the NAD^+ reduction and the acetaldehyde reduction gives the overall reaction (R–L). The reaction quotients for each of these half-reactions and the overall reaction are:

$$Q_{NAD^+}=\frac{[NADH]}{[NAD^+][H^+]}$$

$$Q_{CH_3CHO}=\frac{[CH_3CH_2OH]}{[CH_3CHO][H^+]^2}$$

$$Q_{net}=\frac{Q_{NAD^+}}{Q_{CH_3CHO}}=\frac{[CH_3CHO][NADH][H^+]}{[CH_3CH_2OH][NAD^+]}$$

E5.20 R: $O_2(g)+4\,H^+(aq)+4\,e^-\rightarrow 2\,H_2O(l)$

L: $\text{cystine}(aq)+2\,H^+(aq)+2\,e^-\rightarrow 2\,\text{cysteine}(aq)$

The overall reaction, which must not contain free electrons, is obtained as R – 2 × L and is:

$4\,\text{cysteine}(aq)+O_2(g)\rightarrow 2\,\text{cystine}(aq)+2\,H_2O(l)\qquad \nu=4$

E5.21 Overall reaction: $NADP^+(aq) + 2\,H^+(aq) + 2\,fd_{red}(aq) \rightarrow NADPH(aq) + 2\,fd_{ox}(aq)$ $\boxed{\nu = 2}$

The reduction half-reactions are:

R: $NADP^+(aq) + H^+(aq) + 2\,e^- \rightarrow NADPH(aq)$

L: $2\,fd_{ox}(aq) + 2\,e^- \rightarrow 2\,fd_{red}(aq) + H^+(aq)$

The overall reaction is obtained as R–L with 2 electrons transferred.

E5.22 The half-reactions are

R: $O_2(g) + 4\,H^+(aq) + 4\,e^- \rightarrow 2\,H_2O(l)$ $\quad E_R^{\oplus} = +0.82\text{ V}$

L: $NAD^+(aq) + H^+(aq) + 2\,e^- \rightarrow NADH(aq)$ $\quad E_L^{\oplus} = -0.32\text{ V}$

The overall reaction, which must not contain free electrons, is obtained as R − 2 × L and is:

$2\,NADH(aq) + O_2(g) + 2\,H^+(aq) \rightarrow 2\,NAD^+(aq) + 2\,H_2O(l)$ $\boxed{\nu = 4}$

The biological standard potential and Gibbs energy for the overall reaction are then

$$E_{cell}^{\oplus} = E_R^{\oplus} - E_L^{\oplus}\,[5.17b] = +0.82\text{ V} - (-0.32\text{ V}) = \boxed{+1.14\text{ V}}$$

$$\Delta_r G^{\oplus} = -\nu F E_{cell}^{\oplus}\,[5.12] = -4 \times (96.485\text{ C mol}^{-1}) \times (+1.14\text{ V}) = \boxed{-440\text{ kJ mol}^{-1}}$$

E5.23 R: $O_2(g) + 4\,H^+(aq) + 4\,e^- \rightarrow 2\,H_2O(l)$ $\quad E_R^{\oplus} = +0.81\text{ V}$ [Table 5.2]

L: $Fe^{3+}(\text{Cyt } c) + e^- \rightarrow Fe^{2+}(\text{Cyt } c)$ $\quad E_L^{\oplus} = +0.25\text{ V}$ [Table 5.2]

The overall reaction, which must not contain free electrons, is obtained as R − 4 × L and is:

$O_2(g) + 4\,H^+(aq) + 4\,Fe^{2+}(\text{Cyt } c) \rightarrow 2\,H_2O(l) + 4\,Fe^{3+}(\text{Cyt } c)$ $\boxed{\nu = 4}$

$$E_{cell}^{\oplus} = E_R^{\oplus} - E_L^{\oplus}\,[5.17b] = +0.81\text{ V} - (+0.25\text{ V}) = \boxed{+0.56\text{ V}}$$

$$\Delta_r G^{\oplus} = -\nu F E_{cell}^{\oplus}\,[5.12] = -4 \times (96.485\text{ kC mol}^{-1}) \times (+0.56\text{ V}) = \boxed{-21\bar{6}\text{ kJ mol}^{-1}}\ [1\text{ kC V} = 1\text{ kJ}]$$

$$K = e^{-\Delta_r G^{\oplus}/RT}\,[4.10b] = e^{-(-21\bar{6}\text{ kJ mol}^{-1})/\{(0.0083145\text{ kJ K}^{-1}\text{ mol}^{-1})\times(298\text{ K})\}} = \boxed{7.3 \times 10^{37}}$$

E5.24 (a) Overall cell reaction: $H_2(g) + \frac{1}{2}\,O_2(g) \rightarrow H_2O(l)$ $\quad \nu = 2$

$\Delta_r G^{\ominus} = \Delta_f G^{\ominus}(H_2O, l) = -237.13\text{ kJ mol}^{-1}$ [*Resource section 3*: *Data*]

$$E_{cell}^{\ominus} = -\frac{\Delta_r G^{\ominus}}{\nu F}\,[5.12] = \frac{+237.13\text{ kJ mol}^{-1}}{2 \times (96.485\text{ kC mol}^{-1})} = \boxed{+1.23\text{ V}}$$

(b) Overall cell reaction: $C_4H_{10}(g) + \frac{13}{2}\,O_2(g) \rightarrow 4\,CO_2(g) + 5\,H_2O(l)$

$$\begin{aligned}\Delta_f G^{\ominus} &= 4\Delta_f G^{\ominus}(CO_2, g) + 5\Delta_f G^{\ominus}(H_2O, l) - \Delta_f G^{\ominus}(C_4H_{10}, g)\\ &= [(4) \times (-394.36) + (5) \times (-237.13) - (-17.03)]\text{ kJ mol}^{-1}\ [\textit{Resource section 3}: \textit{Data}]\\ &= -2746.06\text{ kJ mol}^{-1}\end{aligned}$$

In this reaction the number of electrons transferred, ν, is not immediately apparent as in part (a). To find ν we break the cell reaction down into half-reactions as follows.

R: $\frac{13}{2}\,O_2(g) + 26\,e^- + 26\,H^+(aq) \rightarrow 13\,H_2O(l)$

L: $4\,CO_2(g) + 26\,e^- + 26\,H^+(aq) \rightarrow C_4H_{10}(g) + 8\,H_2O(l)$

Overall (R–L): $C_4H_{10}(g) + \frac{13}{2}\,O_2(g) \rightarrow 4\,CO_2(g) + 5\,H_2O(l)$

Hence, $\nu = 26$.

Therefore, $E^{\ominus}_{\text{cell}} = -\dfrac{\Delta_r G^{\ominus}}{\nu F} = \dfrac{+2746.06\ \text{kJ mol}^{-1}}{(26)\times(96.485\ \text{kC mol}^{-1})} = \boxed{+1.09\ \text{V}}$

An alternative, and perhaps faster, method for finding ν involves comparing the oxidation number of carbon in butane with that of carbon in carbon dioxide. In C_4H_{10} each carbon has an oxidation number of $-\frac{5}{2}$ while the carbon of CO_2 has an oxidation number of +4. This is a change of $+\frac{13}{2}$ for each carbon or +26 for the four carbons of the overall reaction. Thus, a mole of butane has lost the equivalent of 26 moles of electrons and $\nu = 26$.

E5.25 The half-reaction and standard potential for the electrode are: $2\,H^+(aq) + 2\,e^- \rightarrow H_2(g)$ and $E^{\ominus} = 0$.

The Nernst equation applies to half-cells as well as cells so at any hydrogen partial pressure and HBr(aq) concentration c we have the relation:

$$\begin{aligned} E &= E^{\ominus} - \frac{RT}{\nu F}\ln Q\ [5.13] \\ &= E^{\ominus} - \frac{25.693\ \text{mV}}{\nu}\ln Q \\ &= E^{\ominus} - \frac{25.693\ \text{mV}}{2}\ln(p_{H_2}/a_{H^+}^2) = E^{\ominus} - (12.85\ \text{mV})\times\ln(p_{H_2}/a_{H^+}^2) \\ &= E^{\ominus} - (12.85\ \text{mV})\times\ln(p_{H_2}/c^2) \quad \text{because } a_{H^+} = \gamma_{H^+}c \simeq c. \end{aligned}$$

Thus, at constant p_{H_2} the difference between the potential when $c_i = 5.0$ mmol dm^{-3} and the potential when $c_f = 25.0$ mmol dm^{-3} is:

$$\begin{aligned} \Delta E &= E_f - E_i \\ &= \{E^{\ominus} - (12.85\ \text{mV})\times\ln(p_{H_2}/c_f^2)\} - \{E^{\ominus} - (12.85\ \text{mV})\times\ln(p_{H_2}/c_i^2)\} \\ &= (12.85\ \text{mV})\times\ln(c_f/c_i)^2 = (25.693\ \text{mV})\times\ln(c_f/c_i) \\ &= (25.693\ \text{mV})\times\ln(25.0/5.0) = \boxed{41\ \text{mV}} \end{aligned}$$

E5.26 Hydrogen electrode: $\text{Pt}|H_2(\text{g, 1 bar})|H^+(aq)$ $\quad H^+(aq) + e^- \rightleftharpoons \frac{1}{2}H_2(\text{g, 1 bar}) \quad \nu = 1$

$$\begin{aligned} E &= E^{\ominus}_{\text{SHE}} - \frac{RT}{\nu F}\ln Q = -\frac{RT}{F}\ln Q\ [E^{\ominus}_{\text{SHE}} = 0,\ \nu = 1] \\ &= -\frac{RT}{F}\ln\frac{1}{a_{H^+}} = -\left(\frac{RT}{F}\right)\times(\ln 10)\times\log\frac{1}{a_{H^+}} = -\left(\frac{RT}{F}\right)\times(\ln 10)\times\text{pH} \\ &= -(59.16\ \text{mV})\text{pH} \end{aligned}$$

A drop of 1 pH unit causes the potential of the hydrogen electrode to increase by 59.16 mV.

Lactic acid is a weak acid ($K_a = 8.4\times10^{-4}$, *Resource section 3: Data*) so an equilibrium computation is required to find the pH of a solution that has a lactic acid preparation concentration of c. We make the approximation that activity coefficients equal 1.

$$\text{Lactic acid(aq)} \rightleftharpoons \text{lactate(aq)} + H^+(aq)$$

$$K_a = \frac{[\text{lactate}][H^+]}{[\text{lactic acid}]} = \frac{[H^+]^2}{c - [H^+]}$$

$$[H^+]^2 + K_a[H^+] - K_a c = 0$$

$$[H^+] = \tfrac{1}{2}\left(-K_a + \sqrt{K_a^2 + 4K_a c}\right)$$

We now draw a table of computations of $[H^+]$, pH, and E at each value of the formal lactic acid concentration c. (Remember that $c^{\ominus} = 1$ mol dm^{-3}).

c/mmole dm^{-3}	5	25
$[H^+]$/mmole dm^{-3}	1.67	4.18
pH	2.78	2.38
E/mV	−164	−141

Thus, the hydrogen electrode potential changes by $\{(-141)-(-164)\}\text{mV} = \boxed{+24\text{ mV}}$ when the lactic acid concentration changes from 5.0 mmol dm^{-3} to 25.0 mmol dm^{-3}.

E5.27 (a) R: $2\,H^+(aq) + 2\,e^- \rightarrow H_2(g, p_R)$

L: $2\,H^+(aq) + 2\,e^- \rightarrow H_2(g, p_L)$

R–L: $H_2(g, p_L) \rightarrow H_2(g, p_R)$

(b) R: $Br_2(l) + 2\,e^- \rightarrow 2\,Br^-(aq)$

L: $Cl_2(g) + 2\,e^- \rightarrow 2\,Cl^-(aq)$

R–L: $Br_2(l) + 2\,Cl^-(aq) \rightarrow Cl_2(g) + 2\,Br^-(aq)$

(c) R:

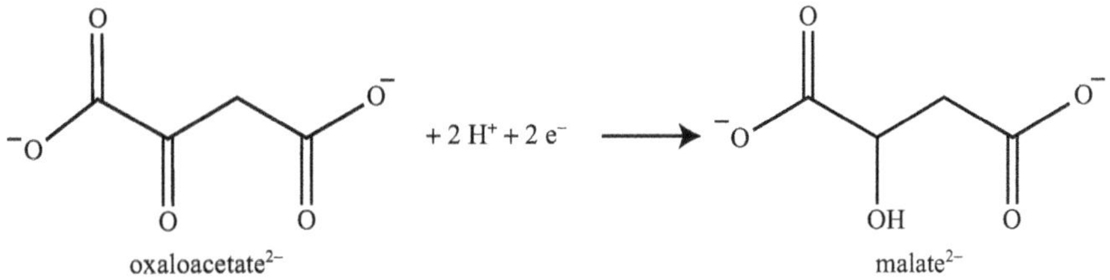

L: $NAD^+(aq) + H^+(aq) + 2\,e^- \rightarrow NADH(aq)$

R−L: oxaloacetate^{2-}(aq) + NADH(aq) + H^+(aq) → malate^{2-}(aq) + NAD^+(aq)

(d) R: $MnO_2(s) + 4\,H^+(aq) + 2\,e^- \rightarrow Mn^{2+}(aq) + 2\,H_2O(l)$

L: $Fe^{2+}(aq) + 2\,e^- \rightarrow Fe(s)$

R–L: $Fe(s) + MnO_2(s) + 4\,H^+(aq) \rightarrow Fe^{2+}(aq) + Mn^{2+}(aq) + 2\,H_2O(l)$

E5.28 (a) $E_{cell} = E^{\ominus}_{cell} - \dfrac{RT}{2F}\ln\dfrac{p_R}{p_L}$, where $E^{\ominus}_{cell} = E^{\ominus}_R - E^{\ominus}_L = \boxed{0}$ (Same electrode on right and left.)

In the following Nernst equations involving ions in aqueous solution we have replaced activities with molar concentrations

(b) $E_{cell} = E^{\ominus}_{cell} - \dfrac{RT}{2F}\ln\left(\dfrac{p_{Cl_2}[Br^-]^2}{[Cl^-]^2}\right)$, where $E^{\ominus}_{cell} = E^{\ominus}_R - E^{\ominus}_L = 1.09\text{ V} - 1.36\text{ V} = \boxed{-0.27\text{ V}}$

This is a non-spontaneous cell reaction under standard conditions.

(c) $E_{cell} = E^{\ominus}_{cell} - \dfrac{RT}{2F}\ln\left(\dfrac{[\text{malate}^{2-}][NAD^+]}{[\text{oxaloacetate}^{2-}][NADH][H^+]}\right)$

where $E^{\ominus}_{cell} = E^{\ominus}_R - E^{\ominus}_L = (-0.17\text{ V}) - (-0.32\text{ V}) = \boxed{+0.15\text{ V}}$

(d) $E_{\text{cell}} = E^{\circ}_{\text{cell}} - \frac{RT}{2F}\ln\left(\frac{[Fe^{2+}][Mn^{2+}]}{[H^+]^4}\right)$, where $E^{\circ}_{\text{cell}} = E^{\circ}_{R} - E^{\circ}_{L} = 1.23\ \text{V} - (-0.44\ \text{V}) = \boxed{+1.67\ \text{V}}$

E5.29 (a) R: $NAD^+ + 2\,e^- + H^+ \rightarrow NADH$

L: $CH_3CHO + 2\,e^- + 2\,H^+ \rightarrow CH_3CH_2OH$

Overall (R−L): $CH_3CH_2OH(aq) + NAD^+(aq) \rightarrow CH_3CHO(aq) + NADH(aq) + H^+(aq)$ $\boxed{\nu = 2}$

The cell arrangement is

$Pt|CH_3CH_2OH(aq), CH_3CHO(aq), H^+(aq)||H^+(aq), NAD^+(aq), NADH(aq)|Pt$

(b) The activity coefficient of the $H^+(aq)$ ion should change as the solution concentration of $MgATP^{2-}$ is changed and this could provide insight into the association of Mg^{2+} and ATP^{4-}.

R: $H^+(aq,c^{\oplus}_{R}) + MgATP^{2-}(c_R) + e^- \rightarrow \frac{1}{2}H_2(g,p^{\circ}) + MgATP^{2-}(c_R)$

L: $H^+(aq,c^{\oplus}_{L}) + MgATP^{2-}(c_L) + e^- \rightarrow \frac{1}{2}H_2(g,p^{\circ}) + MgATP^{2-}(c_L)$

Overall (R−L): $H^+(aq,c^{\oplus}_{R}) \rightarrow H^+(aq,c^{\oplus}_{L})$ $\boxed{\nu = 1}$

The cell arrangement is

$Pt|H_2(p^{\circ})|c^{\oplus}_{H^+}, MgATP^{2-}(c_L)||MgATP^{2-}(c_R), c^{\oplus}_{H^+}|H_2(p^{\circ})|Pt$

(c) R: $CH_3COCO_2^-(aq) + 2\,e^- + 2\,H^+(aq) \rightarrow CH_3CH(OH)CO_2^-(aq)$

L: cyt-c(ox, aq) + $e^- \rightarrow$ cyt-c(red, aq)

The overall reaction is obtained from R − 2 × L. $\boxed{\nu = 2}$

The cell arrangement:

$Pt|$cyt-c(red, aq), cyt-c(ox, aq)$||H^+(aq), CH_3CH(OH)CO_2^-(aq), CH_3COCO_2^-(aq)|Pt$

E5.30 The standard potentials are found in the *Reference section 3: Data.*

(a) $E^{\oplus}_{\text{cell}} = E^{\oplus}_{NAD^+/NADH} - E^{\oplus}_{CH_3CHO/CH_3CH_2OH} = (-0.32\ \text{V}) - (-0.20\ \text{V}) = \boxed{-0.11\ \text{V}}$

The Nernst equation can be used to find the standard potential when the biological standard potential is known. Simply evaluate Q giving all substances other than H^+ an activity of 1 while H^+ is assigned an activity of 1×10^{-7}; call this $Q^{\oplus}$.

$$E^{\oplus}_{\text{cell}} = E^{\circ}_{\text{cell}} - \frac{RT}{\nu F}\ln Q^{\oplus} \quad \text{[Nernst equation evaluated at the biological standard state.]}$$

Thus, at 298 K

$$E^{\circ}_{\text{cell}} = E^{\oplus}_{\text{cell}} + \frac{RT}{2F}\ln a_{H^+} = -0.11\ \text{V} + \frac{0.025693\ \text{V}}{2}\ln 10^{-7} = \boxed{-0.32\ \text{V}}$$

(b) $E^{\oplus}_{\text{cell}} = E^{\circ}_{\text{cell}} = E^{\circ}_{R} - E^{\circ}_{L} = \boxed{0}$ (Hydrogen electrode at both R and L.)

(c) $E^{\oplus}_{\text{cell}} = E^{\oplus}_{CH_3COCO_2^-/CH_3CH(OH)CO_2^-} - E^{\oplus}_{\text{Cyt-c(ox)/Cyt-c(red)}} = (-0.18\ \text{V}) - (+0.25\ \text{V}) = \boxed{-0.43\ \text{V}}$

At 298 K,

$$E^{\circ}_{\text{cell}} = E^{\oplus}_{\text{cell}} + \frac{RT}{2F}\ln\frac{1}{a^2_{H^+}} = -0.43\ \text{V} + \frac{0.025693\ \text{V}}{2}\ln\frac{1}{(10^{-7})^2} = \boxed{-0.02\ \text{V}}$$

E5.31 $MnO_4^- + 8\,H^+ + 5\,e^- \rightarrow Mn^{2+} + 4\,H_2O \quad E^\ominus = 1.51\text{ V at }25°C$

(a) Using the Nernst equation to determine the reduction potential for pH = 6.00, keeping $a_{MnO_4^-}$ and $a_{Mn^{2+}} = 1.00$.

$$E_{cell} = E^\ominus_{cell} - \frac{25.693\text{ mV}}{\nu}\ln Q\,[5.13], \quad \text{where} \quad \ln Q = \ln\frac{1}{a^8_{H^+}} = 8\times\ln(10)\times\text{pH}$$

Thus, $E_{cell} = 1.51\text{ V} - \dfrac{8\times(59.160\text{ mV})\times 6.00}{5} = \boxed{+0.94\text{ V}}$

(b) In general, $\boxed{E_{cell}/\text{V} = 1.51 - 0.094656\times\text{pH}}$

E5.32 (a) $E_{cell} = E^\ominus_{cell} - \dfrac{RT}{2F}\ln\dfrac{p_R}{p_L}$; E_{cell} $\boxed{\text{increases}}$ as p_L increases.

(b) $E_{cell} = E^\ominus_{cell} - \dfrac{RT}{2F}\ln\left(\dfrac{p_{Cl_2}[Br^-]^2}{[Cl^-]^2}\right)$; E_{cell} $\boxed{\text{increases}}$ as [HCl] increases because $[Cl^-]$ increases.

(c) $E_{cell} = E^\ominus_{cell} - \dfrac{RT}{2F}\ln\left(\dfrac{[\text{malate}^{2-}][NAD^+]}{[\text{oxaloacetate}^{2-}][NADH][H^+]}\right)$; E_{cell} $\boxed{\text{increases}}$ as acid is added to both R and L.

(d) $E_{cell} = E^\ominus_{cell} - \dfrac{RT}{2F}\ln\left(\dfrac{[Fe^{2+}][Mn^{2+}]}{[H^+]^4}\right)$; E_{cell} $\boxed{\text{increases}}$ as acid is added to R.

E5.33 (a) The Nernst equation is

$$E_{cell} = E^\ominus_{cell} - \frac{RT}{2F}\ln\frac{[CH_3CHO][NADH][H^+]}{[CH_3CH_2OH][NAD^+]}$$

Increasing pH implies decreasing $[H^+]$; therefore the cell potential $\boxed{\text{decreases}}$.

(b) The Nernst equation is

$$E_{cell} = E^\ominus_{cell} - \frac{RT}{2F}\ln\frac{[MgATP^{2-}]}{[ATP^{4-}][Mg^{2+}]}$$

Increasing $[Mg^{2+}]$ from $MgSO_4$ $\boxed{\text{increases}}$ the cell potential.

(c) The Nernst equation is

$$E_{cell} = E^\ominus_{cell} - \frac{RT}{2F}\ln\frac{[\text{Cyt-c(ox)}][CH_3CH(OH)CO_2^-]}{[\text{Cyt-c(red)}][CH_3COCO_2^-][H^+]^2}$$

Increasing $[CH_3CH(OH)CO_2^-]$ $\boxed{\text{decreases}}$ the cell potential.

E5.34 R: $Tl^+(aq) + e^- \rightarrow Tl(s) \quad E^\ominus = -0.34\text{ V}$

L: $Hg^{2+}(aq) + 2\,e^- \rightarrow Hg(l) \quad E^\ominus = +0.86\text{ V}$

(a) $E^\ominus_{cell} = E^\ominus_R - E^\ominus_L = \boxed{-1.20\text{ V}}$

(b) Overall (2R−L): $2\,Tl^+(aq) + Hg(l) \rightarrow 2\,Tl(s) + Hg^{2+}(aq)$

Replacing activities by molar concentrations, we have $Q = \frac{[Hg^{2+}]}{[Tl^+]^2}$

$$\begin{aligned} E_{cell} &= E^{\ominus}_{cell} - \frac{RT}{\nu F}\ln Q = E^{\ominus} - \frac{25.693\text{ mV}}{\nu}\ln Q \\ &= -1.20\text{ V} - \frac{25.693\text{ mV}}{2} \times \ln\frac{0.150}{(0.93)^2} \\ &= -1.20\text{ V} + 0.023\text{ V} = \boxed{-1.18\text{ V}} \end{aligned}$$

E5.35 (a) $2\,\text{NADH(aq)} + O_2\text{(g)} + 2\,H^+\text{(aq)} \rightarrow 2\,\text{NAD}^+\text{(aq)} + 2\,H_2O\text{(l)} \quad E^{\oplus} = +1.14\text{ V}$

The oxidation number of each oxygen atom changes from 0 on the left to −2 on the right. Since there are two oxygen atoms undergoing this change, we deduce that $\nu = 4$.

$$\begin{aligned} \Delta_r G^{\oplus} &= -\nu F E^{\oplus}_{cell}\ [5.14] \\ &= -4 \times (96.485\text{ kC mol}^{-1}) \times (+1.14\text{ V}) \\ &= \boxed{-440\text{ kJ mol}^{-1}} \end{aligned}$$

(b) $\text{Malate(aq)} + \text{NAD}^+\text{(aq)} \rightarrow \text{oxaloacetate(aq)} + \text{NADH} + H^+\text{(aq)} \quad E^{\oplus} = -0.154\text{ V}$

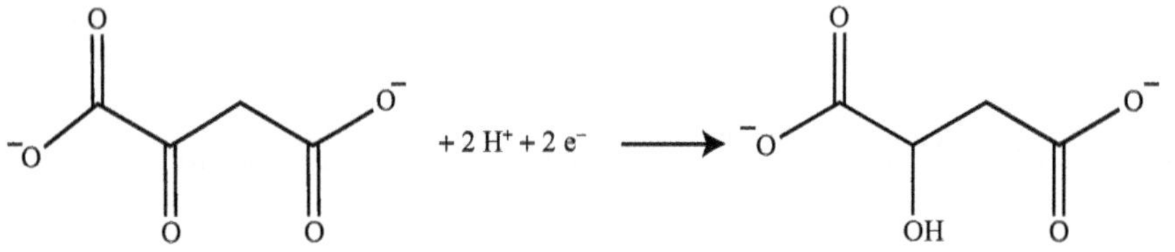

The half-reaction for the reduction of oxaloacetate^{2-} to malate^{2-} shows that $\nu = 2$. As an alternative method for finding that $\nu = 2$ we see that NAD$^+$ gains H$^-$, which is equivalent to 2 e$^-$, to become NADH.

$$\begin{aligned} \Delta_r G^{\oplus} &= -\nu F E^{\oplus}_{cell}\ [5.14] \\ &= -2 \times (96.485\text{ kC mol}^{-1}) \times (-0.154\text{ V}) \\ &= \boxed{+29.7\text{ kJ mol}^{-1}} \end{aligned}$$

(c) $O_2\text{(g)} + 4\,H^+\text{(aq)} + 4\,e^- \rightarrow 2\,H_2O\text{(l)} \quad E^{\oplus} = +0.81\text{ V} \quad \nu = 4$

$$\begin{aligned} \Delta_r G^{\oplus} &= -\nu F E^{\oplus}_{cell}\ [5.14] \\ &= -4 \times (96.485\text{ C mol}^{-1}) \times (+0.81\text{ V}) \\ &= \boxed{-313\text{ kJ mol}^{-1}} \end{aligned}$$

E5.36 (a) $\text{AgCl(s)} + e^- \rightarrow \text{Ag(s)} + \text{Cl}^-\text{(aq)}$

(b) R: $\text{Ag}^+\text{(aq)} + e^- \rightarrow \text{Ag(s)} \qquad E^{\ominus}(\text{Ag}^+/\text{Ag}) = +0.80\text{ V}$

L: $\text{AgCl(s)} + e^- \rightarrow \text{Ag(s)} + \text{Cl}^-\text{(aq)} \qquad E^{\ominus}(\text{AgCl/Ag, Cl}^-) = +0.22\text{ V}$

Overall (R−L): $\text{Ag}^+\text{(aq)} + \text{Cl}^-\text{(aq)} \rightarrow \text{AgCl(s)} \quad E^{\ominus}_{cell} = E^{\ominus}_R - E^{\ominus}_L = +0.58\text{ V} \quad \nu = 1$

The potential under non-standard conditions is calculated with the Nernst equation.

$$\begin{aligned} E_{cell} &= E^{\ominus}_{cell} - \frac{RT}{\nu F}\ln\left(\frac{1}{b_{Ag^+,R}\,b_{Cl^-,L}}\right)\ [5.13] = E^{\ominus}_{cell} + \frac{RT}{\nu F}\ln(b_{Ag^+,R}\,b_{Cl^-,L}) \\ &= +0.58\text{ V} + (0.025693\text{ V}) \times \ln(\{0.010\} \times \{0.025\}) = \boxed{+0.37\text{ V}} \end{aligned}$$

E5.37 R: $O_2(g) + 4\,H^+(aq) + 4\,e^- \rightarrow 2\,H_2O$ $E_R^{\ominus} = +0.81$ V at 298.15 K [Table 5.2]

L: $\text{cystine(aq)} + 2\,H^+(aq) + 2\,e^- \rightarrow 2\,\text{cysteine(aq)}$ $E_L^{\ominus} = -0.34$ V

R−L: $4\,\text{cysteine(aq)} + O_2(g) \rightarrow 2\,\text{cystine(aq)} + 2\,H_2O(l)$ $E_{cell}^{\ominus} = E_R^{\ominus} - E_L^{\ominus} = \boxed{+1.15\text{ V}}$ $\nu = 4$

(a) At 25°C, $\Delta_r G^{\ominus} = -\nu F E^{\ominus}$ [5.14] $= -4 \times (96.485\text{ kC mol}^{-1}) \times (1.15\text{ V}) = \boxed{-444\text{ kJ mol}^{-1}}$

Assuming that the aqueous phase reaction entropy changes balance to a value that is negligibly small compared to the entropy change of gas consumption, the standard reaction entropy is given by

$$\begin{aligned}\Delta_r S^{\ominus} &= 2\,S_m^{\ominus}(\text{cystine, aq}) + 2\,S_m^{\ominus}(H_2O, l) - 4\,S_m^{\ominus}(\text{cysteine, aq}) - S_m^{\ominus}(O_2, g) \quad [2.9]\\ &\approx -S_m^{\ominus}(O_2, g)\\ &\approx -205\text{ J K}^{-1}\text{ mol}^{-1} \quad [\textit{Resource section 3: Data}]\end{aligned}$$

$$\begin{aligned}\Delta_r H^{\ominus} &= \Delta_r G^{\ominus} + T\Delta_r S^{\ominus} \; [5.19]\\ &\approx -444\text{ kJ mol}^{-1} + (298.15\text{ K}) \times (-205\text{ J K}^{-1}\text{ mol}^{-1}) \approx \boxed{-505\text{ kJ mol}^{-1}}\end{aligned}$$

(b) Upon a temperature increase from 25°C to 35°C, we do not expect $\Delta_r S^{\ominus}$ to change significantly, so the change in $\Delta_r G^{\ominus}$ is given by

$$\begin{aligned}\text{Change in } \Delta_r G^{\ominus} &= -(\text{change in } T) \times \Delta_r S^{\ominus} \; [4.17]\\ &= -(10\text{ K}) \times (-205\text{ J K}^{-1}\text{ mol}^{-1}) = 2.05\text{ kJ mol}^{-1}\end{aligned}$$

$$\begin{aligned}\Delta_r G^{\ominus}(308.15\text{ K}) &= \Delta_r G^{\ominus}(298.15\text{ K}) + \text{change in } \Delta_r G^{\ominus}\\ &\approx (-444\text{ kJ mol}^{-1}) + (2.05\text{ kJ mol}^{-1}) \approx \boxed{-442\text{ kJ mol}^{-1}}\end{aligned}$$

Since the overall reaction equilibrium is independent of pH,

$$\Delta_r G^{\oplus}(308.15\text{ K}) = \Delta_r G^{\ominus}(298.15\text{ K}) \approx \boxed{-442\text{ kJ mol}^{-1}}$$

E5.38 The pyruvic acid/lactic acid couple (HP/HL) is

$$CH_3COCOOH + 2\,H^+ + 2\,e^- \rightarrow CH_3CH(OH)COOH \quad E^{\ominus} = -0.19\text{ V} \quad \nu = 2$$

The Nernst equation for the couple is

$$E_{cell} = E_{cell}^{\ominus} - \frac{RT}{2F}\ln\frac{a_{HL}}{a_{HP}a_{H^+}^2} = E_{cell}^{\ominus} - \frac{RT}{2F}\ln\frac{a_{HL}}{a_{HP}} + \frac{RT}{F}\ln a_{H^+} \quad \text{or} \quad E_{cell}^{\ominus} = E_{cell} + \frac{RT}{2F}\ln\frac{a_{HL}}{a_{HP}} - \frac{RT}{F}\ln a_{H^+}.$$

At the biological standard state $a_{HL} = a_{HP} = 1$ and $a_{H^+} = 1 \times 10^{-7}$ (pH = 7) so evaluation of the Nernst equation at the state $E = E^{\ominus}$ gives

$$\begin{aligned}E^{\oplus} &= E^{\ominus} + \frac{RT}{2F}\ln 1 - \frac{RT}{F}\ln(1 \times 10^{-7})\\ &= -0.19\text{ V} - (0.025693\text{ V})\ln(1 \times 10^{-7})\\ &= \boxed{+0.22\text{ V}}\end{aligned}$$

E5.39 R: $O_2(g) + 4\,H^+(aq) + 4\,e^- \rightarrow 2\,H_2O$ $E_R^{\ominus} = +0.81$ V at 298 K [Table 5.2]

L: $\text{cystine(aq)} + 2\,H^+(aq) + 2\,e^- \rightarrow 2\,\text{cysteine(aq)}$ $E_L^{\ominus} = -0.34$ V

R−L: $4\,\text{cysteine(aq)} + O_2(g) \rightarrow 2\,\text{cystine(aq)} + 2\,H_2O(1)$ $E_{cell}^{\ominus} = E_R^{\ominus} - E_L^{\ominus} = +1.15$ V $\nu = 4$

At 25°C, $\Delta_r G^{\ominus} = -\nu F E_{cell}^{\ominus}$ [5.14] $= -4 \times (96.485\text{ kC mol}^{-1}) \times (1.15\text{ V}) = -444\text{ kJ mol}^{-1}$

Assuming that the aqueous phase reaction entropy changes balance to a value that is negligibly small compared to the entropy change of gas consumption, and recognizing that the reaction entropy is independent of pH, the standard reaction entropy is given by

$$\begin{aligned}\Delta_r S^{\oplus} = \Delta_r S^{\ominus} &= 2\,S^{\ominus}_m(\text{cystine, aq}) + 2\,S^{\ominus}_m(H_2O, l) - 4\,S^{\ominus}_m(\text{cysteine, aq}) - S^{\ominus}_m(O_2, g) \quad [2.9]\\ &\approx -S^{\ominus}_m(O_2, g)\\ &\approx -205\ \text{J K}^{-1}\ \text{mol}^{-1}\ [\textit{Resource section 3: Data}]\end{aligned}$$

Upon a temperature increase from 25°C to 35°C, we do not expect $\Delta_r S^{\ominus}$ to change significantly, so the change in $\Delta_r G^{\ominus}$ is given by

$$\begin{aligned}\text{Change in } \Delta_r G^{\oplus} &= -(\text{change in } T) \times \Delta_r S^{\oplus}\ [4.18]\\ &= -(310\ \text{K} - 298.15\ \text{K}) \times (-205\ \text{J K}^{-1}\ \text{mol}^{-1}) = +2.43\ \text{kJ mol}^{-1}\end{aligned}$$

$$\begin{aligned}\Delta_r G^{\oplus}(310\ \text{K}) &= \Delta_r G^{\oplus}(298.15\ \text{K}) + \text{change in } \Delta_r G^{\oplus}\\ &\approx (-444\ \text{kJ mol}^{-1}) + (2.43\ \text{kJ mol}^{-1}) \approx -442\ \text{kJ mol}^{-1}\end{aligned}$$

Thus, at 310 K

$$E^{\oplus}_{\text{cell}} = -\Delta_r G^{\oplus}/\nu F\ [5.14] = -(-442\ \text{kJ mol}^{-1})/(4 \times 96.485\ \text{kC mol}^{-1}) = \boxed{+1.15\ \text{V}}$$

We have found that there is no significant difference between the values of $\Delta_r G^{\oplus}$ and $E^{\oplus}_{\text{cell}}$ at 298.15 K and the values at 310 K.

To find the value of $E^{\ominus}_L$, we recognize that $E^{\oplus}_{\text{cell}} = E^{\oplus}_R - E^{\oplus}_L = E^{\ominus}_R - E^{\ominus}_L$ because the pH dependence of the electrodes vanishes in the subtraction in this particular example. Thus,

$$E^{\ominus}_L = E^{\ominus}_R - E^{\oplus}_{\text{cell}} = (+1.23\ \text{V}) - (+1.15\ \text{V}) = \boxed{+0.08\ \text{V}}\ [\text{Table 5.1}]$$

The Nernst equation can be used to find the biological standard potentials when the standard potentials are known. Simply evaluate Q giving all substances other than H^+ an activity of 1 while H^+ is assigned an activity of 1×10^{-7}; call this $Q^{\oplus}$.

$$E^{\oplus} = E^{\ominus} - \frac{RT}{\nu F}\ln Q^{\oplus} \quad [\text{Nernst equation evaluated at the biological standard state.}]$$

Thus, at 298 K

$$E^{\oplus}_R = E^{\ominus}_R - \frac{RT}{4F}\ln\frac{1}{a^4_{H^+}} = +1.23\ \text{V} - \frac{0.025693\ \text{V}}{4}\ln\frac{1}{(10^{-7})^4} = \boxed{+0.82\ \text{V}}$$

$$E^{\oplus}_L = E^{\ominus}_L - \frac{RT}{2F}\ln\frac{1}{a^2_{H^+}} = +0.08\ \text{V} - \frac{0.025693\ \text{V}}{2}\ln\frac{1}{(10^{-7})^2} = \boxed{-0.33\ \text{V}}$$

E5.40 (a) R: coenzyme Q + 2 H^+ + 2 e^- → coenzyme QH_2 $\quad E^{\oplus}_R = +0.04$ V at 298 K [Table 5.2]

L: FAD + 2 H^+ + 2 e^- → $FADH_2$ $\quad E^{\oplus}_L = -0.22$ V

$$E^{\oplus}_{\text{cell}} = E^{\oplus}_R - E^{\oplus}_L = +0.26\ \text{V}$$

Since $E^{\oplus}_{\text{cell}} > 0$, $FADH_2$ has a thermodynamic tendency to reduce coenzyme Q under standard conditions.

(b) R: Fe^{3+}(cyt *b*) + e^- → Fe^{2+}(cyt *b*) $\quad E^{\oplus}_R = +0.08$ V at 298 K [Table 5.2]

L: Fe^{3+}(cyt *f*) + e^- → Fe^{2+}(cyt *f*) $\quad E^{\oplus}_R = +0.36$ V

$$E^{\oplus}_{\text{cell}} = E^{\oplus}_R - E^{\oplus}_L = -0.28\ \text{V}$$

Since $E^{\oplus}_{\text{cell}} < 0$, Fe^{3+}(cyt *b*) does not have a thermodynamic tendency to oxidize Fe^{2+}(cyt *f*) under standard conditions. The reverse reaction is spontaneous.

E5.41 The more positive the difference $E(R\bullet/R{:}) - E(\text{antioxidant}_{ox}/\text{antioxidant}_{red})$, the greater the thermodynamic tendency to reduce the radical R•. Thus, the most negative value of the biological standard potential of the antioxidant group is used to identify the best antioxidant from the thermodynamic view. The biological standard potentials for the couples involving ascorbic acid, reduced glutathione, reduced lipoic acid, and reduced coenzyme Q are +0.08, −0.23, −0.29, and +0.04 V, respectively, so $\boxed{\text{reduced lipoic acid}}$ is the best antioxidant of this group.

E5.42 pyruvate$^-$ + succinate^{2-} → lactate$^-$ + fumarate^{2-} $\quad \nu = 2$

$E_r^\oplus = E^\oplus(\text{pyruvate/lactate}) - E^\oplus(\text{fumarate/succinate}) = (-0.19\text{ V}) - (+0.03\text{ V}) = -0.22\text{ V}$

$\ln K = \nu F E_r^\oplus/RT$ [5.16] $= 2\times(-0.22\text{ V})/(0.025693\text{ V}) = -17.\bar{1}$ at 25°C

$K = e^{-17.\bar{1}} = \boxed{3.7\times10^{-8}}$

E5.43 R: $Fe^{3+}(\text{cyt }c) + e^- \rightarrow Fe^{2+}(\text{cyt }c)$ $\quad E_R^\oplus = +0.25\text{ V}$ at 298 K [Table 5.2]

L: $^-O_2CCHO(\text{glyoxylate}) + 2\,H^+ + 2\,e^- \rightarrow {}^-O_2CCH_2OH(\text{glycolate})$ $\quad E_L^\oplus = ?$

Overall (2R−L): 2 cyt-c(ox) + glycolate$^-$(aq) → 2 cyt-c(red) + glyoxylate$^-$(aq) + 2 H^+(aq) $\quad \nu = 2$

$K_{298\text{ K}} = 2.14\times10^{11}$ at pH = 7

(a) $E_\text{cell}^\oplus = (RT/\nu F)\ln K$ [5.16]
$= (0.025693\text{ V}/2)\ln(2.14\times10^{11})$
$= \boxed{+0.34\text{ V}}$

(b) $E_\text{cell}^\oplus = E_R^\oplus - E_L^\oplus = E^\oplus(\text{cyt-}c(\text{ox})/\text{cyt-}c(\text{ox})) - E^\oplus(\text{glyoxylate/glycolate})$

$E^\oplus(\text{glyoxylate/glycolate}) = E^\oplus(\text{cyt-}c(\text{ox})/\text{cyt-}c(\text{ox})) - E_\text{cell}^\oplus$
$= (+0.25\text{ V}) - (+0.34\text{ V})$ [*Resource section 3: Data*]
$= \boxed{-0.09\text{ V}}$

E5.44 (a) L: $HCO_3^-(aq) + e^- \rightarrow CO_3^{2-}(aq) + \tfrac{1}{2}H_2(g)$ $\quad \nu = 1$

$\Delta_L G^\ominus = \Delta_f G^\ominus(CO_3^{2-}, aq) + \Delta_f G^\ominus(H_2, g) - \Delta_f G^\ominus(HCO_3^-, aq)$
$= (-527.81\text{ kJ mol}^{-1}) + 0 - (-586.77\text{ kJ mol}^{-1})$
$= +58.96\text{ kJ mol}^{-1}$

$$E_L^\ominus = -\frac{\Delta_L G^\ominus}{\nu F} = -\frac{58.96\text{ kJ mol}^{-1}}{1\times(96.485\text{ kC mol}^{-1})}$$
$= \boxed{-0.6111\text{ V}}$

(b) We now wish to calculate the standard potential of a cell in which the cell net ionic reaction is

$CO_3^{2-}(aq) + H_2O(l) \rightarrow HCO_3^-(aq) + OH^-(aq)$.

This is the **base dissociation reaction** of CO_3^{2-} and it is the net ionic reaction for the combined L half-reaction presented in part (a) and the half-reaction

R: $H_2O(l) + e^- \rightarrow \tfrac{1}{2}H_2(g) + OH^-(aq)$ $\quad E_R^\ominus = -0.83\text{ V}$ $\quad \nu = 1$.

Thus, the cell standard potential is

$E_\text{cell}^\ominus = E_R^\ominus - E_L^\ominus = -0.83\text{ V} - (-0.61) = \boxed{-0.22\text{ V}}$.

Notice that we can now calculate $\Delta_R G^\ominus$ and the basicity constant K_b of CO_3^{2-}.

$$\Delta_r G^{\ominus} = -\nu F E^{\ominus}_{\text{cell}}$$
$$= -1 \times (96.485\ \text{kC mol}^{-1}) \times (-0.22\ \text{V})$$
$$= +21\ \text{kJ mol}^{-1}$$

$$K_b = e^{-\Delta_r G^{\ominus}/RT} = e^{-(21\times 10^3\ \text{J mol}^{-1})/\{(8.31447\ \text{J K}^{-1}\ \text{mol}^{-1})\times(298.15\ \text{K})\}} = 2.1 \times 10^{-4}$$

Also, $pK_b = -\log K_b = -\log(2.1 \times 10^{-4}) = 3.68$.

This value of pK_b is used in part (e) to calculate pK_a of HCO_3^-.

(c) The Nernst equation for the cell of part (b) is

$$E_{\text{cell}} = E^{\ominus}_{\text{cell}} - \frac{RT}{\nu F} \ln Q\ [5.13].$$

Thus,

$$\boxed{E_{\text{cell}} = E^{\ominus}_{\text{cell}} - \frac{RT}{F} \ln\left(\frac{a_{HCO_3^-} a_{OH^-}}{a_{CO_3^{2-}}}\right)}.$$

(d) At the biological standard state $a_{HCO_3^-} = a_{CO_3^{2-}} = 1$ and $a_{OH^-} = a_{H^+} = 1 \times 10^{-7}$ (i.e. pOH = pH = 7). Thus, evaluation of the above Nernst equation at $E_{\text{cell}} = E^{\oplus}_{\text{cell}}$ gives

$$E^{\oplus}_{\text{cell}} = E^{\ominus}_{\text{cell}} - \frac{RT}{F} \ln a_{H^+}$$

$$E^{\oplus}_{\text{cell}} - E^{\ominus}_{\text{cell}} = -\frac{RT}{F} \ln a_{H^+}$$
$$= -(0.025693\ \text{V}) \ln(1 \times 10^{-7})$$
$$= \boxed{+0.41}.$$

(e) Within the acid–base discussion of text Chapter 8 it is shown that for a conjugate acid–base pair:

$$pK_a + pK_b = pK_w.$$

In part (b) it is shown that $pK_b = 3.67$ for the HCO_3^-/CO_3^{2-} conjugate pair. Thus, the pK_a of HCO_3^- is

$$pK_a = pK_w - pK_b = 14.00 - 3.68 = \boxed{10.32}.$$

E5.45 The half-reaction is

$$Cr_2O_7^{2-}(aq) + 14\ H^+(aq) + 6\ e^- \rightarrow 2\ Cr^{3+}(aq) + 7\ H_2O(l).$$

The reaction quotient is

$$Q = \frac{a^2_{Cr^{3+}}}{a_{Cr_2O_7^{2-}} a^{14}_{H^+}} \quad \nu = 6.$$

Hence,

$$E_{\text{cell}} = E^{\ominus}_{\text{cell}} - \frac{RT}{6F} \ln\left(\frac{a^2_{Cr^{3+}}}{a_{Cr_2O_7^{2-}} a^{14}_{H^+}}\right).$$

E5.46 R: $2\,AgCl(s) + 2\,e^- \rightarrow 2\,Ag(s) + 2\,Cl^-(aq)$ $E_R^\ominus = +0.22\text{ V}$

L: $2\,H^+(aq) + 2\,e^- \rightarrow H_2(g)$ $E_L^\ominus = 0$

R−L: $2\,AgCl(s) + H_2(g) \rightarrow 2\,Ag(s) + 2\,Cl^-(aq) + 2\,H^+(aq)$ $E_{cell}^\ominus = E_R^\ominus - E_L^\ominus = +0.22\text{ V}$

For this cell $a_{H^+} = a_{Cl^-}$, $\nu = 2$, and $E_{cell} = 0.312$ V. The Nernst equation can be used to calculate the pH under these conditions.

$$E_{cell} = E_{cell}^\ominus - \frac{RT}{2F}\ln(a_{H^+}^2 a_{Cl^-}^2)$$
$$= E_{cell}^\ominus - \frac{RT}{2F}\ln a_{H^+}^4 = E_{cell}^\ominus - \frac{2RT}{F}\ln a_{H^+} = E_{cell}^\ominus - \frac{2RT\ln 10}{F}\log a_{H^+}$$
$$= E_{cell}^\ominus + \frac{2RT\ln 10}{F}\text{pH}$$

Solving for pH gives

$$\text{pH} = \frac{F}{2RT\ln 10}\times(E_{cell} - E_{cell}^\ominus) = \frac{0.312\text{ V} - 0.22\text{ V}}{0.1183\text{ V}}$$
$$= \boxed{0.78}.$$

E5.47 The method of the solution is first to determine $\Delta_r G^\ominus$, $\Delta_r H^\ominus$, and $\Delta_r S^\ominus$ for the cell reaction

$$\tfrac{1}{2}H_2(g) + AgCl(s) \rightarrow Ag(s) + HCl(aq) \qquad \nu = 1$$

and then, from the values of these quantities and the known values of $\Delta_f G^\ominus$, $\Delta_f H^\ominus$, and $S^\ominus$, for all the species other than $Cl^-(aq)$, to calculate $\Delta_f G^\ominus$, $\Delta_f H^\ominus$, and $S^\ominus$ for $Cl^-(aq)$.

$$\Delta_r G^\ominus = -\nu F E_{cell}^\ominus$$

At 298.15 K (25.00°C)

$$E_{cell}^\ominus/\text{V} = (0.23659) - (4.8564\times10^{-4})\times(25.00) - (3.4205\times10^{-6})\times(25.00)^2$$
$$+ (5.869\times10^{-9})\times(25.00)^3$$
$$= +0.22240\text{ V}.$$

Therefore, $\Delta_r G^\ominus = -(96.485\text{ kC mol}^{-1})\times(0.22240\text{ V}) = -21.46\text{ kJ mol}^{-1}$

$$\Delta_r S^\ominus = -\left(\frac{\partial \Delta_r G^\ominus}{\partial T}\right)_p = \nu F\left(\frac{\partial E_{cell}^\ominus}{\partial T}\right)_p = \nu F\left(\frac{\partial E_{cell}^\ominus}{\partial \theta}\right)_p \frac{°\text{C}}{\text{K}} \quad [d\theta/°\text{C} = dT/\text{K}] \quad (a)$$

$$(\partial E_{cell}^\ominus/\partial\theta)_p/\text{V} = (-4.8564\times10^{-4}/°\text{C}) - (2)\times(3.4205\times10^{-6}\theta/°\text{C}^2) + (3)\times(5.869\times10^{-9}\theta^2/°\text{C}^3)$$

$$(\partial E_{cell}^\ominus/\partial\theta)_p/(\text{V}°\text{C}^{-1}) = (-4.8564\times10^{-4}) - (6.8410\times10^{-6}(\theta/°\text{C})) + (1.7607\times10^{-8}(\theta/°\text{C})^2)$$

Therefore, at 25°C,

$$(\partial E_{cell}^\ominus/\partial\theta)_p = -6.4566\times10^{-4}\text{ V/°C}$$

and

$$(\partial E_{cell}^\ominus/\partial\theta)_p = (-6.4566\times10^{-4}\text{ V/°C})\times(°\text{C/K}) = -6.4566\times10^{-4}\text{ V K}^{-1}$$

Hence, from equation (a)

$$\Delta_r S^\ominus = (-96.485\text{ kC mol}^{-1})\times(6.4566\times10^{-4}\text{ V K}^{-1}) = -62.30\text{ J K}^{-1}\text{ mol}^{-1}$$

and $\Delta_r H^\ominus = \Delta_r G^\ominus + T\Delta_r S^\ominus$
$$= -(21.46\text{ kJ mol}^{-1}) + (298.15\text{ K})\times(-62.30\text{ J K}^{-1}\text{ mol}^{-1}) = -40.03\text{ kJ mol}^{-1}$$

For the cell reaction $\frac{1}{2}\,H_2(g) + AgCl(s) \rightarrow Ag(s) + HCl(aq)$

$$\begin{aligned}\Delta_r G^\circ &= \Delta_f G^\circ(H^+) + \Delta_f G^\circ(Cl^-) - \Delta_f G^\circ(AgCl) \\ &= \Delta_f G^\circ(Cl^-) - \Delta_f G^\circ(AgCl) \quad [\Delta_f G^\circ(H^+) = 0]\end{aligned}$$

Hence, $\Delta_f G^\circ(Cl^-) = \Delta_r G^\circ + \Delta_f G^\circ(AgCl) = (-21.46 - 109.79)\ \text{kJ mol}^{-1}$

$$= \boxed{-131.25\ \text{kJ mol}^{-1}}$$

Similarly, $\Delta_f H^\circ(Cl^-) = \Delta_r H^\circ + \Delta_f H^\circ(AgCl) = (-40.03 - 127.07)\ \text{kJ mol}^{-1}$

$$= \boxed{-167.10\ \text{kJ mol}^{-1}}$$

For the entropy of Cl^- in solution we use

$$\Delta_r S^\circ = S^\circ(Ag) + S^\circ(H^+) + S^\circ(Cl^-) - \tfrac{1}{2}S^\circ(H_2) - S^\circ(AgCl)$$

with $S^\circ(H^+) = 0$. Then,

$$\begin{aligned}S^\circ(Cl^-) &= \Delta_r S^\circ - S^\circ(Ag) + \tfrac{1}{2}S^\circ(H_2) + S^\circ(AgCl) \\ &= \{(-62.30) - (42.55) + \tfrac{1}{2}\times(130.68) + (96.2)\}\ \text{J K}^{-1}\ \text{mol}^{-1} = \boxed{+56.7\ \text{J K}^{-1}\ \text{mol}^{-1}}\end{aligned}$$

E5.48 $\Delta G_m = F\Delta\phi - (RT\ln 10)\Delta\text{pH}$ [5.21]

The normal pH difference across the inner mitochondrial membrane is $\Delta\text{pH} = \text{pH}_{in} - \text{pH}_{out} = -1.4$ and $\Delta\phi = 70$ mV. Thus,

$$\begin{aligned}\Delta G_m &= (9.6485\times10^4\ \text{C mol}^{-1})\times(0.070\ \text{V}) - (8.3145\ \text{J K}^{-1}\ \text{mol}^{-1})\times(310\ \text{K})\times\ln 10\times(-1.4) \\ &= +15.\bar{1}\ \text{kJ mol}^{-1}\end{aligned}$$

For 4 mol H^+, $\Delta G = (4\ \text{mol})\times(+15.\bar{1}\ \text{kJ mol}^{-1}) = +60.\bar{4}\ \text{kJ}$

Since 31 kJ mol^{-1} is needed for ADP phosphorylation, the amount of ATP that could be synthesized is

$$\frac{60.\bar{4}\ \text{kJ}}{31\ \text{kJ mol}^{-1}} = 1.9\ \text{mol} \approx \boxed{2\ \text{mol}}$$

E5.49 $\Delta G_m = F\Delta\phi - (RT\ln 10)\Delta\text{pH}$ [5.21]

The normal pH difference across the inner mitochondrial membrane is $\Delta\text{pH} = \text{pH}_{in} - \text{pH}_{out} = -1.4$ and $\Delta\phi = 70$ mV. With stress that increases pH_{in} by 0.1 units the pH difference becomes $\Delta\text{pH} = \text{pH}_{in} - \text{pH}_{out} = -1.3$. Thus,

$$\begin{aligned}\Delta G_m &= (9.6485\times10^4\ \text{C mol}^{-1})\times(0.070\ \text{V}) - (8.3145\ \text{J K}^{-1}\ \text{mol}^{-1})\times(310\ \text{K})\times\ln 10\times(-1.3) \\ &= +14.\bar{5}\ \text{kJ mol}^{-1}\end{aligned}$$

For 2 mol H^+, $\Delta G = (2\ \text{mol})\times(+14.\bar{5}\ \text{kJ mol}^{-1}) = +29\ \text{kJ}$

Since 31 kJ mol^{-1} is needed for ADP phosphorylation, the amount of ATP that could be synthesized is

$$\frac{29\ \text{kJ}}{31\ \text{kJ mol}^{-1}} = 0.93\ \text{mol} \approx \boxed{1\ \text{mol}}$$

E5.50 Yes, a bacterium can evolve to utilize the ethanol/nitrate pair to exergonically release the free energy needed for ATP synthesis. Ethanol (the reductant) may yield any of the following products.

$$\underset{\text{ethanol}}{CH_3CH_2OH} \rightarrow \underset{\text{ethanal}}{CH_3CHO} \rightarrow \underset{\text{ethanoic acid}}{CH_3COOH} \rightarrow CO_2 + H_2O$$

Nitrate (the oxidant) receives electrons to yield any of the following products.

$$\begin{array}{ccccccc} NO_3^- & \rightarrow & NO_2^- & \rightarrow & N_2 & \rightarrow & NH_3 \\ \text{nitrate} & & \text{nitrite} & & \text{dinitrogen} & & \text{ammonia} \end{array}$$

Oxidation of two ethanol molecules to carbon dioxide and water can transfer 8 electrons to nitrate during the formation of ammonia. The half-reactions and net reaction are:

$$2\,[CH_3CH_2OH(l) \rightarrow 2\,CO_2(g) + H_2O(l) + 4\,H^+(aq) + 4\,e^-]$$

$$NO_3^-(aq) + 9\,H^+(aq) + 8\,e^- \rightarrow NH_3(aq) + 3\,H_2O(l)$$

$$2\,CH_3CH_2OH(l) + H^+(aq) + NO_3^-(aq) \rightarrow 4\,CO_2(g) + 5\,H_2O(l) + NH_3(aq)$$

$\Delta_r G^\ominus = -2331.29$ kJ for the reaction as written (a data-table calculation). Of course, enzymes must evolve that couple this exergonic redox reaction to the production of ATP, which would then be available for carbohydrate, protein, lipid, and nucleic acid synthesis.

E5.51 R: $Fe^{3+}(\text{cyt}\,f) + e^- \rightarrow Fe^{2+}(\text{cyt}\,f)$ $\quad E_R^\ominus = +0.36$ V at 298 K [Table 5.2]

L: $Fe^{3+}(\text{cyt}\,b) + e^- \rightarrow Fe^{2+}(\text{cyt}\,b)$ $\quad E_L^\ominus = +0.08$ V

Overall (R−L): cyt-f(ox) + cyt-b(red) → cyt-f(red) + cyt-b(ox) $\quad E_{cell}^\ominus = +0.28$ V $\quad \nu = 1$

(a) $\Delta_r G^\ominus = -\nu F E_{cell}^\ominus$ [5.14] $= -1 \times (96.485\text{ kC mol}^{-1}) \times (+0.28\text{ V}) = \boxed{-27\text{ kJ mol}^{-1}}$

(b) Since four protons must be transferred across the membrane and 50 kJ mol^{-1} is needed for ADP phosphorylation in the chloroplast, the amount of ATP that could be synthesized from the passage of four protons is

$$\frac{27\text{ kJ}}{50\text{ kJ mol}^{-1}} = 0.54\text{ mol}$$

and the transfer of at least $\boxed{\text{eight}}$ protons is required for the phosphorylation of one ADP.

Solutions to projects

P5.52 The half-reactions involved are:

R: cyt-c_{ox} + e$^-$ → cyt-c_{red} $\quad E_{cyt}^\ominus$

L: $D_{ox} + e^- \rightarrow D_{red}$ $\quad E_D^\ominus$

The overall cell reaction is:

R − L: cyt-c_{ox} + D_{red} ⇌ cyt-c_{red} + D_{ox} $\quad E_{cell}^\ominus = E_{cyt}^\ominus - E_D^\ominus \quad \nu = 1$

(a) The Nernst equation for the cell reaction is

$$E_{cell} = E_{cell}^\ominus - \frac{RT}{F}\ln\frac{[\text{cyt-}c_{red}][D_{ox}]}{[\text{cyt-}c_{ox}][D_{red}]}.$$

$E_{cell} = 0$ at equilibrium. Therefore,

$$\ln\frac{[\text{cyt}_{red}]_{eq}[D_{ox}]_{eq}}{[\text{cyt}_{ox}]_{eq}[D_{red}]_{eq}} = \frac{F}{RT}(E_{cyt}^\ominus - E_D^\ominus)$$

$$\ln\left(\frac{[D_{ox}]_{eq}}{[D_{red}]_{eq}}\right) = \ln\left(\frac{[\text{cyt}_{ox}]_{eq}}{[\text{cyt}_{red}]_{eq}}\right) + \frac{F}{RT}(E_{cyt}^\ominus - E_D^\ominus).$$

Therefore, a plot of $\ln([D_{ox}]_{eq}/[D_{red}]_{eq})$ against $\ln([cyt_{ox}]_{eq}/[cyt_{red}]_{eq})$ is linear with a slope of one and an intercept of $F(E^{\circ}_{cyt} - E^{\circ}_{D})/RT$.

(b) Draw up the following table.

$\ln([D_{ox}]_{eq}/[D_{red}]_{eq})$	−5.882	−4.776	−3.661	−3.002	−2.593	−1.436	−0.6274
$\ln([cyt_{ox}]_{eq}/[cyt_{red}]_{eq})$	−4.547	−3.772	−2.415	−1.625	−1.094	−0.2120	−0.3293

The plot of $\ln([D_{ox}]_{eq}/[D_{red}]_{eq})$ against $\ln([cyt_{ox}]_{eq}/[cyt_{red}]_{eq})$ is shown in Figure 5.2. The figure insert reports the linear least-squares regression fit from which we see that the intercept is −1.2124. In part (a) it is shown that $F(E^{\circ}_{cyt} - E^{\circ}_{D})/RT$ equals the intercept so, given that $E^{\circ}_{D} = +0.237$ V, the intercept value allows the calculation of E°_{cyt}.

$$\begin{aligned} E^{\circ}_{cyt} &= (RT/F) \times (\text{intercept}) + E^{\circ}_{D} \\ &= (0.025693\ \text{V}) \times (-1.2124) + 0.237\ \text{V} \\ &= \boxed{+0.206\ \text{V}} \end{aligned}$$

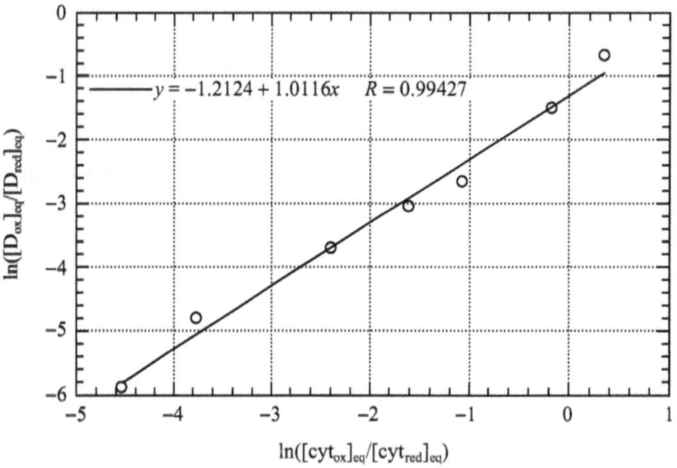

Figure 5.2

PART 2 The kinetics of life processes

6 The rates of reactions

Answers to discussion questions

D6.1 The time scales of atomic processes are rapid indeed: according to the following table, a nanosecond is an eternity. Note that the times given here are in some way typical values for times that may vary over two or three orders of magnitude. A large number of time scales for physical, chemical, and biological processes on the atomic and molecular scale are reported in Figure 2 of A.H. Zewail, Femtochemistry: atomic-scale dynamics of the chemical bond. *J. Phys. Chem. A* **104**, 5660 (2000).

Proton-transfer reactions occur on a time scale of about 10^{-10} to 10^{-9}s. Case Study 12.2: *Vision* describes several events in vision, including the 200-fs photoisomerization that gets the process started. Case Study 12.3: Harvesting of light during plant *Photosynthesis* lists time scales of several energy-transfer and electron-transfer steps in photosynthesis. Initial energy transfer (to a nearby pigment) has a time scale of around 10^{-13} to 10^{-11}s, with longer-range transfer (to the reaction center) taking about 10^{-10}s. Immediate electron transfer is also very fast (about 3 ps), with ultimate transfer (leading to oxidation of water and reduction of plastoquinone) taking from 10^{-10} to 10^{-3}s. Case Study 7.1 discusses helix–coil transitions, including experimental measurements of time scales of tens or hundreds of microseconds (10^{-5} to 10^{-4} s) for formation of tightly packed cores. The rate-determining step for the helix–coil transition of small polypeptides has a relaxation time of about 160 ns in contrast to the faster 50 ns relaxation time of large protein.

Process	t/ns	Reference
proton transfer (in water)	2×10^{-5}	Zewail 2000
initial chemical reaction of vision*	1×10^{-4}	Case Study 12.2
energy transfer in photosynthesis†	1×10^{-3}	Case Study 12.3
electron transfer in photosynthesis	3×10^{-3}	Case Study 12.3
polypeptide helix–coil transition	2×10^{2}	Case Study 7.1

*photoisomerization of retinal from 11-*cis* to all-*trans*
†time from absorption until electron transfer to adjacent pigment

D6.2 This report on one of the listed topics is left for the student to write.

D6.3 The determination of a rate law is simplified by the **isolation method** in which the concentrations of all the reactants except one are in large excess. If B is in large excess, for example, then to a good approximation its concentration is constant throughout the reaction. Although the true rate law might be rate = k_r[A][B], we can approximate [B] by $[B]_0$ and write

$$\text{rate} = k_{\text{eff}}[A], \text{ where } k_{\text{eff}} = k[B]_0$$

which has the form of a first-order rate law. Because the true rate law has been forced into first-order form by assuming that the concentration of B is constant, it is called a **pseudofirst-order** rate law.

[A] has been isolated. The dependence of the rate on the concentration of each of the reactants may be found by isolating them in turn (by having all the other substances present in large excess), and so constructing a picture of the overall rate law.

In the **method of initial rates**, which is often used in conjunction with the isolation method, the rate is measured at the beginning of the reaction for several different initial concentrations of reactants. We shall suppose that the rate law for a reaction with A isolated is rate $= k_r'[A]^a$; then its initial rate, rate$_0$, is given by the initial values of the concentration of A, and we write rate$_0 = k_r'[A]_0^a$. Taking logarithms gives:

$$\log(\text{rate}_0) = \log k_r' + a \log [A]_0 \quad [6.10]$$

For a series of initial concentrations, a plot of the logarithms of the initial rates against the logarithms of the initial concentrations of A should be a straight line with slope a.

The method of initial rates might not reveal the full rate law, for the products may participate in the reaction and affect the rate. For example, products participate in the synthesis of HBr, where the full rate law depends on the concentration of HBr. To avoid this difficulty, the rate law should be fitted to the data throughout the reaction. The fitting may be done, in simple cases at least, by using a proposed rate law to predict the concentration of any component at any time, and comparing it with the data.

Because rate laws are differential equations, we must integrate them if we want to find the concentrations as a function of time. Even the most complex rate laws may be integrated numerically. However, in a number of simple cases analytical solutions are easily obtained, and prove to be very useful. A first-order rate law shows linearity when the logarithm of concentration is plotted against time while a second-order rate law exhibits linearity in a plot of inverse concentration against time. The slope of the linear plot equals the second-order rate constant in the latter case and the negative of the first-order rate constant in the former case.

D6.4 Consider a rate law of the form: rate $= k_r[A]^m[B]^n$ where the concentration orders m and n equal either zero or a positive integer. If the sum $m + n$ equals zero, the rate is zeroth order and the rate is independent of species concentration. If the sum equals either 1 or 2, the rate is first order or second order, respectively. In the case for which $m = 1$ and $n \neq 0$, the reaction order will appear to be 1 if the concentration of B is a large excess and [B] remains basically unchanged during the course of reaction. This is the pseudofirst-order reaction rate for which

$$\text{rate} = k_r[A][B]^n = (k_r[B]^n)[A] = k_{\text{eff}}[A]$$

where $k_{\text{eff}} = k[B]^n =$ pseudofirst-order rate constant.

The apparent order of a reaction changes whenever a concentration varies to make one term in the rate law smaller, or larger, relative to another term. For example, a typical rate law for the action of an enzyme E on a substrate S is (see Chapter 8 and Section 6.3)

$$\text{rate} = \frac{k_r[E][S]}{[S] + K_M}$$

where K_M is a constant. This rate law does not have a definite order overall. If the substrate concentration is so low that $[S] \ll K_M$, the rate becomes

$$\text{rate} = \frac{k_r}{K_M}[E][S]$$

which is first order in S, first order in E, and second order overall. If the substrate concentration is so large that $[S] \gg K_M$, the rate becomes

$$\text{rate} = k_r[\text{E}]$$

which is zero-order in S, first-order in E, and first-order overall.

D6.5 The **Arrhenius equation,** $k_r = Ae^{-E_a/RT}$ [6.19c] or $\ln k_r = \ln A - \frac{E_a}{RT}$ [6.19b], provides for the variation of the reaction rate constant with temperature, $k_r(T)$. The constants A and E_a, which are determined through experimental effort, are called the **Arrhenius parameters**. More specifically, the parameter A is the **pre-exponential factor** and E_a is the **activation energy**. The pre-exponential factor is proportional to the rate at which reactant molecules collide, while the activation energy is the minimum kinetic energy required for a collision to result in a reaction. Determination of the Arrhenius parameters involves preparation of a plot of experimentally determined values of ln $k_r(T)$ against $1/T$. The Arrhenius equation predicts that the plot will be linear with a slope equal to $-E_a/R$ and an extrapolated intercept equal to ln A. Thus, if the plot is linear, a linear least-square regression fit of the data plot yields values for the slope and intercept from which the Arrhenius parameters are calculated with the relations $E_a = -R \times \textit{slope}$ and $A = e^{\textit{intercept}}$.

The Arrhenius equation describes the variation of the reaction rate constant with temperature whenever the activation energy is temperature independent. There are instances for which the plot of $k_r(T)$ against $1/T$ is not Arrhenius-like, in the sense that a straight line is not obtained. However, it is still possible to define a general activation energy as

$$E_a = RT^2\left(\frac{\mathrm{d}\ln k_r}{\mathrm{d}T}\right)$$

This definition accounts for the temperature-dependent activation energy. It reduces to the Arrhenius equation (as the slope of a straight line) for a temperature-independent activation energy. However, this definition is more general, because it allows E_a to be obtained from the slope (at the temperature of interest) of a plot of ln k_r against $1/T$ even if the Arrhenius plot is not a straight line. Non-Arrhenius behavior is sometimes a sign that quantum-mechanical tunneling is playing a significant role in the reaction.

D6.6 The Arrhenius equation is $k_r = Ae^{-E_a/RT}$ [6.19c]. The pre-exponential factor, A, is proportional to the rate at which reactant molecules collide, while the activation energy, E_a, is the minimum kinetic energy required for a collision to result in a reaction.

Solutions to exercises

E6.7

$$\frac{I}{I_0} = 10^{-\varepsilon[\text{substance}]L} \quad [6.1a]$$

Substituting the given data we have

$$\frac{I}{I_0} = 10^{-855\ \text{dm}^3\ \text{mol}^{-1}\ \text{cm}^{-1}\times 3.25\ \text{mmol dm}^{-3}\times 2.55\ \text{mm}}$$

After conversion of the units of the concentration and cell length to the units of the molar absorption coefficient this becomes

$$\frac{I}{I_0} = 10^{-855\times 3.25\times 10^{-3}\times 0.255} = 0.984$$

The percentage reduction in intensity is then $100\% - 20.2\% = \boxed{79.8\%}$.

E6.8 Define $A = \log\left(\frac{I_0}{I}\right) = -\log\left(\frac{I}{I_0}\right) = -\log(0.398) = 0.400$, then

$$[\text{cyt P450}] = \frac{A}{\varepsilon l} \quad [6.1b]$$

$$= \frac{0.400}{(291\ \text{dm}^3\ \text{mol}^{-1}\ \text{cm}^{-1})(0.65\ \text{cm})} = 2.1\times 10^{-3}\ \text{mol dm}^{-3} = \boxed{2.1\ \text{mmol dm}^{-3}}$$

E6.9 (a) Because the rate of formation of C is known, the reaction stoichiometry can be used to determine the rates of consumption and formation of the other participants in the reaction. The reaction is:

$2\text{A} + \text{B} \rightarrow 3\text{C} + 2\text{D}$ and the rate of formation of C equals $2.2\ \text{mol dm}^{-3}\ \text{s}^{-1}$.

$$\text{rate of consumption of A} = \frac{2}{3}\times \text{rate of formation of C} = \frac{2}{3}\times(2.2\ \text{mol dm}^{-3}\ \text{s}^{-1})$$

$$= \boxed{1.5\ \text{mol dm}^{-3}\ \text{s}^{-1}}$$

$$\text{rate of consumption of B} = \frac{1}{3}\times \text{rate of formation of C}$$

$$= \boxed{0.73\ \text{mol dm}^{-3}\ \text{s}^{-1}}$$

$$\text{rate of formation of D} = \frac{2}{3}\times \text{rate of formation of C}$$

$$= \boxed{1.5\ \text{mol dm}^{-3}\ \text{s}^{-1}}$$

(b) $v = k_r[\text{A}][\text{B}][\text{C}]$

The reaction rate has units of concentration per unit time ($\text{mol dm}^{-3}\ \text{s}^{-1}$), hence

$$\text{mol dm}^{-3}\ \text{s}^{-1} = [\text{units of } k]\times(\text{mol dm}^{-3})^3.$$

Therefore, $[\text{units of } k] = \frac{\text{mol dm}^{-3}\ \text{s}^{-1}}{(\text{mol dm}^{-3})^3} = \boxed{\text{mol}^{-2}\ \text{dm}^6\ \text{s}^{-1}}$.

E6.10 (a) concentrations in (molecules m^{-3}) $= N\ \text{m}^{-3}$

(i) second order

$$\text{rate} = N\ \text{m}^{-3}\ \text{s}^{-1} = [k]\times(N\ \text{m}^{-3})^2$$

where $[k]$ = units of k and N = number of molecules

Then, $[k] = \frac{N\ \text{m}^{-3}\ \text{s}^{-1}}{(N\ \text{m}^{-3})^2} = N^{-1}\ \text{m}^3\ \text{s}^{-1}$

Because N is unitless, $[k] = \boxed{\text{m}^3\ \text{s}^{-1}}$,

though loosely we may say molecules^{-1} $\text{m}^3\ \text{s}^{-1}$

(ii) third order $[k] = \frac{N\ \text{m}^{-3}\ \text{s}^{-1}}{(N\ \text{m}^{-3})^3} = N^{-2}\ \text{m}^6\ \text{s}^{-1}$

or $[k] = \boxed{\text{m}^6\ \text{s}^{-1}}$

(b) pressures in kilopascals

(i) second order

rate = kPa s^{-1} = $[k] \times (\text{kPa})^2$

$$[k] = \frac{\text{kPa s}^{-1}}{(\text{kPa})^3} = \boxed{\text{kPa}^{-2}\ \text{s}^{-1}}$$

(ii) third order

$$[k] = \frac{\text{kPa s}^{-1}}{(\text{kPa})^3} = \boxed{\text{kPa}^{-2}\ \text{s}^{-1}}$$

E6.11 See Figure 6.1.

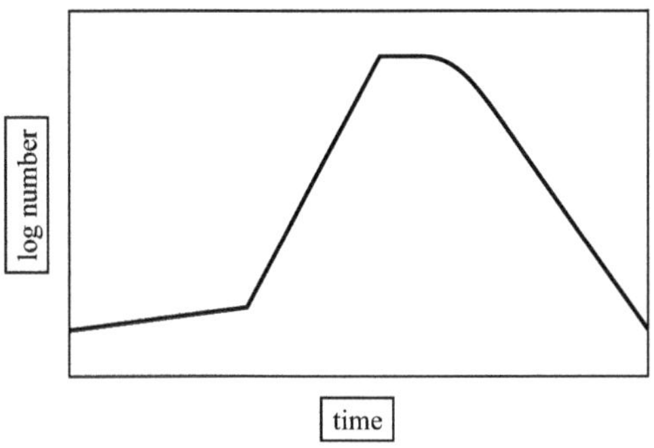

Figure 6.1

E6.12 $\text{Mb} + \text{CO} \rightarrow \text{MbCO} \quad k_r = 5.8 \times 10^5\ \text{dm}^3\ \text{mol}^{-1}\ \text{s}^{-1}$

Since the carbon monoxide concentration (400 mmol dm^{-3}) is much larger than the initial myoglobin concentration (10 mmol dm^{-3}), we assume that Mb is isolated. Additionally, although the exercise suggests the rate to be pseudofirst-order in Mb, we assume that the reaction rate is first-order in the CO. Thus, $v = -\text{d[Mb]}/\text{d}t = k_r[\text{CO}][\text{Mb}] = k_{r,\text{eff}}[\text{Mb}]$ where

$$k_{r,\text{eff}} = (5.8 \times 10^5\ \text{dm}^3\ \text{mol}^{-1}\ \text{s}^{-1}) \times (400\ \text{mmol dm}^{-3}) = 2.3 \times 10^5\ \text{s}^{-1}.$$

The integrated form of this rate law is $\ln([\text{Mb}]_0/[\text{Mb}]) = k_{r,\text{eff}}t$ or $[\text{Mb}] = [\text{Mb}]_0 e^{-k_{r,\text{eff}}t}$. The latter equation is used to prepare the plot of [Mb] against t shown in Figure 6.2. The time range for the computation is chosen by calculating the pseudofirst-order time constant.

$$\tau_{\text{eff}} = 1/k_{r,\text{eff}}\ [6.14] = 1/(2.3 \times 10^5\ \text{s}^{-1}) = 4.3 \times 10^{-6}\ \text{s} = 4.3\ \mu\text{s}$$

The range is chosen to be 0–10 μs or about $2.5\tau_{\text{eff}}$.

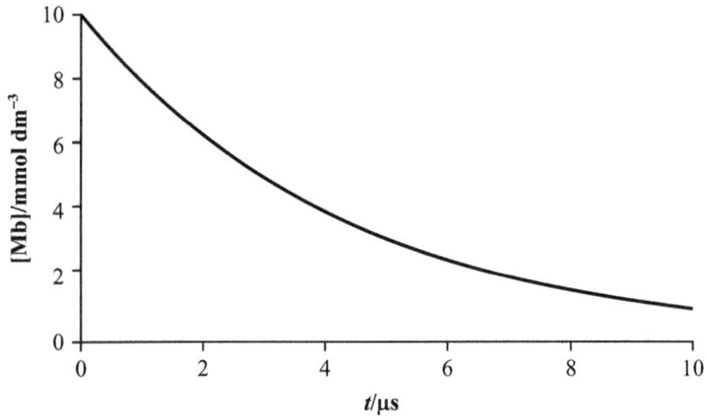

Figure 6.2

E6.13 The rate expression for a zero-order reaction is $\frac{d[A]}{dt} = -k[A]^0 = -k$. Then

$$\int_{[A]_0}^{[A]} d[A] = -k\int_0^t dt \text{ giving}$$

$$[A] - [A]_0 = -kt$$

When the concentration of ethanol drops to 50% we have $[A] = [A]_0/2$ and

$$\frac{[A]_0}{2} - [A]_0 = -kt \text{ or}$$

$$\frac{[A]_0}{2t} = k \text{ or}$$

$$k = \frac{[A]_0}{2t} = \frac{1.5\ \text{g dm}^{-3}}{2\times\left(\frac{49\ \text{min}}{60\ \text{min/h}}\right)} = \boxed{0.92\ \text{g dm}^{-3}\ \text{h}^{-1}}$$

E6.14 Use $\ln\frac{[A]_0}{[A]} = k_r t$ [6.12b] and solve for k_r.

$$k_r = \frac{\ln(220/56.0)}{1.22\times10^4\ \text{s}} = \boxed{1.12\times10^{-4}\ \text{s}^{-1}}$$

E6.15 The reaction is $CH_3COCO_2^-(aq) + H^+(aq) \rightarrow CH_3CHO(aq) + CO_2(g)$.

Data provided: $V_{gas} = 250\ \text{cm}^3$, $V_{solution} = 100\ \text{cm}^3$, $T = 293\ \text{K}$, $[P]_0 = 3.23\ \text{mmol dm}^{-3}$, $p_{CO_2}(t) = 100\ \text{Pa}$ at $t = 522\ \text{s}$, and $p_{CO_2}(0) = 0$.

We can assume that the concentration of pyruvate (P) decreases in proportion to the increase in p_{CO_2}. For the $CO_2(g)$ formed, we may write the perfect gas relation.

$$p_{CO_2}V_{gas} = n_{CO_2}RT$$

$$n_{CO_2} = \frac{V_{gas}}{RT}p_{CO_2} = \Delta n_{CO_2}$$

$$\Delta n_{CO_2} = \frac{250\times10^{-6}\ \text{m}^3\times100\ \text{Pa}}{8.3145\ \text{J K}^{-1}\ \text{mol}^{-1}\times293\ \text{K}} = 0.0102\bar{6}\ \text{mmol}$$

$$[P] = [P]_0 - \frac{\Delta n_{CO_2}}{V_{solution}} = [P]_0 - \frac{\Delta n_{CO_2}}{0.100\ \text{dm}^3} = 3.23\ \text{mmol dm}^{-3} - 0.102\bar{6}\ \text{mmol dm}^{-3}$$

$$= 3.12\bar{7}\ \text{mmol dm}^{-3}$$

For a first-order reaction:

$$[P] = [P]_0 e^{-k_r t} \quad [6.12d]$$

$$k_r = \frac{\ln\left(\dfrac{[P]_0}{[P]}\right)}{t}$$

$$= \frac{\ln\left(\dfrac{3.23 \text{ mmol dm}^{-3}}{3.12\bar{7} \text{ mmol dm}^{-3}}\right)}{522 \text{ s}}$$

$$= \boxed{6.21 \times 10^{-5} \text{ s}^{-1}}.$$

Note: In the above solution we have focused on the particular values of p_{CO_2}, V_{gas}, $V_{solution}$, $[P]_0$, and T. However, an experimentalist may need a general expression for k_r because these are variables in many experiments. Can you derive the general expression? It is

$$k_r = -\frac{1}{t}\ln\left(1 - \frac{p_{CO_2}V_{gas}}{RT[P]_0 V_{solution}}\right).$$

E6.16 $$CO_2(aq) + H_2O \xrightarrow{\text{carbonic anhydrase}} H_2CO_3(aq)$$

We assume that the reaction rate is pseudo-first-order in carbon dioxide as the enzyme concentration may appear in the general rate expression. Then,

$$\ln\frac{[CO_2]_0}{[CO_2]} = k_{r,eff}t \quad [6.12b]$$

$$k_{r,eff} = \frac{1}{t}\ln\frac{[CO_2]_0}{[CO_2]}$$

$$= \left(\frac{1}{1.22 \times 10^4 \text{ s}}\right) \times \ln\left(\frac{220}{56.0}\right)$$

$$= \boxed{1.12 \times 10^{-4} \text{ s}^{-1}}.$$

E6.17 We have the reaction $NO(g) + \frac{1}{2}Cl_2(g) \rightarrow NOCl(g)$.

Data provided: $p_{NO}(0) = 300$ Pa, $p_{NOCl}(522 \text{ s}) = 100$ Pa, $p_{NOCl}(0) = 0$.

The reaction rate is stated to be a pseudo-second-order reaction in NO. Thus,

$\dfrac{dp_{NO}}{dt} = k_{r,eff}p_{NO}^2$, where $k_{r,eff}$ is the pseudo-second-order rate constant and the integrated form of the rate law is

$$\frac{1}{p_{NO}} - \frac{1}{p_{NO}(0)} = k_{r,eff}t \quad [6.15c].$$

Since the reaction stoichiometry indicates that $p_{NOCl} = p_{NO}(0) - p_{NO}$, we substitute $p_{NO} = p_{NO}(0) - p_{NOCl}$ and solve for $k_{r,eff}$.

$$k_{r,eff} = \frac{1}{t}\left(\frac{1}{p_{NO}(0) - p_{NOCl}} - \frac{1}{p_{NO}(0)}\right)$$

$$= \left(\frac{1}{522 \text{ s}}\right) \times \left(\frac{1}{(300 - 100) \text{ Pa}} - \frac{1}{300 \text{ Pa}}\right)$$

$$= \boxed{3.19 \times 10^{-6} \text{ Pa}^{-1} \text{ s}^{-1}}$$

E6.18 As explained in Section 6.3 and in more detail in Chapter 8 there is not likely to be a simple specific order with respect to the substrate S in enzyme kinetics. For the purposes of this solution we will assume that the rate law takes the form $\upsilon = k[S]^n[E]^m$ with the most likely values of m and n both being 1 (eqn 6.7b) and k to be interpreted as k_r/K_M with K_M being the Michaelis constant (Chapter 8). With this assumption, for constant enzyme concentrations and variable substrate concentrations, the rate law takes the form $\upsilon_0 = k'[S]_0^n$ with $k' = k[E]_0^m$. Examination of the data for any given enzyme concentration (a), (b), or (c) shows that the rate is directly proportional to $[S]_0$. Thus, $\boxed{\text{the order with respect to } [S]_0 \text{ is first}}$. Examination of the data for a constant substrate concentration (1.00, 2.00, 3.00, or 4.00 mmol dm^{-3} shows that the rate is directly proportional to $[E]_0$, hence $\boxed{\text{the order with respect to } [E]_0 \text{ is first}}$. The rate constant k can be determined from any pair of $[S]_0$ and $[E]_0$. Choosing the pair $[S]_0 = 4.00$ mmol dm^{-3} and $[E]_0 = 10.0$ mmol dm^{-3} with $\upsilon_0 = 238.0$ mmol dm^{-3} s^{-1} and solving for k we obtain $k = 5.95$ dm^3 mmol^{-1} s^{-1}. Other pairs of data will give slightly different values for k. The best value of k taking into account all the data points can be determined by first plotting $\log \upsilon_0 = \log k' + n \log [S]_0$ against $[S]_0$. The order of the reaction and the effective rate constant k' can be determined from the slope and intercept of this plot. See Figure 6.3(a).

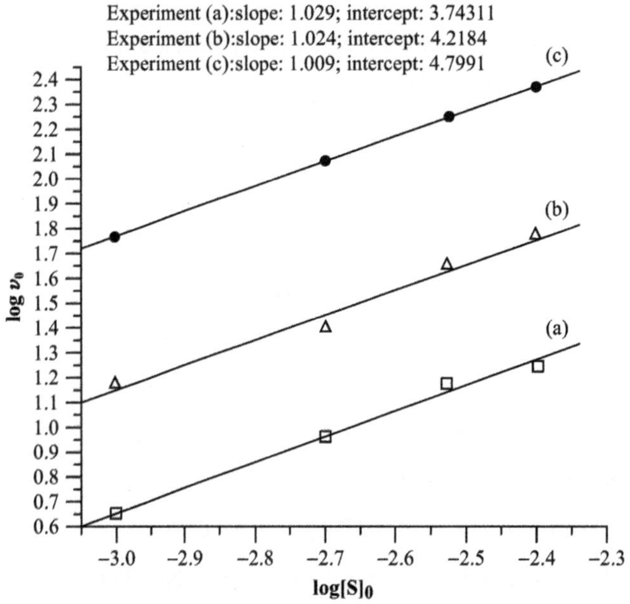

Figure 6.3(a)

The slope of these plots is close to 1 for all enzyme concentrations so we conclude as previously that the order of the reaction with respect to $[S]_0$ is $\boxed{\text{first}}$. The intercepts of the plots determine the effective rate constant, k', for a given $[E]_0$. The order of the reaction with respect to $[E]_0$ and the rate constant k can be determined from the slope and the intercept, respectively, of a plot of

$$\log k' = \log k + m \log [E]_0$$

See Figure 6.3(b).

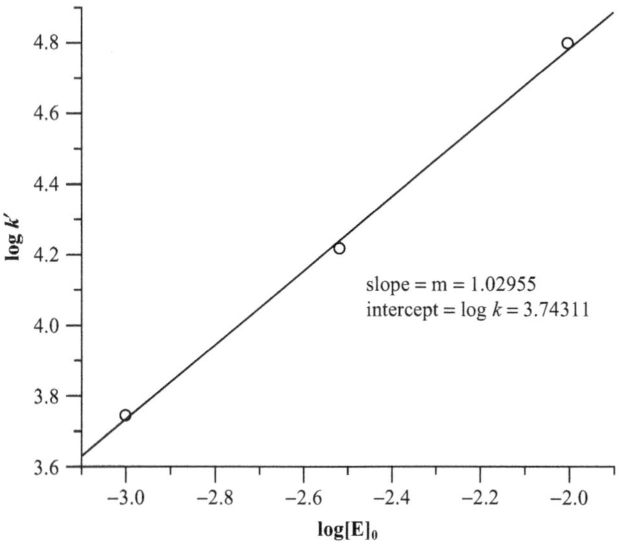

Figure 6.3(b)

The slope of the graph is close to 1, so the order of the reaction with respect to $[E]_0$ is **first**.

The intercept is 3.74311 = log k, so the value of k is $5.5 \times 10^{-3}\ \mathrm{dm^3\ mol^{-1}\ s^{-1}}$ to two significant figures, which is all that is warranted by the data, and is slightly different from the approximate value given above.

E6.19 The data for this experiment do not extend much beyond one half-life. Therefore predicting the order of the reaction from the half-life by use of equations 6.13 or 6.16 cannot be used here. However, a similar method based on three-quarter lives will work. For a first-order reaction, we may write (analogous to the derivation of eqn 6.13, also see the solutions to Exercises 6.33 and 6.34)

$$k_r t_{3/4} = -\ln\frac{\frac{3}{4}[\mathrm{A}]_0}{[\mathrm{A}]_0} = -\ln\frac{3}{4} = \ln\frac{4}{3} = 0.288 \quad \text{or} \quad t_{3/4} = \frac{0.288}{k_r}.$$

Thus the three-quarter life (or any given fractional life) is also independent of concentration for a first-order reaction. Examination of the data shows that the first three-quarters life (time to $[\mathrm{A}] = 0.237\ \mathrm{mol\ dm^{-3}}$) is about 80 min and by interpolation the second (time to $[\mathrm{A}] = 0.178\ \mathrm{mol\ dm^{-3}}$) is also about 80 min. Therefore, the reaction is first order and the rate constant is approximately

$$k_r = \frac{0.288}{t_{3/4}} \approx \frac{0.288}{80\ \mathrm{min}} = 3.6 \times 10^{-3}\ \mathrm{min^{-1}}$$

A least-squares fit of the data to the first-order integrated rate law [6.12c] gives the slightly more accurate result, $k = 3.67 \times 10^{-3}\ \mathrm{min^{-1}}$ An Excel plot of the data is shown in Figure 6.4. Also note that $2 \times t_{3/4} \neq t_{1/2}$.

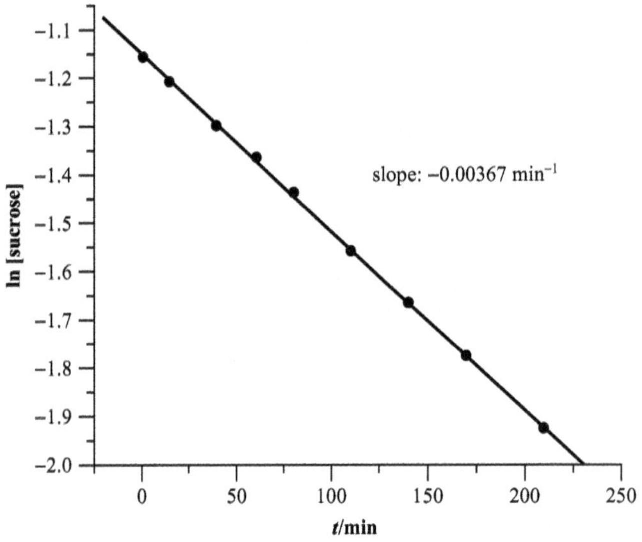

Figure 6.4

E6.20 (a) The initial concentrations given are exactly the same. Hence, the rates of change of the concentrations of both reactants are the same because of the 1:1 stoichiometry.

(b) A good assumption based on the stoichiometry of the reaction is that the reaction is second order. This can be confirmed by plotting 1/[A] (A = N-acetylcysteine) against time to see if a straight line is obtained. See Figure 6.5. The fit is very good; the reaction is second order.

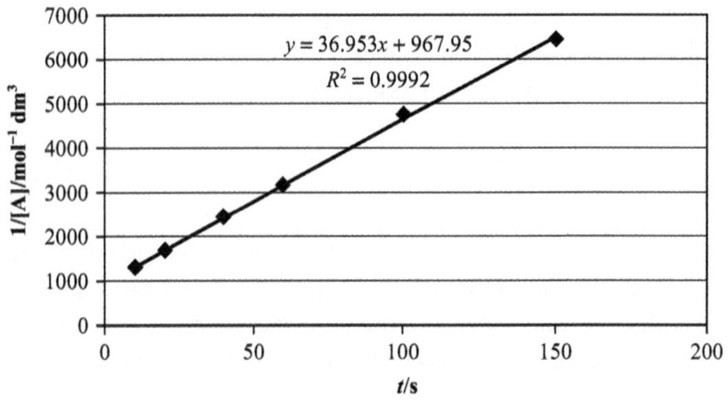

Figure 6.5

(c) From the equation obtained by the fitting process and comparing to eqn 6.15(c) of the text we see that the rate constant is the slope. Thus $k_r = \boxed{36.9\ \mathrm{dm^3\ mol^{-1}\ s^{-1}}}$

E6.21 The solution to Exercise 6.20 demonstrates that the reaction is a second-order reaction. We expect that the order with respect to each reactant is first; if so, then a plot of $\ln\left(\dfrac{[B]/[B]_0}{[A]/[A]_0}\right)$ against time should yield a straight line (eqn 6.17b). In the plot below, Figure 6.6, we designate **B** to be N-acetylcysteine and A to be iodoacetamide.

(a) The plot is linear; hence we conclude that the order of the reaction with respect to each reactant is $\boxed{\text{first.}}$

(b) Eqn 6.17b shows that the slope of the line is

$[B]_0 - [A]_0 \times k_r = -0.03027\ s^{-1}$ giving

$$k_r = \frac{-0.03027\ s^{-1}}{[B]_0 - [A]_0} = \frac{-0.03027\ s^{-1}}{-1.00 \times 10^{-3}\ dm^3\ mol^{-1}} = \boxed{30.27\ dm^3\ mol^{-1}\ s^{-1}}$$

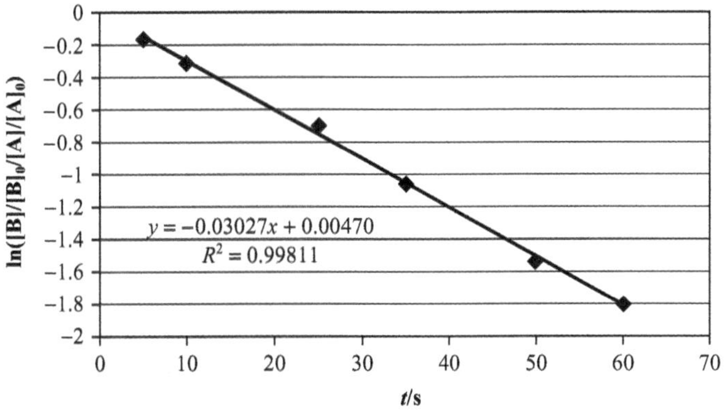

Figure 6.6

E6.22 The stoichiometry of the reaction relates product and reactant concentrations as follows:

$$[A] = [A]_0 - 2[B]$$

When the reaction goes to completion, $[B] = [A]_0/2$; hence $[A]_0 = 0.624\ mol\ dm^{-3}$. We can therefore tabulate [A], and examine its half-life. We see that the half-life of A from its initial concentration is approximately 1200 s, and that its half-life from the concentration at 1200 s is also 1200 s. This indicates a first-order reaction. We confirm this conclusion by plotting the data accordingly (in Figure 6.7), using

$$\ln\frac{[A]_0}{[A]} = k_A t$$

which follows from

$$\frac{d[A]}{dt} = -k_A[A]$$

t/s	0	600	1200	1800	2400
[B]/(mol dm^{-3})	0	0.089	0.153	0.200	0.230
[A]/(mol dm^{-3})	0.624	0.446	0.318	0.224	0.164
$\ln\frac{[A]_0}{[A]}$	0	0.34	0.67	1.02	1.34

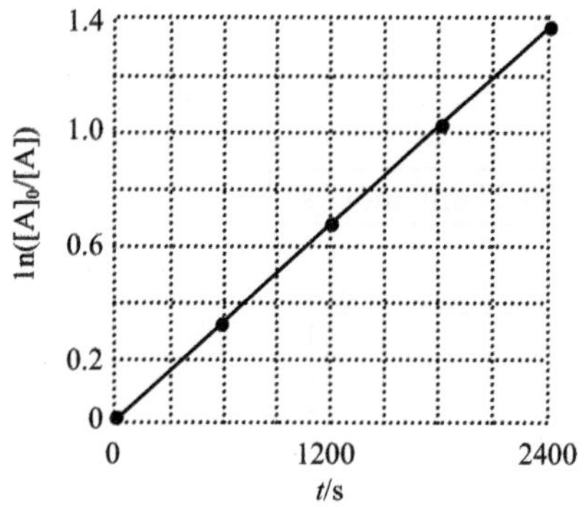

Figure 6.7

The points lie on a straight line, which confirms first-order kinetics. Since the slope of the line is 5.6×10^{-4}, we conclude that $k_A = 5.6 \times 10^{-4}\ s^{-1}$. To express the rate law in the form $v = k_r[A]$ we note that

$$v = -\frac{1}{2}\frac{d[A]}{dt} = -(\tfrac{1}{2}) \times (-k_A[A]) = \tfrac{1}{2}k_A[A]$$

and hence $k_r = \tfrac{1}{2}k_A = \boxed{2.8 \times 10^{-4}\ s^{-1}}$

E6.23 $$\text{rate} = -\frac{d[A]}{dt} = k[A]^3$$

$$\frac{d[A]}{[A]^3} = -k\,dt$$

$$\int_{[A]_0}^{[A]} \frac{d[A]}{[A]^3} = -k\int_0^t dt$$

We now use the standard integral $\int \frac{dx}{x^n} = -\frac{1}{(n-1)x^{n-1}} + \text{constant}$ where $n > 2$

which implies that

$$\int_a^b \frac{dx}{x^n} = \left\{-\frac{1}{(n-1)x^{n-1}} + \text{constant}\right\}\bigg|_b - \left\{-\frac{1}{(n-1)x^{n-1}} + \text{constant}\right\}\bigg|_a$$

$$= -\frac{1}{n-1}\left(\frac{1}{b^{n-1}} - \frac{1}{a^{n-1}}\right)$$

$$-\frac{1}{3-1}\left(\frac{1}{[A]^{3-1}} - \frac{1}{[A]_0^{3-1}}\right) = -kt$$

$$\boxed{\frac{1}{[A]^2} = \frac{1}{[A]_0^2} + 2kt}$$

Confirmation of a third-order reaction rate is provided by a linear plot of $\frac{1}{[A]^2}$ against t. The third-order rate constant equals the linear regression slope divided by 2.

E6.24 $$v=\frac{\mathrm{d}[\mathrm{P}]}{\mathrm{d}t}=k_\mathrm{r}[\mathrm{A}][\mathrm{B}]$$

Let the initial concentrations be $[\mathrm{A}]_0=A_0$, $[\mathrm{B}]_0=B_0$, and $[\mathrm{P}]_0=0$, then, when P is formed in concentration x, the concentration of A changes to A_0-2x and that of B changes to B_0-3x. Therefore,

$$\frac{\mathrm{d}[\mathrm{P}]}{\mathrm{d}t}=\frac{\mathrm{d}x}{\mathrm{d}t}=k_\mathrm{r}(A_0-2x)(B_0-3x)\quad\text{with } x=0 \text{ at } t=0.$$

$$\int_0^t k_\mathrm{r}\,\mathrm{d}t=\int_0^x\frac{\mathrm{d}x}{(A_0-2x)\times(B_0-3x)}$$

$$=\int_0^x\left(\frac{6}{2B_0-3A_0}\right)\times\left(\frac{1}{3(A_0-2x)}-\frac{1}{2(B_0-3x)}\right)\mathrm{d}x$$

$$=\left(\frac{-1}{(2B_0-3A_0)}\right)\times\left(\int_0^x\frac{\mathrm{d}x}{x-(1/2)A_0}-\int_0^x\frac{\mathrm{d}x}{x-(1/3)B_0}\right)$$

$$k_\mathrm{r}t=\left(\frac{-1}{(2B_0-3A_0)}\right)\times\left[\ln\left(\frac{x-\frac{1}{2}A_0}{-\frac{1}{2}A_0}\right)-\ln\left(\frac{x-\frac{1}{3}B_0}{-\frac{1}{3}B_0}\right)\right]$$

$$=\left(\frac{-1}{2B_0-3A_0}\right)\ln\left(\frac{(2x-A_0)B_0}{A_0(3x-B_0)}\right)$$

$$=\boxed{\left(\frac{1}{(3A_0-2B_0)}\right)\ln\left(\frac{(2x-A_0)B_0}{A_0(3x-B_0)}\right)}$$

E6.25 $$\frac{\mathrm{d}[\mathrm{A}]}{\mathrm{d}t}=-2k_\mathrm{r}[\mathrm{A}]^2[\mathrm{B}],\quad 2\mathrm{A}+\mathrm{B}\rightarrow\mathrm{P}$$

(a) Let $[\mathrm{P}]=x$ at t, then $[\mathrm{A}]=A_0-2x$ and $[\mathrm{B}]=B_0-x=\dfrac{A_0}{2}-x$. Therefore,

$$\frac{\mathrm{d}[\mathrm{A}]}{\mathrm{d}t}=-2\frac{\mathrm{d}x}{\mathrm{d}t}=-2k_\mathrm{r}(A_0-2x)^2\times(B_0-x)$$

$$\frac{\mathrm{d}x}{\mathrm{d}t}=k_\mathrm{r}(A_0-2x)^2\times\left(\frac{1}{2}A_0-x\right)=\frac{1}{2}k_\mathrm{r}(A_0-2x)^3$$

$$\frac{1}{2}k_\mathrm{r}t=\int_0^x\frac{\mathrm{d}x}{(A_0-2x)^3}=\frac{1}{4}\times\left[\left(\frac{1}{A_0-2x}\right)^2-\left(\frac{1}{A_0}\right)^2\right]$$

Therefore, $\boxed{k_\mathrm{r}t=\dfrac{2x(A_0-x)}{A_0^2(A_0-2x)^2}}$

(b) Now $B_0=A_0$, so

$$\frac{\mathrm{d}x}{\mathrm{d}t}=k_\mathrm{r}(A_0-2x)^2\times(B_0-x)=k_\mathrm{r}(A_0-2x)^2\times(A_0-x)$$

$$k_\mathrm{r}t=\int_0^x\frac{\mathrm{d}x}{(A_0-2x)^2\times(A_0-x)}$$

We proceed by the method of partial fractions (which is employed in the general case too), and look for the values of α, β, and γ such that

$$\frac{1}{(A_0-2x)^2\times(A_0-x)}=\frac{\alpha}{(A_0-2x)^2}+\frac{\beta}{A_0-2x}+\frac{\gamma}{A_0-x}.$$

This requires that

$$\alpha(A_0-x)+\beta(A_0-2x)\times(A_0-x)+\gamma(A_0-2x)^2=1$$

Expand and gather terms by powers of x:

$$(A_0\alpha+A_0^2\beta+A_0^2\gamma)-(\alpha+3\beta A_0+4\gamma A_0)x+(2\beta+4\gamma)x^2=1$$

This must be true for all x; therefore

$$A_0\alpha+A_0^2\beta+A_0^2\gamma=1$$

$$\alpha+3A_0\beta+4A_0\gamma=0$$

$$2\beta+4\gamma=0$$

These solve to give $\alpha=\dfrac{2}{A_0}$, $\beta=\dfrac{-2}{A_0^2}$, and $\gamma=\dfrac{1}{A_0^2}$

Therefore,

$$kt=\int_0^x\left(\frac{(2/A_0)}{(A_0-2x)^2}-\frac{(2/A_0^2)}{A_0-2x}+\frac{(1/A_0^2)}{A_0-x}\right)\mathrm{d}x$$

$$=\left(\frac{(1/A_0)}{A_0-2x}+\frac{1}{A_0^2}\ln(A_0-2x)-\frac{1}{A_0^2}\ln(A_0-x)\right)\Bigg|_0^x$$

$$=\boxed{\left(\frac{2x}{A_0^2(A_0-2x)}\right)+\left(\frac{1}{A_0^2}\right)\ln\left(\frac{A_0-2x}{A_0-x}\right)}$$

E6.26 Because $(1/2)^6=1/64$, 6 half-lives must pass for the concentration to fall to 1/64th of its initial values for this first-order reaction.

$$t=t_{1/2}\times 6=221\text{ s}\times 6=\boxed{1.33\times 10^3\text{ s}}$$

E6.27 For a first-order reaction: $k_r=\dfrac{\ln 2}{t_{1/2}}$ [6.13]. Therefore,

$$t=\frac{1}{k_r}\ln\frac{[^{14}\mathrm{C}]_0}{[^{14}\mathrm{C}]}\,[6.12\mathrm{b}]=\frac{t_{1/2}}{\ln 2}\ln\frac{[^{14}\mathrm{C}]_0}{[^{14}\mathrm{C}]}$$

$$=\frac{(5730\text{ a})}{\ln 2}\ln\frac{[^{14}\mathrm{C}]_0}{0.69[^{14}\mathrm{C}]_0}=\frac{(5730\text{ a})}{\ln 2}\ln\frac{1}{0.69}$$

$$=\boxed{3067\text{ a}}\pm 100\text{ a}.$$

E6.28 $m_{^{90}\mathrm{Sr}}=m_{^{90}\mathrm{Sr},0}\mathrm{e}^{-k_rt}$ [6.12d] $=m_{^{90}\mathrm{Sr},0}\mathrm{e}^{-t\times(\ln 2)/t_{1/2}}$ [6.13] $=(1.00\ \mu\mathrm{g})\times\mathrm{e}^{-t\times(\ln 2)/(28.1\text{ a})}$

(a) When $t=19$ a,

$$m_{^{90}\mathrm{Sr}}=(1.00\ \mu\mathrm{g})\times\mathrm{e}^{-(19\text{ a})\times(\ln 2)/(28.1\text{ a})}=\boxed{0.63\ \mu\mathrm{g}}$$

(b) When $t=75$ a,

$$m_{^{90}\mathrm{Sr}}=(1.00\ \mu\mathrm{g})\times\mathrm{e}^{-(75\text{ a})\times(\ln 2)/(28.1\text{ a})}=\boxed{0.16\ \mu\mathrm{g}}$$

E6.29 Assume the reaction is first order, then

$$\frac{N}{N_0} = e^{-kt} \quad \text{then} \quad \ln\left(\frac{N}{N_0}\right) = -kt \quad \text{and} \quad k = \frac{\ln 2}{t_{1/2}}$$

We are told

$N_0 = 10^9$ and $N = 10^9 - 1$.

Solving for t from above, we have

$$t = \frac{\ln\left(\frac{N}{N_0}\right)}{-k} = \frac{\ln\left(\frac{10^9 - 1}{10^9}\right)}{-k} = \frac{1 \times 10^{-9}}{k}$$

$$t = \frac{1 \times 10^{-9}\, t_{1/2}}{\ln 2} = \boxed{1.44 \times 10^{-9} t_{1/2}} = 1.44 \times 10^{-9} \times 1.3 \times 10^5 \text{ a}$$

$$t = 1.87 \times 10^{-4} \text{ a} = \boxed{1.64 \text{ min}}$$

E6.30 For first-order kinetics we have the following relations

$$\frac{[A]}{[A]_0} = e^{-k_r t} \text{ [6.12d]} \quad \text{and} \quad k_r = \frac{\ln 2}{t_{1/2}} \quad \text{[6.13]}$$

k_r is given as 4.5 h.

$$\frac{[A]}{[A]_0} = e^{-\frac{\ln 2}{t_{1/2}} t} = e^{-\frac{\ln 2}{4.5\text{ h}} \times 2\text{ h}} = 0.735$$

The mass of phenobarbitol that remains after 2 hours is

$0.735 \times (30 \text{ mg/kg}) \times 15 \text{ kg} = 328 \text{ mg}.$

To restore to the original level of phenobarbitol which is $30 \text{ mg/kg} \times 15 \text{ kg} = 450 \text{ mg}$, about $\boxed{120 \text{ mg}}$ must be re-injected.

E6.31 This is a reaction of the type $A + B \rightarrow P$.

The integrated rate law for this type of reaction is

$$k_r t = \frac{1}{[B]_0 - [A]_0} \ln\left\{\left(\frac{[B]}{[B]_0}\right) \Big/ \left(\frac{[A]}{[A]_0}\right)\right\} \quad \text{[6.17b]}$$

Introducing $[B] = [B]_0 - x$ and $[A] = [A]_0 - x$ and rearranging we obtain

$$kt = \left(\frac{1}{[B]_0 - [A]_0}\right) \ln\left(\frac{[A]_0([B]_0 - x)}{([A]_0 - x)[B]_0}\right)$$

Solving for x yields, after some rearranging,

$$x = \frac{[A]_0[B]_0\{e^{k([B]_0-[A]_0)t} - 1\}}{[B]_0 e^{([B]_0-[A]_0)kt} - [A]_0} = \frac{(0.055) \times (0.150 \text{ mol dm}^{-3}) \times \{e^{(0.150-0.055)\times 0.11 \times t/\text{s}} - 1\}}{(0.150) \times \{e^{(0.150-0.055)\times 0.11 \times t/\text{s}}\} - 0.055}$$

$$= \frac{(0.150 \text{ mol dm}^{-3}) \times (e^{10.45\times 10^{-3} t/\text{s}} - 1)}{2.73 e^{10.45\times 10^{-3} t/\text{s}} - 1}$$

(a) $x = \dfrac{(0.150\ \text{mol dm}^{-3}) \times (e^{0.1568} - 1)}{2.73e^{0.1568} - 1} = 1.21 \times 10^{-2}\ \text{mol dm}^{-3}$

which implies that $[\text{NaOH}] = (0.055 - 0.0121)\ \text{mol dm}^{-3} = \boxed{0.043\ \text{mol dm}^{-3}}$ and

$$[CH_3COOC_2H_5] = (0.150 - 0.0121)\ \text{mol dm}^{-3} = \boxed{0.138\ \text{mol dm}^{-3}}$$

(b) Perform the above calculation with $t = 15\ \text{min} = 900\ \text{s}$.

$$x = \frac{(0.150\ \text{mol dm}^{-3}) \times (e^{9.405} - 1)}{2.73e^{9.405} - 1} = 0.0549\ \text{mol dm}^{-3}$$

Hence, $[\text{NaOH}] = (0.055 - 0.0549)\ \text{mol dm}^{-3} = \boxed{0.0001\ \text{mol dm}^{-3}}$, which is effectively zero to within the significant figures given with the data, and

$$[CH_3COOC_2H_5] = (0.150 - 0.0549)\ \text{mol dm}^{-3} = \boxed{0.0951\ \text{mol dm}^{-3}}$$

E6.32 $2\,A \rightarrow P$

Since the rate law is second order in A, the differential rate expression is

$v = -\frac{1}{2}\,d[A]/dt\ [6.15a] = k_r[A]^2$.

Note the factor $\frac{1}{2}$ that originates from the stoichiometric coefficient of A!

The integrated rate expression is $\dfrac{1}{[A]} - \dfrac{1}{[A]_0} = 2k_r t \quad [6.15b]$.

Solving for t gives

$$t = \frac{1}{2k_r}\left(\frac{1}{[A]} - \frac{1}{[A]_0}\right)$$

$$= \frac{1}{2 \times (1.24 \times 10^{-3}\ \text{dm}^3\ \text{mol}^{-1}\ \text{s}^{-1})}\left(\frac{1}{0.026\ \text{mol dm}^{-3}} - \frac{1}{0.260\ \text{mol dm}^{-3}}\right)$$

$$= \boxed{1.40 \times 10^4\ \text{s.}}$$

E6.33 The rate law $\dfrac{d[A]}{dt} = -k_r[A]^n$ [6.15a] for $n \neq 1$ integrates to

$$k_r t = \left(\frac{1}{n-1}\right) \times \left(\frac{1}{[A]^{n-1}} - \frac{1}{[A]_0^{n-1}}\right)$$

At $t = t_{1/2}$, $k_r t_{1/2} = \left(\dfrac{1}{n-1}\right)\left[\left(\dfrac{2}{[A]_0}\right)^{n-1} - \left(\dfrac{1}{[A]_0}\right)^{n-1}\right]$

At $t = t_{3/4}$, $k_r t_{3/4} = \left(\dfrac{1}{n-1}\right)\left[\left(\dfrac{4}{3[A]_0}\right)^{n-1} - \left(\dfrac{1}{[A]_0}\right)^{n-1}\right]$

Hence, $\dfrac{t_{1/2}}{t_{3/4}} = \boxed{\dfrac{2^{n-1} - 1}{\left(\frac{4}{3}\right)^{n-1} - 1}}$

E6.34 (a) See the solution to Exercise 6.33 above. At $t = t_{1/2}$

$$kt_{1/2} = \left(\frac{1}{n-1}\right)\left[\left(\frac{2}{[\mathrm{A}]_0}\right)^{n-1} - \left(\frac{1}{[\mathrm{A}]_0}\right)^{n-1}\right] = \left(\frac{2^{n-1}-1}{n-1}\right)\times\left(\frac{1}{[\mathrm{A}]_0^{n-1}}\right)$$

$$t_{1/2} = \boxed{\left(\frac{2^{n-1}-1}{n-1}\right)\times\left(\frac{1}{k_\mathrm{r}[\mathrm{A}]_0^{n-1}}\right)}$$ as was to be demonstrated.

(b) Similarly for $t = t_{1/3}$ at which $[\mathrm{A}] = [\mathrm{A}]_0/3$,

$$kt_{1/3} = \left(\frac{1}{n-1}\right)\left[\left(\frac{3}{[\mathrm{A}]_0}\right)^{n-1} - \left(\frac{1}{[\mathrm{A}]_0}\right)^{n-1}\right] = \left(\frac{3^{n-1}-1}{n-1}\right)\times\left(\frac{1}{[\mathrm{A}]_0^{n-1}}\right)$$

and $$t_{1/3} = \boxed{\frac{3^{n-1}-1}{k_\mathrm{r}(n-1)}[\mathrm{A}]_0^{1-n}}$$

E6.35 $$\ln\frac{k_{\mathrm{r},2}}{k_{\mathrm{r},1}} = \frac{E_\mathrm{a}}{R}\left(\frac{1}{T_1} - \frac{1}{T_2}\right) \quad [6.20]$$

Solve the above equation for E_a.

$$E_\mathrm{a} = \frac{R}{\left(\frac{1}{T_1} - \frac{1}{T_2}\right)}\ln\frac{k_{\mathrm{r},2}}{k_{\mathrm{r},1}}$$

$$= \left(\frac{8.3145\ \mathrm{J\ K^{-1}\ mol^{-1}}}{\frac{1}{292\ \mathrm{K}} - \frac{1}{310\ \mathrm{K}}}\right)\times\ln\left(\frac{1.38\times10^{-3}}{1.78\times10^{-4}}\right)$$

$$= \boxed{85.6\ \mathrm{kJ\ mol^{-1}}}$$

For A, use

$$A = k_\mathrm{r}(T)\times \mathrm{e}^{E_\mathrm{a}/RT} \quad [6.19\mathrm{c}]$$

$$= 1.78\times10^{-4}\ \mathrm{mol\ dm^{-3}\ s^{-1}}\times \mathrm{e}^{85600/(8.3145\times292)}$$

$$= \boxed{3.65\times10^{11}\ \mathrm{mol\ dm^{-3}\ s^{-1}}}.$$

E6.36 $$\ln\frac{k_{\mathrm{r},2}}{k_{\mathrm{r},1}} = \frac{E_\mathrm{a}}{R}\left(\frac{1}{T_1} - \frac{1}{T_2}\right) \quad [6.20]$$

$$\ln 1.1 = \left(\frac{408\times10^3\ \mathrm{J\ mol^{-1}}}{8.3145\ \mathrm{J\ K^{-1}\ mol^{-1}}}\right)\times\left(\frac{1}{298.15\ \mathrm{K}} - \frac{1}{T_2}\right)$$

$$0.0953 = 4.91\times10^4\ \mathrm{K}\times\left(\frac{1}{298.15\ \mathrm{K}} - \frac{1}{T_2}\right)$$

Solve for T_2.

$T_2 = \boxed{298.32\ \mathrm{K}}$. A very small temperature increase is required!

E6.37 $k_r(T) = Ae^{-E_a/RT}$ [6.16c]

A rate constant responds more strongly to an increase in temperature when the slope dk_r/dT, or equivalently $d(\ln k_r)/dT$, is larger. Taking the natural logarithm of eqn 6.16c and evaluating the temperature derivative gives

$$\ln k_r = \ln A - \frac{E_a}{RT}$$

$$\frac{d\ln k_r}{dT} = \frac{E_a}{RT^2}.$$

Examination of the right side of the above equation shows that at any particular temperature the reaction that has the greater value of E_a has the greater response to an increase in temperature. That is, the reaction for which $\boxed{E_a = 52\ \text{kJ mol}^{-1}}$ responds more strongly than the reaction for which $E_a = 25\ \text{kJ mol}^{-1}$ even though it may have a smaller rate constant (depending upon relative values of pre-exponential factors).

E6.38 $$\ln \frac{k_{r,2}}{k_{r,1}} = \frac{E_a}{R}\left(\frac{1}{T_1} - \frac{1}{T_2}\right) \quad [6.20]$$

Solve the above equation for E_a.

$$E_a = \frac{R}{\left(\frac{1}{T_1} - \frac{1}{T_2}\right)} \ln \frac{k_{r,2}}{k_{r,1}}$$

$$= \left(\frac{8.3145\ \text{J K}^{-1}\ \text{mol}^{-1}}{\frac{1}{293\ \text{K}} - \frac{1}{300\ \text{K}}}\right) \times \ln(1.23)$$

$$= \boxed{21.6\ \text{kJ mol}^{-1}}$$

E6.39 Use eqns 6.16(a) and (b) of the text. Plot $\ln k_r$ against $1/T$. The slope of the plot is $-E_a/R$. See Figure 6.8. Thus,

$$-3620.8\ \text{K}^{-1} = -\frac{E_a}{R} = -\frac{E_a}{8.3145\ \text{J K}^{-1}\ \text{mol}^{-1}}$$

$$E_a = \boxed{30.1\ \text{kJ mol}^{-1}}$$

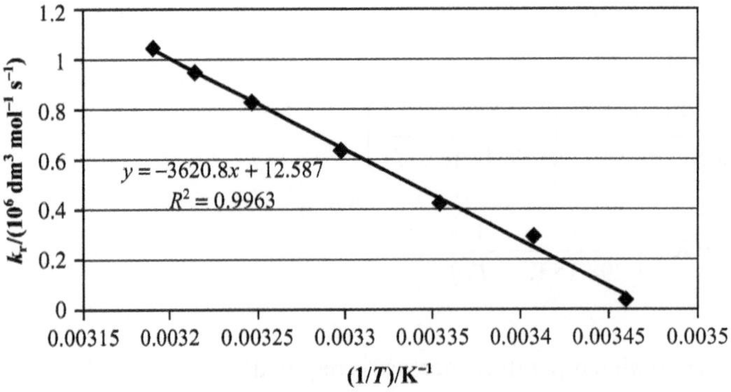

Figure 6.8

E6.40 $$\ln\frac{k_{r,2}}{k_{r,1}}=\frac{E_a}{R}\left(\frac{1}{T_1}-\frac{1}{T_2}\right) \quad [6.20]$$

Solve the above equation for E_a.

$$E_a=\frac{R}{\left(\frac{1}{T_1}-\frac{1}{T_2}\right)}\ln\frac{k_{r,2}}{k_{r,1}}$$

$$=\left(\frac{8.3145\ \text{J K}^{-1}\ \text{mol}^{-1}}{\frac{1}{277\ \text{K}}-\frac{1}{298\ \text{K}}}\right)\times\ln(40)$$

$$=\boxed{121\ \text{kJ mol}^{-1}}$$

E6.41 $$\ln\frac{k_{r,2}}{k_{r,1}}=\frac{E_a}{R}\left(\frac{1}{T_1}-\frac{1}{T_2}\right) \quad [6.20]$$

Solve the above equation for E_a.

$$E_a=\frac{R}{\left(\frac{1}{T_1}-\frac{1}{T_2}\right)}\ln\frac{k_{r,2}}{k_{r,1}}$$

Substitute $k_r=\dfrac{\ln 2}{t_{1/2}}$ [6.13] at each temperature.

$$E_a=\frac{R}{\left(\frac{1}{T_1}-\frac{1}{T_2}\right)}\ln\frac{t_{1/2}(T_1)}{t_{1/2}(T_2)}$$

$$=\left(\frac{8.3145\ \text{J K}^{-1}\ \text{mol}^{-1}}{\frac{1}{293\ \text{K}}-\frac{1}{283\ \text{K}}}\right)\times\ln\left(\frac{1}{2}\right)$$

$$=\boxed{47.8\ \text{kJ mol}^{-1}}$$

Solution to project

P6.42 The reaction is

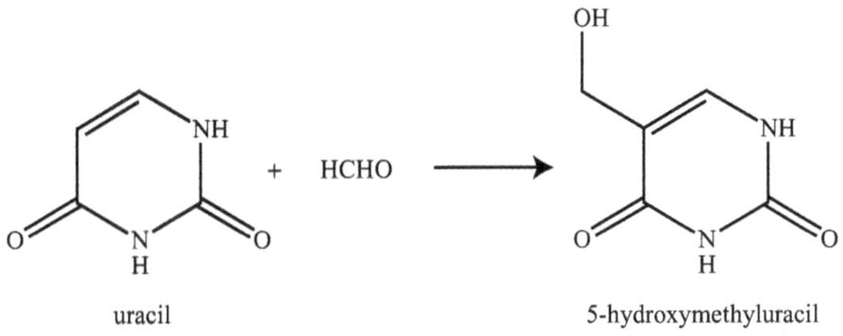

Properties at pH = 7:

$$\log k_r/(\text{dm}^3\,\text{mol}^{-1}\,\text{s}^{-1}) = 11.75 - 5488/(T/\text{K})$$

$$\log K = -1.36 + 1794/(T/\text{K})$$

(a) Plots of k_r and K against T in the range 273 K to 323 K are shown in Figures 6.9 and 6.10.

Figure 6.9 shows that, as expected, an increase in T causes an increase in k_r. Figure 6.10 shows a decrease in K with increasing T, a condition that indicates an exothermic reaction. From a kinetic point of view the reaction becomes more favorable at higher temperatures; from a thermodynamic point of view it becomes less favorable.

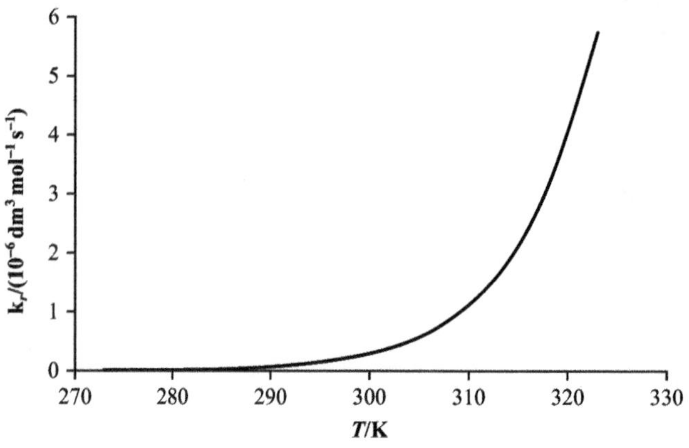

Figure 6.9

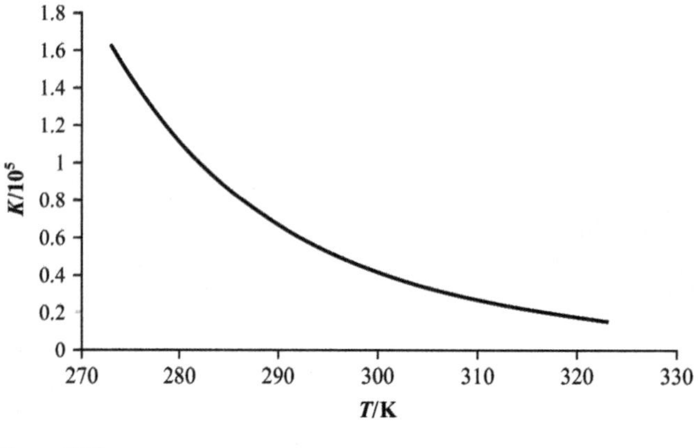

Figure 6.10

(b) Examination of the antilog of the above empirical rate constant relation,

$$k_r/(\text{dm}^3\,\text{mol}^{-1}\,\text{s}^{-1}) = 10^{11.75-5488/(T/\text{K})} = 10^{11.75} \times 10^{-5488/(T/\text{K})} = (5.62 \times 10^{11}) \times 10^{-5488/(T/\text{K})},$$

reveals that it has the Arrhenius form

$$k_r = A\,e^{-E_a/RT} \quad [6.19c]$$

with the pre-exponential factor $A = 5.62 \times 10^{11}$ dm^3 mol^{-1} s^{-1}. The activation energy is deduced by equating the exponentials and solving for E_a.

$$e^{-E_a/RT} = 10^{-5488/(T/\text{K})}$$

$$-E_a/RT = \ln(10^{-5488/(T/\text{K})})$$

$$-E_a/RT = (-5488/(T/\text{K})) \times \ln 10$$

$$E_a = (\ln 10) \times (5488\ \text{K}^{-1}) \times R = (\ln 10) \times (5488\ \text{K}^{-1}) \times (8.3145\ \text{J K}^{-1}\ \text{mol}^{-1})$$

$$= \boxed{105\ \text{kJ mol}^{-1}}$$

The standard biological Gibbs energy (i.e. pH = 7 and other activities equal to 1 in the biological standard state; see Section 4.2) is related to the above empirical expression for K.

$$\Delta_r G^{\oplus} = -RT \ln K \text{ [see Sections 4.3 and 4.7]} = -RT \ln(10) \log K$$

At 298.15 K,

$$\Delta_r G^{\oplus} = -(8.3145\ \text{J K}^{-1}\ \text{mol}^{-1}) \times (298.15\ \text{K}) \times (\ln 10) \times \left(-1.36 + \frac{1794}{298.15}\right)$$

$$= \boxed{-26.6\ \text{kJ mol}^{-1}}.$$

To find a relation between $\Delta_r H^{\oplus}$ and log K, consider the relation

$$\ln K = \frac{-\Delta_r G^{\oplus}}{RT} = -\frac{\Delta_r H^{\oplus}}{RT} + \frac{\Delta_r S^{\oplus}}{R}.$$

We use the approximation that the standard reaction enthalpy and entropy are independent of temperature over the range of interest and take the derivative w/r/t $1/T$. Solving the result for $\Delta_r H^{\oplus}$ yields a very useful relation:

$$\Delta_r H^{\oplus} = -R\frac{\mathrm{d}\ln(K)}{\mathrm{d}(1/T)} = -R\ln(10)\frac{\mathrm{d}\log(K)}{\mathrm{d}(1/T)}.$$

Substitution of the above empirical relation for log K gives the value of $\Delta_r H^{\oplus}$.

$$\Delta_r H^{\oplus} = -R\ln(10)\frac{\mathrm{d}}{\mathrm{d}(1/T)}\left(-1.36 + \frac{1794}{T/\text{K}}\right) = -(8.3145\ \text{J K}^{-1}\ \text{mol}^{-1}) \times (\ln 10) \times (1794\ \text{K})$$

$$= \boxed{-34.3\ \text{kJ mol}^{-1}}$$

(c) The equations for the rate constant k_r and the equilibrium constant K were obtained under conditions corresponding to the biological standard state of pH = 7. Thus the values of $\Delta_r G$ calculated from the equation for K are $\Delta_r G^{\oplus}$ values which can differ significantly from $\Delta_r G^{\ominus}$ for which pH = 0. Prebiotic conditions are more likely to be near pH = 7 than pH = 0 so we expect the part (a) plot of K against T to be relevant to the prebiotic environment. The plot shows that the reaction will be $\boxed{\text{thermodynamically favorable}}$ ($K \gg 1$). Because $\Delta_r G = \Delta_r G^{\oplus} + RT \ln Q$ [4.2] and since we might expect $Q < 1$ in a prebiotic environment, $\Delta_r G < \Delta_r G^{\oplus}$. But, as shown in the calculation above, $\Delta_r G^{\oplus}$ is rather large and negative (–26.6 kJ mol^{-1}), so we expect it will still be large and negative under the prebiotic conditions; hence the reaction will be spontaneous for these conditions. We also expect that $\Delta_r H \approx \Delta_r H^{\oplus}$ under prebiotic conditions because enthalpy changes largely reflect bond-breakage and bond-formation energies.

7 Accounting for the rate laws

Answers to discussion questions

D7.1 Figure 7.1 sketches the concentration variations for second-order reversible steps with an assumed equilibrium constant equal to 1. This figure and Figure 7.2 of the text are qualitatively comparable. The quantitative difference is that text Figure 7.2 illustrates a reaction for which the equilibrium constant is greater than 1.

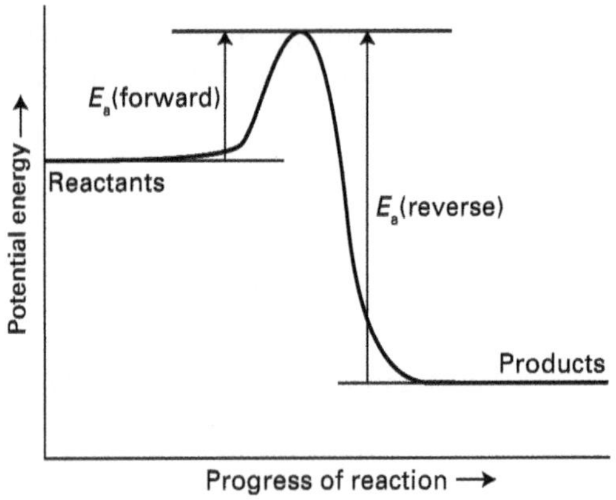

Figure 7.1

D7.2 Heme oxygenase (HO) plays a key role in heme degradation by catalyzing the oxidation of iron protoporphyrin IX, producing biliverdin, CO, and free Fe^{2+}. In humans, biliverdin reductase subsequently reduces biliverdin to bilirubin, which is amenable to excretion. HO is a potent antioxidant capable of protecting tumor cells from chemical therapies and HO inhibitors may provide a route to anticancer, antibacterial, and antifungal therapies. HO inhibitors have been developed for suppression of neonatal jaundice, a common condition in newborns caused by inefficient bilirubin elimination.

HO binding and dissociation rate constants for the association with isopropyl, *n*-butyl, and benzyl isocyanides have been investigated with stopped-flow kinetic measurements at 20°C and pH = 7.4. Stopped-flow concentrations being measured spectrophotometrically with a diode array detector. The dissociation constant is found to be $K_D = 6 \times 10^{-6}$ for the Fe^{3+}-benzyl isocyanide complex and $K_D = 0.03 \times 10^{-6}$ for the Fe^{2+}-benzyl isocyanide complex. Observed binding rates to Fe^{2+}-HO as a function of isocyanide concentration are shown in Figure 7.2. Benzyl isocyanide binds the most tightly and rapidly of the three inhibitors suggesting an HO active site that is more easily accessed by more hydrophobic ligands.

Reference: J.P. Evans, S. Kandel, and P.R. Ortiz de Montellano, *Biochemistry*, 2009, **48**, 8920–8928.

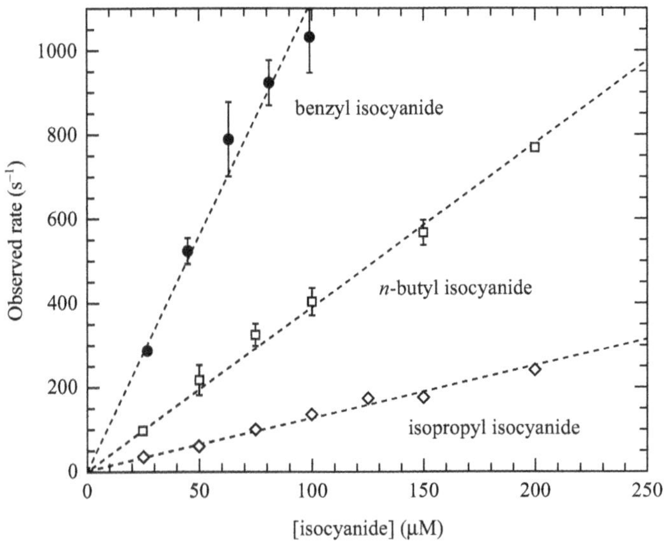

Figure 7.2

D7.3 The **rate-determining step** is not just the slowest step: it must be slow *and* be a crucial gateway for the formation of products. If a faster reaction can also lead to products, then the slowest step is irrelevant because the slow reaction can then be side-stepped. The rate-determining step is like a slow ferry crossing between two fast highways: the overall rate at which traffic can reach its destination is determined by the rate at which it can make the ferry crossing.

If the first step in a mechanism is the slowest step with the highest activation energy, then it is rate determining, and the overall reaction rate is equal to the rate of the first step because all subsequent steps are so fast that once the first intermediate is formed it results immediately in the formation of products. Once over the initial barrier, the intermediates cascade into products. However, a rate-determining step may also stem from the low concentration of a crucial reactant or catalyst and need not correspond to the step with highest activation barrier. A rate-determining step arising from the low activity of a crucial enzyme can sometimes be identified by determining whether or not the reactants and products for that step are in equilibrium: if the reaction is not at equilibrium it suggests that the step may be slow enough to be rate determining.

D7.4 The **steady-state approximation** is the assumption that the concentrations of all intermediates remain constant and small throughout the reaction (except right at the beginning and right at the end). The mathematical form of the approximation for intermediate I is

$$\text{net rate of formation of I} = 0$$

or with the symbols of calculus

$$(\mathrm{d}[\mathrm{I}]/\mathrm{d}t)_{\mathrm{net}} = 0$$

A **pre-equilibrium approximation** is similar in that it is a good approximation when the rate of formation of the intermediate from the reactants and the rate of its reversible decay back to the reactants are both very fast in comparison to the rate of formation of the product from the intermediate. This results in the intermediate being in approximate equilibrium with the reactants over relatively long time periods. Hence, the concentration of the intermediate remains approximately constant over the time period that the equilibrium can be considered to be maintained. This allows one to relate the rate constants and concentrations to each other through a constant (the pre-equilibrium constant).

To illustrate the two approximations, consider the symbolic generalization of the gas-phase mechanism:

$$A + A \underset{k_a'}{\overset{k_a}{\rightleftharpoons}} I$$
$$I + B \xrightarrow{k_b} P$$

Application of the steady-state approximation to intermediate I gives

$$\text{net rate of formation of I} = k_a[A]^2 - k_a'[I] - k_b[I][B] = 0$$

which implies that

$$[I] = \frac{k_a[A]^2}{k_a' - k_b[B]}$$

It follows that the rate of formation of product is

$$\text{rate of formation of P} = k_b[I][B] = \frac{k_a k_b[A]^2[B]}{k_a' - k_b[B]} \quad \text{(steady-state approx.)}$$

In contrast, the pre-equilibrium approximation assumes the equilibrium condition that the formation rate of I equals the rate at which I decomposes to reactants. Thus, $k_a[A]^2 = k_a'[I]$ so that

$$\frac{[I]}{[A]^2} = \frac{k_a}{k_a'} = K$$

which implies that

$$[I] = K[A]^2$$

It follows that the rate of formation of product is

$$\text{rate of formation of P} = k_b[I][B] = k_b K[A]^2[B] \quad \text{(pre-equilibrium approx.)}$$

By comparing the rate of formation of P in the steady-state approximation with the rate in the pre-equilibrium approximation, we see that in general they give very different predictions about the concentration dependence of the rate of product formation. They do, however, agree in a special case. When the rate at which I decomposes to reactants is much faster than the rate at which it forms product, $k_a' \gg k_b[B]$ and the second term in the denominator of the steady-state approximation is negligibly small. In this case, the steady-state approximation simplifies to the form provided by the pre-equilibrium approximation. Thus, we see that the steady-state approximation describes a greater range of concentrations and rates than that provided by the pre-equilibrium approximation.

D7.5 A reaction in solution can be regarded as the outcome of two stages: one is the encounter of two reactant species; this is followed by their reaction in the second stage, if they acquire their activation energy. If the rate-determining step is the former, then the reaction is said to be diffusion controlled. If the rate-determining step is the latter, then the reaction is activation controlled. For a reaction of the form $A + B \rightarrow P$ that obeys the second-order rate law $v = k_r[A][B]$, in the diffusion-controlled regime,

$$k_r = k_d = 4\pi R^* D N_A$$

where D is the sum of the diffusion coefficients of the two reactant species and R^* is the distance at which reaction occurs. A further approximation is that each molecule obeys the Stokes–Einstein relation and Stokes' law, and then

$$k_d \approx \frac{8RT}{3\eta} \quad [7.27]$$

where η is the viscosity of the medium. The result suggests that k_d is independent of the radii of the reactants. It also suggests that the rate constant depends only weakly on temperature, so the activation energy is small.

When both diffusion and activation are important to a bimolecular step, the rate constant is given by

$$k_r = \frac{k_a k_d}{k_a + k_d'} \quad [7.25]$$

In the activation-controlled limit for which $k_d' \gg k_a$, eqn 7.25 reduces to

$$k_r = \frac{k_a k_d}{k_d'}$$

and the reaction rate depends on the rate at which energy accumulates in the encounter pair (as expressed by k_a).

D7.6

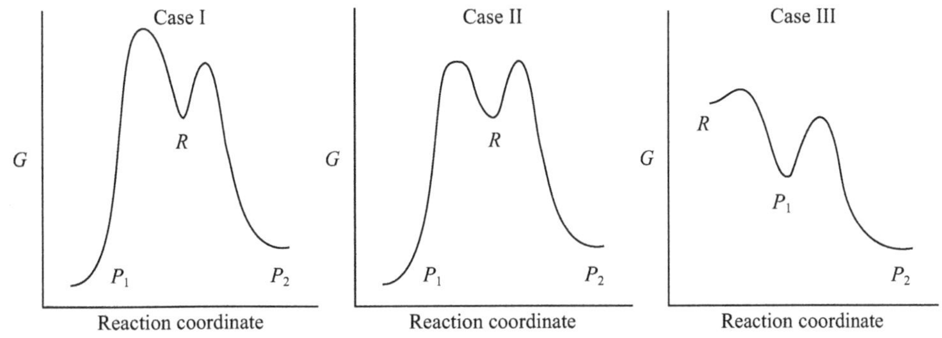

Simple diagrams of Gibbs energy against reaction coordinate are useful for distinguishing between kinetic and thermodynamic control of a reaction. For the simple parallel reactions $R \rightarrow P_1$ and $R \rightarrow P_2$, shown above as Cases I and II, the product P_1 is thermodynamically favored because the Gibbs energy decreases to a greater extent for its formation. However, the rate at which each product appears does not depend upon thermodynamic favorability. Rate constants depend upon activation energy. In Case I the activation energy for the formation of P_1 is much larger than that for formation of P_2. At low and moderate temperature the large activation energy may not be readily available and P_1 either cannot form or forms at a slow rate. The much smaller activation energy for P_2 formation is available and, consequently, P_2 is produced even though it is not the thermodynamically favored product. This is **kinetic control**. In this case, $[P_2]/[P_1] = k_{r2}/k_{r1} > 1$ [7.28].

The activation energies for the parallel reactions are equal in Case II and, consequently, the two products appear at identical rates. If the reactions are irreversible, $[P_2]/[P_1] = k_{r2}/k_{r1} = 1$ at all times. The results are very different for reversible reactions. The activation energy for $P_1 \rightarrow R$ is much larger than that for $P_2 \rightarrow R$ and P_1 accumulates as the more rapid $P_2 \rightarrow R \rightarrow P_1$ occurs. Eventually, the ratio $[P_2]/[P_1]$ approaches the equilibrium value for which

$$\left(\frac{[P_2]}{[P_1]}\right)_{eq} = e^{-(\Delta G_2 - \Delta G_1)/RT} < 1$$

This is **thermodynamic control**.

Case III above represents an interesting consecutive reaction series $R \rightarrow P_1 \rightarrow P_2$. The first step has relatively low activation energy and P_1 rapidly appears. However, the relatively large activation energy for the second step is not available at low and moderate temperatures. By using low or moderate temperatures and short reaction times it is possible to produce more of the thermodynamically less favorable P_1. This is kinetic control. High temperatures and long reaction times will yield the thermodynamically favored P_2.

The ratio of reaction products is determined by relative reaction rates in kinetic-controlled reactions. Favorable conditions include short reaction times, lower temperatures, and irreversible reactions. Thermodynamic control is favored by long reaction times, higher temperatures, and reversible reactions. The ratio of products depends on the relative stability of products for thermodynamically controlled reactions.

D7.7 Bimolecular mechanistic step: $A + B \xrightarrow{k_r}$ products

Description provided by the Eyring activated complex theory:

$$A + B \underset{}{\overset{K=[C^\ddagger]/[A][B]}{\rightleftharpoons}} C^\ddagger \rightarrow \text{products}$$

where $C^\ddagger$ is the **activated complex** at the **transition state** of the mechanistic step.

Eyring equation for the rate constant k_r: $k_r = \kappa(k^\ddagger T/h)K/c^\ominus$ [7.32]

The activated complex theory of bimolecular reaction postulates an equilibrium between reactants and an activated complex $C^\ddagger$ formed by the partial bonding of reactant molecules at the top of the potential energy profile for reaction. The activated complex has the requisite activation energy for reaction but energy flow must occur to cause bond lengthening and distortions, which characterize the transition state. Interactions of $C^\ddagger$ with solvent molecules may also be important to the transition state. The transition state either decomposes into reactant molecules or moves beyond the energy profile barrier to form product. The term kT/h within the Eyring equation arises from consideration of the motions of atoms that lead to the decay of $C^\ddagger$ into products, as specific bonds are broken and formed. The dependence upon temperature indicates that an increase in temperature enhances the rate by causing more vigorous motion in the activated complex, facilitating the rearrangement of atoms and the formation of new bonds. The **transmission coefficient** κ ($0 < \kappa \leqslant 1$) accounts for the fact that the activated complex may not pass through to the transition state. In the absence of information to the contrary, κ is assumed to be about 1.

D7.8 As discussed in Section 7.3(d), it is possible for the activation energy of a composite reaction to be negative. For example, should the rate constant for a net reaction have the form $k_r = k_{r1}k_{r2}/k_{r3}$, where k_{r1}, k_{r2}, and k_{r3} are the rate constants of three different mechanistic steps, the temperature dependence of k_r will exhibit an activation energy equivalent to $E_{a,r1} + E_{a,r2} - E_{a,r3}$. Therefore, it will be negative if $E_{a,r1} + E_{a,r2} < E_{a,r3}$. An important consequence of this discussion is that we have to be very cautious about making predictions about the effect of temperature on reactions that are the outcome of several steps. Enzyme-catalyzed reactions may also exhibit a strongly non-Arrhenius temperature dependence if the enzyme denatures at high temperatures and ceases to function.

D7.9 $\log k_r = \log k_r^\circ + 2Az_Az_BI^{1/2}$ [7.35]

Equation 7.35 expresses the dilute solution **kinetic salt effect**, the variation of the rate constant of a reaction between ions with the ionic strength of the solution. If the reactant ions have the same sign (as in a reaction between cations or between anions), then increasing the ionic strength by the addition of inert ions increases the rate constant. The formation of a single, highly charged ionic complex from two less highly charged ions is favored by a high ionic strength because the new ion has a denser ionic atmosphere and interacts with that atmosphere more strongly. Conversely, ions of opposite charge react more slowly in solutions of high ionic strength. Now, the charges cancel and the complex has a less favorable interaction with its atmosphere than the separated ions.

Solutions to exercises

E7.10 $$\mathrm{E} + \mathrm{S} \underset{k_r'}{\overset{k_r}{\rightleftharpoons}} \mathrm{ES}$$

At equilibrium, the forward and reverse rates are equal:

$$k_r[\mathrm{E}]_{eq}[\mathrm{S}]_{eq} = k_r'[\mathrm{ES}]_{eq} \quad \text{or} \quad \frac{[\mathrm{ES}]_{eq}}{[\mathrm{E}]_{eq}[\mathrm{S}]_{eq}} = \frac{k_r}{k_r'}$$

The latter is closely related to the equilibrium constant.

$$K = \frac{[\mathrm{ES}]_{eq}/c^{\ominus}}{([\mathrm{E}]_{eq}/c^{\ominus})([\mathrm{S}]_{eq}/c^{\ominus})} = \frac{[\mathrm{ES}]_{eq}}{[\mathrm{E}]_{eq}[\mathrm{S}]_{eq}}c^{\ominus} = \frac{k_r}{k_r'}c^{\ominus} \quad [7.1]$$

Solve this expression for k_r'.

$$k_r' = \frac{k_r c^{\ominus}}{K} = \frac{(7.4\times10^7\ \mathrm{dm^3\ mol^{-1}\ s^{-1}})\times(1\ \mathrm{mol\ dm^{-3}})}{235} = \boxed{3.1\times10^5\ \mathrm{s^{-1}}}$$

E7.11 $$\mathrm{A} \underset{k_r'}{\overset{k_r}{\rightleftharpoons}} \mathrm{B}$$

$$\frac{\mathrm{d}[\mathrm{A}]}{\mathrm{d}t} = -k_r[\mathrm{A}] + k_r'[\mathrm{B}] \quad \frac{\mathrm{d}[\mathrm{B}]}{\mathrm{d}t} = -k_r'[\mathrm{B}] + k_r[\mathrm{A}]$$

$[\mathrm{A}]+[\mathrm{B}] = [\mathrm{A}]_0 + [\mathrm{B}]_0$ at all times. Therefore, $[\mathrm{B}] = [\mathrm{A}]_0 + [\mathrm{B}]_0 - [\mathrm{A}]$.

$$\frac{\mathrm{d}[\mathrm{A}]}{\mathrm{d}t} = -k_r[\mathrm{A}] + k_r'\{[\mathrm{A}]_0 + [\mathrm{B}]_0 - [\mathrm{A}]\} = -(k_r + k_r')[\mathrm{A}] + k_r'([\mathrm{A}]_0 + [\mathrm{B}]_0)$$

Rearrangement to place concentration functions to the left of the equality, time functions to the right, and to produce unitless expressions on both sides of the equality gives

$$\frac{\mathrm{d}[\mathrm{A}]}{[\mathrm{A}] - k_r'([\mathrm{A}]_0 + [\mathrm{B}]_0)/(k_r + k_r')} = -(k_r + k_r')\mathrm{d}t$$

Integration of both sides between the concentration conditions at $t = 0$ and $t = t$ gives

$$\int_{[\mathrm{A}]_0}^{[\mathrm{A}]} \frac{\mathrm{d}[\mathrm{A}]}{[\mathrm{A}] - k_r'([\mathrm{A}]_0 + [\mathrm{B}]_0)/(k_r + k_r')} = -(k_r + k_r')\int_0^t \mathrm{d}t$$
$$= -(k_r + k_r')t$$

The remaining integral is a standard integral of the form

$$\int \frac{\mathrm{d}x}{ax+b} = \frac{1}{a}\ln(ax+b) \text{ [found in a maths handbook]}$$

Application to our working equation with the transformations $a = 1$ and $b = -k_r'([\mathrm{A}]_0 + [\mathrm{B}]_0)/(k_r + k_r')$ gives

$$\ln([\mathrm{A}] - k_r'([\mathrm{A}]_0 + [\mathrm{B}]_0)/(k_r + k_r'))\Bigg|_{[\mathrm{A}]=[\mathrm{A}]_0}^{[\mathrm{A}]=[\mathrm{A}]} = -(k_r + k_r')t \quad \text{[Evaluate difference; use } \ln x - \ln y = \ln(x/y)\text{]}$$

$$\ln\left(\frac{[\mathrm{A}] - k_r'([\mathrm{A}]_0 + [\mathrm{B}]_0)/(k_r + k_r')}{[\mathrm{A}]_0 - k_r'([\mathrm{A}]_0 + [\mathrm{B}]_0)/(k_r + k_r')}\right) = -(k_r + k_r')t \quad \text{[Exponentiate each side]}$$

$$\frac{[\mathrm{A}] - k_r'([\mathrm{A}]_0 + [\mathrm{B}]_0)/(k_r + k_r')}{[\mathrm{A}]_0 - k_r'([\mathrm{A}]_0 + [\mathrm{B}]_0)/(k_r + k_r')} = \mathrm{e}^{-(k_r + k_r')t} \quad \text{[Solve for [A]]}$$

$$[A] = k_r'([A]_0 + [B]_0)/(k_r + k_r') + \{[A]_0 - k_r'([A]_0 + [B]_0)/(k_r + k_r')\}e^{-(k_r+k_r')t} \quad \text{[Simplify]}$$

$$[A] = \frac{k_r'([A]_0 + [B]_0) + \{[A]_0(k_r + k_r') - k_r'([A]_0 + [B]_0)\}e^{-(k_r+k_r')t}}{k_r + k_r'}$$

$$\boxed{[A] = \frac{k_r'([A]_0 + [B]_0) + \{k_r[A]_0 - k_r'[B]_0\}e^{-(k_r+k_r')t}}{k_r + k_r'}}$$

Reaction stoichiometry provides the expression $[B] = [A]_0 + [B]_0 - [A]$, so we find that

$$[B] = [A]_0 - \frac{k_r'([A]_0 + [B]_0 + [B]_0) + \{k_r[A]_0 + [B]_0 - k_r'[B]_0\}e^{-(k_r+k_r')t}}{k_r + k_r'} \quad \text{[Simplify]}$$

$$= \frac{(k_r + k')[A]_0 - k_r'([A]_0 + [B]_0) - \{k_r[A]_0 - k_r'[B]_0\}e^{-(k_r+k_r')t} + [B]_0(k_r + k_r')}{k_r + k_r'}$$

$$= \frac{k_r[A]_0 + k_r[B]_0 - \{k_r[A]_0 - k_r'[B]_0\}e^{-(k_r+k_r')t}}{k_r + k_r'}$$

$$\boxed{[B] = \frac{k_r[A]_0(1 - e^{-(k_r+k_r')t}) + [B]_0(k_r + k_r'e^{-(k_r+k_r')t})}{k_r + k_r'}}$$

In the special case for which $[B]_0 = 0$, these general expressions reduce to eqns 7.2a and 7.2b of the text.

E7.12 $$H_2O(l) \underset{k_r'}{\overset{k_r}{\rightleftharpoons}} H^+(aq) + OH^-(aq)$$

$pK_w = 14.01$ at 298 K (thus, $[H^+]_{eq} = [OH^-]_{eq} = 10^{-14.01/2}$ mol dm^{-3}); $\tau = 37$ μs

The net rate of water formation after a dissociation perturbation, perhaps caused by a temperature jump, is

$$\frac{d[H_2O]}{dt} = -k_r[H_2O] + k_r'[H^+][OH^-]$$
$$= -k_r[H_2O] + k_r'[H^+]^2 \quad \text{[because } [H^+] = [OH^-]\text{]}$$

Define x with the expressions $[H_2O] = [H_2O]_{eq} - x$, and $[H^+] = [H^+]_{eq} + x$, recognize that $\frac{d[H_2O]}{dt} = \frac{d\{[H_2O]_{eq} - x\}}{dt} = \frac{d[H_2O]_{eq}}{dt} - \frac{dx}{dt} = -\frac{dx}{dt}$, and substitutions into the working equation gives

$$-\frac{dx}{dt} = -k_r\{[H_2O]_{eq} - x\} + k_r'\{[H^+]_{eq} + x\}^2$$
$$= -k_r\{[H_2O]_{eq} - x\} + k_r'\{[H^+]_{eq}^2 + 2[H^+]_{eq}x + x^2\}$$

The forward and reverse rates are equal at equilibrium so

$$k_r[H_2O]_{eq} = k_r'[H^+]_{eq}^2 \quad \text{[because } [H^+]_{eq} = [OH^-]_{eq}\text{]}$$

and the working equation becomes

$$-\frac{dx}{dt} = k_r x + k_r'\{2[H^+]_{eq}x + x^2\}$$
$$= (k_r + 2k_r'[H^+]_{eq})x + k_r'x^2$$

x is a small fraction of $[H^+]_{eq}$. Consequently, the x^2 term is negligibly smaller than the x term and can be discarded. The working equation becomes

$$\frac{dx}{dt} = -\frac{1}{\tau}x, \quad \text{where} \quad \tau^{-1} = k_r + 2k_r'[H^+]_{eq} = k_r + k_r'([H^+]_{eq} + [OH^-]_{eq})$$

Rearrangement with x on one side of the equality and t on the other followed by integration gives

$$\frac{\mathrm{d}x}{x} = -\frac{1}{\tau}\mathrm{d}t$$

$$\int_{x=x_0}^{x=x} \frac{\mathrm{d}x}{x} = -\frac{1}{\tau}\int_{t=0}^{t} \mathrm{d}t$$

$$\ln x - \ln x_0 = -\frac{t}{\tau} \quad [\text{use } \ln a - \ln b = \ln(a/b)]$$

$$\ln\frac{x}{x_0} = -\frac{t}{\tau} \quad [\text{exponentiate}]$$

$$x = x_0 \mathrm{e}^{-t/\tau}$$

This shows that the perturbation from equilibrium decays exponentially with the relaxation time given by

$$\boxed{\tau^{-1} = k_\mathrm{r} + k_\mathrm{r}'([\mathrm{H}^+]_\mathrm{eq} + [\mathrm{OH}^-]_\mathrm{eq})} = k_\mathrm{r} + 2k_\mathrm{r}'[\mathrm{H}^+]_\mathrm{eq}$$

(b) Evaluate the first equation of part (a) at equilibrium, where $(\mathrm{d}[\mathrm{H_2O}]/\mathrm{d}t)_\mathrm{eq} = 0$, to find

$$K_\mathrm{w} = [\mathrm{H}^+]_\mathrm{eq}[\mathrm{OH}^-]_\mathrm{eq}/(c^\ominus)^2 = \frac{k_\mathrm{r}[\mathrm{H_2O}]_\mathrm{eq}}{k_\mathrm{r}'(c^\ominus)^2} \quad \text{or} \quad k_\mathrm{r}' = \frac{k_\mathrm{r}[\mathrm{H_2O}]_\mathrm{eq}}{K_\mathrm{w} \times (c^\ominus)^2}$$

Substitution of this latter expression into the relaxation time expression of part (a) and solving for k_r gives

$$\frac{1}{\tau} = k_\mathrm{r} + 2k_\mathrm{r}'[\mathrm{H}^+]_\mathrm{eq} = k_\mathrm{r} + 2\left(\frac{k_\mathrm{r}[\mathrm{H_2O}]_\mathrm{eq}}{K_\mathrm{w} \times (c^\ominus)^2}\right)[\mathrm{H}^+]_\mathrm{eq} = k_\mathrm{r} + 2\left(\frac{k_\mathrm{r}[\mathrm{H_2O}]_\mathrm{eq}}{[\mathrm{H}^+]_\mathrm{eq}^2}\right)[\mathrm{H}^+]_\mathrm{eq} = k_\mathrm{r} + 2\left(\frac{k_\mathrm{r}[\mathrm{H_2O}]_\mathrm{eq}}{[\mathrm{H}^+]_\mathrm{eq}}\right)$$

$$k_\mathrm{r} = \frac{1}{\tau\{1 + 2[\mathrm{H_2O}]_\mathrm{eq}/[\mathrm{H}^+]_\mathrm{eq}\}}$$

Since $[\mathrm{H_2O}]_\mathrm{eq} = 55.6\ \mathrm{mol\ dm^{-3}}$ for pure water and $[\mathrm{H}^+]_\mathrm{eq} = [\mathrm{OH}^-]_\mathrm{eq} = 10^{-14.01/2}\ \mathrm{mol\ dm^{-3}}$, we find

$$k_\mathrm{r} = \frac{1}{(37 \times 10^{-6}\ \mathrm{s}) \times \{1 + 2 \times 55.6/10^{-14.01/2}\}} = \boxed{2.4\bar{0} \times 10^{-5}\ \mathrm{s^{-1}}}$$

$$k_\mathrm{r}' = \frac{k_\mathrm{r}[\mathrm{H_2O}]_\mathrm{eq}}{K_\mathrm{w} \times (c^\ominus)^2} = \frac{(2.4\bar{0} \times 10^{-5}\ \mathrm{s^{-1}}) \times (55.6\ \mathrm{mol\ dm^{-3}})}{10^{-14.01} \times (1\ \mathrm{mol\ dm^{-3}})^2} = \boxed{1.3\bar{7} \times 10^{11}\ \mathrm{mol^{-1}\ dm^3\ s^{-1}}}$$

The recombination rate constant for the hydrogen ion and hydroxide ion (k_r') is extremely large, making it the fastest reaction in water. Another example of a fast reaction is combination of the ammonium ion and the hydroxide ion in the dissociation reaction

$$\mathrm{NH_3(aq)} + \mathrm{H_2O(l)} \underset{k_\mathrm{r}'}{\overset{k_\mathrm{r}}{\rightleftharpoons}} \mathrm{NH_4^+(aq)} + \mathrm{OH^-(aq)}$$

where it is found that $k_\mathrm{r} = 1.2\bar{8} \times 10^4\ \mathrm{mol^{-1}\ dm^3\ s^{-1}}$ and $k_\mathrm{r}' = 4.0\bar{0} \times 10^{10}\ \mathrm{mol^{-1}\ dm^3\ s^{-1}}$.

E7.13 $2\,A \underset{k_r'}{\overset{k_r}{\rightleftharpoons}} A_2$

The net rate of A formation is

$$\frac{d[A]}{dt} = 2 \times (\text{rate of decomposition}) \text{ of } A_2 + 2 \times (\text{rate of formation of A}) = -2k_r[A]^2 + 2k_r'[A_2]$$

where the factor "2" originates with the reaction coefficient of A; 2 moles of A react in each step.

Suppose that a jump perturbation diminishes the equilibrium concentration of A_2 by x. Then, by the reaction stoichiometry:

$$[A_2] = [A_2]_{eq} - x \quad \text{and} \quad [A] = [A]_{eq} + 2x$$

Substituting these definitions into the equation for the net rate of A formation and recognizing that

$$\frac{d[A]}{dt} = \frac{d\{[A]_{eq} + 2x\}}{dt} = \frac{d[A]_{eq}}{dt} + 2\frac{dx}{dt} = 2\frac{dx}{dt} \text{ gives}$$

$$2\frac{dx}{dt} = -2k_r\{[A]_{eq} + 2x\}^2 + 2k_r'\{[A_2]_{eq} - x\}$$

$$\frac{dx}{dt} = -k_r\{[A]_{eq}^2 + 4[A]_{eq}x + 4x^2\} + k_r'\{[A_2]_{eq} - x\}$$

$$= -k_r\{4[A]_{eq}x + 4x^2\} - k_r'x \quad [\text{because } k_r[A]_{eq}^2 = k_r'[A_2]_{eq}]$$

x is a small fraction of $[A]_{eq}$. Consequently, the x^2 term is negligibly smaller than the x term and can be discarded. The working equation becomes

$$\frac{dx}{dt} = -\frac{1}{\tau}x, \quad \text{where} \quad \tau^{-1} = 4k_r[A]_{eq} + k_r'$$

Rearrangement with x on one side of the equality and t on the other followed by integration gives

$$\frac{dx}{x} = -\frac{1}{\tau}dt$$

$$\int_{x=x_0}^{x=x} \frac{dx}{x} = -\frac{1}{\tau}\int_{t=0}^{t} dt$$

$$\ln x - \ln x_0 = -\frac{t}{\tau} \quad [\text{use } \ln a - \ln b = \ln(a/b)]$$

$$\ln\frac{x}{x_0} = -\frac{t}{\tau} \quad [\text{exponentiate}]$$

$$x = x_0 e^{-t/\tau}$$

This shows that the perturbation from equilibrium decays exponentially with the relaxation time given by

$$\boxed{\tau^{-1} = 4k_r[A]_{eq} + k_r'}$$

E7.14 $2\,A \underset{k_r'}{\overset{k_r}{\rightleftharpoons}} A_2$

The net rate of A formation is

$$\frac{d[A]}{dt} = 2 \times (\text{rate of decomposition}) \text{ of } A_2 + 2 \times (\text{rate of formation of A})$$

$$= -2k_r[A]^2 + 2k_r'[A_2]$$

(a) At equilibrium, the rate of decomposition equals the rate of formation, so

$$k_r[A]^2_{eq} = k'_r[A_2]_{eq} \quad \text{and} \quad [A_2]_{eq} = k_r[A]^2_{eq}/k'_r$$

Furthermore, $[A]_{tot} = [A] + 2[A_2]$ is constant throughout relaxation, so we may write

$$[A]_{eq} = [A]_{tot} - 2[A_2]_{eq}$$

Now, the relaxation time τ is found in E7.13 to be

$$\tau^{-1} = 4k_r[A]_{eq} + k'_r$$

Squaring this expression, substitution of $[A]_{eq} = [A]_{tot} - 2[A_2]_{eq}$ in the term linear in $[A]_{eq}$, using the substitution $[A_2]_{eq} = k_r[A]^2_{eq}/k'_r$, and simplification gives

$$\begin{aligned}\tau^{-2} &= 16k_r^2[A]^2_{eq} + 8k_r k'_r[A]_{eq} + k'^2_r \\ &= 16k_r^2[A]^2_{eq} + 8k_r k'_r\{[A]_{tot} - 2[A_2]_{eq}\} + k'^2_r \\ &= 16k_r^2[A]^2_{eq} + 8k_r k'_r\{[A]_{tot} - 2k_r[A]^2_{eq}/k'_r\} + k'^2_r \\ &= \boxed{8k_r k'_r[A]_{tot} + k'^2_r}\end{aligned}$$

(b) Plot τ^{-2} against $[A]_{tot}$. The plot should be linear with an $[A]_{tot} = 0$ intercept that equals k'^2_r and a slope that equals $8k_r k'_r$. Thus, $k'_r = \sqrt{intercept}$ and $k_r = \dfrac{slope}{8\sqrt{intercept}}$.

E7.15 $2\,P \underset{k'_r}{\overset{k_r}{\rightleftharpoons}} P_2$

Draw up the following table.

$[P]_{tot}/(\text{mol dm}^{-3})$	0.500	0.352	0.251	0.151	0.101
τ/ns	2.3	2.7	3.3	4.0	5.3
$1/(\tau/\text{ns})^2$	0.189	0.137	0.092	0.062	0.036

It is shown in E7.14 that a plot τ^{-2} against $[P]_{tot}$ should be linear with an $[P]_{tot} = 0$ intercept that equals k'^2_r and a slope that equals $8k_r k'_r$. Thus, $k'_r = \sqrt{intercept}$ and $k_r = \dfrac{slope}{8\sqrt{intercept}}$. The plot is seen to be linear in Figure 7.3 and the linear regression fit is shown in the figure.

$$\begin{aligned}k'_r &= \sqrt{intercept} \\ &= \sqrt{2.\bar{9} \times 10^{-4}\ \text{ns}^{-2}} = \sqrt{2.\bar{9} \times 10^{14}\ \text{s}^{-2}} \\ &= \boxed{1.\bar{7} \times 10^{7}\ \text{s}^{-1}}\end{aligned}$$

$$\begin{aligned}k_r &= \frac{slope}{8\sqrt{intercept}} = \frac{0.38\bar{0}\ \text{mol}^{-1}\ \text{dm}^3\ \text{ns}^{-2}}{8\sqrt{2.\bar{9} \times 10^{-4}\ \text{ns}^{-2}}} = \frac{0.38\bar{0} \times 10^{18}\ \text{mol}^{-1}\ \text{dm}^3\ \text{s}^{-2}}{8\sqrt{2.\bar{9} \times 10^{14}\ \text{s}^{-2}}} \\ &= \boxed{2.8 \times 10^{9}\ \text{mol}^{-1}\ \text{dm}^3\ \text{s}^{-1}}\end{aligned}$$

At equilibrium the rate of decomposition equals the rate of formation, so

$$k_r[P]^2_{eq} = k'_r[P_2]_{eq} \quad \text{and} \quad K = \frac{[P_2]_{eq}c^{\ominus}}{[P]^2_{eq}} = \frac{k_r c^{\ominus}}{k'}$$

Subsequently,

$$K = \frac{(2.8 \times 10^{9}\ \text{mol}^{-1}\ \text{dm}^3\ \text{s}^{-1}) \times (1\ \text{mol dm}^{-3})}{1.\bar{7} \times 10^{7}\ \text{s}^{-1}} = \boxed{1.\bar{6} \times 10^{2}}$$

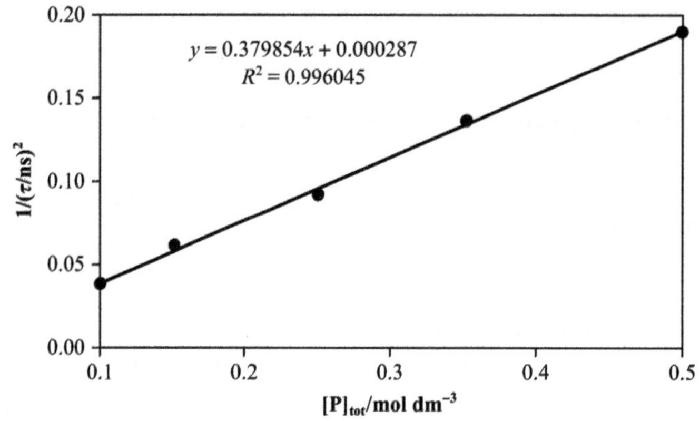

Figure 7.3

E7.16 $A \xrightarrow{k_a} I \xrightarrow{k_b} P$

The differential and integrated rate expressions for consecutive first-order steps are discussed in text Section 7.3. The differential expressions are

$$d[A]/dt = -k_a[A],\ d[I]/dt = k_a[A] - k_b[I], \quad \text{and} \quad d[P]/dt = k_b[I]$$

The integrated rate expressions are

$$[A] = [A]_0 e^{-k_a t},\ [I] = \frac{k_a}{k_b - k_a}(e^{-k_a t} - e^{-k_b t})[A]_0, \quad \text{and} \quad [P] = \left(1 + \frac{k_a e^{-k_b t} - k_b e^{-k_a t}}{k_b - k_a}\right)[A]_0$$

To confirm the integrated expressions, we differentiate each in turn with respect to time. If the result is a match to the differential expression, the integrated form is confirmed. We begin by differentiating the [A] expression.

$$\frac{d[A]}{dt} = \frac{d}{dt}([A]_0 e^{-k_a t})$$
$$= [A]_0 \frac{d}{dt} e^{-k_a t} = -k_a[A]_0 e^{-k_a t}$$
$$= -k_a[A]$$

This is a match to the accepted differential expression, so the integrated expression for [A] is confirmed. We now differentiate the [I] expression.

$$\frac{d[I]}{dt} = \frac{d}{dt}\left(\frac{k_a}{k_b - k_a}(e^{-k_a t} - e^{-k_b t})[A]_0\right)$$
$$= \frac{k_a[A]_0}{k_b - k_a}\frac{d}{dt}(e^{-k_a t} - e^{-k_b t}) = \frac{k_a}{k_b - k_a}(-k_a[A]_0 e^{-k_a t} + k_b[A]_0 e^{-k_b t})$$
$$= \frac{k_a}{k_b - k_a}(-k_a[A] + k_b[A]_0 e^{-k_b t}) = \frac{k_a}{k_b - k_a}\left(-k_a[A] + k_b[A]_0\left\{e^{-k_a t} - \frac{(k_b - k_a)[I]}{k_a[A]_0}\right\}\right)$$
$$= \frac{k_a}{k_b - k_a}\left(-k_a[A] + k_b[A] - \frac{k_b}{k_a}\{k_b - k_a\}[I]\right)$$
$$= k_a[A] - k_b[I]$$

This is a match to the accepted differential expression, so the integrated expression for [I] is confirmed. We now differentiate the [P] expression.

$$\frac{d[P]}{dt} = \frac{d}{dt}\left(1 + \frac{k_a e^{-k_b t} - k_b e^{-k_a t}}{k_b - k_a}\right)[A]_0$$
$$= \frac{[A]_0}{k_b - k_a}(-k_a k_b e^{-k_b t} + k_a k_b e^{-k_a t}) = \frac{k_a k_b [A]_0}{k_b - k_a}(e^{-k_a t} - e^{-k_b t})$$
$$= \frac{k_a k_b [A]_0}{k_b - k_a}(e^{-k_a t} - e^{-k_b t}) = \frac{k_a k_b}{k_b - k_a}\left(\frac{k_b - k_a}{k_a}\right)[I]$$
$$= k_b[I]$$

This is a match to the accepted differential expression, so the integrated expression for [P] is confirmed.

E7.17 $\mathrm{X} \xrightarrow{t_{1/2,a} = 22.5\text{ d}} \mathrm{Y} \xrightarrow{t_{1/2,b} = 33.0\text{ d}} \mathrm{Z}$

We use eqn 7.14 after solving for k_a and k_b from the half-lives.

$$k_a = \frac{\ln 2}{t_{1/2,a}}[6.13] = \frac{\ln 2}{22.5\text{ d}} = 3.08 \times 10^{-2}\text{ d}^{-1}$$

$$k_b = \frac{\ln 2}{33.0\text{ d}} = 2.10 \times 10^{-2}\text{ d}^{-1}$$

$$t = \frac{1}{k_a - k_b}\ln\frac{k_a}{k_b} \quad [7.14]$$
$$= \frac{1}{(3.08 - 2.10) \times 10^{-2}\text{ d}^{-1}}\ln(3.08/2.10)$$
$$= \boxed{39.1\text{ d}}$$

E7.18 $\mathrm{A} \xrightarrow{k_a} \mathrm{I} \xrightarrow{k_b} \mathrm{P}$

Using spreadsheet software to evaluate eqn 7.12 for [I] in the above consecutive series, we draw up a graph shown in Figure 7.4 where the curves represent the concentration of the intermediate [I] as a function of time at various k_a/k_b ratios. $k_b = 1\text{ s}^{-1}$ for all curves so the curve labeled 10 corresponds to $k_a = 10\text{ s}^{-1}$, as specified in part (a) of the problem. As the ratio k_a/k_b gets smaller (or, as the problem puts it, the ratio k_b/k_a gets larger), the concentration profile for I becomes lower, broader, and flatter; that is, [I] becomes more nearly constant over a longer period of time. This is the nature of the $\boxed{\text{steady-state approximation}}$, which becomes more and more valid as consumption of the intermediate becomes fast compared with its formation.

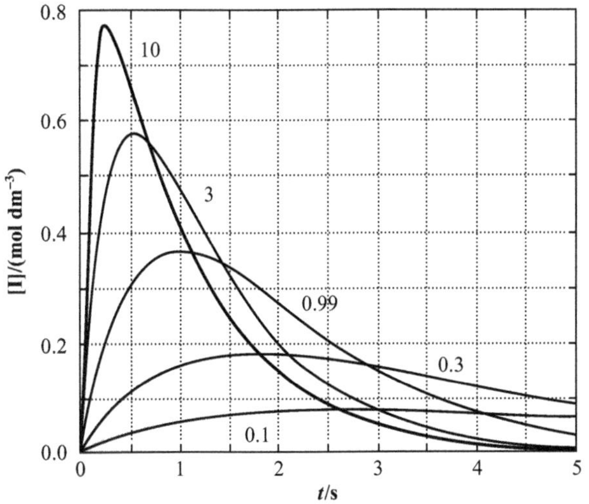

Figure 7.4

E7.19 The first step is rate determining; hence rate $= k_r[H_2O_2][Br^-]$

The reaction is $\boxed{\text{first order in } H_2O_2 \text{ and in } Br^-}$, and $\boxed{\text{second order overall}}$.

E7.20

$$A_2 \underset{k_a'}{\overset{k_a}{\rightleftharpoons}} A + A \quad \text{(fast)}$$
$$A + B \xrightarrow{k_b} P \quad \text{(slow)}$$

We assume pre-equilibrium for the fast, reversible step as the slow step is rate determining, and write

$$K = \frac{[A]^2}{[A_2]c^\ominus} = \frac{k_a}{k_a' c^\ominus}, \text{ which implies that } [A] = (K[A_2]c^\ominus)^{1/2}$$

The rate-determining step then gives

$$\text{rate of product formation} = k_b[A][B] = k_b(Kc^\ominus)^{1/2}[A_2]^{1/2}[B]$$
$$= \boxed{k_r[A_2]^{1/2}[B], \text{ where } k_r = k_b(Kc^\ominus)^{1/2}}$$

E7.21

$$A + B \underset{k_a'}{\overset{k_a}{\rightleftharpoons}} \text{unstable helix} \quad \text{(fast)}$$
$$\text{unstable helix} \xrightarrow{k_b} \text{stable double helix} \quad \text{(slow)}$$

(i) Rate analysis with pre-equilibrium assumption for formation of unstable helix

$$K = \frac{[\text{unstable helix}]c^\ominus}{[A][B]} = \frac{k_a c^\ominus}{k_a'}, \text{ indicating that } [\text{unstable helix}] = K[A][B]/c^\ominus$$

The slow rate-determining step then gives

$$\text{rate of double-helix formation} = k_b[\text{unstable helix}] = k_b K[A][B]/c^\ominus$$
$$= k_r[A][B], \quad \text{where} \quad k_r = k_b K/c^\ominus \quad \text{(pre-equilibrium approx.)}$$

(ii) Rate analysis with the steady-state assumption for the unstable helix intermediate

$$\text{net rate of unstable helix formation} = k_a[A][B] - (k_a' + k_b)[\text{unstable helix}] = 0$$

So,

$$[\text{unstable helix}] = k_a[A][B]/(k_a' + k_b)$$

and

$$\text{rate of double-helix formation} = k_b[\text{unstable helix}] = k_a k_b[A][B]/(k_a' + k_b)$$
$$= k_r[A][B], \quad \text{where} \quad k_r = k_a k_b/(k_a' + k_b) \quad \text{(steady-state approx.)}$$

Both the pre-equilibrium approx. and the steady-state approx. predict that the reaction is first order in A, first order in B, and second order overall. However, the predictions for k_r are generally different functions of the mechanistic rate constants. In the case for which $k_a' \gg k_b$, the predictions for k_r are identical.

E7.22

$$O_3 \underset{k_a'}{\overset{k_a}{\rightleftharpoons}} O_2 + O$$
$$O + O_3 \xrightarrow{k_b} O_2 + O_2$$

The net reaction for the decomposition of atmospheric ozone is $2\,O_3 \rightarrow 3\,O_2$ and the reaction rate, v, is proportional to the net rate of ozone decomposition:

$$v = -\frac{1}{2}\frac{d[O_3]}{dt}$$

In order to simplify the kinetic analysis, we assume that the steady-state approximation applies to [O] (but see the question below). Then,

$$\text{rate of atomic oxygen formation, } \frac{d[O]}{dt} = k_a[O_3] - k_a'[O][O_2] - k_b[O][O_3] = 0$$

Solving for [O],

$$[O] = \frac{k_a[O_3]}{k_a'[O_2] + k_b[O_3]}$$

and

$$\text{net rate of ozone decomposition, } \frac{d[O_3]}{dt} = -k_a[O_3] + k_a'[O][O_2] - k_b[O][O_3]$$

Substituting for [O] from above

$$\frac{d[O_3]}{dt} = -k_a[O_3] + \frac{k_a[O_3](k_a'[O_2] - k_b[O_3])}{k_a'[O_2] + k_b[O_3]}$$
$$= \frac{-k_a[O_3](k_a'[O_2] + k_b[O_3]) + k_a[O_3](k_a'[O_2] - k_b[O_3])}{k_a'[O_2] + k_b[O_3]}$$
$$= \frac{-2k_ak_b[O_3]^2}{k_a'[O_2] + k_b[O_3]}$$

giving

$$\boxed{v = \frac{k_ak_b[O_3]^2}{k_a'[O_2] + k_b[O_3]}}$$

If the second step is slow, then $k_b[O_3] \ll k_a'[O_2]$ and the rate reduces to

$$v = \frac{k_ak_b[O_3]^2}{k_a'[O_2]}$$

which is second order in $[O_3]$ and of order −1 in $[O_2]$.

QUESTION. Can you determine the rate-law expression if the first step of the proposed mechanism is a rapid pre-equilibrium? Under what conditions does the rate expression above reduce to the case of rapid pre-equilibrium?

E7.23

$$AH + B \xrightarrow{k_a} BH^+ + A^-$$
$$A^- + BH^+ \xrightarrow{k_a'} AH + B$$
$$A^- + HA \xrightarrow{k_b} P$$

Apply the steady-state approximation to the carbanion A^-.

Net rate of A^- formation $= k_a[HA][B] - k_a'[A^-][BH^+] - k_b[A^-][HA] = 0$

Therefore, $\boxed{[A^-] = \dfrac{k_a[HA][B]}{k_a'[BH^+] + k_b[HA]}}$

and the rate of formation of product is

Net rate of P formation $= k_b[HA][A^-] = \boxed{\dfrac{k_ak_b[HA]^2[B]}{k_a'[BH^+] + k_b[HA]}}$

E7.24

$$HA + H^+ \underset{k_a'}{\overset{k_a}{\rightleftharpoons}} HAH^+ \quad \text{(fast)}$$

$$HAH^+ + B \xrightarrow{k_b} BH^+ + HA \quad \text{(slow)}$$

The rate of production of the product is

$$\frac{d[BH^+]}{dt} = k_b[HAH^+][B]$$

HAH^+ is an intermediate involved in a rapid pre-equilibrium

$$\frac{[HAH^+]}{[HA][H^+]} = \frac{k_a}{k_a'}, \quad \text{so} \quad [HAH^+] = \frac{k_a[HA][H^+]}{k_a'}$$

$$\text{and } \frac{d[BH^+]}{dt} = \boxed{\frac{k_a k_b}{k_a'}[HA][H^+][B]}$$

This rate law can be made independent of $[H^+]$, if the source of H^+ is the weak acid HA, for then H^+ is given by another equilibrium.

$$\frac{[H^+][A^-]}{[HA]} = K_a = \frac{[H^+]^2}{[HA]}, \quad \text{so} \quad [H^+] = (K_a[HA])^{1/2}$$

$$\text{and } \frac{d[BH^+]}{dt} = \boxed{\frac{k_a k_b K_a^{1/2}}{k_a'}[HA]^{3/2}[B]}$$

E7.25 The Malthus model for the rate of change of the population N: $\frac{dN}{dt} = bN - dN$, where b and d are the rate constants of births and deaths. To find the integrated form of the rate expression, we rearrange it with expressions of N on the left and time expressions on the right. We then integrate using the symbol N_0 to represent the population at $t = 1750$ y, while N represents the population in the year t.

$$\frac{dN}{N} = (b-d)dt$$

$$\int_{N_0}^{N} \frac{dN}{N} = (b-d)\int_{1750\text{ y}}^{t} dt$$

$$\ln N \Big|_{N=N_0}^{N=N} = (b-d)\times(t-1750\text{ y})$$

$$\ln\frac{N}{N_0} = (b-d)\times(t-1750\text{ y}) \quad \text{or} \quad \boxed{N = N_0 e^{(b-d)\times(t-1750\text{ y})}}$$

In this model, the population grows exponentially when $b > d$, it decays exponentially when $d > b$, and it is in a steady state when $b = d$. To test the model against the data, we note that the above integrated expression can be placed in the general linear form: $\ln N$ = intercept + slope × *year*. Thus, we prepare a data plot of $\ln N$ against *year*. If the plot is linear, the Malthus model might possibly be applicable, but not necessarily precisely correct as other models may also predict the linearity of this plot. The plot is shown in Figure 7.5. Inspection of the plot indicates that it is approximately linear so $\boxed{\text{the Malthus model does seem to describe the data as an exponential growth}}$. The linear regression fit, shown in the figure, indicates that the rate constant for the exponential growth is $\boxed{k_r = b - d = 0.0095\text{ y}^{-1}}$. Extrapolation (a risky proposition) of the regression fit to the year 2050, yields a predicted population of 7.7×10^9.

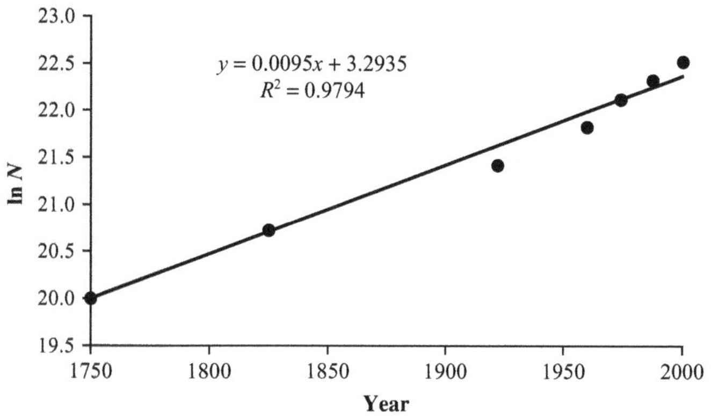

Figure 7.5

E7.26 The rate constant for diffusion control is

$$k_r = \frac{8RT}{3\eta} \quad [7.27]$$
$$= \frac{8 \times (8.3145\ \text{J K}^{-1}\ \text{mol}^{-1}) \times (298\ \text{K})}{3 \times (1.06 \times 10^{-3}\ \text{kg m}^{-1}\ \text{s}^{-1})} \; [1\ \text{J} = 1\ \text{kg m}^2\ \text{s}^{-2})$$
$$= 6.23 \times 10^6\ \text{mol}^{-1}\ \text{m}^3\ \text{s}^{-1}$$
$$= \boxed{6.23 \times 10^9\ \text{mol}^{-1}\ \text{dm}^3\ \text{s}^{-1}}$$

E7.27 The fraction of collisions that occur with at least the energy E_a is given by $f = e^{-E_a/RT}$ [7.30]

(a) (i) $E_a = 10\ \text{kJ mol}^{-1}$, $T = 300\ \text{K}$, $f = e^{-(10\times10^3\ \text{J mol}^{-1})/\{(8.3145\ \text{J K}^{-1}\ \text{mol}^{-1})\times(300\ \text{K})\}} = \boxed{0.018}$

(ii) $E_a = 10\ \text{kJ mol}^{-1}$, $T = 1000\ \text{K}$, $f = e^{-(10\times10^3\ \text{J mol}^{-1})/\{(8.3145\ \text{J K}^{-1}\ \text{mol}^{-1})\times(1000\ \text{K})\}} = \boxed{0.30}$

(b) (i) $E_a = 100\ \text{kJ mol}^{-1}$, $T = 300\ \text{K}$, $f = e^{-(100\times10^3\ \text{J mol}^{-1})/\{(8.3145\ \text{J K}^{-1}\ \text{mol}^{-1})\times(300\ \text{K})\}} = \boxed{3.9 \times 10^{-18}}$

(ii) $E_a = 100\ \text{kJ mol}^{-1}$, $T = 1000\ \text{K}$, $f = e^{-(100\times10^3\ \text{J mol}^{-1})/\{(8.3145\ \text{J K}^{-1}\ \text{mol}^{-1})\times(1000\ \text{K})\}} = \boxed{6.0 \times 10^{-6}}$

These calculations demonstrate that an increase in temperature causes an increase in a particular rate constant (fixed activation energy) by increasing the fraction of collisions that have a kinetic energy larger than the activation energy. They also show that very, very few collisions (as a fraction of all collisions) have the energy needed for reaction at ordinary temperatures when the activation energy is about the order of a bond strength.

E7.28 The fraction of collisions that occur with at least the energy E_a is given by $f(T) = e^{-E_a/RT}$ [7.30] so the percentage increase in $f(T)$ upon a temperature increase of $\Delta T = 10$ K (i.e. 10°C) is given by the expression:

$$\text{percentage increase} = \frac{f(T+\Delta T) - f(T)}{f(T)} 100\% = (f(T)^{-1} f(T+\Delta T) - 1)100\%$$
$$= (e^{E_a/RT} e^{-E_a/R(T+\Delta T)} - 1)100\% = \left(e^{\frac{E_a}{R}\left\{\frac{1}{T} - \frac{1}{T+\Delta T}\right\}} - 1 \right) 100\%$$

(a) (i) $E_a = 10\ \text{kJ mol}^{-1}$, $T = 300\ \text{K}$

$$\text{percentage increase} = \left(e^{\frac{10\times10^3\ \text{J mol}^{-1}}{8.3145\ \text{J K}^{-1}\ \text{mol}^{-1}}\left\{\frac{1}{300\ \text{K}} - \frac{1}{310\ \text{K}}\right\}} - 1 \right) 100\% = \boxed{13.8\%}$$

(ii) $E_a = 10\ \text{kJ mol}^{-1}$, $T = 1000\ \text{K}$

$$\text{percentage increase} = \left(e^{\frac{10\times10^3\ \text{J mol}^{-1}}{8.3145\ \text{J K}^{-1}\ \text{mol}^{-1}}\left\{\frac{1}{1000\ \text{K}}-\frac{1}{1010\ \text{K}}\right\}} - 1\right)100\% = \boxed{1.2\%}$$

(b) (i) $E_a = 100\ \text{kJ mol}^{-1}$, $T = 300\ \text{K}$

$$\text{percentage increase} = \left(e^{\frac{100\times10^3\ \text{J mol}^{-1}}{8.3145\ \text{J K}^{-1}\ \text{mol}^{-1}}\left\{\frac{1}{300\ \text{K}}-\frac{1}{310\ \text{K}}\right\}} - 1\right)100\% = \boxed{264\%}$$

These values relate to the lab rule that each 10°C increase in temperature approximately doubles the reaction rate.

(ii) $E_a = 100\ \text{kJ mol}^{-1}$, $T = 1000\ \text{K}$

$$\text{percentage increase} = \left(e^{\frac{100\times10^3\ \text{J mol}^{-1}}{8.3145\ \text{J K}^{-1}\ \text{mol}^{-1}}\left\{\frac{1}{1000\ \text{K}}-\frac{1}{1010\ \text{K}}\right\}} - 1\right)100\% = \boxed{12.6\%}$$

E7.29 $k_r = \dfrac{kT}{h}e^{-\Delta^{\ddagger}G/RT}$ [7.34 with $\kappa = 1$]

The ratio of the rates is then given by

$$\frac{k_{\text{cat}}}{k_{\text{uncat}}} = \frac{e^{-\Delta^{\ddagger}G_{\text{cat}}/RT}}{e^{-\Delta^{\ddagger}G_{\text{uncat}}/RT}} = e^{\Delta^{\ddagger}G_{\text{uncat}}/RT-\Delta^{\ddagger}G_{\text{uncat}}/RT} = e^{(\Delta^{\ddagger}G_{\text{uncat}}-\Delta^{\ddagger}G_{\text{uncat}})/RT}$$
$$= e^{(100-10)\text{kJ mol}^{-1}/(8.3145\times10^{-3}\ \text{kJ mol}^{-1}\ \text{K}^{-1}\times310\ \text{K})}$$
$$= \boxed{1.5\times10^{15}}$$

E7.30 $C_2H_4(g) + H_2(g) \xrightarrow{k_r = Ae^{-E_a/RT}} C_2H_6(g)$

The pre-exponential factor according to kinetic theory is given by

$$A = \sigma\left(\frac{8kT}{\pi\mu}\right)^{1/2} N_A \quad [7.31] \quad \text{where} \quad \mu = \frac{m_A m_B}{m_A + m_B} = \frac{M_A M_B}{M_A + M_B}\times\frac{1}{N_A}$$

The reduced mass for ethene/hydrogen collisions is

$$\mu = \frac{M_{H_2}M_{C_2H_4}}{M_{H_2}+M_{C_2H_4}}\times\frac{1}{N_A}$$
$$= \frac{(2.016\ \text{g mol}^{-1})\times(0.028052\ \text{kg mol}^{-1})}{(2.016\ \text{g mol}^{-1}+28.052\ \text{g mol}^{-1})\times(6.02214\times10^{23}\ \text{mol}^{-1})}$$
$$= 3.123\times10^{-27}\ \text{kg}$$

In the note at the end of this exercise we give the justification for computing the collision cross-section with the expression

$$\sigma = (\sigma_{H_2}^{1/2} + \sigma_{C_2H_4}^{1/2})^2$$
$$= \{(0.27)^{1/2} + (0.64)^{1/2}\}^2\ \text{nm}^2 \quad [\text{Table 7.1}]$$
$$= 1.7\bar{4}\ \text{nm}^2 = 1.7\bar{4}\times10^{-18}\ \text{m}^2$$

Thus, the pre-exponential factor is

$$A = (1.7\bar{4}\times10^{-18}\ \text{m}^2)\left(\frac{8\times(1.3807\times10^{-23}\ \text{J K}^{-1})\times(673\ \text{K})}{\pi\times(3.123\times10^{-27}\ \text{kg})}\right)^{1/2}\times(6.02214\times10^{23}\ \text{mol}^{-1})$$
$$= 2.9\times10^{9}\ \text{m}^3\ \text{mol}^{-1}\ \text{s}^{-1} = \boxed{2.9\times10^{12}\ \text{dm}^3\ \text{mol}^{-1}\ \text{s}^{-1}}$$

The pre-exponential factor should include a **steric factor *P*** that accounts for orientational requirements for reaction between collision pairs. For the ethene/hydrogen pair P is found to be 1.7×10^{-6}. Thus, a much better estimate is given by

$$(1.7 \times 10^{-6}) \times (2.9 \times 10^{12}\ \text{dm}^3\ \text{mol}^{-1}\ \text{s}^{-1}) = 4.9 \times 10^{6}\ \text{dm}^3\ \text{mol}^{-1}\ \text{s}^{-1}$$

Note: The collision cross-section is given by $\sigma = \pi\{r_A + r_B\}^2$, where r_A and r_B are the effective radii computed from individual collision cross-sections. That is: $\sigma_A = \pi r_A^2$ and $\sigma_B = \pi r_B^2$. Thus,

$$\sigma = \pi\{(\sigma_A/\pi)^{1/2} + (\sigma_B/\pi)^{1/2}\}^2 = (\sigma_A^{1/2} + \sigma_B^{1/2})^2.$$

E7.31 $\mathrm{A + B} \underset{k_a'}{\overset{k_a}{\rightleftharpoons}} \mathrm{I}\ \text{(fast)} \quad \text{and} \quad \mathrm{I} \xrightarrow{k_b} \mathrm{P}$

The rate of reaction is

$$v = \frac{d[P]}{dt} = k_b[I]$$

Applying the pre-equilibrium approximation yields

$$\frac{[I]}{[A][B]} = K = \frac{k_a}{k_a'}, \quad \text{so} \quad [I] = \frac{k_a[A][B]}{k_a'}$$

$$\text{and } v = \frac{k_a k_b[A][B]}{k_a'} = k_r[A][B] \quad \text{with} \quad k_r = \frac{k_a k_b}{k_a'}$$

Substituting the Arrhenius expression $k_{r,\text{step}} = A_{\text{step}}e^{-E_{a,\text{step}}/RT}$ [6.19c] for the rate constant of each individual step and for the overall rate constant ($k_r = Ae^{-E_a/RT}$) gives the composite Arrhenius parameters.

$$\begin{aligned} Ae^{-E_a/RT} &= \frac{(A_{\text{step}}e^{-E_{a,\text{step}}/RT})_a(A_{\text{step}}e^{-E_{a,\text{step}}/RT})_b}{(A_{\text{step}}e^{-E_{a,\text{step}}/RT})_{a'}} \\ &= \frac{(A_a e^{-E_{a,a}/RT})(A_b e^{-E_{a,b}/RT})}{(A_{a'}e^{-E_{a,a'}/RT})} \\ &= \frac{A_a A_b}{A_{a'}} e^{-(E_{a,a}+E_{a,b}-E_{a,a'})/RT} \end{aligned}$$

By equating the pre-exponentials and exponents on each side of the equality we find the composite parameters

$$A = \frac{A_a A_b}{A_{a'}} \quad \text{and} \quad E_a = E_{a,a} + E_{a,b} - E_{a,a'}$$

Thus, with the stepwise activation energies provided in this exercise the composite acitivation energy is:

$$E_a = (25 + 10 - 38)\ \text{kJ mol}^{-1} = \boxed{-3\ \text{kJ mol}^{-1}}$$

This exercise demonstrates that, while activation energies of individual steps are positive, the activation energy of a composite rate constant can be negative.

E7.32 For bovine rhodopsin the ratio of transition half-lives is

$$\frac{t_{1/2}(310\ \text{K})}{t_{1/2}(273\ \text{K})} = \frac{600 \times 10^{-6}\ \text{s}}{1\ \text{s}} = 600 \times 10^{-6}$$

while the same ratio for a frog rhodopsin is 1/6. To find whether the difference is related to activation energies for the transition we relate this ratio at the two temperatures $T_2 > T_1$ with rate constants and Arrhenius parameters.

$$\frac{t_{1/2}(T_2)}{t_{1/2}(T_1)} = \frac{(\ln 2)/k_r(T_2)}{(\ln 2)/k_r(T_1)} [6.13] = \frac{k_r(T_1)}{k_r(T_2)} = \frac{A\mathrm{e}^{-E_a/RT_1}}{A\mathrm{e}^{-E_a/RT_2}} \quad [6.19c]$$

$$= \mathrm{e}^{-\frac{E_a}{R}\times\left(\frac{1}{T_1}-\frac{1}{T_2}\right)}$$

Solving for E_a gives

$$E_a = -\frac{R}{T_1^{-1} - T_2^{-1}} \ln\frac{t_{1/2}(T_2)}{t_{1/2}(T_1)}$$

$$E_{a,\text{bovine}} = -\frac{8.3145 \times 10^{-3}\ \text{kJ K}^{-1}\ \text{mol}^{-1}}{(273\ \text{K})^{-1} - (310\ \text{K})^{-1}} \ln(600 \times 10^{-6})$$

$$= 141\ \text{kJ mol}^{-1}$$

$$E_{a,\text{frog}} = -\frac{8.3145 \times 10^{-3}\ \text{kJ K}^{-1}\ \text{mol}^{-1}}{(273\ \text{K})^{-1} - (310\ \text{K})^{-1}} \ln(1/6)$$

$$= 34\ \text{kJ mol}^{-1}$$

The bovine activation energy is 107 kJ mol^{-1} larger than the frog activation energy. The low activation energy of the frog rhodopsin means that there is greater activity of the protein under photo conditions of night-light. With the nocturnal life style of the frog, the enhanced vision improves both the pursuit of insect food and awareness of predators.

E7.33 $(NH_2)_2CO(aq) + 2\ H_2O(l) \xrightarrow{k_r} 2\ NH_4^+(aq) + CO_3^{2-}(aq)$

Pseudo-first-order reaction rate = $k_r[(NH_2)_2CO(aq)]$

$$k_r = \kappa\frac{kT}{h}K^{\ddagger}\ [7.32 \text{ and } 7.34] = \frac{\kappa kT}{h}\mathrm{e}^{-\Delta^{\ddagger}G/RT}$$

$$= \frac{kT}{h}\mathrm{e}^{-\Delta^{\ddagger}G/RT} \quad \text{[assume that the transmission coefficient equals 1]}$$

Solving for $\Delta^{\ddagger}G$ gives

$$\Delta^{\ddagger}G = RT\ln\frac{kT}{hk_r}.$$

At 333 K,

$$\Delta^{\ddagger}G = (8.3145\ \text{J K}^{-1}\ \text{mol}^{-1}) \times (333\ \text{K}) \ln\frac{(1.381 \times 10^{-23}\ \text{J K}^{-1}) \times (333\ \text{K})}{(6.626 \times 10^{-34}\ \text{J s}) \times (1.2 \times 10^{-7}\ \text{s}^{-1})} = 12\bar{6}\ \text{kJ mol}^{-1}.$$

At 343 K,

$$\Delta^{\ddagger}G = (8.3145\ \text{J K}^{-1}\ \text{mol}^{-1}) \times (343\ \text{K}) \ln\frac{(1.381 \times 10^{-23}\ \text{J K}^{-1}) \times (343\ \text{K})}{(6.626 \times 10^{-34}\ \text{J s}) \times (4.6 \times 10^{-7}\ \text{s}^{-1})} = 12\bar{6}\ \text{kJ mol}^{-1}.$$

We conclude that $\Delta^{\ddagger}G = \boxed{12\bar{6}\ \text{kJ mol}^{-1}}$ in this temperature range.

E7.34 $$k_r = \left(\frac{kT}{h}\mathrm{e}^{\Delta^{\ddagger}S/R}\right)\mathrm{e}^{-\Delta^{\ddagger}H/RT} \quad \text{[7.34, assume that the transmission coefficient equals 1]}$$

$$= A\mathrm{e}^{-\Delta^{\ddagger}H/RT}, \quad \text{where} \quad A = \frac{kT}{h}\mathrm{e}^{\Delta^{\ddagger}S/R}$$

Let us assume that the pre-exponential factor is essentially a constant over this small temperature range (10 K). Then, the equation for the natural logarithm of the ratio of rate constants at two different temperatures can be solved for $\Delta^{\ddagger}H$ to give

$$\Delta^{\ddagger}H = \frac{R}{\left(\frac{1}{T}-\frac{1}{T'}\right)}\ln\frac{k_r(T')}{k_r(T)}$$

$$= \frac{(8.3145\ \text{J K}^{-1}\ \text{mol}^{-1})}{\left(\frac{1}{333\ \text{K}}-\frac{1}{343\ \text{K}}\right)}\ln\frac{4.6}{1.2} = 12\bar{8}\ \text{kJ mol}^{-1}.$$

To evaluate $\Delta^{\ddagger}S$, solve $\Delta^{\ddagger}G = \Delta^{\ddagger}H - T\Delta^{\ddagger}S$ [7.33] for $\Delta^{\ddagger}S$.

$$\Delta^{\ddagger}S = \frac{\Delta^{\ddagger}H - \Delta^{\ddagger}G}{T}$$

In Exercise 7.33 we found that $\Delta^{\ddagger}G = 12\bar{6}$ kJ mol^{-1} so the precision of this data does not warrant the claim that the value of $\Delta^{\ddagger}H - \Delta^{\ddagger}G = 2$ kJ mol^{-1}. All we really know is that

$\Delta^{\ddagger}H - \Delta^{\ddagger}G \approx 0$ so $\boxed{\Delta^{\ddagger}S \approx 0}$. If the rate constants had an additional significant figure so that we could claim that $\Delta^{\ddagger}H - \Delta^{\ddagger}G = 2$ kJ mol^{-1}, $\Delta^{\ddagger}S$ in this temperature ranges would have the estimate

$$\Delta^{\ddagger}S \approx \frac{2000\ \text{J mol}^{-1}}{333\ \text{K}} = 6\ \text{J K}^{-1}\ \text{mol}^{-1}.$$

E7.35 Eqn 7.34 with $\kappa = 1$ indicates that a plot of $\ln(hk_rc^{\ominus}/kT)$ or $\ln(k_r/T)$ against $1/T$ is linear provided that $\Delta^{\ddagger}S$ and $\Delta^{\ddagger}H$ are constants over the temperature range of interest. To see this, divide by kT/h and take the natural logarithm of both sides of the equation.

$$k_r = \left(\frac{kT}{hc^{\ominus}}\right)e^{\Delta^{\ddagger}S/R}e^{-\Delta^{\ddagger}H/RT} \quad [7.34]$$

$$\frac{hk_rc^{\ominus}}{kT} = e^{\Delta^{\ddagger}S/R}e^{-\Delta^{\ddagger}H/RT}$$

$$\ln\frac{hk_rc^{\ominus}}{kT} = \ln\{e^{\Delta^{\ddagger}S/R}e^{-\Delta^{\ddagger}H/RT}\}$$

$$= \ln e^{\Delta^{\ddagger}S/R} + e^{-\Delta^{\ddagger}H/RT}$$

$$= \frac{\Delta^{\ddagger}S}{R} - \frac{\Delta^{\ddagger}H}{RT}$$

$$= \textit{intercept} + \textit{slope} \times \frac{1}{T}, \quad \text{where} \quad \textit{intercept} = \frac{\Delta^{\ddagger}S}{R} \quad \text{and} \quad \textit{slope} = -\frac{\Delta^{\ddagger}H}{R}$$

Consequently, we prepare a data table of $\ln(hk_rc^{\ominus}/kT)$ and $1/T$ transformations of the experimental data and make a plot of $\ln(hk_rc^{\ominus}/kT)$ against $1/T$. If the plot is linear, we perform a linear regression fit to find the intercept and slope and calculate $\Delta^{\ddagger}S$ and $\Delta^{\ddagger}H$ with the eqns

$$\Delta^{\ddagger}S = R \times \textit{intercept} \quad \text{and} \quad \Delta^{\ddagger}H = -R \times \textit{slope}$$

T/K	289.0	293.5	298.1	303.2	308.0	313.5
$k_r/10^6$ dm^3 mol^{-1} s^{-1}	1.04	1.34	1.53	1.89	2.29	2.84
1000 K/T	3.4602	3.4072	3.3546	3.2982	3.2468	3.1898
$\ln(hk_rc^{\ominus}/kT)$	−15.572	−15.334	−15.217	−15.022	−14.846	−14.649

The plot, shown in Figure 7.6, is seen to be linear, so we perform the linear regression fit and make the desired computations.

$$\Delta^{\ddagger}S = (8.3145\ \text{J K}^{-1}\ \text{mol}^{-1}) \times (-4.06) = \boxed{-33.8\ \text{J K}^{-1}\ \text{mol}^{-1}}$$

$$\Delta^{\ddagger}H = -(8.3145 \times 10^{-3}\ \text{kJ K}^{-1}\ \text{mol}^{-1}) \times (-3.32 \times 10^{3}\ \text{K}) = \boxed{+27.6\ \text{kJ mol}^{-1}}$$

We estimate $\Delta^{\ddagger}G$ over this temperature range using the value at 300 K.

$$\begin{aligned}\Delta^{\ddagger}G &= \Delta^{\ddagger}H - T\Delta^{\ddagger}S \quad [7.33]\\ &= (+27.6\ \text{kJ mol}^{-1}) - (300\ \text{K}) \times (-0.0338\ \text{kJ K}^{-1}\ \text{mol}^{-1})\\ &= \boxed{37.7\ \text{kJ mol}^{-1}}\end{aligned}$$

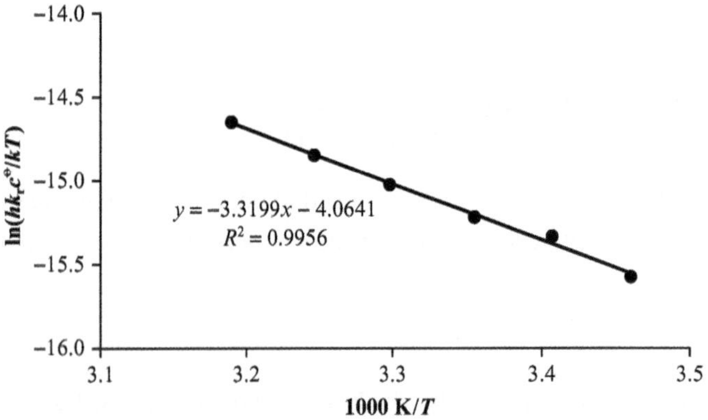

Figure 7.6

E7.36 Empirical rate constant: $k_r = ae^{b/T}$, where $a = 8.72 \times 10^{12}\ \text{dm}^3\ \text{mol}^{-1}\ \text{s}^{-1}$ and $b = 6134\ \text{K}$

We compare the empirical rate constant to the expression of transition state theory, $k_r = \left(\dfrac{kT}{hc^{\ominus}}\right) e^{\Delta^{\ddagger}S/R} e^{-\Delta^{\ddagger}H/RT}$ [7.34], make the pre-exponential and exponential associations

$$\left(\frac{kT}{hc^{\ominus}}\right) e^{\Delta^{\ddagger}S/R} = a \quad \text{and} \quad e^{-\Delta^{\ddagger}H/RT} = e^{b/T} \quad (\text{or } \Delta^{\ddagger}H = -Rb),$$

and solve for $\Delta^{\ddagger}H$ and $\Delta^{\ddagger}S$.

$$\begin{aligned}\Delta^{\ddagger}H &= -(8.3145 \times 10^{-3}\ \text{kJ K}^{-1}\ \text{mol}^{-1}) \times (6134\ \text{K})\\ &= \boxed{-51.0\ \text{kJ mol}^{-1}}\end{aligned}$$

$$\Delta^{\ddagger}S = R\ln\left(\frac{hc^{\ominus}a}{kT}\right)$$

$$= (8.3145\ \text{J K}^{-1}\ \text{mol}^{-1}) \times \ln\left\{\frac{(6.62608 \times 10^{-34}\ \text{J s}) \times (1\ \text{mol dm}^{-3}) \times (8.72 \times 10^{12}\ \text{dm}^3\ \text{mol}^{-1}\ \text{s}^{-1})}{(1.38065 \times 10^{-23}\ \text{J K}^{-1}) \times (298.15\ \text{K})}\right\}$$

$$= \boxed{2.82\ \text{J K}^{-1}\ \text{mol}^{-1}}$$

E7.37 $^{-}O_2CH{=}CHCO_2^{-}(\text{Fumarate}^{2-},aq) + H_2O(l) \xrightleftharpoons{\text{fumarase}} {}^{-}O_2CH(OH)CH_2CO_2^{-}\ (\text{malate}^{2-},aq)$

(a) The standard enthalpies of (i)–(iv) are shown in Figure 7.7.

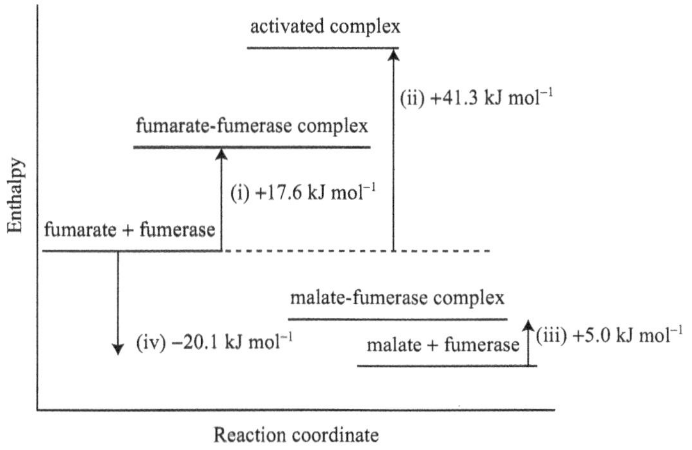

Figure 7.7

(b) Inspection of Figure 7.7 indicates that the enthalpy of activation of the reverse reaction is $\boxed{+61.4 \text{ kJ mol}^{-1}}$.

E7.38 Text Figure 7.21 shows that $\Delta^{\ddagger}H$ is a linear function of $\Delta^{\ddagger}S$ for the reaction catalyzed by myosin APTase in a variety of fish species living in environments of widely differing temperatures. The figure gives an intercept at $\Delta^{\ddagger}S = 0$ of 78 kJ mol^{-1} and a slope of 250 K; all values of the figure are based on reaction rate measurements at 274 K (I.A. Johnson and G. Goldspink, *Nature*, **257**, 620 (1975). Thus, we have the empirical relation:

$$\Delta^{\ddagger}H = a + b\Delta^{\ddagger}S, \quad \text{where} \quad a = 78 \text{ kJ mol}^{-1} \quad \text{and} \quad b = 250 \text{ K}$$

The Gibbs activation energy is related to $\Delta^{\ddagger}S$ by the expression

$$\begin{aligned}\Delta^{\ddagger}G &= \Delta^{\ddagger}H - T_{\text{assay}}\Delta^{\ddagger}S \quad [7.33], \quad &&\text{where} \quad T_{\text{assay}} = 274 \text{ K}\\ &= a + b\Delta^{\ddagger}S - T_{\text{assay}}\Delta^{\ddagger}S\\ &= a - c\Delta^{\ddagger}S, &&\text{where} \quad c = T_{\text{assay}} - b = 24 \text{ K}\end{aligned}$$

The extremes of $\Delta^{\ddagger}S$ are as high as about +200 J K^{-1} mol^{-1} for fish of hot-spring environments and as low −175 J K^{-1} mol^{-1} for fish of Antarctic environments. Placing these value into the $\Delta^{\ddagger}G$ equation gives values as low as 73 kJ mol^{-1} for the hot-spring species and as high as 82 kJ mol^{-1} for the Antarctic species. This 9 kJ mol^{-1} variation between extremes is small and, consequently, $\Delta^{\ddagger}G$ is approximately constant for fish of different environments. In contrast, the figure reports $\Delta^{\ddagger}H$ extremes from about 130 kJ mol^{-1} for the hot-spring species to 25 kJ mol^{-1} for the Antarctic species; this is a large, 5-fold variation.

But does the activation energy, E_a, show a variation between these species? To answer this question we calculate the extremes of the activation energy using an eqn that is part of transition-state theory of a reaction in solution ($E_a = \Delta^{\ddagger}H + RT$).

$$E_a = \Delta^{\ddagger}H + RT_{\text{assay}} = \Delta^{\ddagger}H + 2 \text{ kJ mol}^{-1}$$

Thus, the E_a extremes are from about 132 kJ mol^{-1} for the hot-spring species to 27 kJ mol^{-1} for the Antarctic species. Apparently, the high activation energy of myosin ATPase provides survival benefits for the hot-spring species but low activation energy is required by cold-water species; the variation between extremes is large.

E7.39 The rate constant at zero ionic strength, k_r°, is given by

$$\log\frac{k_r^\circ}{k_r} = -2Az_Az_BI^{1/2} \quad [7.35]$$

$$= -2\times(0.509)\times(+1)\times(+1)\times(0.0241)^{1/2}$$

$$= -0.158$$

$$k_r^\circ = 10^{-0.158}\,k_r = 10^{-0.158}\times(1.55\text{ dm}^6\text{ mol}^{-2}\text{ min}^{-1})$$

$$= \boxed{1.08\text{ dm}^6\text{ mol}^{-2}\text{ min}^{-1}}$$

Solutions to projects

P7.40 (a) $\mathrm{A} \xrightarrow{k_a} \mathrm{B} \xrightarrow{k_b} \mathrm{C}$, where A is the drug at site of administration, B is the drug dispersed in the blood, and C is the eliminated drug.

(i) The peak concentration of B, $[\mathrm{P}]_n$, immediately after administration of the nth dose, each of which is administered at the time interval τ, is given by the sum:

$$\boxed{[\mathrm{P}]_n = [\mathrm{B}]_0 + [\mathrm{B}]_0e^{-k_b\tau} + [\mathrm{B}]_0e^{-2k_b\tau} + \cdots\cdots + [\mathrm{B}]_0e^{-(n-1)k_b\tau} = [\mathrm{B}]_0\sum_{m=0}^{n-1}e^{-mk_b\tau}} > [\mathrm{B}]_0$$

Conc. contribution of nth dose; Remainder of nth − 1 dose; Remainder of nth − 2 dose; Remainder of 1st dose

The residual concentration of B, $[\mathrm{R}]_n$, just before administration of the nth + 1 dose results from the first-order elimination of $[\mathrm{P}]_n$: $\boxed{[\mathrm{R}]_n = [\mathrm{P}]_ne^{-k_b\tau}}$ [6.12 d].

$$[\mathrm{P}]_\infty = \text{limit as } n\to\infty \text{ of } [\mathrm{P}]_n = [\mathrm{B}]_0\sum_{m=0}^{\infty}e^{-mk_b\tau} = [\mathrm{B}]_0(1+x+x^2+\cdots), \text{ where } x = e^{-k_b\tau} < 1$$

This may be simplified using the Taylor series: $1+x+x^2+\cdots = \dfrac{1}{1-x} = \dfrac{1}{1-e^{-k_b\tau}}$ when $x<1$

We conclude that $\boxed{[\mathrm{P}]_\infty = [\mathrm{B}]_0(1-e^{-k_b\tau})^{-1}}$

(ii) Furthermore, $[\mathrm{R}]_\infty = [\mathrm{P}]_\infty e^{-k_b\tau} = \dfrac{[\mathrm{B}]_0e^{-k_b\tau}}{1-e^{-k_b\tau}} = \dfrac{[\mathrm{B}]_0}{e^{k_b\tau}-1}$

$$\boxed{[\mathrm{R}]_\infty = [\mathrm{B}]_0(e^{k_b\tau}-1)^{-1}}$$

$$[\mathrm{P}]_\infty - [\mathrm{R}]_\infty = [\mathrm{B}]_0\{(1-e^{-k_b\tau})^{-1} - (e^{k_b\tau}-1)^{-1}\} = [\mathrm{B}]_0\{(1-e^{-k_b\tau})^{-1} - e^{-k_b\tau}(1-e^{-k_b\tau})^{-1}\}$$
$$= [\mathrm{B}]_0\{(1-e^{-k_b\tau})(1-e^{-k_b\tau})^{-1}\} = [\mathrm{B}]_0$$

$$\boxed{[\mathrm{P}]_\infty - [\mathrm{R}]_\infty = [\mathrm{B}]_0}$$

(b) (i) Solving the equation $[\mathrm{P}]_\infty = [\mathrm{B}]_0(1-e^{-k_b\tau})^{-1}$ for τ gives:

$$\frac{[\mathrm{B}]_0}{[\mathrm{P}]_\infty} = 1-e^{-k_b\tau} \quad\text{or}\quad e^{-k_b\tau} = 1-\frac{[\mathrm{B}]_0}{[\mathrm{P}]_\infty} \quad\text{or}\quad -k_b\tau = \ln\left(1-\frac{[\mathrm{B}]_0}{[\mathrm{P}]_\infty}\right)$$

$$\boxed{\tau = -\frac{1}{k_b}\ln\left(1-\frac{[B]_0}{[P]_\infty}\right)}$$

$$\tau = -\frac{1}{0.0289\ \mathrm{h}^{-1}}\ln\left(1-\frac{1}{10}\right)$$

$$\boxed{\tau = 3.65\ \mathrm{h}}$$

The plots of Figure 7.8 show peak and residual drug concentrations against the number of administrations. The plot of Figure 7.9 shows the concentration variation in time. It clearly demonstrates the peak and residual concentration and the elimination decay between drug administrations.

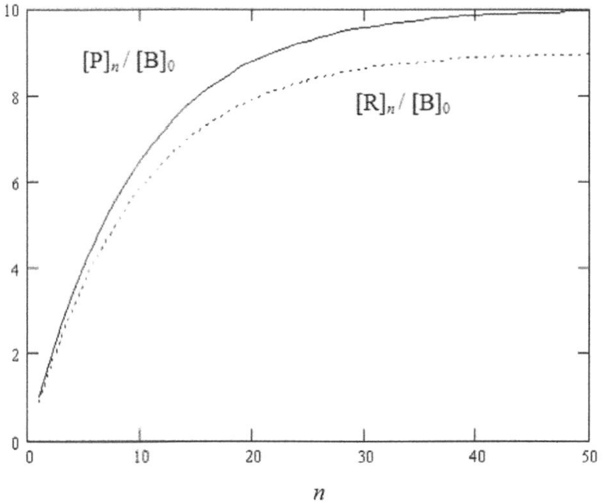

Figure 7.8

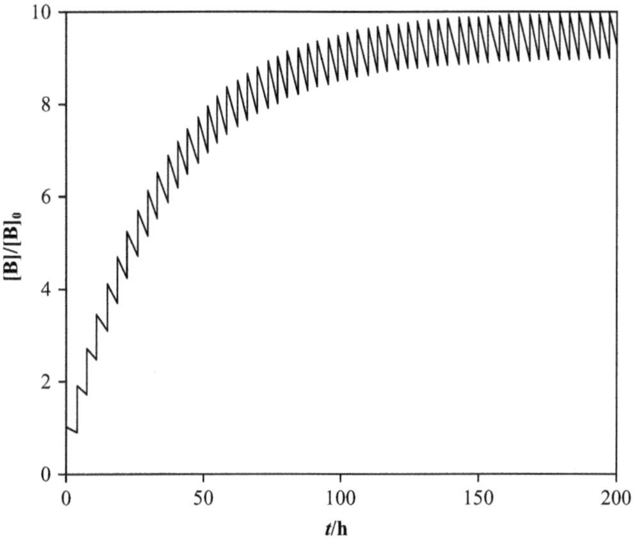

Figure 7.9

(ii) By using the trace function of the Figure 7.8 plot, or by directly reading the graph it is found that $[P]_n$ is 75% of the maximum value when $n = 13$.

$$t_{75\%\,\max} = (n-1)\tau$$
$$= (13-1)(3.65\text{ h})$$
$$= \boxed{43.8\text{ h}}$$

(iii) The magnitude of the variation $[\text{P}]_n - [\text{R}]_n$ may be reduced by reducing the drug dosage $[\text{B}]_0$. However, in order to avoid changing $[\text{P}]_\infty$ it becomes necessary to reduce τ.

(c) For first-order absorption and zero-order elimination of a single dose $[\text{A}]_0$:

$$-\frac{\text{d[A]}}{\text{d}t} = k_\text{a}[\text{A}] \quad \text{and} \quad [\text{A}] = [\text{A}]_0 e^{-k_\text{a}t} \quad [6.12\text{ d}]$$

$$\frac{\text{d[B]}}{\text{d}t} = k_\text{a}[\text{A}] - k_\text{b} = k_\text{a}[\text{A}]_0 e^{-k_\text{a}t} - k_\text{b}$$

$$\int_0^{[\text{B}]} \text{d[B]} = \int_0^t (k_\text{a}[\text{A}]_0 e^{-k_\text{a}t} - k_\text{b})\text{d}t = k_\text{a}[\text{A}]_0 \int_0^t e^{-k_\text{a}t}\text{d}t - k_\text{b}\int_0^t \text{d}t$$

$$\boxed{[\text{B}] = [\text{A}]_0(1 - e^{-k_\text{a}t}) - k_\text{b}t}$$

The Figure 7.10 plot of $[\text{B}]/[\text{A}]_0$ shows rapid absorption of the drug into the blood followed by the slower, linear elimination that corresponds to zeroth-order elimination. Elimination occurs within 25 h with these rate constants ($k_\text{a} = 10\text{ h}^{-1}$ and $k_\text{b} = 4.0 \times 10^{-3}\text{ mmol dm}^{-3}\text{ h}^{-1}$).

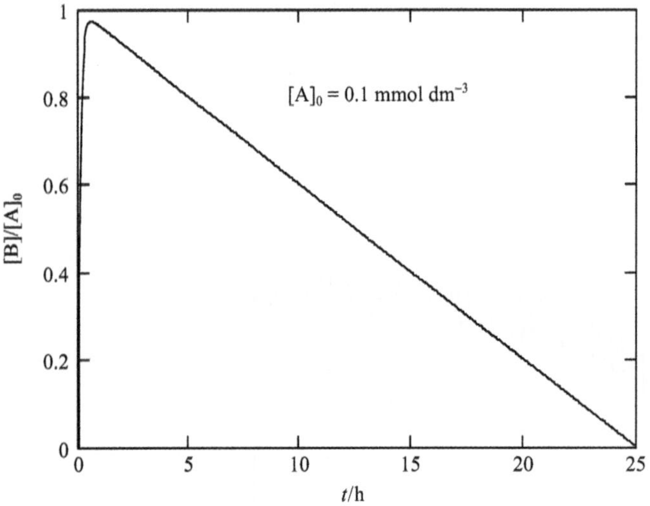

Figure 7.10

(d) Let $\{[\text{B}]_{\max}, t_{\max}\}$ be the maximum of a curve such as that shown in Figure 7.10. To find formulas for this point, we must examine the curve at the point for which $\text{d[B]}/\text{d}t = 0$.

$$\frac{\text{d[B]}}{\text{d}t} = k_\text{a}[\text{A}]_0 e^{-k_\text{a}t_{\max}} - k_\text{b} = 0$$

$$e^{-k_\text{a}t_{\max}} = \frac{k_\text{b}}{k_\text{a}[\text{A}]_0} \quad \text{or} \quad -k_\text{a}t_{\max} = \ln\left(\frac{k_\text{b}}{k_\text{a}[\text{A}]_0}\right)$$

$$\boxed{t_{\max} = \frac{1}{k_\text{a}}\ln\left(\frac{k_\text{a}[\text{A}]_0}{k_\text{b}}\right)}$$

$$[B]_{max} = [A]_0(1 - e^{-k_a t_{max}}) - k_b t_{max}$$

$$= [A]_0\left(1 - \frac{k_b}{k_a[A]_0}\right) - k_b t_{max}$$

$$\boxed{[B]_{max} = [A]_0 - \frac{k_b}{k_a} - k_b t_{max}}$$

P7.41 (a) For the mechanism

$$hhhh\ldots \underset{k_a'}{\overset{k_a}{\rightleftharpoons}} hchh\ldots$$

$$hchh\ldots \underset{k_b'}{\overset{k_b}{\rightleftharpoons}} cccc\ldots$$

the rate equations are

$$\frac{d[hhhh\ldots]}{dt} = -k_a[hhhh\ldots] + k_a'[hchh\ldots]$$

$$\frac{d[hchh\ldots]}{dt} = k_a[hhhh\ldots] - k_a'[hchh\ldots] - k_b[hchh\ldots] + k_b'[cccc\ldots]$$

$$\frac{d[cccc\ldots]}{dt} = k_b[hchh\ldots] - k_b'[cccc\ldots]$$

(b) Apply the steady-state approximation to the intermediate in the above mechanism:

$$\frac{d[hchh\ldots]}{dt} = k_a[hhhh\ldots] - k_a'[hchh\ldots] - k_b[hchh\ldots] + k_b'[cccc\ldots] = 0$$

so $[hchh\ldots] = \dfrac{k_a[hhhh\ldots] + k_b'[cccc\ldots]}{k_a' + k_b}$

Therefore, $\dfrac{d[hhhh\ldots]}{dt} = -\dfrac{k_a k_b}{k_a' + k_b}[hhhh\ldots] + \dfrac{k_a' k_b'}{k_a' + k_b}[cccc\ldots]$

or

$$\boxed{\frac{d[hhhh\ldots]}{dt} = -k_{r,eff}[hhhh\ldots] + k'_{r,eff}[cccc\ldots] \quad \text{where} \quad k_{r,eff} = \frac{k_a k_b}{k_a' + k_b} \quad \text{and} \quad k'_{r,eff} = \frac{k_a' k_b'}{k_a' + k_b}}$$

The simple mechanism

$$hhhh\ldots \underset{k_r'}{\overset{k_r}{\rightleftharpoons}} cccc\ldots$$

has the rate law $\dfrac{d[hhhh\ldots]}{dt} = -k_r[hhhh\ldots] + k_r'[cccc\ldots]$, which is equivalent to the above rate law.

(c) It is difficult to make conclusive inferences about intermediates from kinetic data alone. For example, if rate measurements show formation of coils from helices with a single rate constant, they tell us nearly nothing about the mechanism. The rate law

$$\frac{d[cccc\ldots]}{dt} = k_r[hhhh\ldots]$$

is consistent with a single-step mechanism, with a two-step mechanism with a rate-determining second step, and with a two-step mechanism with a steady-state intermediate. Even if kinetic monitoring of the product shows production with two rate constants, the rate constants could belong to competing paths or to steps of a single reaction path. The best evidence for an intermediate's participation in a reaction is detection of the intermediate, or at least detection of structural features that can belong to a proposed intermediate but not reactant or product.

8 Complex biochemical processes

Answers to discussion questions

D8.1 The **Michaelis–Menten mechanism** of enzyme activity models the enzyme with one active site, that weakly and reversibly binds a substrate in homogeneous solution. It is a three-step mechanism. The first and second steps are the reversible formation of the enzyme–substrate complex (ES). The third step is the decay of the complex into the product. The steady-state approximation is applied to the concentration of the intermediate (ES) and its use simplifies the derivation of the final rate expression. However, the justification for the use of the approximation with this mechanism is suspect, in that both rate constants for the reversible step may not be as large, in comparison to the rate constant for the decay to products, as they need to be for the approximation to be valid. The mechanism clearly indicates that the simplest form of the rate law, $v = v_{max} = k_b[E]_0$, occurs when $[S]_0 \gg K_M$ and the general form of the rate law [8.1] does seem to match the principal experimental features of enzyme-catalyzed reactions. It provides a mechanistic understanding of both the turnover number and catalytic efficiency. The model may be expanded to include multisubstrate reactions and it must be modified to accommodate the effects of competitive and non-competitive inhibition.

D8.2 (a) Enzyme-catalyzed reactions that require minutes or hours may be followed with standard spectroscopic methods including UV-visible, infrared, fluorescence, and NMR techniques. It may even be possible to determine the progression of substrate consumption, or product formation, with a chemical titration or radioactivity assay. Electrical conductivity may be used when ions are reaction participants; pH measurements are used to follow the reaction rate when an acid or base is a participant. An inhibitor that binds very strongly to the active site may be used to quench the reaction at any time, thereby, making it possible to separate substrate and product with a chromatographic method; after which spectroscopic techniques, and application of the Beer–Lambert law, may prove useful in the case for which substrate and product have overlapping spectra.

A stopped-flow technique provides for the rapid mixing that is necessary to follow reactions that require milliseconds or minutes. Standard spectroscopic, conductivity, or pH measurements are used to follow the reaction rate.

Very fast reactions may be followed by disturbing an equilibrium system with the excitation energy of flash photolysis or by a very sudden temperature jump initiated with a large current burst through the reaction solution. A pulsed laser beam is subsequently used to generate absorption, emission, or fluorescence spectra and the evolution of such spectra yields reaction rates.

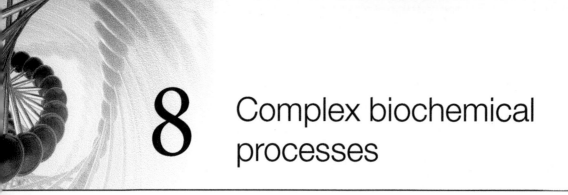

(b) The initial rate of an enzyme-catalyzed reaction is acquired by extrapolation of the time evolution of observed rates to the initial mixing time. When repeated over a range of initial substrate concentrations, it is possible to prepare a double reciprocal, **Lineweaver–Burk plot** of $1/v_0$ against $1/[S]_0$. It is the intercept and slope of this plot that provides the values of the maximum reaction rate and the Michaelis constant (eqns 8.2b and 8.3). The manner in which an enzyme inhibitor alters the slope and intercept provides both evidence for the type of inhibition (competitive, uncompetitive, or non-competitive) and values of inhibitor binding constants.

(c) The molecular shape of a strongly enzyme-binding, competitive inhibitor gives clues about the intermediate enzyme–substrate activated complex because, like the inhibitor, the activated ES complex must bind strongly at the active site in order to initiate reaction. Clues include charge distribution, hydrogen bonding, and hydrophobic interactions. The idea of a transition-state intermediate involves a slight modification of the Michaelis–Menten mechanism:

$$E + S \rightleftharpoons ES \rightleftharpoons ES^* \rightarrow EP \rightleftharpoons E + P$$

The activated transition-state enzyme–substrate complex, ES*, is a very short lived intermediate because it has the activation energy necessary to react. It has been shown that the enzyme has a higher affinity for the transition-state intermediate than for the substrate and will bind the intermediate more strongly. Good inhibitors are generally transition-state analogs.

D8.3 As temperature increases we expect the rate of an enzyme-catalyzed reaction to increase. However, at a sufficiently high temperature the enzyme **denatures** and a decrease in the reaction rate is observed. Temperature-related denaturation is caused by the action of vigorous vibrational motion, which destroys secondary and tertiary protein structure. Electrostatic, internal hydrogen bonding, and van der Waals interactions that hold the protein in its active, folded shape are broken with the protein unfolding into a **random coil**. The active site and enzymatic activity is lost.

The rate of a particular enzyme-catalyzed reaction may also appear to decrease at high temperatures in the special case in which an alternative substrate reaction, which has a relatively slow rate at low temperature, has the faster rate increase with increasing temperature. A temperature may be reached at which the alternative reaction predominates.

D8.4 (a) **Sequential reaction mechanism**:

$$E + S_1 \rightleftharpoons ES_1 \qquad K_{M1} = \frac{[E][S_1]}{[ES_1]}$$

$$ES_1 + S_2 \rightleftharpoons ES_1S_2 \qquad K_{M12} = \frac{[ES_1][S_2]}{[ES_1S_2]}$$

$$ES_1S_2 \rightarrow E + P \qquad v = k_b[ES_1S_2]$$

Ping-pong mechanism:

$$E + S_1 \rightleftharpoons ES_1 \qquad K_{M1} = \frac{[E][S_1]}{[ES_1]}$$

$$ES_1 \rightarrow E^* + P_1 \qquad v_1 = k_{b1}[ES_1]$$

$$E^* + S_2 \rightleftharpoons E^*S_2 \qquad K_{M2} = \frac{[E^*][S_2]}{[E^*S_2]}$$

$$E^*S_2 \rightarrow E + P_2 \qquad v_2 = k_{b2}[E^*S_2]$$

For sequential reactions, the slope of a plot of $1/v$ against $1/[S_1]$ depends on $[S_2]$, so a series of such plots for different values of $[S_2]$ form a family of non-parallel lines. However, for "ping-pong" reactions the lines described by plots of $1/v_2$ against $1/[S_1]$ for different values of $[S_2]$ are parallel because the slopes are independent of $[S_2]$. See text Section 8.2, in particular, Figure 8.5.

(b) Text Figure 8.6 summarizes the important characteristics of the three major modes of enzyme inhibition: competitive inhibition, uncompetitive inhibition, and non-competitive inhibition.

Mathematical models for inhibition, which are the analogs of the Michaelis–Menten and Lineweaver–Burk equations (8.1), are presented in eqns 8.10a, b, and c.

$$\frac{1}{v}=\frac{\alpha'}{v_{\max}}+\left(\frac{\alpha K_{\mathrm{M}}}{v_{\max}}\right)\frac{1}{[\mathrm{S}]_0} \quad [8.9c]$$

where $\alpha = 1 + [\mathrm{I}]/K_{\mathrm{I}}$, $\alpha' = 1 + [\mathrm{I}]/K'_{\mathrm{I}}$, $K_{\mathrm{I}} = [\mathrm{E}][\mathrm{I}]/[\mathrm{EI}]$, and $K'_{\mathrm{I}} = [\mathrm{ES}][\mathrm{I}]/[\mathrm{ESI}]$

In **competitive inhibition** the inhibitor binds only to the active site of the enzyme and thereby inhibits the attachment of the substrate. This condition corresponds to $\alpha > 1$ and $\alpha' = 1$ (because ESI does not form). The slope of the Lineweaver–Burk plot increases by a factor of α relative to the slope for data on the uninhibited enzyme ($\alpha = \alpha' = 1$). The y-intercept does not change as a result of competitive inhibition.

In **uncompetitive inhibition** the inhibitor binds to a site of the enzyme that is removed from the active site, but only if the substrate is already present. The inhibition occurs because ESI reduces the concentration of ES, the active type of the complex. In this case $\alpha = 1$ (because EI does not form) and $\alpha' > 1$. The y-intercept of the Lineweaver–Burk plot increases by a factor of α' relative to the y-intercept for data on the uninhibited enzyme, but the slope does not change.

In **non-competitive inhibition** (also called **mixed inhibition**) the inhibitor binds to a site other than the active site, and its presence reduces the ability of the substrate to bind to the active site. Inhibition occurs at both the E and ES sites. This condition corresponds to $\alpha > 1$ and $\alpha' > 1$. Both the slope and y-intercept of the Lineweaver–Burk plot increase upon addition of the inhibitor. Figure 8.6c shows the special case of $K_{\mathrm{I}} = K'_{\mathrm{I}}$ and $\alpha = \alpha'$, which results in intersection of the lines at the x-axis.

In all cases, the efficiency of the inhibitor may be obtained by determining K_{M} and $v_{\max}$ from a control experiment with uninhibited enzyme and then repeating the experiment with a known concentration of inhibitor. From the slope and y-intercept of the Lineweaver–Burk plot for the inhibited enzyme (eqn 8.9c), the mode of inhibition, the values of α or α', and the values of K_{I}, or K'_{I} may be obtained.

D8.5 (a) Figure 8.1 is a sketch of the enzyme-catalyzed reaction rate against substrate concentration both with and without product inhibition. Inhibition reduces the reaction rate and lowers the maximum achievable reaction rate.

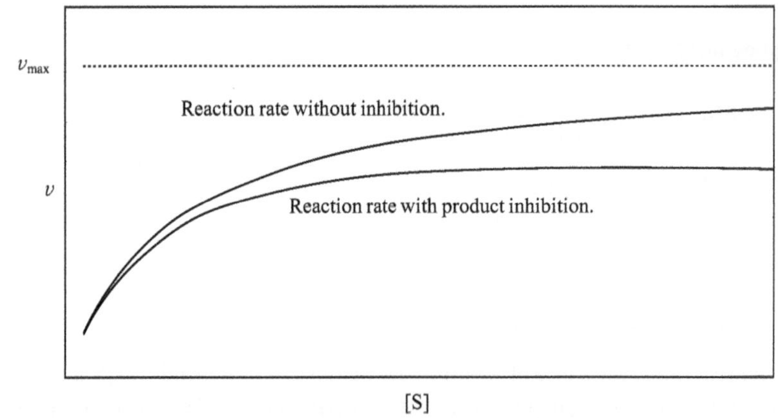

Figure 8.1

(b) Glycolysis occurs in the **cytosol**, the aqueous material encapsulated by the cell membrane, and consists of 10 enzyme-catalyzed reactions. The first step, which is catalyzed by hexokinase, uses an ATP molecule to phosphorylate glucose to glucose-6-phosphate (G6P):

$$\text{glucose(aq)} + \text{ATP(aq)} \xrightarrow{\text{hexokinase}} \text{G6P(aq)} + \text{ADP(aq)} + \text{H}^+\text{(aq)} \quad \Delta_r G^{\ominus} = -17 \text{ kJ mol}^{-1}$$

Once phosphorylated, glucose is trapped within the cell in contrast with unphosphorylated glucose, which easily traverses the plasma membrane. Attachment of the negatively charged phosphate group prevents G6P from crossing the membrane so, among other functions, phosphorylation allows the cell to stock up on glucose while levels are high. The catalyzing enzyme, hexokinase, is allosterically inhibited by the product G6P in a feedback inhibition mechanism that helps control the passage of glucose into the glycolytic pathway. Should the diversion of G6P into either glycogen, the storage form of glucose, or fructose-6-phosphate (F6P) be blocked for some reason, (such as sedentary periods in which ATP usage in muscle and brain tissue is minimal) the buildup of excess G6P inhibits its own formation.

D8.6 Passive diffusion of a molecule through a biological membrane via an embedded ion channel protein is expected to be driven by the concentration gradient of the molecule across the membrane; in contrast to the observations of this exercise, the flux is proportional to the gradient and does not reach a maximum that depends upon the size of the gradient. This is **Fick's first law of diffusion**. To explain the phenomenon of **mediated transport** which exhibits an increased flux at larger gradients until reaching a maximum value at a particular gradient, consider the transport system of Figure 8.2. The transportable molecule locks into a surface hole of a membrane-embedded carrier protein, which undergoes a conformational change to admit the molecule into the hashed protein region in the figure, after which the conformational change squeezes the molecule into the biological cell. The transport may be passive or active. This transport mechanism does provide for a maximum in the flux of the molecule across the membrane. When the gradient is low and the concentration of the transportable molecule outside the cell is low, not all of the carrier proteins in the membrane are engaged as a molecule–protein complex. However, as the concentration of the molecule increases there comes a point at which all carrier proteins in the membrane of the cell are engaged with the transportable molecule. At that point the flux reaches a maximum because a further increase in concentration does not increase the number of molecules being transported.

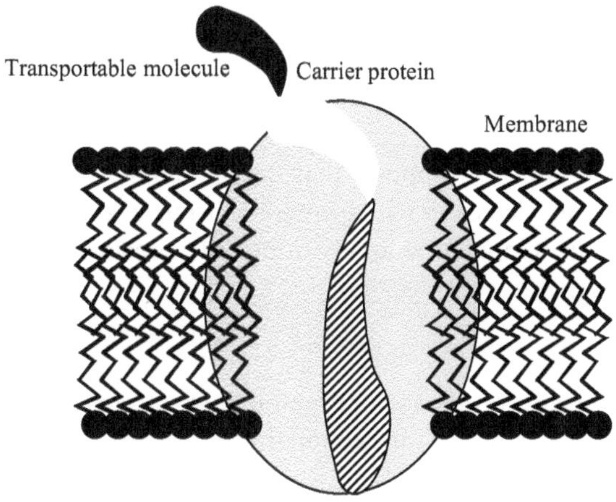

Figure 8.2

D8.7 It is believed that the especially high mobility of the proton in water reflects a very special mechanism for conduction, the **Grotthus mechanism**, in which the proton on one H_2O molecule migrates to its neighbors, the proton on that H_2O molecule migrates to its neighbors, and so on along a chain (text Figure 8.16). The motion is therefore an *effective* motion of a proton, not the actual motion of a single proton.

The mechanism details are still highly contentious. Attention focuses on the $H_9O_4^+$ unit in which the nearly trigonal planar H_3O^+ ion is linked to three strongly solvating H_2O molecules. This cluster of atoms is itself hydrated, but the hydrogen bonds in the secondary sphere are weaker than in the primary sphere. It is envisaged that the rate-determining step is the cleavage of one of the weaker hydrogen bonds of this secondary sphere (Figure 8.3a). After this bond cleavage has taken place, and the released molecule has rotated through a few degrees (a process that takes about 1 ps), there is a rapid adjustment of bond lengths and angles in the remaining cluster, to form a $H_5O_2^+$ cation of structure $H_2O\cdots H^+\cdots OH_2$ (Figure 8.3b). Shortly after this reorganization has occurred, a new $H_9O_4^+$ cluster forms as other molecules rotate into a position where they can become members of a secondary hydration sphere, but now the positive charge is located one molecule to the right of its initial location (Figure 8.3c). According to this model, there is no coordinated motion of a proton along a chain of molecules, simply a very rapid hopping between neighboring sites, with a low activation energy. The model is consistent with the observation that the molar conductivity of protons increases as the pressure is raised, for increasing pressure ruptures the hydrogen bonds in water.

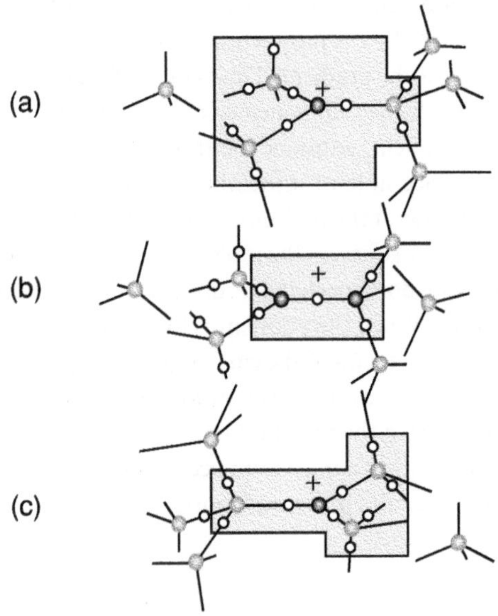

Figure 8.3

D8.8 (a) The **Marcus theory** of electron transfer supposes that the probability that an electron will move from D to A in the transition state decreases with increasing distance between D and A in the DA complex. For given values of the temperature and $\Delta^{\ddagger}G$, the rate constant k_{et} varies with the edge-to-edge distance r as

$$k_{et} \propto e^{-\beta r} \quad [8.28]$$

β is a constant with a value that depends on the medium through which the electron must travel from donor to acceptor. That is, the rate constant for electron transfer decreases exponentially with increasing r.

(b) The DA complex and the medium surrounding it must rearrange spatially as charge is redistributed to form the ions D^+ and A^-. These molecular rearrangements include the relative reorientation of the D and A molecules in DA and the relative reorientation of the solvent molecules surrounding DA. According to the Marcus theory, the expression for the Gibbs energy of activation is

$$\Delta^{\ddagger}G = \frac{(\Delta_r G^{\ominus} + \lambda)^2}{4\lambda} \quad [8.29]$$

where $\Delta_r G^\ominus$ is the standard reaction Gibbs energy for the electron transfer process DA $\rightarrow$ D^+A^- and λ is the **reorganization energy**, the energy change associated with molecular rearrangements that must take place so that DA can take on the equilibrium geometry of D^+A^-. This influences the rate constant of electron transfer by the relation

$$k_{et} \propto e^{-\Delta^\ddagger G/RT} \quad [8.30]$$

at constant r. Thus, electron transfer becomes more efficient as $\Delta_r G^\ominus$ becomes more negative. For example, kinetically efficient oxidation of D requires that its standard reduction potential be lower than the standard reduction potential of A. Also, the electron transfer becomes more efficient as the reorganization energy is matched closely by the standard reaction Gibbs energy. When $\Delta_r G^\ominus = -\lambda$, we see that $\Delta^\ddagger G = 0$ and eqn 8.30 implies that electron transport is not slowed down by an activation energy.

Solutions to exercises

E8.9 $$\mathrm{E} + \mathrm{S} \underset{k_a'}{\overset{k_a}{\rightleftharpoons}} \mathrm{ES} \overset{k_b}{\rightleftharpoons} \mathrm{E} + \mathrm{P}$$

In the pre-equilibrium approximation, the intermediate ES is in equilibrium with the reactants E and S. This requires equality between the rate for formation of ES and the rate of ES dissociation to reactants.

rate of formation of ES = rate of ES dissociation to reactants

$$k_a\,[\mathrm{E}][\mathrm{S}] = k_a'\,[\mathrm{ES}]$$

$$\frac{[\mathrm{ES}]}{[\mathrm{E}][\mathrm{S}]} = \frac{k_a}{k_a'} = K \quad \text{or} \quad [\mathrm{ES}] = K[\mathrm{E}][\mathrm{S}]$$

If $[\mathrm{E}]_0$ is the total enzyme concentration, then $[\mathrm{E}] = [\mathrm{E}]_0 - [\mathrm{ES}]$ by conservation of mass.

$$[\mathrm{ES}] = K[\mathrm{E}][\mathrm{S}] = K([\mathrm{E}]_0 - [\mathrm{ES}])[\mathrm{S}]$$

$$[\mathrm{ES}] + \frac{[\mathrm{ES}]}{K[\mathrm{S}]} = [\mathrm{E}]_0$$

$$[\mathrm{ES}] = \frac{[\mathrm{E}]_0}{1 + \dfrac{1}{K[\mathrm{S}]}} = \frac{[\mathrm{E}]_0[\mathrm{S}]}{[\mathrm{S}] + \dfrac{1}{K}}$$

Substitution in the expression for the rate of formation of P gives:

$$\boxed{\text{rate of formation of P} = k_b[\mathrm{ES}] = \frac{k_b[\mathrm{E}]_0[\mathrm{S}]}{[\mathrm{S}] + \dfrac{1}{K}}} \quad \text{pre-equilibrium approximation, } K = \frac{k_a}{k_a'}$$

$$\text{rate of formation of P} = k_b[\mathrm{ES}] = \frac{k_b[\mathrm{E}]_0[\mathrm{S}]}{[\mathrm{S}] + K_M} \quad \text{steady-state approximation} \quad [8.1]$$

Comparison of the pre-equilibrium approximation with the steady-state approximation shows that the two approximations are the same when $K^{-1} = K_M$.

Since $\dfrac{1}{K} = \dfrac{k_a'}{k_a}$ and $K_M = \dfrac{k_a' + k_b}{k_a}$ [8.1], the two approximations are identical when $\boxed{k_a' \gg k_b}$.

E8.10 $$E+S \underset{k_a'}{\overset{k_a}{\rightleftharpoons}} ES \underset{k_b'}{\overset{k_b}{\rightleftharpoons}} E+P$$

Assume that the steady-state approximation is appropriate for the ES intermediate:

$$\frac{d[ES]}{dt}=k_a[E][S]-k_a'[ES]-k_b[ES]+k_b'[E][P]=0, \quad \text{so} \quad [ES]=\left(\frac{k_a[S]+k_b'[P]}{k_a'+k_b}\right)[E].$$

Define the Michaelis constants $K_M=\frac{k_a'+k_b}{k_a}$ and $K_M'=\frac{k_a'+k_b}{k_b'}$ so that the [ES] expression becomes:

$$[ES]=\left(\frac{[S]}{K_M}+\frac{[P]}{K_M'}\right)[E]$$

Substitution of the mass-conservation expression $[E] = [E]_0 - [ES]$ for [E] and solving for [ES] yields:

$$[ES]=\left(\frac{[S]/K_M+[P]/K_M'}{1+[S]/K_M+[P]/K_M'}\right)[E]_0$$

Now, substitution of this expression for [ES] in the expression for the rate of product formation, v, followed by algebraic simplification, yields:

$$v=k_b[ES]-k_b'[E][P]=k_b[ES]-k_b'([E]_0-[ES])[P]=(k_b+k_b'[P])[ES]-k_b'[E]_0[P]$$
$$=(k_b+k_b'[P])\left(\frac{[S]/K_M+[P]/K_M'}{1+[S]/K_M+[P]/K_M'}\right)[E]_0-k_b'[E]_0[P]$$
$$=\frac{(k_b+k_b'[P])([S]/K_M+[P]/K_M')-k_b'[P](1+[S]/K_M+[P]/K_M')}{1+[S]/K_M+[P]/K_M'}[E]_0$$
$$=\frac{k_b([S]/K_M+[P]/K_M')-k_b'[P]}{1+[S]/K_M+[P]/K_M'}[E]_0=\frac{k_b([S]/K_M+[P]/K_M')-(k_a'+k_b)[P]/K_M'}{1+[S]/K_M+[P]/K_M'}[E]_0$$
$$=\frac{k_b[S]/K_M-k_a'[P]/K_M'}{1+[S]/K_M+[P]/K_M'}[E]_0$$

$$\boxed{v=\frac{v_{max}[S]/K_M-v_{max}'[P]/K_M'}{1+[S]/K_M+[P]/K_M'}, \quad \text{where } v_{max}=k_b[E]_0 \text{ and } v_{max}'=k_a'[E]_0}$$

which finishes the derivation of eqn 8.4a. When [S] is so large that $[S]/K_M \gg [P]/K_M'$, the rate expression reduces to the Michaelis–Menten rate law:

$$\boxed{v=\frac{v_{max}[S]/K_M}{1+[S]/K_M}, \quad \text{where } [S]/K_M \gg [P]/K_M'}$$

When [P] is so large that $[S]/K_M \ll [P]/K_M'$, the rate expression reduces to the Michaelis–Menten rate law:

$$\boxed{v=\frac{-v_{max}'[P]/K_M'}{1+[P]/K_M'}, \quad \text{where } [S]/K_M \ll [P]/K_M'}$$

E8.11 Assume that the steady-state approximation is appropriate for both intermediates ([ES] and [ES′]).

For [ES]:

$$\frac{d[ES]}{dt} = k_a[E][S] - k_a'[ES] - k_b[ES] = 0 \quad \text{and} \quad [ES] = \left(\frac{k_a}{k_a' + k_b}\right)[E][S].$$

For [ES′]:

$$\frac{d[ES']}{dt} = k_b[ES] - k_c[ES'] = 0 \quad \text{and} \quad [ES'] = \left(\frac{k_b}{k_c}\right)[ES].$$

We now have two equations in the three unknowns [E], [ES], and [ES′]. A third is provided by the mass-balance expression $[E]_0 = [E] + [ES] + [ES']$. These three equations may be solved to give expressions for each of the three unknowns in terms of the rate constants, $[E]_0$, and [S]. (For practical purposes the free substrate concentration is replaced by $[S]_0$ because the substrate is typically in large excess relative to the enzyme.) The expression found for [ES′] is

$$[ES'] = \frac{v_{max}/k_c}{1 + K_M/[S]_0}, \quad \text{where} \quad v_{max} = \left(\frac{k_b k_c}{k_b + k_c}\right)[E]_0 \quad \text{and} \quad K_M = \frac{k_c\,(k_a' + k_b)}{k_a\,(k_b + k_c)}.$$

Substitution into the rate expression for product formation yields the desired equation.

$$v = k_c[ES'] = \frac{v_{max}}{1 + K_M/[S]_0}.$$

E8.12 Rate of formation of P, $v = k_r[E]_0 = \frac{k_b[S][E]_0}{K_M + [S]} = \frac{[S]v_{max}}{K_M + [S]}$ [8.1, 8.2b]

Solving the reaction rate expression for v_{max} gives

$$\begin{aligned} v_{max} &= \frac{(K_M + [S])v}{[S]} \\ &= \frac{(0.045\ \text{mol dm}^{-3} + 0.110\ \text{mol dm}^{-3}) \times (1.15\ \text{mmol dm}^{-3}\ \text{s}^{-1})}{0.110\ \text{mol dm}^{-3}} \\ &= \boxed{1.62\ \text{mmol dm}^{-3}\ \text{s}^{-1}} \end{aligned}$$

E8.13 Michaelis–Menten kinetics: $v = \frac{v_{max}}{1 + K_M/[S]}$ [8.1, 8.2b]

By inspection of the Michaelis—Menten reaction rate expression we see that $v = \frac{1}{2}v_{max}$ when $K_M/[S] = 1$ or $\boxed{[S] = K_M}$.

E8.14 Lineweaver−Burk plot for Michaelis−Menten mechanism: $\frac{1}{v} = \frac{1}{v_{max}} + \left(\frac{K_M}{v_{max}}\right)\frac{1}{[S]}$ [8.3]

As illustrated in text Example 8.1, we draw up a data table that transforms experimental data into rows of 1/[S] and $1/v$ values, and prepare a plot of $1/v$ against 1/[S]. If the plot appears to be linear, we perform a linear regression fit for the plot and recognize that the Lineweaver−Burk equation equates the regression intercept and slope to the ratios $1/v_{max}$ and K_M/v_{max}, respectively. Thus, solving for v_{max} and K_M, we find that

$$v_{max} = 1/intercept \quad \text{and} \quad K_M = slope/intercept.$$

[isocitrate]/μmol dm^{-3}	31.8	46.4	59.3	118.5	222.2
v/pmol dm^{-3} s^{-1}	70.0	97.2	116.7	159.2	194.5
100 μmol dm^{-3}/[isocitrate]	3.145	2.155	1.686	0.844	0.450
100 pmol dm^{-3} s^{-1}/v	1.429	1.029	0.857	0.628	0.514

The plot, shown in Figure 8.4, is seen to be linear so we perform a linear regression fit that is summarized within the figure box and calculate v_{max} and K_M.

$$v_{max} = \frac{100 \text{ pmol dm}^{-3}\text{ s}^{-1}}{0.3359} = \boxed{298 \text{ pmol dm}^{-3}\text{ s}^{-1}}$$

$$K_M = \left(\frac{0.3354 \text{ pmol dm}^{-3}\text{ s}^{-1}/100}{\mu\text{mol}^{-1}\text{ dm}^3/100}\right) \Big/ (0.3359 \text{ pmol dm}^{-3}\text{ s}^{-1}/100) = \boxed{99.9 \ \mu\text{mol dm}^{-3}}$$

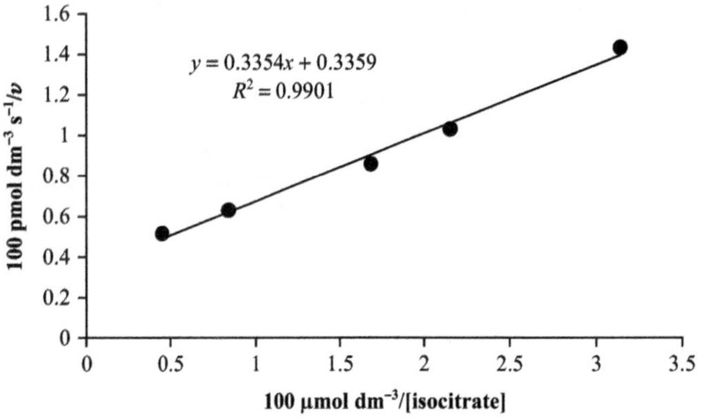

Figure 8.4

E8.15 We fit the data to the Lineweaver–Burk equation: $\frac{1}{v} = \frac{1}{v_{max}} + \left(\frac{K_M}{v_{max}}\right)\frac{1}{[S]}$ [8.3]

Hence, draw up the following table.

[ATP]/μmol dm^{-3}	0.60	0.80	1.4	2.0	3.0
v/μmol dm^{-3} s^{-1}	0.81	0.97	1.30	1.47	1.69
1/[ATP]/μmol^{-1} dm^3	1.67	1.25	0.71	0.50	0.33
1/v/μmol^{-1} dm^3 s	1.23	1.03	0.769	0.680	0.592

A plot of $1/v$ against 1/[S] is shown in Figure 8.5. The plot is linear and the linear least-squares regression fit is shown as a box insert.

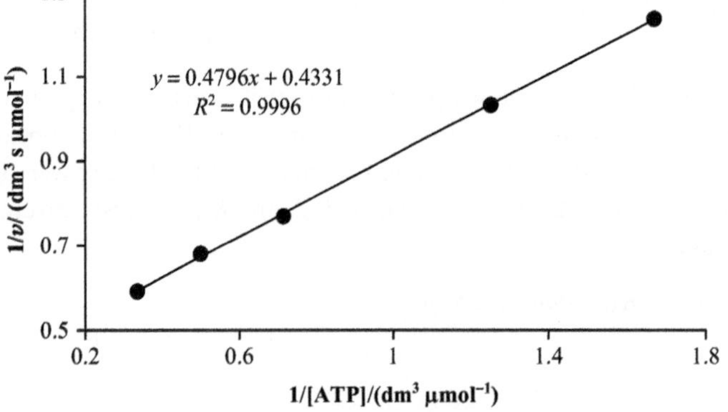

Figure 8.5

The intercept is $1/v_{max} = 0.433\ dm^3\ s\ \mu mol^{-1}$. Therefore, $v_{max} = \boxed{2.31\ \mu mol\ dm^{-3}\ s^{-1}}$

The slope is $K_M/v_{max} = 0.480$ s. Therefore,

$$K_M = (0.480\ s) \times (2.31\ \mu mol\ dm^{-3} s^{-1}) = \boxed{1.11\ \mu mol\ dm^{-3}}$$

The maximum turnover number k_{cat} is

$$k_{cat} = \frac{v_{max}}{[E]_0}\ [8.7] = \frac{2.31\ \mu mol\ dm^{-3}\ s^{-1}}{0.020\ \mu mol\ dm^{-3}} = \boxed{1.1\bar{6} \times 10^2\ s^{-1}}$$

The catalytic efficiency η is

$$\eta = \frac{k_{cat}}{K_M}\ [8.8] = \frac{1.1\bar{6} \times 10^2\ s^{-1}}{1.11\ \mu mol\ dm^{-3}} = \boxed{1.0 \times 10^2\ \mu mol^{-1}\ dm^3\ s^{-1}}$$

E8.16 (a) We start with the Lineweaver–Burk expression: $\frac{1}{v} = \frac{1}{v_{max}} + \left(\frac{K_M}{v_{max}}\right)\frac{1}{[S]}$ [8.3]

Multiply both sides of this equation by $v \times v_{max}$.

$$v_{max} = v + K_M\left(\frac{v}{[S]}\right) \quad \text{or} \quad \boxed{\frac{v}{[S]} = \frac{v_{max}}{K_M} - \frac{v}{K_M}} \quad \text{Eadie–Hofstee relation}$$

(b) Examination of the above Eadie–Hofstee relation reveals that, if the Michaelis–Menten mechanism is valid, a plot of $v/[S]$ against v should be linear with a slope equal to $-1/K_M$ and an intercept equal to v_{max}/K_M. Thus, $K_M = -1/slope$ and $v_{max} = intercept \times K_M$.

(c) An Eadie–Hofstee plot of the appropriately transformed data of Exercise 8.14 is shown in Figure 8.6. The plot is reasonably linear and the linear least-squares regression fit is shown as a box insert.

$$K_M = -1/slope = -1/(-0.0112\ dm^3\ \mu mol^{-1}) = \boxed{89.3\ \mu mol\ dm^{-3}}$$

$$v_{max} = intercept \times K_M = (3.126\ pmol\ dm^{-3}\ s^{-1} \times (\mu mol\ dm^{-3})^{-1} \times (89.3\ \mu mol\ dm^{-3})$$
$$= \boxed{279\ pmol\ dm^{-3}\ s^{-1}}$$

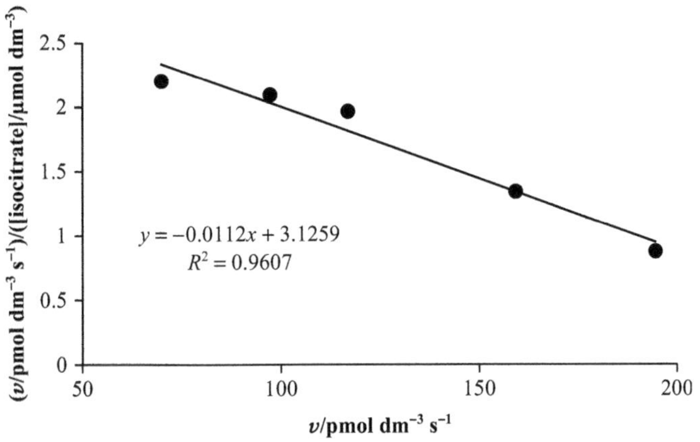

Figure 8.6

E8.17 (a) We start with the Lineweaver–Burk expression: $\frac{1}{v} = \frac{1}{v_{max}} + \left(\frac{K_M}{v_{max}}\right)\frac{1}{[S]}$ [8.3]

Multiplication of the Lineweaver–Burk expression by [S] gives

$$\boxed{\frac{[S]}{v} = \frac{[S]}{v_{max}} + \frac{K_M}{v_{max}}}$$ [Hanes relation]

(b) The above Hanes relation reveals that a plot of $[S]/v$ against [S] should be linear with a slope equal to $1/v_{max}$ and an intercept equal to K_M/v_{max}. Thus, $v_{max} = 1/slope$ and $K_M = intercept \times v_{max}$.

(c) A Hanes plot of the appropriately transformed data of Exercise 8.14 is shown in Figure 8.7. The plot appears to be linear and the linear least-squares regression fit is shown as a box insert.

$$v_{max} = 1/slope = 1/(0.00372 \text{ pmol}^{-1} \text{ dm}^3 \text{ s}) = \boxed{279 \text{ pmol dm}^{-3} \text{ s}^{-1}}$$

$$K_M = intercept \times v_{max} = 0.3095 \text{ µmol dm}^{-3}/(\text{pmol dm}^{-3} \text{ s}^{-1}) \times (279 \text{ pmol dm}^{-3} \text{ s}^{-1})$$
$$= \boxed{86.4 \text{ µmol dm}^{-3}}$$

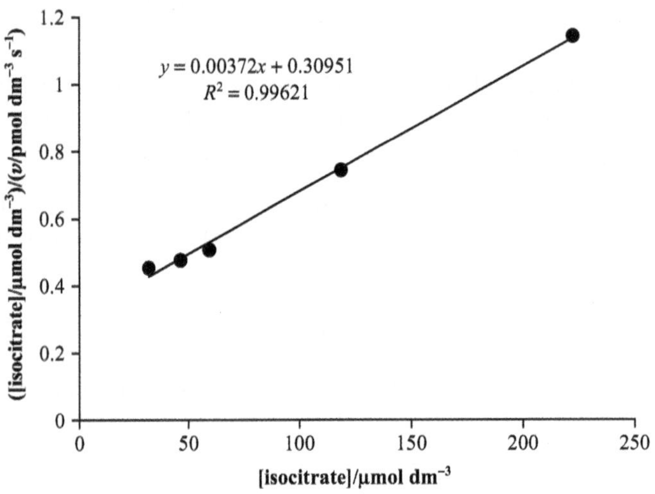

Figure 8.7

E8.18 (a) Figure 8.8 is a sketch that contrasts Michaelis–Menten enzyme activity with the activity of an allosteric enzyme for which the substrate acts as the effector. We see that both approach the maximum reaction rate v_{max} at very high substrate concentration. At low concentrations of the substrate the allosteric enzyme has a slower reaction rate because the effector concentration is insufficient to activate the enzyme subunits. However, the allosteric enzyme shows an activity that is concave-up as [S] increases from zero to somewhat higher values as the substrate effects ever-increasing activity of the subunits. At an intermediate substrate concentration, the allosteric enzyme exhibits a greater reaction rate, which is shown as a "crossing over" of the Michaelis–Menten activity in Figure 8.8. This illustrates the phenomena in which small variations of the effector concentration effectively switch the allosteric enzyme on or off. The allosteric reaction rate has a **sigmoidal**, S-shaped, characteristic; the shape of the Michaelis–Menten reaction rate is said to be **hyperbolic**.

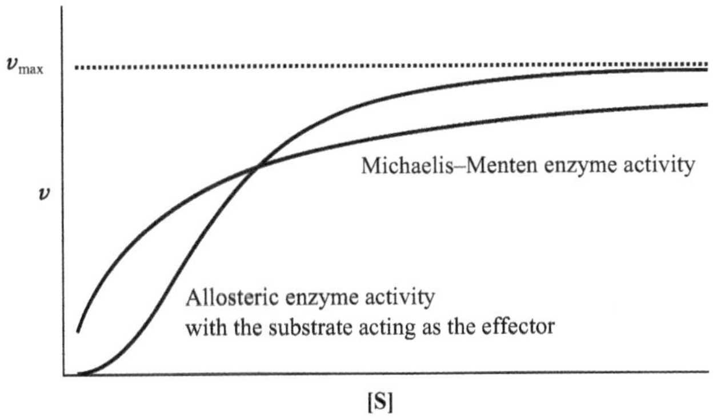

Figure 8.8

(b) $E + n\,S \rightleftharpoons ES_n \rightarrow E + n\,P \quad \dfrac{v}{v_{max}} = \dfrac{1}{1 + (K'/c^{\ominus})/([S]/c^{\ominus})^n}$

A plot of this rate expression is shown in Figure 8.9 for $n = 1, 2, 3$, and 4. All curves use $K' = 5 \times 10^{-4}$ mol dm^{-3}. We see that the $n = 1$ curve is hyperbolic and that the sigmoidal character of the curves grows with increasing n. In some cases, n may equal the number of subunits and active sites in the enzyme. However, the above model is usually an inadequate model of enzymatic activity and experimentation usually finds that n is smaller than the number of subunits in an allosteric enzyme. It is best to consider n as a measure of the sigmoidal character in the plot of v against [S] and recognize that the sigmoidal character is an indicator of the cooperative binding by which the attachment of a substrate molecule to one subunit enhances the affinity of another subunit to bind substrate. Thus, n is called the **interaction coefficient**.

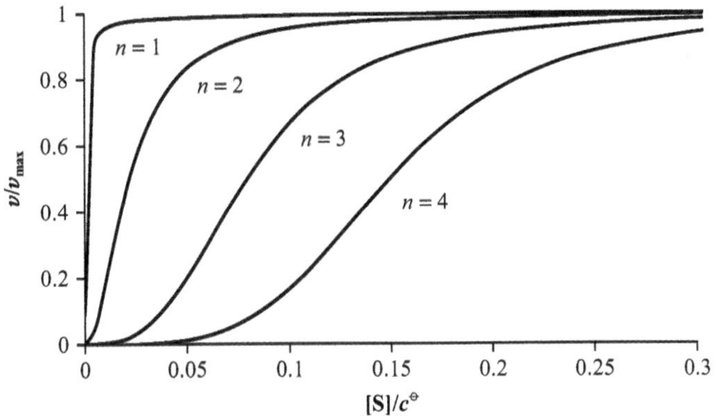

Figure 8.9

E8.19 $E + n\,S \rightleftharpoons ES_n \rightarrow E + n\,P \quad \dfrac{v}{v_{max}} = \dfrac{1}{1 + (K'/c^{\ominus})/([S]/c^{\ominus})^n}$

(a) Invert this rate expression and subtract it from 1 to get

$$\frac{v_{max} - v}{v} = \frac{K'/c^{\ominus}}{([S]/c^{\ominus})^n}$$

Invert it again and take the base-10 logarithm to get (after a little logarithm algebra) the desired relation.

$$\log\frac{v}{v_{max} - v} = \log([S]/c^{\ominus})^n - \log(K'/c^{\ominus}) = \log\left\{\frac{([S]/c^{\ominus})^n}{K'/c^{\ominus}}\right\}$$

$$= \log([S]/c^{\ominus})^n - \log(K'/c^{\ominus}) = n\log([S]/c^{\ominus}) - \log(K'/c^{\ominus})$$

If rate data is described by this model, a plot of $\log\frac{v}{v_{max}-v}$ against $\log([S]/c^{\ominus})$ will be linear with a slope equal to n and an intercept equal to $-\log(K'/c^{\ominus})$. Thus,

$$n = slope \quad \text{and} \quad K' = 10^{-intercept}c^{\ominus}.$$

(b) Make a table containing data transformations to the functions $\log([S]/c^{\ominus})$ and $\log\frac{v}{v_{max}-v}$. Use $v_{max} = 4.17\ \mu\text{mol dm}^{-3}\ \text{s}^{-1}$.

$[S]_0/10^{-5}$ mol dm^{-3}	0.10	0.40	0.50	0.60	0.80	1.0	1.5	2.0	3.0
v/μmol dm^{-3} s^{-1}	0.0040	0.25	0.46	0.75	1.42	2.08	3.22	3.70	4.02
$\log\{[S]_0/c^{\ominus}\}$	−6.00	−5.40	−5.30	−5.22	−5.10	−5.00	−4.82	−4.70	−4.52
$\log\frac{v}{v_{max}-v}$	−3.02	−1.20	−0.907	−0.659	−0.297	−0.002	0.530	0.896	1.43

A plot of $\log\frac{v}{v_{max}-v}$ against $\log([S]/c^{\ominus})$ is shown in Figure 8.10. It is seen to be linear and the linear regression fit is shown as a box insert within the figure. Thus,

$$n = slope = \boxed{3.0}$$

$$K' = 10^{-intercept}c^{\ominus} = 10^{-15.04}\ \text{mol dm}^{-3} = 9.1 \times 10^{-16}\ \text{mol dm}^{-3}$$
$$= \boxed{0.91\ \text{fmol dm}^{-3}}$$

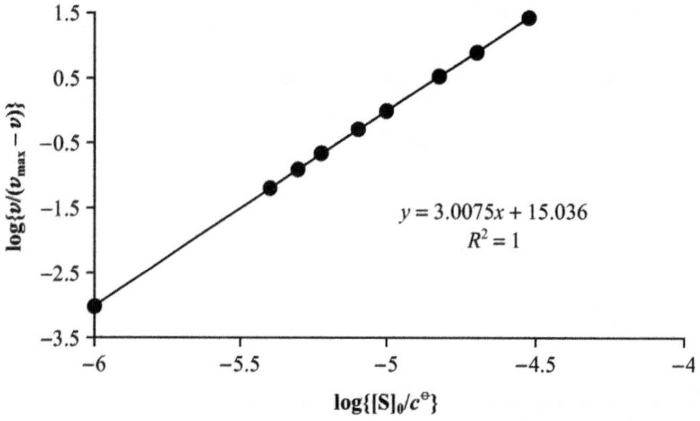

Figure 8.10

E8.20 $$\log\frac{v}{v_{max}-v} = n\log([S]/c^{\ominus}) - \log(K'/c^{\ominus})$$

Evaluation of the above expression at both $v = 0.90\ v_{max}$ and $v = 0.10\ v_{max}$ yields the relations

$$\log 9 = n\log([S]_{90}/c^{\ominus}) - \log(K'/c^{\ominus}) \quad \text{and} \quad \log\frac{1}{9} = n\log([S]_{10}/c^{\ominus}) - \log(K'/c^{\ominus})$$

Subtraction of the two and use of the logarithm property $\log x - \log y = \log(x/y)$ yields the result

$$\log\frac{9}{1/9} = n\log([S]_{90}/[S]_{10})$$

Thus, $\log 81 = \log([S]_{90}/[S]_{10})^n$ which implies that $\boxed{([S]_{90}/[S]_{10})^n = 81}$.

(a) $n = 1$ in the Michaelis–Menten rate law so we conclude that $[S]_{90}/[S]_{10} = 81$.

(b) Solving the general case for $[S]_{90}/[S]_{10}$ yields $[S]_{90}/[S]_{10} = 81^{1/n}$.

(c) Inspection of the E8.19 data reveals that, since $v_{max} = 4.17\ \mu mol\ dm^{-3}\ s^{-1}$, the reaction rate 3.70 $\mu mol\ dm^{-3}\ s^{-1}$ is $0.89\ v_{max}$ and the reaction rate $0.46\ \mu mol\ dm^{-3}\ s^{-1}$ is $0.11\ v_{max}$ so we use the substrate concentrations at these points to estimate n. We easily solve the general expression above for n to find that

$$n \simeq \frac{\log 81}{\log([S]_{89}/[S]_{11})} \simeq \frac{\log 81}{\log(2.00/0.50)} \simeq \boxed{3.2}$$

E8.21 $$CH_3CH_2OH(aq) + NAD^+(aq) \xrightarrow{\text{alcohol dehydrogenase}} CH_3CHO(aq) + NADH(aq) + H^+(aq)$$

As discussed in text Section 8.2, plots of $1/v$ against $1/[S_1]_0$ at different values of constant $[S_2]_0$ for each plot can be used to distinguish a sequential enzymatic mechanism from a ping-pong mechanism. If the plots are linear with identical (i.e. parallel) slopes, the ping-pong mechanism is indicated. But should the linear plots have a slope that progresses toward zero with increasing $[S_2]$, we suspect the sequential mechanism. The data plots with $[S_1] = [EtOH]$ and $[S_2] = [NAD^+]$ are shown in Figure 8.11. The plots are linear with progressively decreasing slope (the linear regression fits are shown as figure inserts) so the $\boxed{\text{sequential mechanism}}$ is indicated:

$$E + S_1 \rightleftharpoons ES_1 \qquad K_{M1} = \frac{[E][S_1]}{[ES_1]}$$

$$ES_1 + S_2 \rightleftharpoons ES_1S_2 \qquad K_{M12} = \frac{[ES_1][S_2]}{[ES_1S_2]}$$

$$ES_1S_2 \rightarrow E + P \qquad v = k_b[ES_1S_2]$$

$$E + S_2 \rightleftharpoons ES_2 \qquad K_{M2} = \frac{[E][S_2]}{[ES_2]}$$

$$ES_2 + S_1 \rightleftharpoons ES_1S_2 \qquad K_{M21} = \frac{[ES_2][S_1]}{[ES_1S_2]}$$

for which the plot slopes and intercepts are related to the mechanism constants by

$$\text{slope} = \frac{K_{M21} + K_{M1}K_{M12}/[S_2]_0}{v_{max}} \qquad y\text{-intercept} = \frac{1 + K_{M12}/[S_2]_0}{v_{max}} \qquad [8.5c]$$

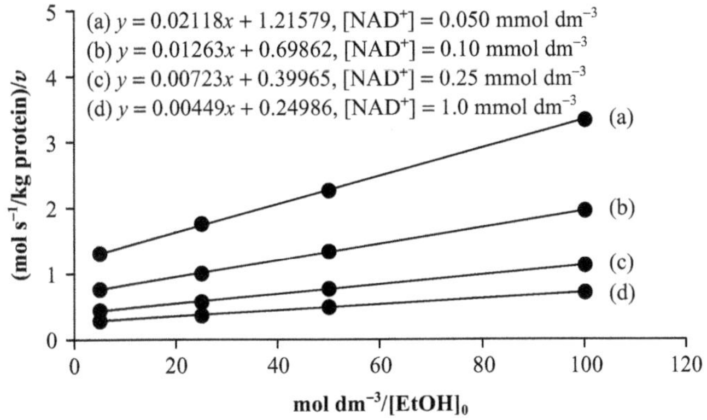

Figure 8.11

Eqn 8.5c implies that a plot of the Figure 8.11 intercepts against $1/[NAD^+]$ should be linear with an intercept equal to $1/v_{max}$ and a slope equal to K_{M12}/v_{max}. This plot is shown in Figure 8.12. It is seen to be linear with a linear regression fit reported within the figure. Thus,

$$v_{max} = (1/intercept_{Fig\ 8.12})\ \text{mol s}^{-1}\ (\text{kg protein})^{-1} = (1/0.196)\ \text{mol s}^{-1}\ (\text{kg protein})^{-1}$$
$$= \boxed{5.10\ \text{mol s}^{-1}\ (\text{kg protein})^{-1}}$$

$$K_{M12} = v_{max} \times slope_{Fig\ 8.12}$$
$$= \{5.10\ \text{mol s}^{-1}\ (\text{kg protein})^{-1}\} \times 0.0508\ \text{mmol dm}^{-3}\ \{\text{mol s}^{-1}\ (\text{kg protein})^{-1}\}^{-1}$$
$$= \boxed{0.259\ \text{mmol dm}^{-3}}$$

Eqn 8.5c also implies that a plot of Figure 8.11 slopes against $1/[NAD^+]$ should be linear with an intercept equal to K_{M21}/v_{max} and a slope equal to $K_{M1}K_{M12}/v_{max}$. This plot is shown in Figure 8.13. It is seen to be linear with a linear regression fit reported within the figure. Thus,

$$K_{M21} = v_{max} \times intercept_{Fig\ 8.13}$$
$$= \{5.10\ \text{mol s}^{-1}\ (\text{kg protein})^{-1}\} \times 0.003705\ \text{mol dm}^{-3}\ \{\text{mol s}^{-1}\ (\text{kg protein})^{-1}\}^{-1}$$
$$= \boxed{0.0189\ \text{mol dm}^{-3}}$$

$$K_{M1} = \frac{v_{max} \times slope_{Fig\ 8.13}}{K_{M12}}$$
$$= \frac{\{5.10\ \text{mol s}^{-1}\ (\text{kg protein})^{-1}\} \times 8.774 \times 10^{-4}\ \text{mol dm}^{-3}\ \{\text{mol s}^{-1}\ (\text{kg protein})^{-1}\}^{-1}\ \text{mmol dm}^{-3}}{0.259\ \text{mmol dm}^{-3}}$$
$$= \boxed{0.0173\ \text{mol dm}^{-3}}$$

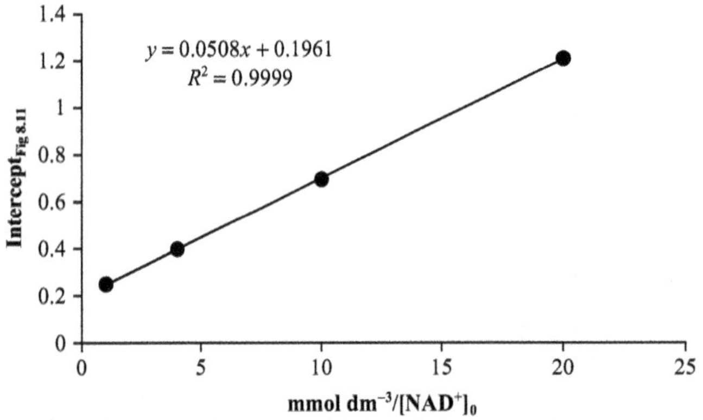

Figure 8.12

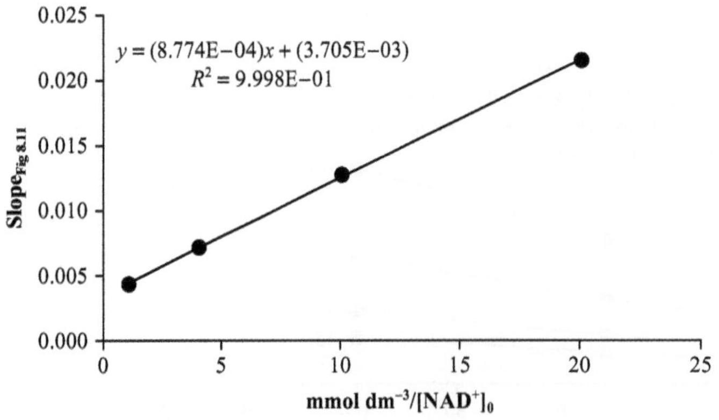

Figure 8.13

E8.22 The catalytic efficiency is given by

$$\eta = \frac{k_{cat}}{K_M} [8.8] = \frac{1.4 \times 10^4 \text{ s}^{-1}}{9.0 \times 10^{-5} \text{ mol dm}^{-3}} = \boxed{1.6 \times 10^8 \text{ dm}^3 \text{ mol}^{-1} \text{ s}^{-1}}$$

This is a very high catalytic efficiency and, since it is in the range of 10^8–10^9 $dm^3\ mol^{-1}\ s^{-1}$ (the diffusion-controlled rate constant is $k_d = 8RT/3\eta_{viscosity}$ [7.27] = $7.4 \times 10^9\ dm^3\ mol^{-1}\ s^{-1}$ for aqueous solution at 25°C), the enzyme is "catalytically perfect". See text Section 8.3.

E8.23 When using reaction rates v, the Lineweaver–Burk plot without inhibition has the form:

$$\frac{1}{v} = \frac{1}{v_{max}} + \left(\frac{K_M}{v_{max}}\right)\frac{1}{[S]_0} \quad [8.3]$$

where the intercept and slope are simple functions of v_{max} and K_M. When using reaction rates relative to a specific, non-inhibited rate ($v_{rel} = v/v_{reference}$), the Lineweaver–Burk plot without inhibition has the same basic form:

$$\frac{1}{v_{rel}} = \frac{1}{v_{max,rel}} + \left(\frac{K_M}{v_{max,rel}}\right)\frac{1}{[S]_0}$$

The linear regression fit of the non-inhibited Lineweaver–Burk data plot is

$$\frac{1}{v_{rel}} = 0.797 + (2.17)\frac{1}{[\text{CBGP}]_0/10^{-2} \text{ mol dm}^{-3}} \qquad R^2 = 0.980.$$

Consequently, $v_{max,rel} = 1/\text{intercept} = 1/0.797 = 1.25$ and

$K_M = \text{slope} \times v_{max,rel} = (2.17 \times 10^{-2} \text{ mol dm}^{-3}) \times (1.25) = 2.71 \times 10^{-2} \text{ mol dm}^{-3}$.

The Lineweaver–Burk plot with inhibition has the basic form:

$$\frac{1}{v_{rel}} = \frac{\alpha'}{v_{max,rel}} + \left(\frac{\alpha K_M}{v_{max,rel}}\right)\frac{1}{[S]_0} \quad [8.9c]$$

The linear regression fit of the Lineweaver–Burk data plot for phenylbutyrate ion inhibition is

$$\frac{1}{v_{rel}} = 1.02 + (6.01)\frac{1}{[\text{CBGP}]_0/10^{-2} \text{ mol dm}^{-3}} \qquad R^2 = 0.972.$$

Therefore, $\alpha' = \textit{intercept} \times v_{max,rel} = 1.02 \times 1.25 = 1.28$ and

$\alpha = \textit{slope} \times v_{max,rel}/K_M = (6.01 \times 10^{-2} \text{ mol dm}^{-3}) \times (1.25)/(2.71 \times 10^{-2} \text{ mol dm}^{-3}) = 2.77$.

Since both $\alpha > 1$ and $\alpha' \sim 1$ (see text Section 8.4), we conclude that phenylbutyrate ion is a competitive inhibitor of carboxypeptidase.

The linear regression fit of the Lineweaver–Burk data plot for benzoate ion inhibition is

$$\frac{1}{v_{rel}} = 3.75 + (3.01)\frac{1}{[\text{CBGP}]_0/10^{-2} \text{ mol dm}^{-3}} \qquad R^2 = 0.999.$$

Therefore, $\alpha' = \textit{intercept} \times v_{max,rel} = 3.75 \times 1.25 = 4.69$ and

$\alpha = \textit{slope} \times v_{max,rel}/K_M = (3.01 \times 10^{-2} \text{ mol dm}^{-3}) \times (1.25)/(2.71 \times 10^{-2} \text{ mol dm}^{-3}) = 1.39$.

Since both $\alpha \sim 1$ and $\alpha' > 1$, we conclude that the benzoate ion is an uncompetitive inhibitor of carboxypeptidase.

E8.24 Let $f = v_{\text{inhibited}}/v_{\text{uninhibited}}$. Then, by eqns 8.1 and 8.9a with $\alpha' = 1$ for a competitive inhibitor (see text Section 8.4):

$$f = \frac{1 + K_M/[S]_0}{1 + \alpha K_M/[S]_0}, \quad \text{where } \alpha = 1 + [I]/K_I \quad [8.9b]$$

Solving for α gives

$$\alpha = \frac{1 - f + K_M/[S]_0}{f K_M/[S]_0}$$

When a competitive inhibitor causes a 50% reduction in the rate of product formation, $f = 0.5$ and α is

$$\alpha_{50} = \frac{0.5 + K_M/[S]_0}{0.5 \times K_M/[S]_0} = \frac{0.5 + 3.0/0.10}{0.5 \times 3.0/0.10} = 2.0\overline{3}$$

Solving eqn 8.9b for $[I]_{50}$ gives

$$[I]_{50} = (\alpha_{50} - 1)K_I = (2.0\overline{3} - 1) \times (20\ \mu\text{mol dm}^{-3}) = \boxed{20.7\ \mu\text{mol dm}^{-3}}$$

E8.25 (a) We add to the Michaelis–Menten mechanism the inhibition by the substrate

$$\text{SES} \rightleftharpoons \text{ES} + \text{S} \quad K_I = [\text{ES}][\text{S}]/[\text{SES}]$$

where the inhibited enzyme, SES, forms when S binds to ES and, thereby, prevents the formation of product. This inhibition might possibly occur when S is at a very high concentration. Enzyme mass balance is written in terms of [ES], K_I, K_M (= [E][S]/[ES]), and [S]. (For practical purposes the free substrate concentration is replaced by $[S]_0$ because the substrate is typically in large excess relative to the enzyme.) By the conservation of enzyme mass

$$\begin{aligned}[E]_0 &= [E] + [ES] + [SES] \\ &= \frac{K_M[ES]}{[S]} + [ES] + \frac{[ES][S]}{K_I} \\ &= \left(1 + \frac{K_M}{[S]} + \frac{[S]}{K_I}\right)[ES]\end{aligned}$$

Thus,

$$[ES] = \frac{[E]_0}{\left(1 + \frac{K_M}{[S]} + \frac{[S]}{K_I}\right)}$$

and the expression for the rate of product formation becomes

$$v = k_b[ES] = \frac{v_{max}}{1 + \frac{K_M}{[S]_0} + \frac{[S]_0}{K_I}} \quad \text{where} \quad v_{max} = k_b[E]_0.$$

The denominator term $[S]_0/K_I$ reflects a reduced reaction rate caused by inhibition as the concentration of S becomes very large.

(b) To examine the effect that substrate inhibition has on the double reciprocal, **Lineweaver–Burk plot** of $1/v$ against $1/[S]_0$ take the inverse of the above rate expression and compare it to the uninhibited expression [8.3]:

$$\frac{1}{v} = \frac{1}{v_{\text{max}}} + \left(\frac{K_{\text{M}}}{v_{\text{max}}}\right)\frac{1}{[\text{S}]_0} \quad [8.3]$$

The inverse of the inhibited rate law is

$$\frac{1}{v} = \frac{1}{v_{\text{max}}} + \left(\frac{K_{\text{M}}}{v_{\text{max}}}\right)\frac{1}{[\text{S}]_0} + \left(\frac{[\text{S}]_0^2}{v_{\text{max}}K_{\text{I}}}\right)\frac{1}{[\text{S}]_0}$$

$$= \frac{1}{v_{\text{max}}} + \left(\frac{K_{\text{M}}}{v_{\text{max}}} + \frac{[\text{S}]_0^2}{v_{\text{max}}K_{\text{I}}}\right)\frac{1}{[\text{S}]_0}.$$

The uninhibited and inhibited line shapes are sketched in Figure 8.14.

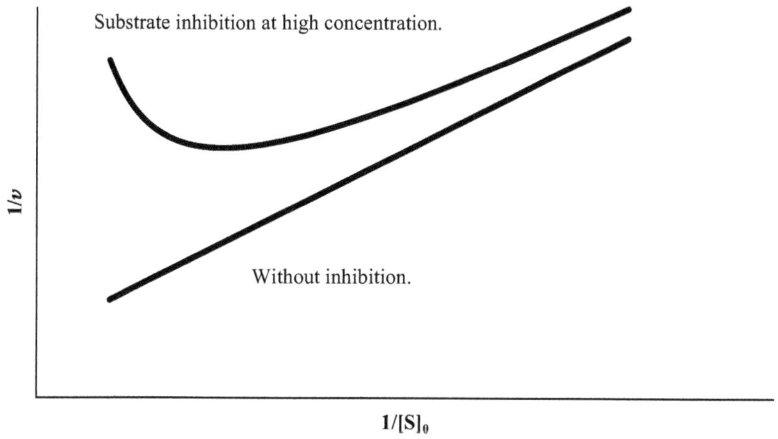

Figure 8.14

Comparing the two expressions, we see that the two curves match at high values of $1/[\text{S}]_0$. However, as the concentration of $[\text{S}]_0$ increases ($1/[\text{S}]_0$ decreases) the $1/v$ curve with inhibition curves upward because the reaction rate is decreasing.

E8.26 (a) $J = -D \times (\text{concentration gradient})$ Fick's first law [8.11]

$$= -(5.22 \times 10^{-10}\ \text{m}^2\ \text{s}^{-1}) \times (0.10\ \text{mol dm}^{-3}\ \text{m}^{-1}) \times \left(\frac{1\ \text{dm}^3}{10^{-3}\ \text{m}^3}\right)$$

$$= \boxed{-5.2 \times 10^{-8}\ \text{mol m}^{-2}\ \text{s}^{-1}}$$

The negative value indicates the flow of mass from high to low concentration.

(b) Let n be the number of moles of molecules that flow down the concentration gradient through area A, which is perpendicular to the concentration gradient, in time Δt. Then, with the gradient expressed in molar concentration per meter:

$$n = JA\Delta t \quad [8.10\text{b}]$$

$$= (5.2 \times 10^{-8}\ \text{mol m}^{-2}\ \text{s}^{-1}) \times (5.0\ \text{mm}^2) \times (60\ \text{s}) \times \left(\frac{10^{-6}\ \text{m}^2}{1\ \text{mm}^2}\right)$$

$$= \boxed{1.6 \times 10^{-11}\ \text{mol}}$$

E8.27 To find an approximate relation between the distance x that a molecule diffuses in time t, consider that the molecule has one-dimensional diffusional speed $v_{\text{d}} = x/t$. At any instant, half of the molecules are diffusing in the positive direction and half are diffusing in the negative direction so it is reasonable to write a quantity diffusional flux of A in the positive direction as

$$J = v_d \times (\tfrac{1}{2}[\mathrm{A}]) = x \times (\tfrac{1}{2}[\mathrm{A}])/t$$

Additionally, the flux is related to the diffusion coefficient by $J = D[\mathrm{A}]_0/x$ [8.17]. Equating the two expressions for the flux and solving for t yields the useful expression

$$\boxed{t = x^2/2D \quad \text{or} \quad x = (2Dt)^{1/2}}$$

(a) $t = \dfrac{(10 \times 10^{-3}\ \mathrm{m})^2}{2(5.22 \times 10^{-10}\ \mathrm{m^2\ s^{-1}})}$ [sucrose, Table 8.1] $= 9.6 \times 10^4\ \mathrm{s} = \boxed{27\ \mathrm{h}}$

(b) $t = \dfrac{(10 \times 10^{-2}\ \mathrm{m})^2}{2(5.22 \times 10^{-10}\ \mathrm{m^2\ s^{-1}})} = \boxed{2.7 \times 10^3\ \mathrm{h}}$

(c) $t = \dfrac{(10\ \mathrm{m})^2}{2(5.22 \times 10^{-10}\ \mathrm{m^2\ s^{-1}})} = \boxed{3.0 \times 10^3\ \mathrm{a}}$

These times are so long that we obviously do not wait for diffusion to move mass across macroscopic distances. Stirring and convective motion have much higher mixing rates on the macroscopic scale.

E8.28 Solving the equation derived in E8.27 for D gives: $\boxed{D = x^2/2t}$

(a) $D = \dfrac{(150 \times 10^{-12}\ \mathrm{m})^2}{2(1.8 \times 10^{-12}\ \mathrm{s})} = \boxed{6.3 \times 10^{-9}\ \mathrm{m^2\ s^{-1}}}$

(b) $D = \dfrac{(75 \times 10^{-12}\ \mathrm{m})^2}{2(1.8 \times 10^{-12}\ \mathrm{s})} = \boxed{1.6 \times 10^{-9}\ \mathrm{m^2\ s^{-1}}}$

E8.29 As shown in E8.27, $x^2 = 2Dt$, gives the diffusional distance traveled in any one dimension in time t. We need the distance traveled from a point in any direction. The distinction here is the distinction between the one-dimensional and three-dimensional diffusion. We can assume that random motions in the x, y, and z directions are indistinguishable so that r, the 3D diffusional distance traveled in time t, is given by: $r^2 = x^2 + y^2 + z^2$ [Pythagorean theorem] $= 3x^2$. Thus,

$\boxed{r^2 = 6Dt \quad \text{or} \quad t = r^2/6D}$ in three dimensions. Therefore,

$$t = \frac{(1.0 \times 10^{-6}\ \mathrm{m})^2}{6(1.0 \times 10^{-11}\ \mathrm{m^2\ s^{-1}})} = \boxed{1.7 \times 10^{-2}\ \mathrm{s}}$$

E8.30 $t = x^2/2D$ [E8.27]

For the plasma membrane: $t = \dfrac{(10 \times 10^{-9}\ \mathrm{m})^2}{2(1.0 \times 10^{-10}\ \mathrm{m^2\ s^{-1}})} = \boxed{500\ \mathrm{ns}}$

For the lipid bilayer: $t = \dfrac{(10 \times 10^{-9}\ \mathrm{m})^2}{2(1.0 \times 10^{-9}\ \mathrm{m^2\ s^{-1}})} = \boxed{50\ \mathrm{ns}}$

E8.31 Since $D \propto M^{-1/2}$, the relation $t = x^2/2D$ [E8.27] may be written in the form: $t \propto M^{1/2}\, x^2$ and the ratio of times for two species A and B to travel the same distance is $t_A/t_B = (M_A/M_B)^{1/2}$. Thus,

$$t_{\text{ribonuclease}}/t_{\text{catalase}} = (13.683/250)^{1/2} = \boxed{0.234}$$

E8.32 Diffusion is only important in the absence of macroscopic fluid flow or convection or turbulence as may be the case when microscopic spatial constraints exist such as passage of gases through lung alveoli or the passage of neurotransmitter molecules across a synaptic gap. An extraordinarily large time is required for a molecule to move across a lake by diffusion alone. A pollutant with a radius of 100 pm (the radius of a water molecule) requires about 90 millennia to travel across a 100 m lake at the mean surface temperature (15°C).

$$D_{15°\text{C}} = \frac{kT}{6\pi\eta a} \quad \text{[Stokes–Einstein relation, 8.14]}$$

$$= \frac{(1.381\times10^{-23}\text{ J K}^{-1})\times(288\text{ K})}{6\pi(1.139\times10^{-3}\text{ kg m}^{-1}\text{ s}^{-1})\times(100\times10^{-12}\text{ m})} \quad \text{[text Figure 8.14]}$$

$$= 1.85\times10^{-9}\text{ m}^2\text{ s}^{-1}$$

$$t_{\text{H}_2\text{O}} \approx \frac{x^2}{2D}\text{[E8.27]} = \frac{(100\text{ m})^2}{2(1.85\times10^{-9}\text{ m}^2\text{ s}^{-1})} = 2.70\times10^{12}\text{ s} = \boxed{8.55\times10^4\text{ a}}$$

E8.33 Let N be the number of one-dimensional steps of length λ that a molecule takes in time t and let τ be the time each step takes; $\tau = t/N$. When the relation derived in E8.27 is applied to an individual step in a journey of duration t we find that

$$D = \frac{\lambda^2}{2\tau}\text{[E8.27]} = \frac{\lambda^2 N}{2t} \quad \text{or} \quad t = \frac{\lambda^2 N}{2D}$$

Now, apply the relation of E8.27 to the total journey in which the diffusional distance x is traversed.

$$x = (2Dt)^{1/2} = \left(\frac{2D\lambda^2 N}{2D}\right)^{1/2} = (\lambda^2 N)^{1/2} \quad \text{or} \quad N = \left(\frac{x}{\lambda}\right)^2$$

For $x = 1000\,\lambda$, $N = \left(\frac{1000\,\lambda}{\lambda^2}\right)^2 = \boxed{1\times10^6\text{ steps}}$

E8.34

$$\eta = \eta_0 e^{E_a/RT} \quad [8.13]$$

$$\frac{\eta_{T_1}}{\eta_{T_2}} = \frac{e^{E_a/RT_1}}{e^{E_a/RT_2}} = e^{\frac{E_a}{R}\left(\frac{1}{T_1}-\frac{1}{T_2}\right)}$$

$$\ln\left(\frac{\eta_{T_1}}{\eta_{T_2}}\right) = \frac{E_a}{R}\left(\frac{1}{T_1}-\frac{1}{T_2}\right) \quad \text{or} \quad E_a = R\left(\frac{1}{T_1}-\frac{1}{T_2}\right)^{-1}\ln\left(\frac{\eta_{T_1}}{\eta_{T_2}}\right)$$

$$E_a = (8.3145\text{ J mol}^{-1}\text{ K}^{-1})\times\left(\frac{1}{293\text{ K}}-\frac{1}{303\text{ K}}\right)^{-1}\ln\left(\frac{1.0019}{0.7982}\right) = \boxed{16.8\text{ kJ mol}^{-1}}$$

E8.35

$$s = u\mathcal{E}\,[8.22] \quad \text{and} \quad \mathcal{E} = \frac{\Delta\phi}{l}$$

Therefore,

$$s = u\left(\frac{\Delta\phi}{l}\right) = (5.19\times10^{-8}\text{ m}^2\text{ s}^{-1}\text{ V}^{-1})\times\left(\frac{12.0\text{ V}}{0.0100\text{ m}}\right)$$

$$= 6.23\times10^{-5}\text{ m s}^{-1} = \boxed{62.3\ \mu\text{m s}^{-1}}$$

E8.36 (a) The calf thymus histone has 4 acidic amino acid residues (1 Asp, 1 Glu, and 2 His), while there are a total of 26 basic residues (11 Lys and 15 Arg). Being overwhelmingly basic, we expect that the histone will bear a net charge that is close to +26 at pH = 7 because of base protonation. We also expect that pI > 7. To estimate pI, we imagine the solution being made progressively more basic, thereby, reducing the total positive charge until the last basic residues, those of Arg, are neutralized in the reaction:

$$NH_3^+\text{—}\cdots(ArgH^+)_{15}\cdots\text{—}CO_2^- \xrightleftharpoons{-15\,H^+,\ K_a=\{K_{a3}(Arg)\}^{15}.\ \text{Table 4.6}} NH_3^+\text{—}\cdots(Arg)_{15}\cdots\text{—}CO_2^-$$

The species on the right is the neutral zwitterion and by this analysis all of the histone molecules will have the zwitterion form above a minimum pH, pH_{min}, which we presume to occur when the ratio $[NH_3^+\text{—}\cdots(ArgH^+)_{15}\cdots\text{—}CO_2^-]/[NH_3^+\text{—}\cdots(Arg)_{15}\cdots\text{—}CO_2^-]$ is about 1/25 (or less). The Henderson–Hasselbalch equation [4.35] is used to calculate pH_{min}.

$$15\,pH_{min} = 15\,pK_{a3}(Arg) - \log\left(\frac{[acid]}{[base]}\right)$$

$$\begin{aligned} pH_{min} &= pK_{a3}(Arg) - \frac{1}{15}\log\left(\frac{[acid]}{[base]}\right) \\ &= 12.10 - (1/15)\times\log(1/25) \\ &= 12.2 \end{aligned}$$

Thus, pI ~ 12.2.

(b) The egg albumin has 58 acidic amino acid residues (Asp, Glu, and 7 His), while there are a total of 35 basic residues (20 Lys and 15 Arg). Being overwhelmingly acidic, we expect that the egg albumin will bear a net charge that is very negative at pH = 7 because of deprotonation of carboxylic acid groups. We also expect that pI < 7.

(c) The pH of the isoelectric points in parts (a) and (b) are so widely separated that the two polypeptides should be separable by gel electrophoresis with isoelectric focusing.

E8.37 (a) $D_{Na^+} = \dfrac{u_{Na^+}RT}{z_{Na^+}F}$

$$D_{Na^+} = \frac{(5.19\times10^{-8}\ m^2\,s^{-1}\,V^{-1})\times(8.3145\ J\,K^{-1}\,mol^{-1})\times(298\ K)}{96485.3\ C\,mol^{-1}} \quad [\text{Table 8.2; } 1\ J = 1\ V\,C]$$

$$= \boxed{1.33\times10^{-9}\ m^2\,s^{-1}}$$

Solving eqn 8.23 for the hydrodynamic radius a gives

$$\begin{aligned} a_{Na^+} &= \frac{z_{Na^+}e}{6\pi\eta u_{Na^+}} \\ &= \frac{1\times(1.6022\times10^{-19}\ C)}{6\pi\times(8.91\times10^{-4}\ kg\,m^{-1}\,s^{-1})\times(5.19\times10^{-8}\ m^2\,s^{-1}\,V^{-1})} \quad [1\ C\,V = 1\ J = 1\ kg\,m^2\,s^{-2}] \\ &= 1.84\times10^{-10}\ m = \boxed{184\ pm} \end{aligned}$$

(b) The hydrodynamic volume is $\frac{4}{3}\pi(a_{Na^+})^3 = \frac{4}{3}\pi(184\times10^{-10}\ cm)^3 = 2.61\times10^{-23}\ cm^3$.

The ion volume is $\frac{4}{3}\pi(r_{Na^+})^3 = \frac{4}{3}\pi(102\times10^{-10}\ cm)^3$ [Table 9.3] $= 0.44\times10^{-23}\ cm^3$.

The difference between these is the volume of water that is part of the hydrodynamic volume of Na^+: $V_{H_2O} = (2.61\times10^{-23}\ cm^3) - (0.44\times10^{-23}\ cm^3) = 2.17\times10^{-23}\ cm^3$.

Assuming that the density of water molecules in the hydrodynamic volume is identical to the density of bulk water (1 g cm^{-3}), the number density of water in the hydrodynamic volume is

$$(1.00\text{ g cm}^{-3}) \times (6.022 \times 10^{23}\text{ H}_2\text{O molecules}/18.015\text{ g}) = 3.34 \times 10^{22}\text{ H}_2\text{O molecules per cm}^3.$$

Multiplication of the number density of water by the volume of water that is part of the hydrodynamic volume gives the number of water molecules in the hydrodynamic volume of a Na^+ cation. It is only about $\boxed{1}$. This minimal value indicates that water does not form a coordination complex with the Na^+ cation like water does with many transition metal cations. A water molecule in the hydration sphere of Na^+ is very labile and readily substituted by a bulk water molecule at a diffusion-controlled rate (> 10^8 s^{-1}).

E8.38 Eqn 8.32 holds for a donor–acceptor pair separated by a constant distance, assuming that the reorganization energy is constant:

$$\ln k_{\text{et}} = -\frac{(\Delta_r G^\ominus)^2}{4\lambda RT} - \frac{\Delta_r G^\ominus}{2RT} + \text{constant} \quad \text{when using molar energies.}$$

$$\ln k_{\text{et}} = -\frac{(\Delta_r G^\ominus)^2}{4\lambda kT} - \frac{\Delta_r G^\ominus}{2kT} + \text{constant} \quad \text{when using energies per molecular event.}$$

Two sets of rate constants and reaction Gibbs energies can be used to generate two equations (eqn 8.32 applied to the two sets) in two unknowns: λ and the constant. Using energies per molecular event, we write

$$\ln k_{\text{et},1} + \frac{(\Delta_r G_1^\ominus)^2}{4\lambda kT} + \frac{\Delta_r G_1^\ominus}{2kT} = \text{constant} = \ln k_{\text{et},2} + \frac{(\Delta_r G_2^\ominus)^2}{4\lambda kT} + \frac{\Delta_r G_2^\ominus}{2kT}$$

so
$$\frac{(\Delta_r G_1^\ominus)^2 - (\Delta_r G_2^\ominus)^2}{4\lambda kT} = \ln\frac{k_{\text{et},2}}{k_{\text{et},1}} + \frac{\Delta_r G_2^\ominus - \Delta_r G_1^\ominus}{2kT}$$

and
$$\lambda = \frac{(\Delta_r G_1^\ominus)^2 - (\Delta_r G_2^\ominus)^2}{4 \times \left(kT \ln\frac{k_{\text{et},2}}{k_{\text{et},1}} + \frac{\Delta_r G_2^\ominus - \Delta_r G_1^\ominus}{2}\right)}$$

$$\lambda = \frac{(-0.665\text{ eV})^2 - (-0.975\text{ eV})^2}{\frac{4 \times (1.381 \times 10^{-23}\text{ J K}^{-1}) \times (298\text{ K})}{1.602 \times 10^{-19}\text{ J eV}^{-1}} \ln\frac{3.33 \times 10^6}{2.02 \times 10^5} - 2 \times (0.975 - 0.665)\text{ eV}} = \boxed{1.53\bar{1}\text{ eV}}$$

As a molar energy,

$$\lambda = (1.53\bar{1}\text{ eV}) \times (1.602 \times 10^{-19}\text{ J eV}^{-1}) \times (6.022 \times 10^{23}\text{ mol}^{-1}) = \boxed{148\text{ kJ mol}^{-1}}$$

E8.39 (a) For the same donor and acceptor at two different distances, eqn 8.31 implies that

$$\ln(k_{\text{et},1}/\text{s}^{-1}) + \beta r_1 = \text{constant} = \ln(k_{\text{et},2}/\text{s}^{-1}) + \beta r_2$$

Solving for β:

$$\beta = \frac{\ln(k_{\text{et},2}/k_{\text{et},1})}{r_1 - r_2} = \frac{\ln(2.02 \times 10^5/2.8 \times 10^4)}{(1.23 - 1.11)\text{ nm}} = \boxed{16.\bar{5}\text{ nm}^{-1}}$$

(b) Either of the data pairs can be used to calculate the constant of eqn 8.31.

$$\begin{aligned}\text{constant} &= \ln(k_{\text{et},1}/\text{s}^{-1}) + \beta r_1\\ &= \ln(2.02\times 10^5) + (16.\overline{5}\ \text{nm}^{-1})\times(1.11\ \text{nm})\\ &= 30.5\overline{3}\end{aligned}$$

Solving eqn 8.31 for k_{et} gives

$$k_{\text{et}}/\text{s}^{-1} = \text{e}^{-\beta r+\text{constant}}$$

$$\begin{aligned}k_{\text{et}} &= \text{e}^{-(16.\overline{5}\ \text{nm}^{-1})\times(1.48\ \text{nm})+30.5\overline{3}}\ \text{s}^{-1}\\ &= 450\ \text{s}^{-1}\end{aligned}$$

E8.40 azurin(red) + cytochrome c(ox) $\rightarrow$ azurin(ox) + cytochrome c(red)

$$E^{\ominus}_{\text{cell}} = E^{\ominus}_{\text{R}} - E^{\ominus}_{\text{L}} = 0.260\ \text{V} - 0.304\ \text{V} = -0.044\ \text{V}$$

$$K = \text{e}^{\frac{\nu F E^{\ominus}_{\text{cell}}}{RT}}\ [5.16] = \text{e}^{\frac{1\times(96485.3\ \text{C mol}^{-1})\times(-0.044\ \text{V})}{(8.31451\ \text{J K}^{-1}\ \text{mol}^{-1})\times(298.15\ \text{K})}} = \text{e}^{-1.713} = 0.180$$

$$k_{\text{obs}} = (k_{\text{DD}}k_{\text{AA}}K)^{1/2}\ [8.34]$$

$$k_{\text{DD}} = \frac{k^2_{\text{obs}}}{k_{\text{AA}}K} = \frac{(1.6\times 10^3\ \text{dm}^3\ \text{mol}^{-1}\ \text{s}^{-1})^2}{(1.5\times 10^2\ \text{dm}^3\ \text{mol}^{-1}\ \text{s}^{-1})\times(0.180)} = \boxed{9.5\times 10^4\ \text{dm}^3\ \text{mol}^{-1}\ \text{s}^{-1}}$$

Solutions to projects

P8.41 (a) $\text{A} + \text{P} \rightarrow \text{P} + \text{P}$ autocatalytic step, $v = k_{\text{r}}\,[\text{A}][\text{P}]$

Let $[\text{A}] = [\text{A}]_0 - x$ and $[\text{P}] = [\text{P}]_0 + x$.

We substitute these definitions into the rate expression, simplify, and integrate.

$$v = -\frac{\text{d}[\text{A}]}{\text{d}t} = k_{\text{r}}[\text{A}][\text{P}]$$

$$-\frac{\text{d}([\text{A}]_0 - x)}{\text{d}t} = k_{\text{r}}([\text{A}]_0 - x)([\text{P}]_0 + x)$$

$$\frac{\text{d}x}{([\text{A}]_0 - x)\,([\text{P}]_0 + x)} = k_{\text{r}}\,\text{d}t$$

$$\frac{1}{[\text{A}]_0 + [\text{P}]_0}\left(\frac{1}{[\text{A}]_0 - x} + \frac{1}{[\text{P}]_0 + x}\right)\text{d}x = k_{\text{r}}\,\text{d}t$$

$$\frac{1}{[\text{A}]_0 + [\text{P}]_0}\int_0^x\left(\frac{1}{[\text{A}]_0 - x} + \frac{1}{[\text{P}]_0 + x}\right)\text{d}x = k_{\text{r}}\int_0^t \text{d}t$$

$$\frac{1}{[\text{A}]_0 + [\text{P}]_0}\left\{\ln\left(\frac{[\text{A}]_0}{[\text{A}]_0 - x}\right) + \ln\left(\frac{[\text{P}]_0 + x}{[\text{P}]_0}\right)\right\} = k_{\text{r}}t$$

$$\ln\left\{\left(\frac{[\text{A}]_0}{[\text{P}]_0}\right)\left(\frac{[\text{P}]_0 + x}{[\text{A}]_0 - x}\right)\right\} = k_{\text{r}}([\text{A}]_0 + [\text{P}]_0)t$$

$$\ln\left\{\left(\frac{[\text{A}]_0}{[\text{P}]_0}\right)\left(\frac{[\text{P}]}{[\text{A}]_0 + [\text{P}]_0 - [\text{P}]}\right)\right\} = k_{\text{r}}([\text{A}]_0 + [\text{P}]_0)t$$

$$\ln\left\{\left(\frac{1}{b}\right)\left(\frac{[P]}{[A]_0+[P]_0-[P]}\right)\right\}=at, \quad \text{where} \quad a=([A]_0+[P]_0)k_r \quad \text{and} \quad b=\frac{[P]_0}{[A]_0}$$

$$\frac{[P]}{[A]_0+[P]_0-[P]}=be^{at}$$

$$[P]=([A]_0+[P]_0)\,be^{at}-be^{at}\,[P]$$

$$(1+be^{at})[P]=[P]_0\left(1+\frac{[A]_0}{[P]_0}\right)be^{at}=[P]_0\left(1+\frac{1}{b}\right)be^{at}=[P]_0(b+1)e^{at}$$

$$\boxed{\frac{[P]}{[P]_0}=(b+1)\frac{e^{at}}{1+be^{at}}}$$

(b) See Figure 8.15.

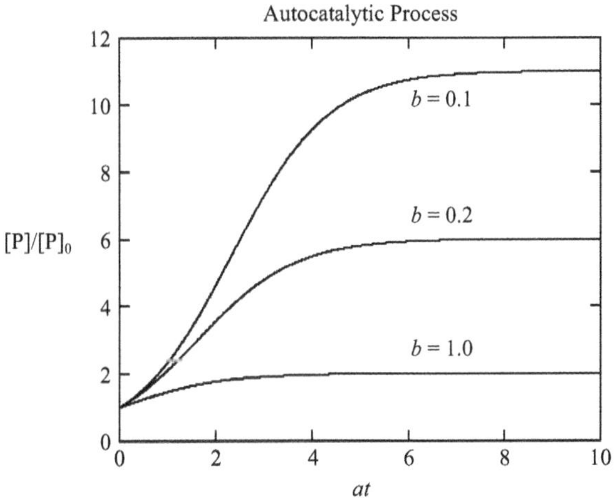

Figure 8.15

The growth to [P] reaches a maximum at very long times. As $t \to \infty$, the exponential term in the denominator of $[P]/[P]_0=(b+1)e^{at}/(1+be^{at})$ becomes so large that the denominator becomes be^{at}. Thus, $([P]/[P]_0)_{max}=(b+1)e^{at}/(be^{at})=(b+1)/b$, where $b=[P]_0/[A]_0$ and this maximum occurs as $t \to \infty$.

The autocatalytic curve $[P]/[P]_0=(b+1)e^{at}/(1+be^{at})$ has a shape that is very similar to that of the first-order process $[P]/[A]_0=1-e^{-k_r t}$. However, $[P]_{max}=[A]_0$ at $t \to \infty$ for the first-order process whereas $[P]_{max}=(1+1/b)[P]_0$ for the autocatalytic mechanism. In a series of experiments at fixed $[A]_0$ and assorted $[P]_0$, only the autocatalytic mechanism will show variation in $[P]_{max}$. Another difference is that the autocatalytic curve is initially concave up, which gives an overall sigmoidal curve, whereas the first-order curve is concave down. See Figure 8.16.

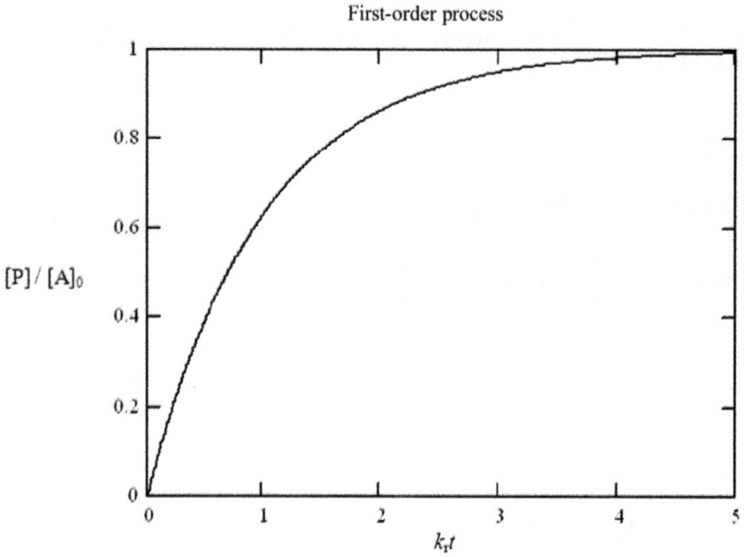

Figure 8.16

(c) Let $[P]_{v_{max}}$ be the concentration of P at which the reaction rate is a maximum and let t_{max} be the corresponding time.

$$v = k_r[A][P] = k_r([A]_0 - x)([P]_0 + x)$$
$$= k_r\{[A]_0[P]_0 + ([A]_0 - [P]_0)x - x^2\}$$

$$\frac{dv}{dt} = k_r([A]_0 - [P]_0 - 2x)$$

The reaction rate is a maximum when $dv/dt = 0$. This occurs when

$$x = [P]_{v_{max}} - [P]_0 = \frac{[A]_0 - [P]_0}{2} \quad \text{or} \quad \frac{[P]_{v_{max}}}{[P]_0} = \frac{b+1}{2b}.$$

Substitution into the final equation of part (a) gives:

$$\frac{[P]_{v_{max}}}{[P]_0} = \frac{b+1}{2b} = (b+1)\frac{e^{at_{max}}}{1 + be^{at_{max}}}.$$

Solving for t_{max}:

$$1 + be^{at_{max}} = 2be^{at_{max}}$$

$$e^{at_{max}} = b^{-1}$$

$$at_{max} = \ln(b^{-1}) = -\ln(b)$$

$$t_{max} = \boxed{-\frac{1}{a}\ln(b)}$$

(d) The description of the progress of infectious diseases can be represented by the mechanism

$S \rightarrow I \rightarrow R$.

(i) Only the $\boxed{\text{formation of I is autocatalytic}}$ as indicated by the first term to the right of the equality in the rate expression: $\frac{dI}{dt} = rSI - aI$

(ii) Whether the infection spreads or dies out is determined by

$$\frac{\mathrm{dI}}{\mathrm{d}t} = r\mathrm{SI} - a\mathrm{I}.$$

At $t = 0$, $\mathrm{I} = \mathrm{I}(0) = \mathrm{I}_0$. Since the process is autocatalytic, $\mathrm{I}(0) \neq 0$.

$$\left(\frac{\mathrm{dI}}{\mathrm{d}t}\right)_{t=0} = (r\mathrm{S}_0 - a)\mathrm{I}_0$$

If $a > r\mathrm{S}_0$, $\left(\frac{\mathrm{dI}}{\mathrm{d}t}\right)_{t=0} < 0$, and the infection dies out. If $a < r\mathrm{S}_0$, $\left(\frac{\mathrm{dI}}{\mathrm{d}t}\right)_{t=0} > 0$ and the infection spreads (an epidemic). Thus,

$$\boxed{\frac{a}{r} < \mathrm{S}_0} \text{ [infection spreads] and } \boxed{\frac{a}{r} > \mathrm{S}_0} \text{ [infection dies out].}$$

(iii) Addition of the three rate equations yields a sum of zero: $\frac{\mathrm{dS}}{\mathrm{d}t} + \frac{\mathrm{dI}}{\mathrm{d}t} + \frac{\mathrm{dR}}{\mathrm{d}t} = 0$

Hence, $\frac{\mathrm{d}}{\mathrm{d}t}(\mathrm{S} + \mathrm{I} + \mathrm{R}) = 0$, which implies that the sum S + I + R is a constant (i.e. "N").

P8.42

$$\mathrm{EH} + \mathrm{S} \underset{k_\mathrm{a}'}{\overset{k_\mathrm{a}}{\rightleftharpoons}} \mathrm{ESH} \xrightarrow{k_\mathrm{b}} \mathrm{E} + \mathrm{P}$$

$$\mathrm{EH} \xrightleftharpoons{K_\mathrm{E,a} = [\mathrm{E}^-][\mathrm{H}^+]/[\mathrm{EH}]} \mathrm{E}^- + \mathrm{H}^+ \qquad \mathrm{EH}_2^+ \xrightleftharpoons{K_\mathrm{E,b} = [\mathrm{EH}][\mathrm{H}^+]/[\mathrm{EH}_2^+]} \mathrm{EH} + \mathrm{H}^+$$

$$\mathrm{ESH} \xrightleftharpoons{K_\mathrm{ES,a} = [\mathrm{ES}^-][\mathrm{H}^+]/[\mathrm{ESH}]} \mathrm{ES}^- + \mathrm{H}^+ \qquad \mathrm{ESH}_2^+ \xrightleftharpoons{K_\mathrm{ES,b} = [\mathrm{ESH}][\mathrm{H}^+]/[\mathrm{ESH}_2^+]} \mathrm{ESH} + \mathrm{H}^+$$

(a) The dissociation equilibrium may be rearranged to give the following relationships.

$$[\mathrm{E}^-] = K_\mathrm{E,a}[\mathrm{EH}]/[\mathrm{H}^+] \qquad [\mathrm{EH}_2^+] = [\mathrm{EH}][\mathrm{H}^+]/K_\mathrm{E,b}$$
$$[\mathrm{ES}^-] = K_\mathrm{ES,a}[\mathrm{ESH}]/[\mathrm{H}^+] \qquad [\mathrm{ESH}_2^+] = [\mathrm{ESH}][\mathrm{H}^+]/K_\mathrm{ES,b}$$

Mass balance provides an equation for [EH].

$$[\mathrm{E}]_0 = [\mathrm{E}^-] + [\mathrm{EH}] + [\mathrm{EH}_2^+] + [\mathrm{ES}^-] + [\mathrm{ESH}] + [\mathrm{ESH}_2^+]$$
$$= \frac{K_\mathrm{E,a}[\mathrm{EH}]}{[\mathrm{H}^+]} + [\mathrm{EH}] + \frac{[\mathrm{EH}][\mathrm{H}^+]}{K_\mathrm{E,b}} + \frac{K_\mathrm{ES,a}[\mathrm{ESH}]}{[\mathrm{H}^+]} + [\mathrm{ESH}] + \frac{[\mathrm{ESH}][\mathrm{H}^+]}{K_\mathrm{ES,b}}$$

$$[\mathrm{EH}] = \frac{[\mathrm{E}]_0 - \left\{1 + \frac{[\mathrm{H}^+]}{K_\mathrm{ES,b}} + \frac{K_\mathrm{ES,a}}{[\mathrm{H}^+]}\right\}[\mathrm{ESH}]}{1 + \frac{[\mathrm{H}^+]}{K_\mathrm{E,b}} + \frac{K_\mathrm{E,a}}{[\mathrm{H}^+]}}$$
$$= \frac{[\mathrm{E}]_0 - c_1[\mathrm{ESH}]}{c_2}, \quad \text{where } c_1 = 1 + \frac{[\mathrm{H}^+]}{K_\mathrm{ES,b}} + \frac{K_\mathrm{ES,a}}{[\mathrm{H}^+]} \text{ and } c_2 = 1 + \frac{[\mathrm{H}^+]}{K_\mathrm{E,b}} + \frac{K_\mathrm{E,a}}{[\mathrm{H}^+]}$$

The steady-state approximation provides an equation for [ESH].

$$\frac{\mathrm{d}[\mathrm{ESH}]}{\mathrm{d}t} = k_\mathrm{a}[\mathrm{EH}][\mathrm{S}] - k_\mathrm{a}'[\mathrm{ESH}] - k_\mathrm{b}[\mathrm{ESH}] = 0$$

$$[\mathrm{ESH}] = \frac{k_\mathrm{a}}{k_\mathrm{a}' + k_\mathrm{b}}[\mathrm{EH}][\mathrm{S}] = K_\mathrm{M}^{-1}[\mathrm{EH}][\mathrm{S}] = K_\mathrm{M}^{-1}[\mathrm{S}]\left\{\frac{[\mathrm{E}]_0 - c_1[\mathrm{ESH}]}{c_2}\right\}$$

$$[\text{ESH}] = \frac{K_M^{-1}[\text{S}][\text{E}]_0/c_2}{1 + \dfrac{K_M^{-1}[\text{S}]c_1}{c_2}} = \frac{[\text{E}]_0/c_1}{1 + \dfrac{K_M c_2}{[\text{S}]c_1}}$$

With the substitutions for [ESH], c_1, and c_2 the rate law becomes:

$$\upsilon = \text{d}[\text{P}]/\text{d}t = k_b[\text{ESH}]$$

$$= \frac{k_b[\text{E}]_0/c_1}{1 + \dfrac{K_M c_2}{[\text{S}]c_1}} = \frac{\upsilon_{max}/c_1}{1 + \dfrac{K_M c_2}{[\text{S}]c_1}}, \quad \text{where} \quad \upsilon_{max} = k_b[\text{E}]_0$$

$$= \boxed{\frac{\upsilon'_{max}}{1 + \dfrac{K'_M}{[\text{S}]_0}}}, \quad \text{where} \quad \upsilon'_{max} = \frac{\upsilon_{max}}{c_1} = \frac{\upsilon_{max}}{1 + \dfrac{[\text{H}^+]}{K_{ES,b}} + \dfrac{K_{ES,a}}{[\text{H}^+]}} \quad \text{and} \quad K'_M = \frac{K_M c_2}{c_1} = \frac{K_M\left(1 + \dfrac{[\text{H}^+]}{K_{E,b}} + \dfrac{K_{E,a}}{[\text{H}^+]}\right)}{1 + \dfrac{[\text{H}^+]}{K_{ES,b}} + \dfrac{K_{ES,a}}{[\text{H}^+]}}$$

(b) $\upsilon_{max} = 1.0 \times 10^{-6}$ mol dm^{-3} s^{-1}, $K_{ES,b} = 1.0 \times 10^{-6}$ mol dm^{-3}, and $K_{ES,a} = 1.0 \times 10^{-8}$ mol dm^{-3}

The plot of υ'_{max} against pH shown in Figure 8.17 indicates a maximum value of υ'_{max} at pH = 7.0 for this set of equilibrium and kinetic constants. A formula for the pH of the maximum can be derived by finding the point at which $\dfrac{\text{d}\upsilon'_{max}}{\text{d}[\text{H}^+]} = 0$. This gives: $[\text{H}^+]_{max} = (K_{ES,a}\, K_{ES,b})^{1/2}$

Thus, for this data

$$[\text{H}^+]_{max} = \sqrt{(1.0 \times 10^{-8}\ \text{mol dm}^{-3})\,(1.0 \times 10^{-6}\ \text{mol dm}^{-3})} = 1.0 \times 10^{-7}\ \text{mol dm}^{-3}$$

which corresponds to $\boxed{\text{pH} = 7.0}$

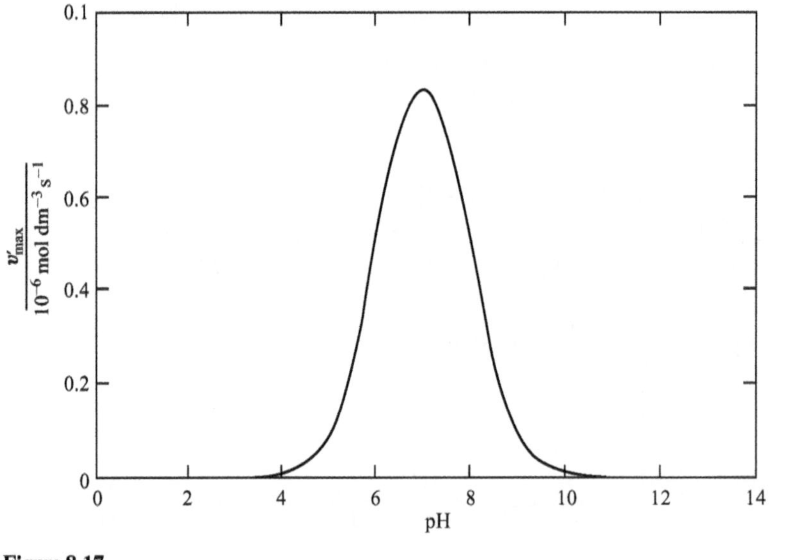

Figure 8.17

(c) $\upsilon_{max} = 1.0 \times 10^{-6}$ mol dm^{-3} s^{-1}, $K_{ES,b} = 1.0 \times 10^{-4}$ mol dm^{-3}, and $K_{ES,a} = 1.0 \times 10^{-10}$ mol dm^{-3}

The plot of υ'_{max} against pH shown in Figure 8.18 indicates that the constants of part (c) give a much broader curve than do the constants of part (b). This reflects the behavior of the term $1 + [\text{H}^+]/K_{ES,b} + K_{ES,a}/[\text{H}^+]$ in the denominator of the υ'_{max} expression. When $K_{ES,b}$ is relatively large, large [H$^+$] values (low pH) cause growth in the values of υ'_{max}. However, when $K_{ES,a}$ is relatively small, very small [H$^+$] values (high pH) cause a decline in the υ'_{max} values.

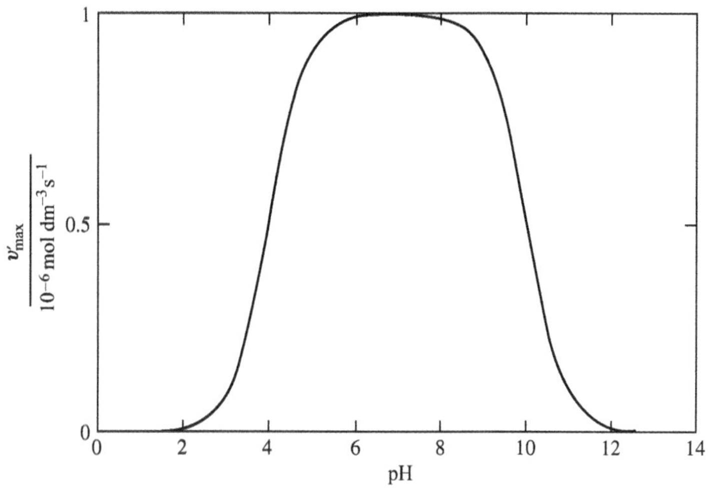

Figure 8.18

P8.43 (a) For a series of reactions with a fixed edge-to-edge distance and reorganization energy, the logarithm of the rate constant for electron transfer depends quadratically on the reaction free energy; eqn 8.32 applies:

$$\ln k_{et} = -\frac{(\Delta_r G^\ominus)^2}{4\lambda RT} - \frac{\Delta_r G^\ominus}{2RT} + \text{constant} \quad \text{when using molar energies.}$$

$$\ln k_{et} = -\frac{(\Delta_r G^\ominus)^2}{4\lambda kT} - \frac{\Delta_r G^\ominus}{2kT} + \text{constant} \quad \text{when using energies per molecular event.}$$

where we have replaced RT by kT since the energies are given in molecular rather than molar units. Draw up the following table and plot $\ln k_{et}$ against $\Delta_r G^\ominus$.

$\Delta_r G^\ominus$/eV	$k_{et}/(10^6\ \text{s}^{-1})$	$\ln(k_{et}/\text{s}^{-1})$
−0.665	0.657	13.4
−0.705	1.52	14.2
−0.745	1.12	13.9
−0.975	8.99	16.0
−1.015	5.76	15.6
−1.055	10.1	16.1

The least squares quadratic fit of the plot, shown as a box insert in Figure 8.19, is used to calculate the reorganization energy. The coefficient of the second-order term gives:

$$-\frac{1}{4\lambda kT} = -\frac{8.48}{\text{eV}^2}$$

$$\lambda = \frac{1}{4 \times (8.48/\text{eV}^2) \times kT}$$

$$= \frac{1.6022 \times 10^{-19}\ \text{J/eV}}{4 \times (8.48/\text{eV}^2) \times (1.381 \times 10^{-23}\ \text{J K}^{-1}) \times (298\ \text{K})}$$

$$= \boxed{1.15\ \text{eV}}$$

As a molar energy,

$$\lambda = (1.15\ \text{eV}) \times (1.602 \times 10^{-19}\ \text{J eV}^{-1}) \times (6.022 \times 10^{23}\ \text{mol}^{-1}) = \boxed{111\ \text{kJ mol}^{-1}}$$

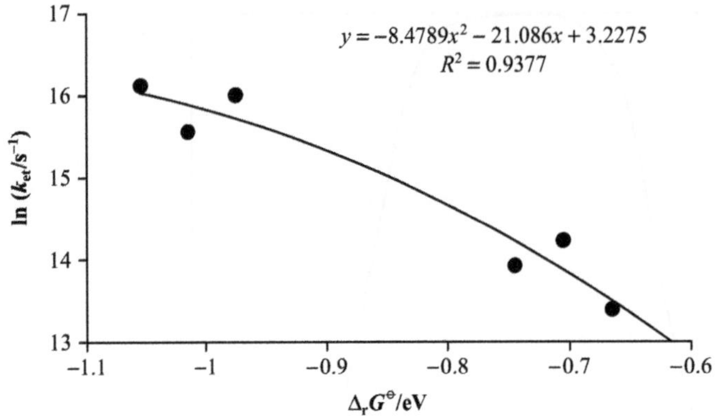

Figure 8.19

(b) Eqn 8.31, $\ln k_{et} = -\beta r + \text{constant}$, predicts that a plot of $\ln k_{et}$ against r should be linear with a slope equal to $-\beta$. Draw up the following table and prepare the required plot, shown in Figure 8.20. It is seen to be linear so a linear regression fit is performed and shown as an insert within the figure. The slope value indicates that $\boxed{\beta = 13.4\ \text{nm}^{-1}}$.

r/nm	k_{et}/s^{-1}	$\ln(k_{et}$/s$^{-1})$
0.48	1.58×10^{12}	28.1
0.95	3.98×10^{9}	22.1
0.96	1.00×10^{9}	20.7
1.23	1.58×10^{8}	18.9
1.35	3.98×10^{7}	17.5
2.24	6.31×10^{1}	4.14

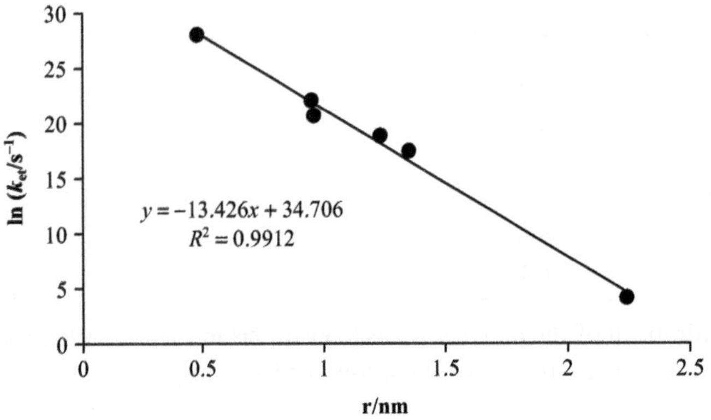

Figure 8.20

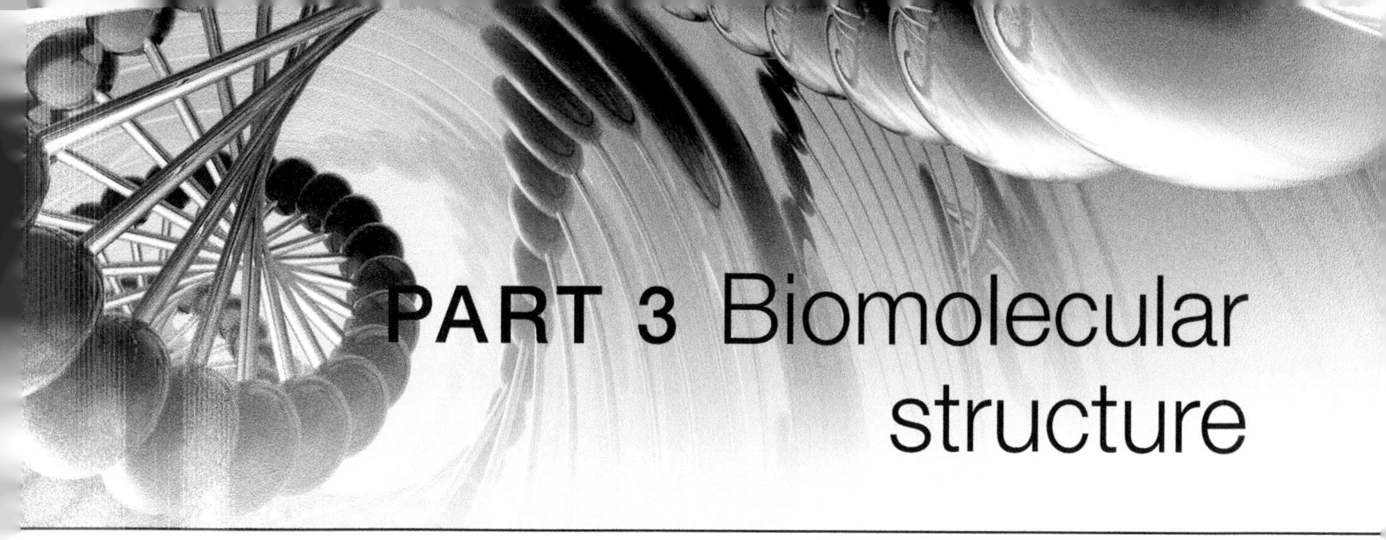

PART 3 Biomolecular structure

9 Microscopic systems and quantization

Answers to discussion questions

D9.1 At the end of the nineteenth century and the beginning of the twentieth, there were many experimental results obtained from studies on the properties of matter and radiation that could not be explained on the basis of established physical principles and theories. Here, we list only some of the most significant.

(1) The photoelectric effect revealed that electromagnetic radiation, classically considered to consist of waves, also exhibits particle-like behavior. These "particles" of radiation came to be called photons. Each photon is a discrete unit, or quantum, of energy that is absorbed during collisions with electrons. Photons are never partially absorbed. They either completely give up their energy or they are not absorbed. The energy of a photon can be calculated if either the radiation frequency or wavelength is known: $\boldsymbol{E_{\text{photon}} = h\nu = hc/\lambda}$.

(2) Absorption and emission spectra indicated that atoms and molecules can only absorb or emit discrete packets of energy (i.e. photons). This implies that an atom or molecule has specific, allowed energy levels and we say that their energies are quantized. During a spectroscopic transition the atom or molecule gains or loses the energy ΔE by either absorption of a photon or emission of a photon, respectively. Thus, spectral lines must satisfy the **Bohr frequency condition:** $\boldsymbol{\Delta E = h\nu}$.

(3) Neutron and electron diffraction studies indicated that entities such as electrons which previously had been thought of as just particles also possess wave-like properties of constructive and destructive interference. The joint particle and wave character of matter and radiation is called **wave–particle duality**. The **de Broglie relation**, $\lambda_{\text{de Broglie}} = h/p$, connects the wave character of a particle ($\lambda_{\text{de Broglie}}$) with its particulate momentum (p).

Evidence that resulted in the development of quantum theory also included:

(4) The energy density distribution of blackbody radiation as a function of wavelength.
(5) The heat capacities of monatomic solids such as copper metal.

D9.2 See the following table.

Particle	Size	Technique
Atom	40–240 pm	Electron microscopy, AFM & STM
Small molecule	300–600 pm	Electron microscopy, AFM & STM
Protein	~3 nm	Electron microscopy, AFM & STM
Ribosome	~20 nm	Electron microscopy, AFM & STM
Bacterium	100–500 nm	Light microscopy & electron microscopy
Animal cell	3–20 μm	Light microscopy & electron microscopy
Plant cell	~50 μm	Light microscopy & electron microscopy

D9.3 The wave–particle duality of quantum theory requires that the wavefunction of a particle has the property that it does not experience destructive interference upon reflection by a barrier or in motion around a closed loop. These are **boundary conditions** on the wavefunction and they are the cause of energy quantization. The criteria of particle existence and the restrictions of the boundary conditions result in quantum conditions that must be satisfied for a wavefunction to be acceptable. The conditions on the wavefunction, ψ, are:

(1) ψ must be single valued at each point.
(2) The probability of finding a particle in a very small subatomic region, $\psi^2\delta V$, cannot exceed 1.
(3) ψ is continuous everywhere.
(4) ψ has a continuous first derivative everywhere.

When applied to a particle of mass m confined to move in a one-dimensional box of length L, these requirements restrict the de Broglie wavelength to $\lambda = 2L/n$, where the **quantum number** n is a non-zero, positive integer. Then, using the relation $E = E_k = p^2/2m$ and the de Broglie relation $\lambda = h/p$, the energy is quantized at $E_n = n^2h^2/8mL^2$ [9.9]. This derivation applies specifically to the particle in a box but the derivation is similar for the particle on a ring; see Section 9.5(a) of the text.

D9.4 The lowest energy level possible for a confined quantum-mechanical system is the **zero-point energy**, and zero-point energy is not necessarily zero energy. This lowest, irremovable energy is consistent with the uncertainty principle, $\Delta x\Delta p_x \geq \frac{1}{2}\hbar$ [9.5]. In this relation, the uncertainties in position and momentum refer to the same direction of motion and these uncertainties are defined with the relations $(\Delta x)^2 = (x^2)_{\text{mean}} - (x)^2_{\text{mean}}$ and $(\Delta p_x)^2 = (p_x^2)_{\text{mean}} - (p_x)^2_{\text{mean}}$. Should a hypothetical quantum state exhibit both $(p_x^2)_{\text{mean}} = 0$ and $(p_x)^2_{\text{mean}} = 0$, so that the momentum uncertainty is zero, a finite value of Δx would give $\Delta x\Delta p_x = 0$ and quantum theory declares that, because the uncertainty principle is violated, the state is invalid and will not be observed in nature. However, should Δx be infinitely large the uncertainty principle may be satisfied even though there is no momentum uncertainty.

The zero-point energy of the particle in a box is the energy of the $n = 1$ quantum state of eqn 9.9: $E_1 = h^2/8mL^2$. For an electron in a 1-nm box, the zero-point energy is calculated to be 6.0×10^{-20} J and 36 kJ mol^{-1}, an energy that remains even after cooling to the absolute zero of temperature. This is consistent with the uncertainty principle as quantum theory does not assign a precise location to the particle, it is in the box but the uncertainty of knowing its position equals the length L of the box. Thus, a hypothetical zero-point energy of zero implies zero kinetic energy so that both $(p_x^2)_{\text{mean}} = 0$ and $(p_x)^2_{\text{mean}} = 0$. Consequently, there is zero uncertainty in knowledge of the momentum giving $\Delta x\Delta p_x = L \times 0 = 0$ in violation of the uncertainty principle and quantum theory declares that the particle in a box cannot have zero energy.

The zero-point energy of the harmonic oscillator is the energy of the $\upsilon = 0$ quantum state of eqn 9.29: $E_0 = \frac{1}{2}h\nu$, where υ (italic vee) is the quantum number and ν (Greek nu) is the frequency of the oscillator. A typical chemical bond has a vibrational frequency of 3.0×10^{13} Hz (corresponding to a wavenumber of 1000 cm^{-1}) so the zero-point energy of molecular vibration is typically about 1×10^{-20} J. In this case, a hypothetical zero-point energy of zero implies precise knowledge of the oscillator position; it is at the bottom of the harmonic potential where the potential is zero and the displacement is exactly zero. Zero energy implies zero uncertainty in knowledge of position. It also implies a precise momentum equal to zero giving $\Delta x\Delta p_x = 0 \times 0 = 0$ in violation of the uncertainty principle. Consequently, quantum theory declares that the harmonic oscillator cannot have zero energy.

The energy levels for a particle confined on a ring of constant potential, see text Section 9.5(a) and Figure 9.33, are given by:

$E_{m_l} = m_l^2\hbar^2/2I$ [9.22], where $m_l = 0, \pm1, \pm2, \ldots$

so the zero-point energy equals zero because it is the state for which $m_l = 0$. This does not violate the uncertainty principle. To see this, we recognize that the complementary variables of this two-dimensional quantum system are the position angle ϕ and the angular momentum J_z and we write the uncertainty principle as $\Delta\phi\Delta J_z \geq \frac{1}{2}\hbar$. As the classical particle travels through one cycle of rotation, the angle sweeps from zero through 2π radians. The second cycle takes the particle through 4π radians, the third cycle through 6π radians, etc. In quantum theory, however, we cannot precisely know either the angle within the first cycle or the cycle of rotation. The angular uncertainty is not 2π. It is infinitely large. Thus, even though J_z is precisely known as zero at the zero point, the uncertainty principle is not violated.

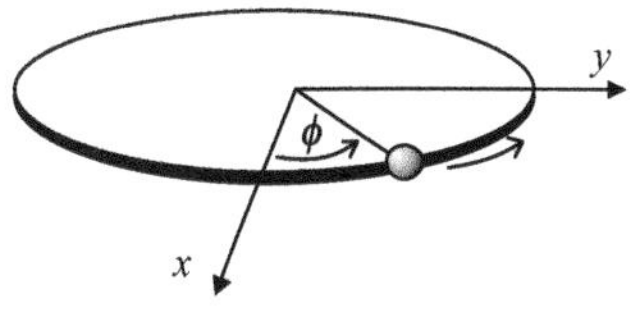

Figure 9.1

D9.5 The physical origin of tunneling is related to the probability density of the particle, which according to the Born interpretation is the square of the wavefunction that represents the particle. This interpretation requires that the wavefunction of the system be everywhere continuous, even at barriers. Therefore, if the wavefunction is non-zero on one side of a barrier it must be non-zero on the other side of the barrier and this implies that the particle has tunneled through the barrier. The transmission probability depends upon the mass of the particle (specifically $m^{1/2}$): the greater the mass the smaller the probability of tunneling. Electrons and protons have small masses, molecular groups large masses; therefore, tunneling effects are more observable in processes involving electrons and protons. An electron tunnels more readily than a proton and a proton tunnels more readily than a deuteron.

The very rapid equilibration of proton transfer reactions is a manifestation of the ability of protons to tunnel through barriers and transfer quickly from an acid to a base. Tunneling of protons between acidic and basic groups is also an important feature of the mechanism of some enzyme-catalysed reactions. Electron tunnelling is one of the factors that determine the rates of electron transfer reactions at electrodes and in biological systems.

D9.6 The time-independent wavefunction in three-dimensional space is a function of x, y, and z so we write $\psi(x,y,z)$ with each variable ranging from $-\infty$ to $+\infty$ in the general case. The time-independent wavefunction is said to be a **stationary state**. It is reasonable to expect that for special quantum systems the probability densities in each of the three independent directions should be mutually independent. This implies that the probability density for the time-independent wavefunction $\psi(x,y,z)$ should be the product of three probability densities, one for each coordinate: $|\psi(x,y,z)|^2 \propto |X(x)|^2 \times |Y(y)|^2 \times |Z(z)|^2$. Subsequently, we see that the wavefunction is the product of three independent wavefunctions and we write $\psi(x,y,z) \propto X(x) \times Y(y) \times Z(z)$. Such a wavefunction is said to exhibit the **separation of variables**. When a wavefunction can be separated in this manner, the solution to the Schrödinger equation is greatly simplified, as demonstrated by the particle in the three-dimensional box.

For a particle free to move within a cube of volume L^3 under the influence of a constant potential, which is assigned the value zero, we may generalize the simple wavefunction provided by the one-dimensional solution discussed in Section 9.4(a) of the text. Using the method of separation of variables, we find that each dimension has a solution analogous to that of the one-dimensional problem: $X(x) \propto \sin(n_x\pi x/L)$, $Y(y) \propto \sin(n_y\pi y/L)$ and $Z(z) \propto \sin(n_z\pi z/L)$. The wavefunction is then given by $\psi(x,y,z) \propto \sin(n_x\pi x/L) \times \sin(n_y\pi y/L) \times \sin(n_z\pi z/L)$ with three independent quantum numbers each of which is an integer that can range between 1 and ∞. Furthermore, the total energy is the sum of the energies that originate from motion along each of the independent variables. The energy along each coordinate is analogous to that of the one-dimensional problem giving:

$$E_{n_x,n_y,n_z} = E_{n_x} + E_{n_y} + E_{n_z} = \frac{n_x^2h^2}{8mL^2} + \frac{n_y^2h^2}{8mL^2} + \frac{n_z^2h^2}{8mL^2} = (n_x^2 + n_y^2 + n_z^2)\frac{h^2}{8mL^2}$$

(Remarkably, when the potential energy term of the hamiltonian is either zero or a constant value throughout space, the time-independent wavefunction does not depend upon the particle mass! Similarly, electrical charge does not appear in the time-independent wavefunction in this particular quantum system.)

The method of separation of variables is not generally valid and a simple analytical solution to the Schrödinger equation does not usually exist. This means that an attempt must generally be made to find solutions with rather difficult numerical methods. The question then becomes "When is a solution by separation of variables possible?" The answer is: When motion along a particular variable does not change the potential energy of the particle, that variable may be separated from the others.

For example, consider an isolated hydrogen atom. This quantum system has a proton nucleus and one electron. The Cartesian coordinate system is centered on the nucleus and we want to know whether the wavefunction for the electron, $\psi(x,y,z)$, is amenable to the method of separation of variables in Cartesian coordinates. To address this question, we remember that the electron and proton exhibit an electrostatic attraction that has a potential energy that is inversely proportion to the distance r between them: $V \propto 1/r$, where $r = (x^2 + y^2 + z^2)^{1/2}$. Examination of the potential reveals that when the electron's x position, or y position, or z position change, the electron potential changes and we conclude that none of these variables may be separated. However, we remember that r is one of the three independent variables of the spherical coordinate system and we write the wavefunction as $\psi(r,\theta,\phi)$. The electron potential does not change when either of the angles changes so functions for each of the angles may be separated giving a wavefunction that is the product of functions for each variable. It has the form: $\psi(r,\theta,\phi) \propto R(r) \times \Theta(\theta) \times \Phi(\phi)$, where the functions R, Θ, and Φ are each functions of a single variable only. This provides the complete separation of variables. It greatly simplifies the solution to the Schrödinger equation and an analytical, non-numeric, solution becomes possible for the hydrogen atom.

D9.7 (1) The **principal quantum number n** determines the energy of the atomic orbitals in a hydrogenic shell through eqn 9.31. The shells K, L, M, and N correspond to the principal quantum numbers $n = 1$, 2, 3, and 4. Successive shells are further away from the nucleus on average and successively higher in energy. The permitted orbital energies approach zero as n becomes very large because this is defined to be the minimum energy at which the electron and nucleus are infinitely separated.

(2) The **orbital angular momentum quantum number l**, also called the azimuthal quantum number, determines the magnitude of the orbital angular momentum of a hydrogenic atomic orbital through the formula $\{l(l+1)\}^{1/2}\hbar$. The permitted values of l are 0, 1, 2, 3, . . . , $n - 1$ for the nth shell and these correspond to the s, p, d, f, . . . subshells. The degeneracy of a subshell is $2l + 1$ because this is the number of orbitals in each subshell. Thus, the s, p, d, and f subshells consist of 1, 3, 5, and 7 orbitals, respectively. In many-electron atoms a maximum of two electrons can be in an orbital; this is the **Pauli exclusion principle.**

(3) The **magnetic quantum number m_l** determines the z-component of the angular momentum of a hydrogenic orbital through the formula $m_l\hbar$. The permitted values for subshell l are l, $l - 1$, $l - 2$, . . . , $-l$, which accounts for the orbital degeneracy of the subshell.

(4) The **spin quantum number s** determines the magnitude of the electron spin angular momentum through the formula $\{s(s+1)\}^{1/2}\hbar$. For hydrogenic atomic orbitals, s can only be 1/2.

(5) The **spin quantum number m_s** determines the z-component of the spin angular momentum through the formula $m_s\hbar$. m_s can only be $\pm\frac{1}{2}$. $m_s = +\frac{1}{2}$ corresponds to the α or $\uparrow$ spin; $m_s = -\frac{1}{2}$ corresponds to the β or $\downarrow$ spin.

D9.8 (a) A **boundary surface** for a hydrogenic orbital is drawn to contain most (say 90%) of the probability density of an electron in that orbital. Its shape varies from orbital to orbital because the electron density distribution is different for different orbitals. Example boundary surfaces are shown in text Figures 9.45 (s orbital), 9.48 (p orbitals), and 9.49 (d orbitals).

(b) The **radial distribution function** gives the probability that the electron will be found anywhere within a shell of radius r around the nucleus (see text Figure 9.46). It gives a better picture of where the electron is likely to be found with respect to the nucleus than the probability density, which is the square of the wavefunction. The radial distribution function for an s orbital is

$$P(r) = 4\pi r^2 \psi_{ns}^2 \quad [9.34]$$

The more general form, which also applies to orbitals that depend on angle, is

$$P(r) = r^2 R_{n,l}(r)^2, \quad \text{where } R_{n,l}(r) \text{ is the radial wavefunction.}$$

The usefulness of the radial distribution function is illustrated in text Figure 9.47 which shows the relative penetration of the inner core by subshell orbitals of the M shell ($n = 3$). The order of penetration is 3s > 3p > 3d and, consequently, electrons in these subshells have the same order of relative effective nuclear charge attracting them to the nucleus. High effective nuclear charge means lower orbital energy, so the order of subshell energy is 3s < 3p < 3d, a fact that is very important when using the building-up principle to determine the ground electron configuration of many-electron atoms.

D9.9 In the crudest form of the **orbital approximation**, the many-electron wavefunction for an atom is represented as a simple product of one-electron wavefunctions (see eqn 9.36), each of which has the form of a hydrogenic atomic orbital. This is said to be the independent-electron model. For example, the orbital approximation for the lithium atom ground-state wavefunction is the product of the orbitals for each of the three lithium electrons:

$$\psi_{\text{Li}} = \psi_{1s}(1)\alpha(1) \times \psi_{1s}(2)\beta(2) \times \psi_{2s}(3)\alpha(3)$$

This is synonymous with the ground-state electronic configuration given by the building-up principle. For the lithium atom it is $1s^2 2s$. The simplest form of this approximation neglects electron repulsions in many-electron atoms so it does not give a very good estimate of the atomic energy. It does, however, provide concepts for the quick analysis of a great many atomic and molecular problems in chemistry and biochemistry.

At a somewhat more sophisticated level, the many-electron wavefunctions are written as linear combinations of such simple product functions that explicitly satisfy the Pauli exclusion principle. Relatively good one-electron functions are generated by the Hartree–Fock self-consistent field method (see the authors' *Physical chemistry*) in which an electron moves in the average electron repulsion potential field of all other electrons. We can in principle obtain exact energies and wavefunctions with such numerical methods; however, there are significant numerical challenges.

D9.10 Electronic configurations of neutral, 4th period transition atoms in the ground state are summarized in the following table along with observed, positive oxidation states. The most common, positive oxidation states are indicated with bright boxing.

Group	3	4	5	6	7	8	9	10	11	12
Oxidation state	Sc	Ti	V	Cr	Mn	Fe	Co	Ni	Cu	Zn
0	$3d4s^2$	$3d^24s^2$	$3d^34s^2$	$3d^54s$	$3d^54s^2$	$3d^64s^2$	$3d^74s^2$	$3d^84s^2$	$3d^{10}4s$	$3d^{10}4s^2$
+1			☺	☺	☺		☺	☺	☺	
+2		☺	☺	☺	☺	☺	☺	☺	☺	☺
+3	☺	☺	☺	☺	☺	☺	☺	☺		
+4		☺	☺	☺	☺	☺	☺	☺		
+5			☺	☺	☺	☺				
+6				☺	☺	☺				
+7					☺					

Toward the middle of the first transition series (Cr, Mn, and Fe) elements exhibit the widest ranges of oxidation states. This phenomenon is related to the availability of both electrons and orbitals favorable for bonding. Elements to the left (Sc and Ti) of the series have few electrons and relatively low effective nuclear charge leaves d orbitals at high energies that are relatively unsuitable for bonding. To the far right (Cu and Zn) effective nuclear charge may be higher but there are few, if any, orbitals available for bonding. Consequently, it is more difficult to produce a range of compounds that promote a wide range of oxidation states for elements at either end of the series. At the middle and right of the series the +2 oxidation state is very commonly observed because normal reactions can provide the requisite ionization energies for the removal of 4s electrons. The readily available +2 and +3 oxidation states of Mn, Fe, and the +1 and +2 oxidation states of Cu make these cations useful in electron-transfer processes occurring in chains of specialized protein within biological cells. The special size and charge of the Zn^{2+} cation makes it useful for the function of some enzymes. The tendency of Fe^{2+} and Cu^+ to bind oxygen proves very useful in hemoglobin and electron transport (respiratory) chains, respectively.

Solutions to exercises

E9.11 (a) For an electronic transition of known frequency the transition quantum is the corresponding photon energy. Thus,

$$E_{\text{photon}} = h\nu$$
$$= (6.626 \times 10^{-34}\ \text{J s}) \times (1.0 \times 10^{15}\ \text{s}^{-1}) = \boxed{6.6 \times 10^{-19}\ \text{J}}$$

and for a mole of photons

$$E_{\text{m}} = N_{\text{A}} h\nu$$
$$= (6.022 \times 10^{23}\ \text{mol}^{-1}) \times (6.626 \times 10^{-34}\ \text{J s}) \times (1.0 \times 10^{15}\ \text{s}^{-1}) = \boxed{4.0 \times 10^{2}\ \text{kJ mol}^{-1}}$$

(b) The harmonic oscillator is used as the model for the quantum motion of molecular vibration and eqn 9.29, along with Figure 9.38, indicates that quantum states are separated by the energy quantum $\Delta E = h\nu = h/T$ where the period T is defined to be the inverse of frequency ($T = 1/\nu$).

$$\Delta E = h/T$$
$$= (6.626 \times 10^{-34}\ \text{J s})/(20 \times 10^{-15}\ \text{s}) = \boxed{3.3 \times 10^{-20}\ \text{J}}$$

$$\Delta E_{\text{m}} = N_{\text{A}} E$$
$$= (6.022 \times 10^{23}\ \text{mol}^{-1}) \times (3.3 \times 10^{-20}\ \text{J}) = \boxed{20.\ \text{kJ mol}^{-1}}$$

(c) The harmonic oscillator is also used as the model for the quantum states of pendulum motion. So, like part (b) eqn 9.29 indicates that quantum states are separated by the energy quantum $\Delta E = h\nu = h/T$.

$$\Delta E = h/T$$
$$= (6.626 \times 10^{-34}\ \text{J s})/(0.50\ \text{s}) = \boxed{1.3 \times 10^{-33}\ \text{J}}$$

$$\Delta E_{\text{m}} = N_{\text{A}}E$$
$$= (6.022 \times 10^{23}\ \text{mol}^{-1}) \times (1.3 \times 10^{-33}\ \text{J}) = \boxed{7.8 \times 10^{-13}\ \text{kJ mol}^{-1}}$$

This extraordinarily small separation is caused by the macroscopic, large mass characteristics of a pendulum. The energy levels are so close together that the pendulum energies appear as a continuum of values that are successfully described by the classical laws of physics.

E9.12 The definition of power P is $P = E/t$ and the energy transported by N photons in time t is $E = Nh\nu = Nhc/\lambda$. Thus, $P = Nhc/\lambda t$.

(a) $$P = \frac{Nhc}{\lambda t} = \frac{8 \times 10^7 \times 6.626 \times 10^{-34}\ \text{J s} \times 2.998 \times 10^8\ \text{m s}^{-1}}{470 \times 10^{-9}\ \text{m} \times 3.8 \times 10^{-3}\ \text{s}} = \boxed{8.9 \times 10^{-9}\ \text{W}}$$

(b) $$P = \frac{Nhc}{\lambda t} = \frac{8 \times 10^7 \times 6.626 \times 10^{-34}\ \text{J s} \times 2.998 \times 10^8\ \text{m s}^{-1}}{780 \times 10^{-9}\ \text{m} \times 3.8 \times 10^{-3}\ \text{s}} = \boxed{5.3 \times 10^{-9}\ \text{W}}$$

E9.13 $$\lambda = \frac{h}{p} = \frac{h}{mv} \quad [9.3]$$

(a) $$\lambda = \frac{(6.626 \times 10^{-34}\ \text{J s})}{(1.00\ \text{m s}^{-1}) \times (1.0 \times 10^{-3}\ \text{kg})} = \boxed{6.6 \times 10^{-31}\ \text{m}}$$

(b) $$\lambda = \frac{(6.626 \times 10^{-34}\ \text{J s})}{(1.0 \times 10^{8}\ \text{m s}^{-1}) \times (1.0 \times 10^{-3}\ \text{kg})} = \boxed{6.6 \times 10^{-39}\ \text{m}}$$

(c) $$\lambda = \frac{(6.626 \times 10^{-34}\ \text{J s})}{4.003 \times (1.6605 \times 10^{-27}\ \text{kg}) \times (1.0 \times 10^{3}\ \text{m s}^{-1})} = \boxed{99.7\ \text{pm}}$$

(d) $m = 85$ kg $\quad v = 8.0$ km h^{-1}

$$\lambda = \frac{h}{p} = \frac{h}{mv} \quad \text{de Broglie relation [9.3]}$$
$$= \frac{6.626 \times 10^{-34}\ \text{J s}}{(85\ \text{kg}) \times (8.0 \times 10^3\ \text{m h}^{-1})}\left(\frac{3600\ \text{s}}{1\ \text{h}}\right) = \boxed{3.5 \times 10^{-36}\ \text{m}}$$

(e) This extraordinarily small wavelength calculated in part (d) is much, much smaller than the diameter of a hydrogen nucleus and that calculation illustrates the hopelessness of measuring the de Broglie wavelength of a macroscopic object. The de Broglie wavelength does increase as the speed of an object decreases and, according to the quantum behavior of a particle in a one-dimensional box of length L, the de Broglie wavelength may be as long as $2L$. For yourself at rest, the de Broglie wavelength would increase to infinity, but what meaning could be attached to this result in unclear.

E9.14 $$p = \frac{h}{\lambda}\ [9.3]\quad E = h\nu\ [9.1] = \frac{hc}{\lambda}$$

$$hc = (6.6261 \times 10^{-34}\ \text{J s}) \times (2.99792 \times 10^8\ \text{m s}^{-1}) = 1.986 \times 10^{-25}\ \text{J m}$$

$$N_A hc = (6.02214 \times 10^{23}\ \text{mol}^{-1}) \times (1.986 \times 10^{-25}\ \text{J m})$$
$$= 0.1196\ \text{J m mol}^{-1}$$

We can therefore draw up the following table

λ	p/kg m s^{-1}	E/J	E/(kJ mol^{-1})
(a) 600 nm	1.10×10^{-27}	3.31×10^{-19}	199
(b) 550 nm	1.20×10^{-27}	3.61×10^{-19}	218
(c) 400 nm	1.66×10^{-27}	4.97×10^{-19}	299
(d) 200 nm	3.31×10^{-27}	9.93×10^{-19}	598
(e) 150 pm	4.41×10^{-24}	1.32×10^{-15}	7.98×10^{5}
(f) 1.00 cm	6.6×10^{-32}	1.99×10^{-23}	0.012

E9.15 **(a)** The shortest wavelength as estimated without relativistic correction is calculated using eqn. 9.3:

$$\lambda_{\text{non-relativistic}} = \frac{h}{p} = \frac{h}{(2m_e E_k)^{1/2}} = \frac{h}{(2m_e eV)^{1/2}}$$
$$= \frac{6.626 \times 10^{-34}\ \text{J s}}{\{2(9.109 \times 10^{-31}\ \text{kg}) \times (1.602 \times 10^{-19}\ \text{C}) \times (50.0 \times 10^{3}\ \text{V})\}^{1/2}}$$
$$= 5.48\ \text{pm}$$

$$\lambda_{\text{relativistic}} = \frac{h}{\left\{2m_e eV\left(1+\dfrac{eV}{2m_e c^2}\right)\right\}^{1/2}}$$
$$= \frac{\lambda_{\text{non-relativistic}}}{\left(1+\dfrac{eV}{2m_e c^2}\right)^{1/2}} = \frac{5.48\ \text{pm}}{\left\{1+\dfrac{(1.602 \times 10^{-19}\ \text{C})(50.0 \times 10^{3}\ \text{V})}{2(9.109 \times 10^{-31}\ \text{kg})(3.00 \times 10^{8}\ \text{m s}^{-1})^2}\right\}^{1/2}}$$
$$= \boxed{5.35\ \text{pm}}$$

(b) For an electron accelerated through 50 kV the non-relativistic de Broglie wavelength is calculated to be high by 2.4%. This error may be insignificant for many applications. However, should an accuracy of 1% or better be required, use the relativistic equation at accelerations through a potential above 20.4 V, as demonstrated in the following calculation.

$$\frac{\lambda_{\text{non-relativistic}} - \lambda_{\text{relativistic}}}{\lambda_{\text{relativistic}}} = \frac{\lambda_{\text{non-relativistic}}}{\lambda_{\text{relativistic}}} - 1 = \left(1+\frac{eV}{2m_e c^2}\right)^{1/2} - 1$$
$$= \cancel{1} + \frac{1}{2}\left(\frac{eV}{2m_e c^2}\right) - \frac{1}{2\cdot 4}\left(\frac{eV}{2m_e c^2}\right)^2 + \frac{1\cdot 3}{2\cdot 4\cdot 6}\left(\frac{eV}{2m_e c^2}\right)^3 - \cdots \cancel{-1}$$
$$\simeq \frac{1}{2}\left(\frac{eV}{2m_e c^2}\right) \quad \text{because 2nd- and 3rd-order terms are very small.}$$

The largest value of V for which the non-relativistic equation yields a value that has less than 1% error:

$$V \simeq 2\left(\frac{2m_e c^2}{e}\right) \times \left(\frac{\lambda_{\text{non-relativistic}} - \lambda_{\text{relativistic}}}{\lambda_{\text{relativistic}}}\right) = 2\left(\frac{2m_e c^2}{e}\right)(0.01) = 20.4\ \text{kV}$$

E9.16 The momentum per photon of wavelength 650 nm is

$$p_{\text{photon}} = \frac{h}{\lambda}[9.3] = \frac{6.626 \times 10^{-34}\ \text{J s}}{650 \times 10^{-9}\ \text{m}} = 1.02 \times 10^{-27}\ \text{kg m s}^{-1}$$

and this is also the change of momentum per photon absorbed by the fabric. The laser produces a hefty N_A photons per second and all photons are absorbed by the spacecraft sail. The power P of this 650 nm laser is

$$P = (N_A\ \text{s}^{-1}) \times E_{\text{photon}} = (N_A\ \text{s}^{-1}) \times hc/\lambda$$

$$= (6.022 \times 10^{23}\ \text{s}^{-1}) \times (6.626 \times 10^{-34}\ \text{J s}) \times (2.998 \times 10^{8}\ \text{m s}^{-1})/(650 \times 10^{-9}\ \text{m}) = 184\ \text{kW}$$

(a) The force F in SI units on the sail is the change in momentum experienced by the sail per second. This is equal to the photon flux, $N_A\ \text{s}^{-1}$, multiplied by the momentum lost by a photon.

$$F = (N_A\ \text{s}^{-1}) \times p_{\text{photon}}$$

$$= (6.022 \times 10^{23}\ \text{s}^{-1}) \times (1.02 \times 10^{-27}\ \text{kg m s}^{-1}) = \boxed{6.14 \times 10^{-4}\ \text{N}}$$

(b) The pressure exerted by the radiation equals the force F divided by the sail area A.

$$F/A = (6.14 \times 10^{-4}\ \text{N})/(1.0 \times 10^{6}\ \text{m}^2) = \boxed{614\ \text{pPa}}$$

(c) $$t = \left(\frac{mv}{F}\right)_{\text{spacecraft}} = \frac{(1.0\ \text{kg}) \times (1.0\ \text{m s}^{-1})}{6.14 \times 10^{-4}\ \text{N}} = 1.63 \times 10^{3}\ \text{s} = \boxed{0.452\ \text{h}}$$

E9.17 $$\Delta p = 1.00 \times 10^{-4} p\ [\text{i.e. } 0.0100\% \text{ of } p] = 1.00 \times 10^{-4} m_p v$$

$$\Delta x = \frac{\hbar}{2\Delta p}[9.5] = \frac{\hbar}{2 \times (1.00 \times 10^{-4}) \times m_p v}$$

$$= \frac{(1.055 \times 10^{-34}\ \text{J s})}{2 \times (1.00 \times 10^{-4}) \times (1.673 \times 10^{-27}\ \text{kg}) \times (3.5 \times 10^{5}\ \text{m s}^{-1})}$$

$$= 9.0 \times 10^{-10}\ \text{m, or } \boxed{0.90\ \text{nm}}$$

E9.18 The minimum uncertainty in position is $\boxed{100\ \text{pm}}$. Therefore, because $\Delta x \Delta p \geq \frac{1}{2}\hbar[9.5]$

$$\Delta p \geq \frac{\hbar}{2\Delta x} = \frac{1.0546 \times 10^{-34}\ \text{J s}}{2(100 \times 10^{-12}\ \text{m})} = 5.3 \times 10^{-25}\ \text{kg m s}^{-1}$$

$$\Delta v = \frac{\Delta p}{m_e} = \frac{5.3 \times 10^{-25}\ \text{kg m s}^{-1}}{9.11 \times 10^{-31}\ \text{kg}} = \boxed{5.8 \times 10^{5}\ \text{m s}^{-1}}$$

E9.19 The Born interpretation (Section 9.2(b)) of a normalized wavefunction states that the probability, P, of finding a particle in a very small region equals $\psi^2\delta V$. When consideration is given to a particle in a one-dimensional box, this becomes $P = \psi^2\delta x$ where ψ is evaluated at the mid-point of the δx range. This method is an estimate for which improvements require the use of calculus.

$$P = \psi^2\delta x = \left\{\left(\frac{2}{L}\right)^{1/2} \sin\left(\frac{2\pi x}{L}\right)\right\}^2 \delta x = \left(\frac{2}{L}\right) \sin^2\left(\frac{2\pi x}{L}\right)\delta x$$

(a) $P=\left(\frac{2}{10\text{ nm}}\right)\times\sin^2\left(\frac{2\pi\times 0.15\text{ nm}}{10\text{ nm}}\right)\times(0.2\text{ nm}-0.1\text{ nm})=\boxed{1.\overline{77}\times10^{-4}}$

(b) $P=\left(\frac{2}{10\text{ nm}}\right)\sin^2\left(\frac{2\pi\times 5.05\text{ nm}}{10\text{ nm}}\right)(5.2\text{ nm}-4.9\text{ nm})=\boxed{5.\overline{92}\times10^{-5}}$

E9.20 When the Born interpretation (Section 9.29b) describes the infinitesimally small probability, $\mathrm{d}P$, of finding a particle in an infinitely small region $\mathrm{d}V$, it is written as the differential equation $\mathrm{d}P=\psi^2\mathrm{d}V$, where the coordinates of ψ are those of the infinitesimal volume $\mathrm{d}V$. To find the probability that the particle will be found in the region of V, integration must be performed over the region.

$$P=\int_{\text{region}}\mathrm{d}V=\int_{\text{region}}\psi^2\,\mathrm{d}V$$

When consideration is given to a particle in a one-dimensional box, where the region is between x_1 and x_2, this becomes

$$P=\int_{x_1}^{x_2}\psi^2\,\mathrm{d}x,\quad\text{where }\psi=\left(\frac{2}{L}\right)^{1/2}\sin\left(\frac{2\pi x}{L}\right)$$

$$P=\int_{x_1}^{x_2}\left\{\left(\frac{2}{L}\right)^{1/2}\sin\left(\frac{2\pi x}{L}\right)\right\}^2\mathrm{d}x=\left(\frac{2}{L}\right)\int_{x_1}^{x_2}\sin^2\left(\frac{2\pi x}{L}\right)\mathrm{d}x$$

Using the standard integral $\int\sin^2(ax)\,\mathrm{d}x=\frac{x}{2}-\frac{\sin(2ax)}{4a}$, the working equation becomes

$$P=\left(\frac{2}{L}\right)\left[\frac{x}{2}-\frac{\sin\left(2\left(\frac{2\pi}{L}\right)x\right)}{4\left(\frac{2\pi}{L}\right)}\right]_{x_1}^{x_2}=\left[\frac{x}{L}-\frac{1}{4\pi}\sin\left(\frac{4\pi x}{L}\right)\right]_{x_1}^{x_2}$$

(a) $P=\left[\frac{x}{10\text{ nm}}-\frac{1}{4\pi}\sin\left(\frac{4\pi x}{10\text{ nm}}\right)\right]_{0.1\text{ nm}}^{0.2\text{ nm}}=1.\overline{84}\times10^{-4}$

Error of the Exercise 9.19 approximation: $\frac{1.\overline{84}\times10^{-4}-1.\overline{77}\times10^{-4}}{1.\overline{84}\times10^{-4}}\times100=\boxed{12.9\%}$

(b) $P=\left[\frac{x}{10\text{ nm}}-\frac{1}{4\pi}\sin\left(\frac{4\pi x}{10\text{ nm}}\right)\right]_{4.9\text{ nm}}^{5.2\text{ nm}}=2.\overline{36}\times10^{-4}$

Error of the Exercise 9.19 approximation: $\left|\frac{2.\overline{36}\times10^{-4}-5.\overline{92}\times10^{-5}}{2.\overline{36}\times10^{-4}}\right|\times100=\boxed{74.9\%}$

E9.21 $P=\int_{x_1}^{x_2}\psi^2\,\mathrm{d}x,\quad\text{where }\psi=\left(\frac{2}{L}\right)^{1/2}\sin\left(\frac{\pi x}{L}\right)$

$$P=\int_{x_1}^{x_2}\left\{\left(\frac{2}{L}\right)^{1/2}\sin\left(\frac{\pi x}{L}\right)\right\}^2\mathrm{d}x=\left(\frac{2}{L}\right)\int_{x_1}^{x_2}\sin^2\left(\frac{\pi x}{L}\right)\mathrm{d}x$$

Using the standard integral $\int \sin^2(ax)\,dx = \frac{x}{2} - \frac{\sin(2ax)}{4a}$, the working equation becomes

$$P = \left(\frac{2}{L}\right)\left[\frac{x}{2} - \frac{\sin\left(2\left(\frac{\pi}{L}\right)x\right)}{4\left(\frac{\pi}{L}\right)}\right]_{x_1}^{x_2} = \left[\frac{x}{L} - \frac{1}{2\pi}\sin\left(\frac{2\pi x}{L}\right)\right]_{x_1}^{x_2}$$

(a) $$P = \left[\frac{x}{L} - \frac{1}{2\pi}\sin\left(\frac{2\pi x}{L}\right)\right]_{0*L}^{L/3} = \left[x - \frac{1}{2\pi}\sin(2\pi x)\right]_0^{1/3} = \boxed{0.196}$$

(b) $$P = \left[\frac{x}{L} - \frac{1}{2\pi}\sin\left(\frac{2\pi x}{L}\right)\right]_{L/3}^{2L/3} = \left[x - \frac{1}{2\pi}\sin(2\pi x)\right]_{1/3}^{2/3} = \boxed{0.609}$$

(c) $$P = \left[\frac{x}{L} - \frac{1}{2\pi}\sin\left(\frac{2\pi x}{L}\right)\right]_{2L/3}^{L} = \left[x - \frac{1}{2\pi}\sin(2\pi x)\right]_{2/3}^{1} = \boxed{0.196}$$

Note that the probabilities sum to 1.

E9.22 $$\int_{-\infty}^{\infty} \psi^2 dx = \int_0^L \psi^2 dx = \int_0^L A^2 dx = A^2\int_0^L dx = A^2 x\Big|_0^L = A^2 L = 1 \text{ [the normalization condition]}$$

Therefore, $A = \left(\frac{1}{L}\right)^{1/2}$ and the normalized wavefunction is $\boxed{\psi = \left(\frac{1}{L}\right)^{1/2}}$.

E9.23 (a) The energy levels are given by:

$$E_n = \frac{h^2 n^2}{8mL^2},$$

and we are looking for the energy difference between $n = 6$ and $n = 7$:

$$\Delta E = \frac{h^2(7^2 - 6^2)}{8mL^2}.$$

Since there are 12 atoms on the conjugated backbone, the length of the box is 11 times the bond length:

$$L = 11(140 \times 10^{-12}\text{ m}) = 1.54 \times 10^{-9}\text{ m},$$

so $$\Delta E = \frac{(6.626 \times 10^{-34}\text{ J s})^2(49 - 36)}{8(9.11 \times 10^{-31}\text{ kg})(1.54 \times 10^{-9}\text{ m})^2} = \boxed{3.30 \times 10^{-19}\text{ J}}.$$

(b) The relationship between energy and frequency is:

$$\Delta E = h\nu, \quad \text{so} \quad \nu = \frac{\Delta E}{h} = \frac{3.30 \times 10^{-19}\text{ J}}{6.626 \times 10^{-34}\text{ J s}} = \boxed{4.95 \times 10^{-14}\text{ s}^{-1}}.$$

This frequency corresponds to a wavelength of about 600 nm, which is in the orange region of the spectrum.

E9.24 The rate of tunneling is proportional to the transmission probability, so a ratio of tunneling rates is equal to the corresponding ratio of transmission probabilities. The desired factor is T_1/T_2, where the subscripts denote the tunneling distances in nanometers:

$$\frac{T_1}{T_2}=\frac{1+\dfrac{(e^{\kappa L_2}-e^{-\kappa L_2})^2}{16\varepsilon(1-\varepsilon)}}{1+\dfrac{(e^{\kappa L_1}-e^{-\kappa L_1})^2}{16\varepsilon(1-\varepsilon)}}.$$ (See *Physical Chemistry* 2010 for the full formula used here).

If $\dfrac{(e^{\kappa L_2}-e^{-\kappa L_2})^2}{16\varepsilon(1-\varepsilon)}\gg 1$, and similarly for L_1.

then $\dfrac{T_1}{T_2}\approx\dfrac{(e^{\kappa L_2}-e^{-\kappa L_2})^2}{(e^{\kappa L_1}-e^{-\kappa L_1})^2}\approx e^{2\kappa(L_2-L_1)}=e^{2(7/\text{nm})(2.0-1.0)\text{nm}}=\boxed{1.2\times10^6}$.

That is, the tunneling rate increases about a million-fold. Note: if the first approximation does not hold, we need more information, namely $\varepsilon=E/V$. If the first approximation is valid, then the second is also likely to be valid, namely that the negative exponential is negligible compared to the positive one.

E9.25 (a) $\upsilon=Ae^{-d/l}=(5\times10^{14}\text{ s}^{-1})e^{-(750\text{ pm})/(70\text{ pm})}=\boxed{1.\bar{1}\times10^{10}\text{ s}^{-1}}$

(b) $\dfrac{\upsilon(d_2)}{\upsilon(d_1)}=\dfrac{Ae^{-d_2/l}}{Ae^{-d_1/l}}=e^{-(d_2-d_1)/l}=e^{-(850\text{ pm}-750\text{ pm})/(70\text{ pm})}=\boxed{0.24}$

Thus, the current is reduced by about a factor of $\boxed{4}$.

E9.26 With 10 electrons, the five lowest states will be occupied by two electrons each. The energy levels are (eqn. 9.13b)

$$E_{n_1,n_2}=\left(\frac{n_1^2}{L_1^2}+\frac{n_2^2}{L_2^2}\right)\frac{h^2}{8m_e}$$

$$=\left(\frac{n_1^2}{(L_1/\text{pm})^2}+\frac{n_2^2}{(L_2/\text{pm})^2}\right)\frac{(6.626\times10^{-34}\text{ J s})^2}{8\times9.11\times10^{-31}\text{ kg}\times(10^{-12}\text{ m})^2}$$

$$=\left(\frac{n_1^2}{280^2}+\frac{n_2^2}{450^2}\right)\times6.02\times10^{-14}\text{ J.}$$

The seven lowest energy levels are shown in the table below:

n_1,n_2	1,1	1,2	2,1	1,3	2,2	1,4	2,3
$E/10^{-18}$ J	1.07	1.96	3.37	3.45	4.26	5.53	5.75

(a) The highest occupied level is (2,2); its energy is $\boxed{4.26\times10^{-18}\text{ J}}$.

(b) The energy of the photon is equal to the difference in energy levels, in this case between levels (2,2) and (1,4):

(2) $\psi_{1,2}$

$n_x := 1$

$n_y := 2$

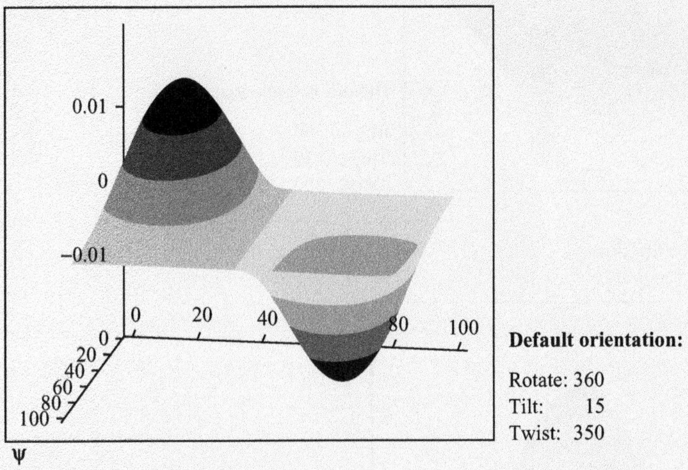

Figure 9.2(c) Wavefunction

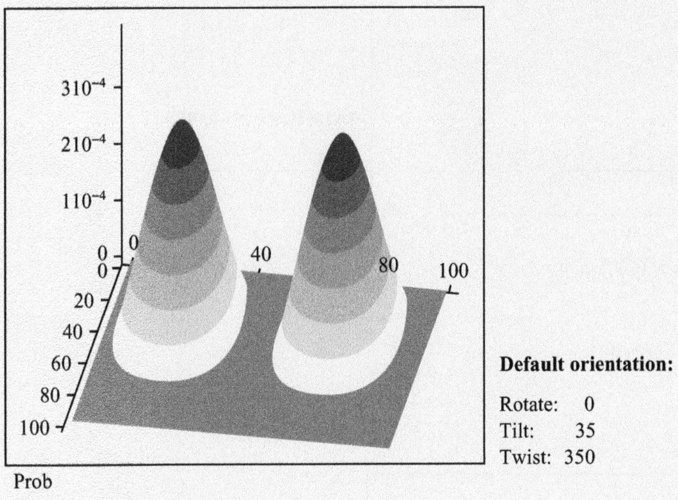

Figure 9.2(d) Probability density

$\psi_{2,1}$

$n_x := 2$

$n_y := 1$

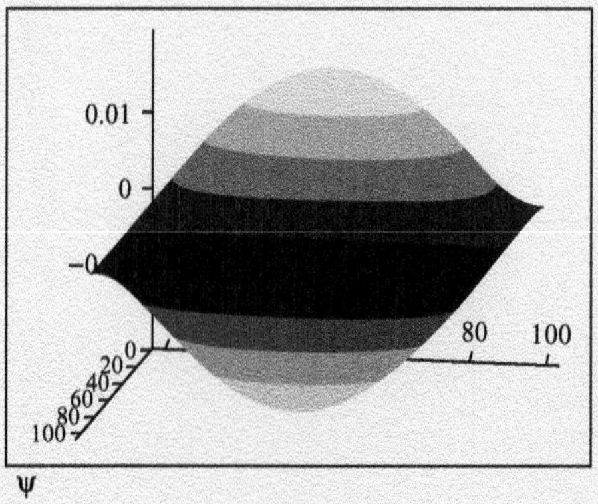

Default orientation:

Rotate: 360
Tilt: 15
Twist: 350

Figure 9.2(e) Wavefunction

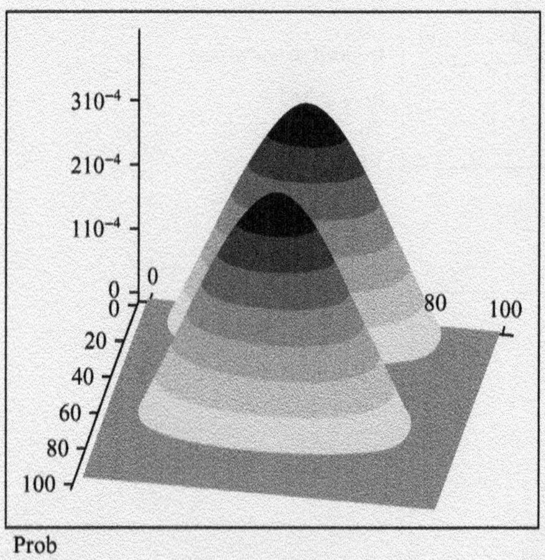

Default orientation:

Rotate: 0
Tilt: 35
Twist: 350

Figure 9.2(f) Probability density

$\psi_{2,1}$

$n_x := 2$

$n_y := 2$

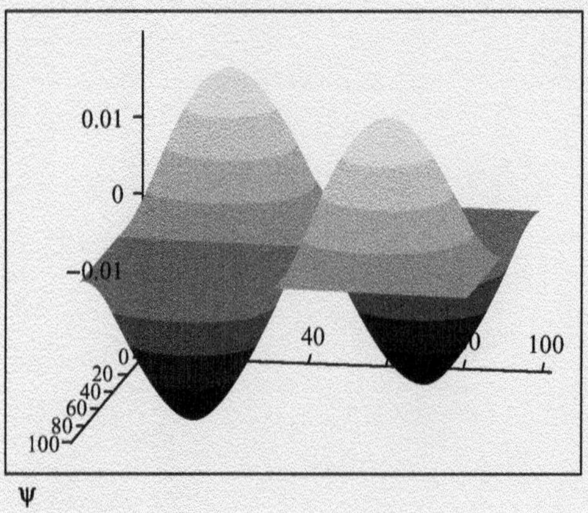

Default orientation:

Rotate: 360
Tilt: 15
Twist: 350

Figure 9.2(g) Wavefunction

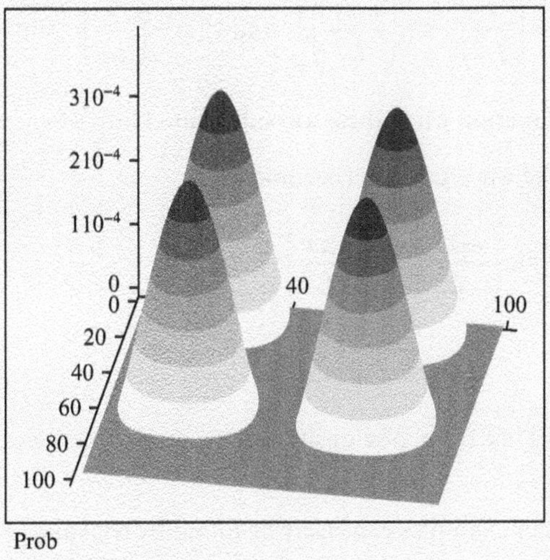

Figure 9.2(h) Probability density

E9.29 (a) The wavefunctions in question are solutions of the Schrödinger equation. The hamiltonian inside the box is

$$\hat{H} = -\frac{\hbar^2}{2m}\left(\frac{\partial^2}{\partial x^2} + \frac{\partial^2}{\partial y^2} + \frac{\partial^2}{\partial z^2}\right),$$

so the equation to be solved is

$$-\frac{\hbar^2}{2m}\left(\frac{\partial^2\psi(x,y,z)}{\partial x^2} + \frac{\partial^2\psi(x,y,z)}{\partial y^2} + \frac{\partial^2\psi(x,y,z)}{\partial z^2}\right) = E\psi(x,y,z).$$

We look for solutions that factor into functions of one variable. That is, let

$$\psi(x,y,z) = X(x) \times Y(y) \times Z(z),$$

and see under what conditions (if any) the differential equation can be solved. In substituting our trial separated solution into the differential equation, note that each partial derivative acts only on one factor:

$$-\frac{\hbar^2}{2m}\left(Y(y)Z(z)\frac{d^2X(x)}{dx^2} + X(x)Z(z)\frac{d^2Y(y)}{dy^2} + X(x)Y(y)\frac{d^2Z(z)}{dz^2}\right) = EX(x)Y(y)Z(z)$$

Divide both sides of the equation by the trial wavefunction and by $-h^2/2m$:

$$\frac{1}{X(x)}\frac{d^2X(x)}{dx^2} + \frac{1}{Y(y)}\frac{d^2Y(y)}{dy^2} + \frac{1}{Z(z)}\frac{d^2Z(z)}{dz^2} = -\frac{2mE}{\hbar^2}.$$

The right side of the equation is a constant, while the left side is a sum of three terms, each of which depends only on one of the three independent variables. Because the three terms depend on independent variables, the only way that they can sum to a constant value is if *each* is equal to a constant value. Thus,

$$\frac{1}{X(x)}\frac{d^2X(x)}{dx^2} = -\frac{2mE_X}{\hbar^2},\quad \frac{1}{Y(y)}\frac{d^2Y(y)}{dy^2} = -\frac{2mE_Y}{\hbar^2},\ \text{and}\ \frac{1}{Z(z)}\frac{d^2Z(z)}{dz^2} = -\frac{2mE_Z}{\hbar^2},$$

where $E_X + E_Y + E_Z = E$. Each of these equations rearranges to an ordinary differential equation, the Schrödinger equation for a one-dimensional particle in a box. Thus, each factor in the three-dimensional wavefunction has the form given in eqn 9.7:

$$X(x) = \left(\frac{2}{L_1}\right)^{1/2} \sin\left(\frac{n_1\pi x}{L_1}\right), \quad Y(y) = \left(\frac{2}{L_2}\right)^{1/2} \sin\left(\frac{n_2\pi y}{L_2}\right), \quad \text{and } Z(z) = \left(\frac{2}{L_3}\right)^{1/2} \sin\left(\frac{n_3\pi z}{L_3}\right),$$

yielding the three-dimensional wavefunction when these are substituted into $\psi(x, y, z)$ above.

(b) For a cubical box with sides L the wavefunctions become

$$\psi_{n_1,n_2,n_3}(x, y, z) = \left(\frac{8}{L^3}\right)^{1/2} \sin\left(\frac{n_1\pi x}{L}\right) \sin\left(\frac{n_2\pi x}{L}\right) \sin\left(\frac{n_3\pi x}{L}\right)$$

and the energies become

$E_{n_1,n_2,n_3} = (n_1^2 + n_2^2 + n_3^2) \times \left(\dfrac{h^2}{8mL^2}\right)$. This form obviously admits of degeneracies, for example, $E_{2,1,1} = E_{1,2,1} = E_{1,1,2} = \dfrac{6h^2}{8mL^2}$. Many other examples can easily be found by trial and error.

E9.30 (a) $I = m_{\rm H}r^2$ [text Section 9.5(a)]

$$= (1.008\ \text{u}) \times (1.6605 \times 10^{-27}\ \text{kg/u}) \times (161 \times 10^{-12}\ \text{m})^2$$

$$= \boxed{4.34 \times 10^{-47}\ \text{kg m}^2}$$

(b) $E_{m_l} = \dfrac{m_l^2\hbar^2}{2I}$ [9.22], where $m_l = 0, \pm1, \pm2, \ldots$

$$\Delta E = E_1 - E_0 = \frac{h^2}{8\pi^2 I} \quad \text{and} \quad \Delta E = h\nu = \frac{hc}{\lambda}$$

$$\lambda = \frac{hc}{\Delta E} = \frac{8\pi^2 cI}{h}$$

$$= \frac{8\pi^2 \times (2.998 \times 10^8\ \text{m s}^{-1}) \times (4.34 \times 10^{-47}\ \text{kg m}^2)}{6.626 \times 10^{-34}\ \text{J s}}$$

$$= 1.55 \times 10^{-3}\ \text{m} = \boxed{1.55\ \text{mm}}$$

This wavelength is in the microwave region of the electromagnetic spectrum.

E9.31 (a) $\nu = \dfrac{1}{2\pi}\left(\dfrac{k_{\rm f}}{m}\right)^{1/2}$ [9.29]

$$= \frac{1}{2\pi}\left(\frac{314\ \text{N m}^{-1}}{(1.0079\ \text{u}) \times (1.6605 \times 10^{-27}\ \text{kg u}^{-1})}\right)^{1/2} = \boxed{6.89 \times 10^{13}\ \text{s}^{-1}}$$

(b) $\lambda = \dfrac{c}{\nu} = \dfrac{2.998 \times 10^8\ \text{m s}^{-1}}{6.89 \times 10^{13}\ \text{s}^{-1}} = 4.35 \times 10^{-6}\ \text{m} = \boxed{4.35\ \mu\text{m}}$

(c) The D–I bond and the H–I bond are expected to have almost identical bond strengths and identical bonding force constants, k, because bonding is an electronic, not a mass/isotopic, property. However, the vibrational frequency does have a mass dependence.

$$\nu = \frac{1}{2\pi}\left(\frac{k_{\rm f}}{m}\right)^{1/2} \quad [9.29]$$

$$\frac{\nu_{\mathrm{DI}}}{\nu_{\mathrm{HI}}}=\frac{\dfrac{1}{2\pi}\left(\dfrac{k_{\mathrm{f}}}{m_{\mathrm{D}}}\right)^{1/2}}{\dfrac{1}{2\pi}\left(\dfrac{k_{\mathrm{f}}}{m_{\mathrm{H}}}\right)^{1/2}}=\left(\frac{m_{\mathrm{H}}}{m_{\mathrm{D}}}\right)^{1/2}=\left(\frac{1}{2}\right)^{1/2}=0.707$$

When hydrogen-1 is replaced by hydrogen-2 (deuterium) in H–I, the vibrational frequency decreases by a factor of 0.707.

E9.32 (a) $\psi = N\mathrm{e}^{-ax^2/2} \quad -\infty < x < \infty$

The normalization constant is found as follows:

$$N^2\int_{-\infty}^{\infty}(\mathrm{e}^{-ax^2/2})^2\,\mathrm{d}x = 1 \quad \text{(See Justification 9.1 for a similar example)}$$

$$N^2\int_{-\infty}^{\infty}\mathrm{e}^{-ax^2}\,\mathrm{d}x = 1$$

Using the standard, definite integral $\int_{-\infty}^{\infty}\mathrm{e}^{-ax^2}\,\mathrm{d}x = \left(\dfrac{\pi}{a}\right)^{1/2}$, we find that

$$N^2\left(\frac{\pi}{a}\right)^{1/2} = 1 \quad \text{or} \quad \boxed{N=\left(\frac{a}{\pi}\right)^{1/4}}$$

The normalized wavefunction is $\psi = \left(\dfrac{a}{\pi}\right)^{1/4}\mathrm{e}^{-ax^2/2}$.

(b) The function $\psi = \left(\dfrac{a}{\pi}\right)^{1/4}\mathrm{e}^{-ax^2/2}$ is a "bell" or "Gaussian" function with a maximum at $x = 0$.

Consequently, ψ^2 has a maximum value at the displacement $x = 0$ and the Born interpretation of ψ^2 (see Section 9.2(b)) indicates that the displacement $x = 0$ is the most probable displacement. It is instructive to use analytic geometry to find the maximum. This requires identification of the displacement for which $\mathrm{d}\psi/\mathrm{d}x = 0$ and showing that $\mathrm{d}\psi/\mathrm{d}x > 0$ before the maximum and $\mathrm{d}\psi/\mathrm{d}x < 0$ after the maximum

$$\frac{\mathrm{d}\psi}{\mathrm{d}x}=\frac{\mathrm{d}}{\mathrm{d}x}\left\{\left(\frac{a}{\pi}\right)^{1/4}\mathrm{e}^{-ax^2/2}\right\}=\left(\frac{a}{\pi}\right)^{1/4}\frac{\mathrm{d}}{\mathrm{d}x}\mathrm{e}^{-ax^2/2}=-\left(\frac{a}{\pi}\right)^{1/4}ax\,\mathrm{e}^{-ax^2/2}$$

The factor of x in the first derivative indicates that the derivative equals zero when $x = 0$. Furthermore, the formula for the derivative clearly shows that the derivative is positive when $x < 0$ and the derivative is negative when $x > 0$. The function is a maximum at $x = 0$.

E9.33 $\nu = \dfrac{1}{2\pi}\left(\dfrac{k_{\mathrm{f}}}{\mu}\right)^{1/2}$ [9.29], where $\mu = m_{\mathrm{A}}m_{\mathrm{B}}/(m_{\mathrm{A}}+m_{\mathrm{B}}) = M_{\mathrm{A}}M_{\mathrm{B}}/\{N_{\mathrm{A}}(M_{\mathrm{A}}+M_{\mathrm{B}})\}$

(a) $\mu_{^{12}\mathrm{C}^{16}\mathrm{O}} = \dfrac{(12.00)\times(16.00)\times10^{-3}\,\mathrm{kg}}{(6.0221\times10^{23})\times(28.00)} = 1.139\times10^{-26}\,\mathrm{kg}$

$$\nu_{^{12}\mathrm{C}^{16}\mathrm{O}} = \frac{1}{2\pi}\left(\frac{1860\,\mathrm{N\,m^{-1}}}{1.139\times10^{-26}\,\mathrm{kg}}\right)^{1/2} = \boxed{6.432\times10^{13}\,\mathrm{s^{-1}}}$$

(b) $\tilde{\nu}_{^{12}C^{16}O} = 1/\lambda_{^{12}C^{16}O} = \dfrac{\nu_{^{12}C^{16}O}}{c}$

$$= \frac{6.432 \times 10^{13}\ \text{s}^{-1}}{2.9979 \times 10^{8}\ \text{m s}^{-1}} = 2.146 \times 10^{5}\ \text{m}^{-1} = \boxed{2146\ \text{cm}^{-1}}$$

(c) Computations like those of part (a) and part (b) can be repeated for additional isotopes of carbon monoxide. We will take a short-cut by recognizing that these isotopic CO molecules have identical force constants and differ only in their reduced mass. Consequently, since both the frequency and wavenumber are inversely proportional to $\mu^{1/2}$, we write

$$\mu_{^{13}C^{16}O} = \frac{(13.00) \times (16.00) \times 10^{-3}\ \text{kg}}{(6.0221 \times 10^{23}) \times (29.00)} = 1.191 \times 10^{-26}\ \text{kg}$$

$$\tilde{\nu}_{^{13}C^{16}O} = \left(\frac{\mu_{^{12}C^{16}O}}{\mu_{^{13}C^{16}O}}\right)^{1/2} \tilde{\nu}_{^{12}C^{16}O}$$

$$= \left(\frac{1.139}{1.191}\right)^{1/2} \times (2146\ \text{cm}^{-1})$$

$$= \boxed{2099\ \text{cm}^{-1}}$$

Similarly,

$$\mu_{^{12}C^{18}O} = 1.196 \times 10^{-26}\ \text{kg} \quad \text{and} \quad \boxed{\tilde{\nu}_{^{12}C^{18}O} = 2094\ \text{cm}^{-1}}$$

$$\mu_{^{13}C^{18}O} = 1.253 \times 10^{-26}\ \text{kg} \quad \text{and} \quad \boxed{\tilde{\nu}_{^{13}C^{18}O} = 2046\ \text{cm}^{-1}}$$

E9.34 The reduced mass is nearly identical for hydrogenic atoms so the Rydberg constant for all of them may be estimated to equal R_H. Also, the ionization energy is equivalent to the energy of the transition $n = 1 \rightarrow n = \infty$ so we write

$$I = E_{n=\infty} - E_{n=1} = -hcR_H Z^2\left(\frac{1}{\infty^2} - \frac{1}{1^2}\right) = hcR_H Z^2$$

Since $I_{He^+} = 4hcR_H$, we may write $I_{Li^{2+}} = 9hcR_H = \frac{9}{4}I_{He^+}$. With the ionization of He^+ given as 54.36 eV, the ionization energy of Li^{2+} is $\frac{9}{4} \times (54.36\ \text{eV}) = \boxed{122.31\ \text{eV}}$.

E9.35 $n = 4$ for the N shell

$n^2 = 4^2 = \boxed{16\ \text{orbitals}}$

E9.36 (a) The probability density varies as

$$\psi^2 = \frac{1}{\pi a_0^3} e^{-2r/a_0}$$

The maximum value is at $r = 0$ and ψ^2 is 25 per cent of the maximum when $e^{-2r/a_0} = 0.25$, so that $r = 1/2\ a_0 \ln(0.25)$, which is at $\boxed{r = 0.693\ a_0}$, which corresponds to 36.7 pm.

(b) The radial distribution function varies as

$$P = 4\pi r^2\psi^2 = \frac{4r^2}{a_0^3} e^{-2r/a_0}$$

The maximum value of P occurs at $r = a_0$ because

$$\frac{\mathrm{d}P}{\mathrm{d}r} \propto \left(2r - \frac{2r^2}{a_0}\right)\mathrm{e}^{-2r/a_0} = \text{at } r = a_0 \text{ and } P_{\max} = \frac{4}{a_0}\mathrm{e}^{-2}$$

P falls to a fraction f of its maximum when

$$f = \frac{\dfrac{4r^2}{a_0^3}\mathrm{e}^{-2r/a_0}}{\dfrac{4}{a_0}e^{-2}} = \frac{r^2}{a_0^2}\mathrm{e}^2\mathrm{e}^{-2r/a_0}$$

Therefore, solve

$$\frac{f^{1/2}}{\mathrm{e}} = \left(\frac{r}{a_0}\right)\mathrm{e}^{-r/a_0}$$

$$f = 0.25$$

solves to $r = 2.6783\ a_0$ or $0.2320\ a_0 = \boxed{142 \text{ pm or } 12 \text{ pm}}$

(c) The most probable distance of a 1s electron from the nucleus occurs when the first derivative of the radial distribution function equals zero.

$$P_{1s} = 4\pi r^2\psi_{1s}^2\ [13.8a] = 4\pi r^2\,(N\mathrm{e}^{-r/a_0})^2\ [13.7] = 4\pi N^2(r^2\mathrm{e}^{-2r/a_0})$$

$$\frac{\mathrm{d}P_{1s}}{\mathrm{d}r} = 4\pi N^2\frac{\mathrm{d}(r^2\mathrm{e}^{-2r/a_0})}{\mathrm{d}r} = 4\pi N^2\left\{2r\mathrm{e}^{-2r/a_0} + r^2\left(-\frac{2}{a_0}\mathrm{e}^{-2r/a_0}\right)\right\} = 8\pi N^2\left\{1 - \frac{r}{a_0}\right\}r\mathrm{e}^{-2r/a_0}$$

The derivative equals zero when the factor $1 - r/a_0$ equals zero. Therefore, $\boxed{r_{\max} = a_0}$.

E9.37 There are 2 lobes to a p-orbital so the probability that a p-orbital electron will be found in one or the other lobe is $\boxed{\frac{1}{2}}$. However, there are 3 degenerate orbitals in a p subshell so the probability of finding a p subshell electron in one or another p-orbital lobe of the subshell is $\frac{1}{6}$.

E9.38 (a) We will make the assumption that ψ^2 is a constant within this very small volume. Then, Probability $= \int\psi^2(r)\delta V \approx \psi^2\delta V$ with $\delta V = 1.0\ \mathrm{pm}^3$

$$\psi^2 = \frac{1}{32\pi a_0^3}\left(2 - \frac{r}{a_0}\right)^2\mathrm{e}^{-r/a_0} = 6.72\times10^{-8}\ \mathrm{pm}^{-3}\left(2 - \frac{r}{a_0}\right)^2\mathrm{e}^{-r/a_0}$$

(i) $\psi^2 = 6.72\times10^{-8}\ \mathrm{pm}^{-3}\times2^2\times1 = 2.7\times10^{-7}\ \mathrm{pm}^{-3}$

$\psi^2\delta V = \boxed{2.7\times10^{-7}}$

(ii) $\psi^2 = 6.72\times10^{-8}\ \mathrm{pm}^{-3}\times1\times\mathrm{e}^{-1} = 2.47\times10^{-8}\ \mathrm{pm}^3$

$\psi^2\delta V = \boxed{2.5\times10^{-8}}$

(iii) $\psi^2 = 0$, $\psi^2\delta V = \boxed{0}$. This is a radial node.

(b) The most probable distance of a 2*s* electron from the nucleus may be determined by plotting the radial distribution function against r/a_0 and using the trace function of the plotting software to evaluate the coordinates of the maximum. The following function is plotted in Figure 9.3. The plot reveals that $\boxed{r_{max} = 5.235\ a_0}$.

$$P_{2s} = 4\pi r^2 \psi_{2s}^2 \,[13.7a] = 4\pi r^2 \left\{ \left(\frac{1}{32\pi a_0^3} \right)^{1/2} \left(2 - \frac{r}{a_0} \right) e^{-r/2a_0} \right\}^2$$

$$= \frac{1}{8a_0^3} \left(2 - \frac{r}{a_0} \right)^2 r^2 e^{-r/a_0} = \boxed{\frac{1}{8a_0}(2-x)^2 x^2 e^{-x} \quad \text{where} \quad x = r/a_0}$$

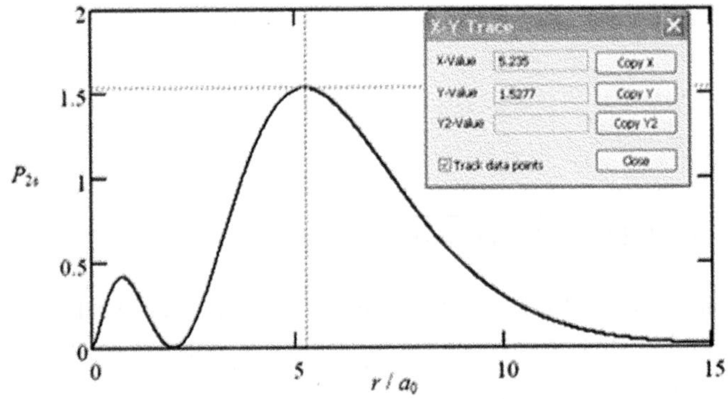

Figure 9.3

(c) Let $x = r/a_0$, then

$$P_{2s} = \frac{1}{8a_0}\{x^2(2-x)^2 e^{-x}\} = \frac{1}{8a_0}(4x^2 - 4x^3 + x^4)e^{-x}$$

$$\frac{dP_{2s}}{dr} = \frac{dx}{dr}\frac{dP_{2s}}{dx} = \frac{1}{8a_0^2}\frac{d\{(4x^2-4x^3+x^4)e^{-x}\}}{dx}$$

$$= \frac{1}{8a_0^2}\{(8x - 12x^2 + 4x^3)e^{-x} + (4x^2 - 4x^3 + x^4)(-e^{-x})\} = \frac{1}{8a_0^2}\{8 - 16x + 8x^2 - x^3\}xe^{-x}$$

$$= -\frac{1}{8a_0^2} \times (x-2) \times (x^2 - 6x + 4)xe^{-x}$$

The derivative equals zero when $x = r/a_0 = 0, 3 - 5^{1/2}, 2, 3 + 5^{1/2}$, and ∞. These correspond to the radial distribution function being a minimum, a maximum, a minimum, a maximum, and a minimum, respectively. The following ratio identifies the maximum that is most probable.

$$\frac{P_{2s}(x = 3 + 5^{1/2})}{P_{2s}(x = 3 - 5^{1/2})} = \frac{[(4x^2 - 4x^3 + x^4)e^{-x}]_{x=3+5^{1/2}}}{[(4x^2 - 4x^3 + x^4)e^{-x}]_{x=3-5^{1/2}}} = \frac{1.528}{0.415} = 3.68$$

The most probable distance is $\boxed{(3 + 5^{1/2})a_0}$.

E9.39 There are two methods that can be used to locate the radial nodes of a hydrogen atom orbital. We could find a textbook plot of the radial function $R_{n,l}$ against r/a_0 and read the node values directly from the plot. Alternatively, we could search an advanced textbook on physical chemistry to find the mathematical form for $R_{n,l}$ and analyze the function for its nodes.

(a) Figure 9.4 is a typical plot for the 3s hydrogen orbital. The nodes occur when the radial function passes through zero. These occur when $r = \boxed{1.9\ a_0 \text{ and } 7.1\ a_0}$.

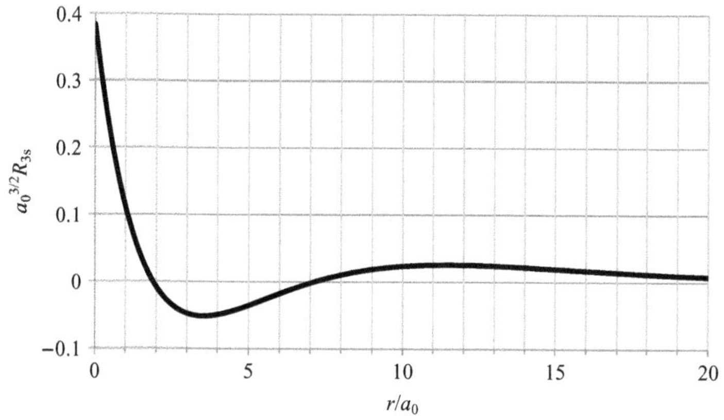

Figure 9.4

(b) Finding the nodes of the 4s orbital of a hydrogen atom is a greater challenge. Few textbooks display a plot of this radial wavefunction so it becomes necessary to find the mathematical form of this function. Eqn 10.14 of P. Atkins and J. de Paula, *Physical Chemistry*, 8th edn, Freeman, 2006 reports that the radial function is

$$R_{n,l} = N_{n,l}\rho^l L_{n+1}^{2l+1}(\rho)e^{-\rho/2} \quad \text{where} \quad \rho = \frac{2r}{na_0}$$

$L_{n+1}^{2l+1}(\rho)$ is the associated Laguerre polynomial and examination of the above radial function indicates that the radial nodes occur when $L_{n+1}^{2l+1}(\rho)$ equals zero. Thus, we must solve for the zeros of the associated Laguerre polynomial for the 4s orbital. L. Pauling and E. Wilson Jr., *Introduction to Quantum Mechanics with Applications to Chemistry*, McGraw-Hill, New York (1935) report the following form for the associated Laguerre polynomial.

$$\frac{L_{n+1}^{2l+1}(\rho)}{\{(n+l)!\}^2} = \sum_{k=0}^{n-l-1} \frac{(-1)^{k+1}\rho^k}{(n-l-1-k)!(2l+1+k)!k!} \quad \text{where } 0! = 1 \text{ and } m! = m\times(m-1)\times(m-2)\times\cdots\times 1$$

For the 4s orbital we use $n = 4$, $l = 0$, and $\rho = \dfrac{r}{2a_0}$.

$$\frac{L_5^1(\rho)}{\{4!\}^2} = \sum_{k=0}^{3} \frac{(-1)^{k+1}\rho^k}{(3-k)!(1+k)!k!}$$

$$= \frac{(-1)^1\rho^0}{(3)!(1)!0!} + \frac{(-1)^2\rho^1}{(2)!(2)!1!} + \frac{(-1)^3\rho^2}{(1)!(3)!2!} + \frac{(-1)^4\rho^3}{(0)!(4)!3!}$$

$$= -\frac{1}{6} + \frac{\rho}{4} - \frac{\rho^2}{12} + \frac{\rho^3}{144} = \frac{1}{144}(\rho^3 - 12\rho^2 + 36\rho - 24)$$

Thus, the nodes of the 4s orbital occur when

$$\rho^3 - 12\rho^2 + 36\rho - 24 = 0$$

The roots of this polynomial can be found with either the root function or the numeric solver of the modern scientific calculator. The roots found are $\rho = 0.936$, 3.305, and 7.759. Since $r = 2a_0\rho$, the nodes are at

$$r = \boxed{1.87\ a_0,\ 6.61\ a_0,\ \text{and}\ 15.5\ a_0}$$

The plot of R_{4s} against r/a_0, shown in Figure 9.5, confirms this result.

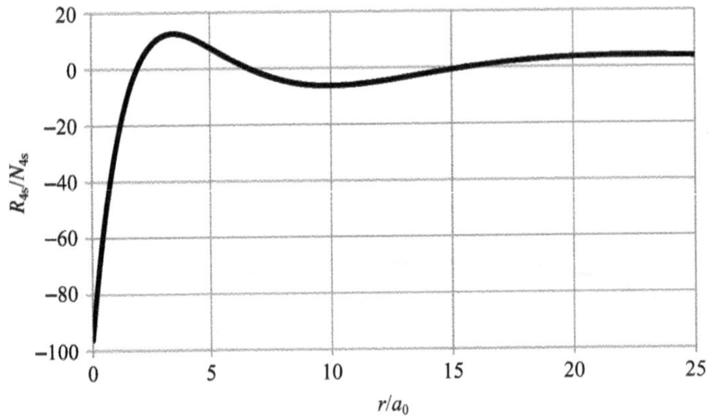

Figure 9.5

E9.40 Look for values of θ for which sin θ or cos θ go to zero. Sin θ goes to zero at $\theta = \boxed{0° \text{ and } 180°}$; cos θ at $\boxed{90° \text{ and } 270°}$.

E9.41 Identify l and use angular momentum $= \{l(l+l)\}^{1/2}\hbar$

(a) $l = 0$, so $\boxed{\text{ang. mom.} = 0}$

(b) $l = 0$, so $\boxed{\text{ang. mom.} = 0}$

(c) $l = 2$, so $\boxed{\text{ang. mom.} = \sqrt{6}\,\hbar}$

(d) $l = 1$, so $\boxed{\text{ang. mom.} = \sqrt{2}\,\hbar}$

(e) $l = 1$, so $\boxed{\text{ang. mom.} = \sqrt{2}\,\hbar}$

The total number of nodes is equal to $n - 1$, and the number of angular nodes is equal to l; hence the number of radial nodes is equal to $n - 1 - l$. We can draw up the following table:

	1*s*	3*s*	3*d*	2*p*	3*p*
n, l	1,0	3,0	3,2	2,1	3,1
Ang. nodes	0	0	2	1	1
Rad. nodes	0	2	0	0	1

E9.42 For a given l there are $2l + 1$ values of m_l and hence $2l + 1$ orbitals. Each orbital may be occupied by two electrons. Therefore, the maximum occupancy is $2(2l + 1)$.

	l	$2(2l+1)$
(a)	0	2
(b)	3	14
(c)	5	22

E9.43 (a) Periodic table

1*s*	H	He										
2*s*	Li	Be	B	C	N	O	F	Ne	Na	Mg	2*p*
3s	Al	Si	P	S	Cl	Ar	K	Ca	Sc	Ti	3*p*
4s	V	Cr								↑	

(b) "Noble gases"

(c) Probably N because it has 5 electrons in its valence shell which corresponds to half-filled, like carbon.

E9.44 The electron configurations of the ions are:

Fe^{2+} [Ar]$3d^6$

Fe^{3+} [Ar]$3d^5$

Fe^{2+} is expected to be larger. The reason is the repulsion between electrons in the same outer sub-shell. The more electrons, the greater the overall repulsions, leading to a greater average distance from the nucleus for the outer electrons and hence increased ionic size.

E9.45 See Figure 9.6 for the ionization trends of Group 13.

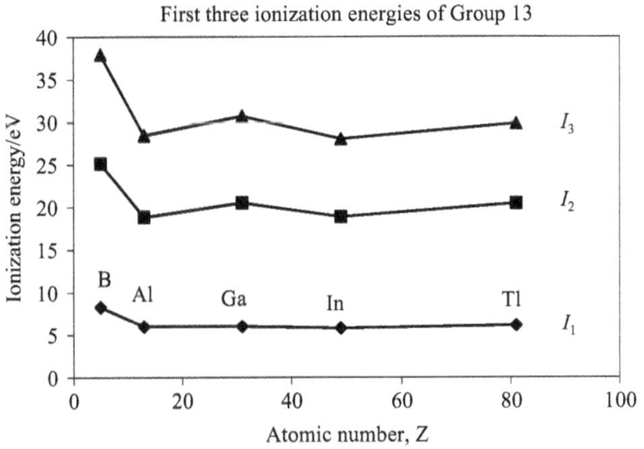

Figure 9.6

Trends:

(i) $I_1 < I_2 < I_3$ because of decreased nuclear shielding as each successive electron is removed.

(ii) The ionization energies of boron are much larger than those of the remaining group elements because the valence shell of boron is very small and compact with little nuclear shielding. The boron atom is much smaller than the aluminum atom.

(iii) The ionization energies of Al, Ga, In, and Tl are comparable, even though successive valence shells are further from the nucleus because the ionization energy decrease expected from large atomic radii is balanced by an increase in effective nuclear charge.

E9.46 By analogy with equation [9.37] the ionization energy of an anion, I_1^-, is for the reaction

$$X^-(g) \rightarrow X(g) + e^-(g) \quad I_1^- = E(X) - E(X^-)$$

while the electron affinity of equation [9.38] is for the reaction

$$X(g) + e^-(g) \rightarrow X^-(g) \quad E_{ea} = -\{E(X^-) - E(X)\} = E(X) - E(X^-)$$

The above expression for the electron affinity E_{ea} is $-\{E(X^-) - E(X)\}$ because the **electron affinity is the energy released** when an electron attaches to a gas-phase atom. Inspection of the expressions for $I_{\bar{1}}$ and E_{ea} indicates that $\boxed{I_{\bar{1}} = E_{ea}}$. However, the reaction equation for the electron affinity is the reverse of the ionization reaction of X^-.

E9.47 The increasing magnitude of the equilibrium constant for the binding of the cations to Lewis bases in going from Ba^{2+} to Mg^{2+} parallels the decreasing size of the cations from Ba^{2+} to Mg^{2+}. Because the nature of the interaction is coulombic its magnitude increases with decreasing distance between the ions, resulting in larger equilibrium constants for the binding.

Solutions to projects

P9.48 (a) In effect, we are looking for the vibrational frequency of an O atom bound, with a force constant equal to that of free CO, to an infinitely massive and immobile protein complex. The angular frequency is

$$\omega = \left(\frac{k_f}{m}\right)^{1/2},$$

where m is the mass of the O atom

$$m = (16.0\text{u})(1.66 \times 10^{-27}\ \text{kg u}^{-1}) = 2.66 \times 10^{-26}\ \text{kg},$$

and k is the same force constant as in exercise 9.33, namely 1860 N m^{-1}:

$$\omega = \left(\frac{1860\ \text{N m}^{-1}}{2.66 \times 10^{-26}\ \text{kg}}\right)^{1/2} = \boxed{2.64 \times 10^{14}\ \text{s}^{-1}}.$$

(b) Assuming that one can identify the CO peak in the infrared spectrum of the CO–myoglobin complex, taking infrared spectra of each of the isotopic variants of CO–myoglobin complexes can show which atom binds to the heme group and determine the C≡O force constant. Compare isotopic variants to $^{12}C^{16}O$ as the standard; when an isotope changes but the vibrational frequency does not, then the atom whose isotope was varied is the atom that binds to the heme. See table below, which includes predictions of the wavenumber of all isotopic variants compared to that of $\tilde{\nu}(^{12}C^{16}O)$. (As usual, the better the experimental results agree with the whole set of predictions, the more confidence one would have in the conclusion.)

Wavenumber for isotopic variant	If O binds	If C binds
$\tilde{\nu}(^{12}C^{18}O) =$	$\tilde{\nu}(^{12}C^{16}O)$†	$(16/18)^{1/2}\tilde{\nu}(^{12}C^{16}O)$
$\tilde{\nu}(^{13}C^{16}O) =$	$(12/13)^{1/2}\tilde{\nu}(^{12}C^{16}O)$	$\tilde{\nu}(^{12}C^{16}O)$†
$\tilde{\nu}(^{13}C^{18}O) =$	$(12/13)^{1/2}\tilde{\nu}(^{12}C^{16}O)$	$(16/18)^{1/2}\tilde{\nu}(^{12}C^{16}O)$

†That is, no change compared to the standard.

The wavenumber is related to the force constant as follows:

$$\omega = 2\pi c\tilde{\nu} = \left(\frac{k_f}{m}\right)^{1/2}, \quad \text{so} \quad k_f = m(2\pi c\tilde{\nu})^2,$$

$$k_f = m(1.66 \times 10^{-27}\ \text{kg u}^{-1})[(2\pi)(2.998 \times 10^{10}\ \text{cm s}^{-1})\tilde{\nu}(^{12}\text{C}^{16}\text{O})]^2,$$

and $k_f/(\text{kg s}^{-1}) = (5.89 \times 10^{-5})(m/\text{u})[\tilde{\nu}(^{12}\text{C}^{16}\text{O})/\text{cm}^{-1}]^2$.

Here, m is the mass of the atom that is not bound, i.e. 12 u if O is bound and 16 u if C is bound. (Of course, one can compute k from any of the isotopic variants, and take k to be a mean derived from all the relevant data.)

P9.49 (a) We are told to assume that the difference in activation energy, $\Delta E_a = E_a(\text{C–D}) - E_a(\text{C–H})$, is due only to the difference in zero-point energies for vibration. The activation energies are the differences between ground-state energy of the activated complex (the upper energy level, assumed to be small and unchanged upon deuteration) and that of the reactant (the lower energy level). Thus, the difference in activation energy is

$$\Delta E_a = E_a(\text{C–D}) - E_a(\text{C–H}) = E_0(\text{C–H}) - E_0(\text{C–D}).$$

Use eqn 9.29:

$$\Delta E_a = E_0(\text{C–H}) - E_0(\text{C–D}) = \tfrac{1}{2}\hbar\omega(\text{C–H}) - \tfrac{1}{2}\hbar\omega(\text{C–D}).$$

Since $\omega = 2\pi\nu = \left(\dfrac{k_f}{\mu}\right)^{1/2}$ [9.29],

with the same force constant for both isotopes, we can write

$$\omega(\text{C–D}) = \omega(\text{C–H})\left(\frac{\mu_{\text{CH}}}{\mu_{\text{CD}}}\right)^{1/2}$$

and $\Delta E_a = \tfrac{1}{2}\hbar\omega(\text{C–H})\left\{1 - \left(\dfrac{\mu_{\text{CH}}}{\mu_{\text{CD}}}\right)^{1/2}\right\}$.

Two more cosmetic substitutions give us the desired expression. First, note that the expressions for zero-point energy hold for single oscillators, while the activation energy is a molar quantity; thus, the activation energy involves Avogadro's number of oscillators. Secondly, we use wavenumber rather than frequency, using $\hbar\omega = hc\tilde{\nu}$. Therefore,

$$\Delta E_a = \boxed{\tfrac{1}{2}N_A hc\tilde{\nu}(\text{C–H})\left\{1 - \left(\frac{\mu_{\text{CH}}}{\mu_{\text{CD}}}\right)^{1/2}\right\}}.$$

(b) (i) $\dfrac{k_r(\text{C–D})}{k_r(\text{C–H})} = \dfrac{e^{-E_a(\text{C–D})/RT}}{e^{-E_a(\text{C–H})/RT}} = e^{-\Delta E_a/RT} = e^{-\lambda}$

where $\lambda = \dfrac{N_A hc\tilde{\nu}(\text{C–H})}{2RT}\left\{1 - \left(\dfrac{\mu_{\text{CH}}}{\mu_{\text{CD}}}\right)^{1/2}\right\} = \boxed{\dfrac{hc\tilde{\nu}(\text{C–H})}{2kT}\left\{1 - \left(\dfrac{\mu_{\text{CH}}}{\mu_{\text{CD}}}\right)^{1/2}\right\}}$.

(ii) To determine how the ratio of rate constants varies with temperature, note that $\lambda > 0$. We know this because $\mu_{\text{CH}} < \mu_{\text{CD}}$. (Recall that $\mu_{\text{AB}} = \dfrac{m_A m_B}{m_A + m_B}$. In the limit where one mass is considerably lower than the other, μ approximates the lesser mass; thus $\mu_{\text{CH}} \approx m_H$, $\mu_{\text{CD}} \approx m_D = 2m_H$, and $\mu_{\text{CH}} < \mu_{\text{CD}}$.) As T decreases, λ increases, and $e^{-\lambda}$ decreases; thus, the ratio $\boxed{\text{decreases}}$.

(c) $$\lambda = \frac{hc\tilde{\nu}(\text{C–H})}{2kT}\left\{1-\left(\frac{\mu_{\text{CH}}}{\mu_{\text{CD}}}\right)^{1/2}\right\}$$

$$= \frac{6.626\times10^{-34}\ \text{J s}\times2.998\times10^{10}\ \text{cm s}^{-1}\times3000\ \text{cm}^{-1}}{2\times1.381\times10^{-23}\ \text{J K}^{-1}\times298\ \text{K}}\left\{1-\left(\frac{m_{\text{H}}}{2m_{\text{H}}}\right)^{1/2}\right\} = 2.12.$$

The ratio is $e^{-\lambda} = e^{-2.12} = \boxed{0.12}$.

(d) Sometimes the ratio $k(\text{C–D})/k(\text{C–H})$ is even lower than this model would account for. Tunneling can explain this discrepancy. The C–H rate constant can be greater than expected because of tunneling; that is, tunneling can account for extension of the C–H bond beyond the classical limit with a non-zero probability. The smaller the mass, the greater the effect of tunneling. Thus, tunneling would enhance the rate of C–H scission preferentially to that of C–D scission.

10 The chemical bond

Answers to discussion questions

D10.1 Our comparison of the two theories will focus on the manner of construction of the trial wavefunctions for the hydrogen molecule in the simplest versions of both theories. In the valence-bond method, the trial function is a linear combination of two simple product wavefunctions, in which one electron resides totally in an atomic orbital (AO) on atom A, and the other totally in an orbital on atom B. See text Figure 10.2. There is no contribution to the wavefunction from products in which both electrons reside on either atom A or B.

$$\psi_{\text{H-H}}(1,2) = \psi_{\text{A}}(1)\psi_{\text{B}}(2) + \psi_{\text{A}}(2)\psi_{\text{B}}(1) \quad [10.1]$$

So the valence-bond approach undervalues, by totally neglecting, any ionic contribution to the trial function. It is a totally covalent function.

The modern one-electron molecular orbital (MO) extends throughout the molecule and is written as a linear combination of atomic orbitals (LCAO).

$$\psi_{\text{MO}}(1) = c_{\text{A}}\psi_{\text{A}}(1) + c_{\text{B}}\psi_{\text{B}}(1) \quad [10.4\text{a}]$$

The squares of the coefficients give the relative proportions of the AO contributing to the MO.

The two-electron molecular orbital function for the hydrogen molecule is a product of two one-electron MOs. That is

$$\begin{aligned}\psi &= [c_{\text{A}}\psi_{\text{A}}(1) + c_{\text{B}}\psi_{\text{B}}(1)] \times [c_{\text{A}}\psi_{\text{A}}(2) + c_{\text{B}}\psi_{\text{B}}(2)] \\ &= c_{\text{A}}^2\psi_{\text{A}}(1)\psi_{\text{A}}(2) + c_{\text{B}}^2\psi_{\text{B}}(1)\psi_{\text{B}}(2) + c_{\text{A}}c_{\text{B}}\psi_{\text{A}}(1)\psi_{\text{B}}(2) + c_{\text{A}}c_{\text{B}}\psi_{\text{A}}(2)\psi_{\text{B}}(1)\end{aligned}$$

The first two terms are ionic forms for which both electrons are on either atom A or on atom B. The molecular orbital approach greatly overvalues the ionic contributions. At these crude levels of approximation, the valence-bond method gives dissociation energies closer to the experimental values. However, more sophisticated versions of the molecular orbital approach are the method of choice for obtaining quantitative results on both diatomic and polyatomic molecules.

D10.2 Consider the case of the carbon atom. Mentally we break the process of hybridization into two major steps. The first is promotion, in which we imagine that one of the electrons in the 2s orbital of carbon ($2s^22p^2$) is promoted to the empty 2p orbital giving the configuration $2s2p^3$. In the second step we mathematically mix the four orbitals by way of the specific linear combinations in eqns 10.2a corresponding to the sp^3 hybrid orbitals. There is a principle of conservation of orbitals that enters here. If we mix four unhybridized atomic orbitals we must end up with four hybrid orbitals. In the construction of the sp^2 hybrids we start with the 2s orbital and two of the 2p orbitals, and after mixing we end up with three sp^2 hybrid orbitals. In the sp case we start with the 2s orbital and one of the 2p orbitals. Equations 10.2b and 10.2c present the LCAO for the sp^2 and sp hybrid orbitals; text Figures 10.7 and 10.9 describe the orientations of sp^3 and sp^2 hybrid orbitals. The justification for all of this is in a sense the First Law of Thermodynamics. Energy is a state function and therefore its value is determined only by the final state of the system, not by the path taken to achieve that state, and the path can even be imaginary.

D10.3 Both the Pauling and Mulliken methods for measuring the attracting power of atoms for electrons seem to make good chemical sense.

Pauling electronegativity scale:

$$|\chi_A - \chi_B| = 0.102 \times (\Delta E/\text{kJ mol}^{-1})^{1/2} \quad [10.8a], \text{ where } \Delta E = E(\text{A—B}) - \tfrac{1}{2}\{E(\text{A—A}) + E(\text{B—B})\}$$

Mulliken electronegativity scale:

$$\chi = \tfrac{1}{2}(I + E_{ea}) \quad [10.8c]$$

If we look at the Pauling scale, we see that if E(A—B) were equal to $\frac{1}{2}[E(\text{A—A}) + E(\text{B—B})]$ the calculated electronegativity difference would be zero, as expected for completely non-polar bonds. Hence, any increased strength of the A—B bond over the average of the A—A and B—B bonds, can reasonably be thought of as being due to the polarity of the A—B bond, which in turn is due to the difference in electronegativity of the atoms involved. Therefore, this difference in bond strengths can be used as a measure of electronegativity difference. To obtain numerical values for individual atoms, a reference state (atom) for electronegativity must be established. The value for fluorine is arbitrarily set at 4.0.

The Mulliken scale may be more intuitive than the Pauling scale because we are used to thinking of ionization energies and electron affinities as measures of the electron attracting powers of atoms. The choice of factor $\frac{1}{2}$, however, is arbitrary, though reasonable, and no more arbitrary than the specific form that defines the Pauling scale.

D10.4 The ground electronic configurations of the valence electrons are found in text Figures 10.30, 10.31, 10.35.

N_2	$1\sigma_g^2 1\sigma_u^2 1\pi_u^4 2\sigma_g^2$	$b = 3$	$2S + 1 = 1$
O_2	$1\sigma_g^2 1\sigma_u^2 2\sigma_g^2 1\pi_u^4 1\pi_g^2$	$b = 2$	$2S + 1 = 3$
NO	$1\sigma^2 2\sigma^2 3\sigma^2 1\pi^4 2\pi^1$	$b = 2\frac{1}{2}$	$2S + 1 = 2$

The following figures show HOMOs of each. Shaded versus unshaded AO lobes represent opposite signs of the wave functions. A relatively large AO represents the major contribution to the MO.

N_2 2σ MO

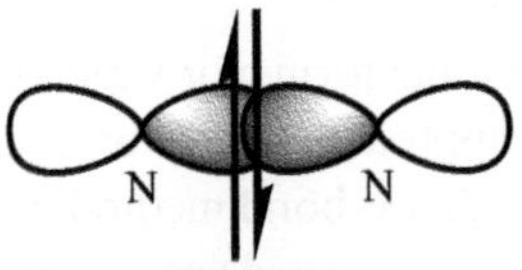

O_2 $1\pi_g$ MO, doubly degenerate

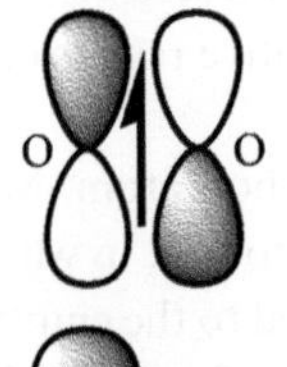

NO 2π

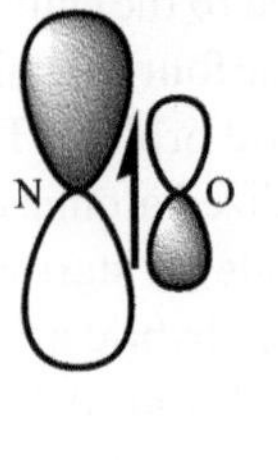

Dinitrogen with a bond order of three and paired electrons in relatively low energy MOs is very unreactive. Special biological, or industrial, processes are needed to channel energy for promotion of 2π electrons into high-energy, reactive states. The high-energy $1\pi_g$ LUMO is not expected to form stable complexes with electron donors.

Dinitrogen is very stable in most biological organisms, and as a result the task of converting plentiful atmospheric N_2 to the fixed forms of nitrogen that can be incorporated into proteins is a difficult one. The fact that N_2 possesses no unpaired electrons is itself an obstacle to facile reactivity, and the great strength (large dissociation energy) of the N_2 bond is another obstacle. Molecular orbital theory explains both of these obstacles by assigning N_2 a configuration that gives rise to a high bond order (triple bond) with all electrons paired.

Dioxygen is kinetically stable because of a bond order equal to two and a high effective nuclear charge that causes the MOs to have relatively low energy. But two electrons are in the high-energy $1\pi_g$ HOMO level, which is doubly degenerate. These two electrons are unpaired and can contribute to bonding of dioxygen with other species such as the atomic radicals Fe(II) of haemoglobin and Cu(II) of the electron-transport chain. When sufficient, though not excessively large, energy is available, biological processes can channel an electron into this HOMO to produce the reactive superoxide anion of bond order $1\frac{1}{2}$. As a result, O_2 is very reactive in biological systems in ways that promote function (such as respiration) and in ways that disrupt it (damaging cells).

Although the bond order of nitric oxide is $2\frac{1}{2}$, the nitrogen nucleus has a smaller effective nuclear charge than an oxygen atom would have. Thus, the one electron of the 2π HOMO is a high-energy, reactive radical compared to the HOMO of dioxygen. Additionally, the HOMO, being antibonding and predominantly centered on the nitrogen atom, is expected to bond through the nitrogen. Oxidation can result from the loss of the radical electron to form the nitrosyl ion, NO^+, which has a bond order equal to 3. Even though it has a rather high bond order, NO is readily converted to the damagingly reactive peroxynitrite ion ($ONOO^-$) by reaction with O_2^-—without breaking the NO linkage.

D10.5 The following list identifies, and justifies, the approximations used in the simple Hückel theory of hydrocarbon π-electron systems.

- Only the carbon p_z valence orbitals of sp^2-hybridized carbon atoms contribute to the LCAO of the π system. This is justified to an extent because the hybridization approximation gives reasonable estimates in many instances and p_z orbitals do not overlap with sp^2-hybridized orbitals.
- All terms of the form S_{AA} are set equal to 1 and terms of the form S_{AB} are set equal to zero. Overlap integrals between neighbors have small values (even smaller for nonadjacent neighbors) and their neglect eases the mathematics so that an indication of the molecular orbital energy-level diagram can be obtained.
- All terms of the form H_{AA} equal α (a negative quantity). The electronic environment of each sp^2-hybridized carbon is very similar, thereby, making all p_z valence orbitals equal in size and energy.
- All terms of the form H_{AB} equal β (a negative quantity) if the atoms are neighbors and to zero otherwise. In addition to a justification similar to that for H_{AA}, when A and B are not neighbors the p_{zA} orbital overlap with the p_{zB} orbital is negligibly small.

These approximations are obviously very severe, but they let us calculate at least a general picture of the molecular orbital energy levels with very little work.

D10.6 Intermediate values of ionization energy and electron affinity make carbon's bonding primarily covalent rather than ionic. A low ionization energy would favor cation formation (particularly in interactions with elements whose electron affinity is high), and a high electron affinity would favor anion formation (particularly in interactions with elements whose ionization energy is low). Intermediate values of these quantities also mean that the energy of the valence atomic orbitals in carbon is not terribly different from the valence orbital energies of other non-metals; this in turn implies that the molecular orbitals it can form with them are significantly different in energy from the component atomic orbitals. That is, the atomic orbitals of carbon can interact with those of both moderately electronegative and electropositive elements to produce molecular orbitals with considerable bonding character – leading to strong covalent bonds. The fact that carbon has four valence electrons is intimately connected to its ability to form four bonds – a connection that has been rationalized by pre-quantum notions such as the octet rule, as well as by quantum chemical theories such as valence-bond and molecular orbital theory. The ability of carbon to make four bonds is in turn directly related to the incredible variety and complexity of organic chemistry and biochemistry. That carbon can make four bonds allows it to form long chains and rings (requiring two bonds for all but the terminal carbons for the C–C σ skeleton) and still have valences left over for a variety of functional groups. Finally, the strength of C–C bonds relative to C–N bonds and C–O bonds permits large organic molecules to persist in an atmosphere of oxygen and nitrogen.

D10.7 The electron configuration of Zn^{2+} is $1s^22s^22p^63s^23p^63d^{10}$. Zn^{2+} has no unfilled orbitals at $n = 3$. On the other hand, the Fe^{2+} ion has the configuration $1s^22s^22p^63s^23p^63d^6$ and thus has unfilled 3d orbitals available for occupancy by electrons from O_2. As discussed in more detail in Case Study 10.4, the availability of these unfilled 3d orbitals plays an important role in the binding of O_2 to haemoglobin. If the Fe^{2+} ion in haemoglobin is replaced with Zn^{2+} ions the electrons from O_2 have no place to go at the $n = 3$ level. They could in theory occupy empty orbitals at $n = 4$, but this is energetically very unfavorable.

D10.8 In *ab initio* methods an attempt is made to evaluate the Schrödinger equation numerically without employing empirical information. Approximations are employed, but these are mainly associated with the construction of the wavefunctions involved in the integrals of the computation. *Ab initio* computations are iterative in that each cycle of the computation gives an improved estimate of the energy and wavefunction that is used in the next calculation cycle. The computation is **self-consistent** in that the iteration of energy and wavefunction are repeated (after an initial approximation is made about the mathematical form of the wavefunction) until the energy and wavefunction, are unchanged to within some acceptable tolerance.

In semiempirical methods, many of the integrals are expressed in terms of spectroscopic data or physical properties. Semiempirical methods exist at several levels. At some levels, in order to simplify the calculations, many of the integrals are set equal to zero.

Density functional theory methods have a lot in common with *ab initio* methods in that the Schrödinger equation is solved iteratively and self-consistently, but the central focus of DFT is the electron density, ρ, rather than the wavefunction, ψ. The first step in DFT methods is to guess the electron density, rather than the construction of wavefunctions. DFT methods are an approach to electronic structure calculations that are in some sense in between *ab initio* methods and cruder semiempirical methods.

Solutions to exercises

E10.9 The three valence-bond wavefunctions for N_2 are of the form described by eqn 10.1.

$$\psi_1(\sigma\text{-bond}) = \psi_{2p_zA}(1)\,\psi_{2p_zB}(2) + \psi_{2p_zA}(2)\,\psi_{2p_zB}(1)$$

$$\psi_2(\pi\text{-bond}) = \psi_{2p_xA}(1)\,\psi_{2p_xB}(2) + \psi_{2p_xA}(2)\,\psi_{2p_xB}(1)$$

$$\psi_3(\pi\text{-bond}) = \psi_{2p_yA}(1)\,\psi_{2p_yB}(2) + \psi_{2p_yA}(2)\,\psi_{2p_yB}(1)$$

E10.10 The repulsion between two nuclei at $R = 74.1$ pm and $Z = 1$ is given by Coulomb's law.

$$V_{\text{nuc,nuc}} = \frac{e^2}{4\pi\varepsilon_0 R} = \frac{(1.602\times10^{-19}\ \text{C})^2}{(1.113\times10^{-10}\ \text{J}^{-1}\ \text{C}^2\ \text{m}^{-1})\times(74.1\times10^{-12}\ \text{m})}$$
$$= 3.11\times10^{-18}\ \text{J}$$

The molar value is

$$N_A V_{\text{nuc,nuc}} = (6.022\times10^{23}\ \text{mol}^{-1})\times(3.11\times10^{-18}\ \text{J}) = \boxed{1.87\times10^{6}\ \text{J mol}^{-1}}$$

E10.11 SO_2

There are two localized S—O σ-bonds formed from S(sp^2) and O(p_z) orbitals. There is a π-bond that exhibits resonance and that can be described as the following superposition of wavefunctions:

$$\psi(\pi\text{-bond}) = (\psi_{2p_{xS}} + \psi_{2p_{xOA}}) + (\psi_{2p_{xS}} + \psi_{2p_{xOB}})$$

The sulfur has a lone pair.

SO_3

There are three localized S—O σ-bonds from S(sp^2) and O(p_z) orbitals. There is a π-bond that exhibits resonance and that can be described as the following superposition of wavefunctions:

$$\psi(\pi\text{-bond}) = (\psi_{2p_{xS}} + \psi_{2p_{xOA}}) + (\psi_{2p_{xS}} + \psi_{2p_{xOB}}) + (\psi_{2p_{xS}} + \psi_{2p_{xOC}})$$

In the above OA, OB, and OC refer to oxygen atoms A, B, and C.

E10.12 The ion has 24 valence electrons as shown in the Lewis structure. The hybridizations are (from left to right) sp^2 for the first O atom and for the N and sp^3 for the next two O atoms. The first bond is a double bond whose σ component arises from the overlap of sp^2 orbitals and whose π component comes from overlap of unhybridized p orbitals. The next bond is a σ bond involving Nsp^2 and Osp^3 orbitals. The O–O bond is a σ bond involving sp^3 orbitals.

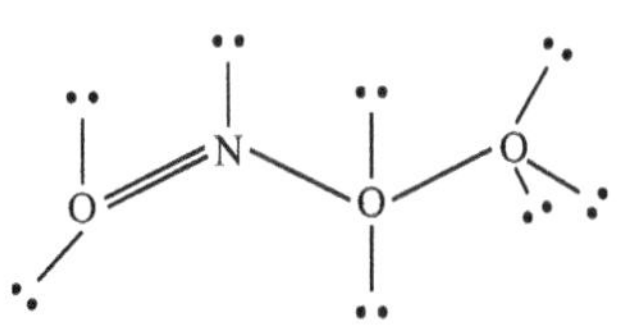

E10.13 Refer to the structure shown below in Figure 10.1 for the numbering of the carbon atoms in cis-retinal. Carbon atoms 5–15 each have three sp^2 hybrid atomic orbitals which form σ bonds with their neighboring atoms. There are six conjugated π bonds between these 11 C atoms and the one O atom. These six π bonds are formed from 12 p_x atomic orbitals, one on each of the 12 atoms. They are resonance hybrids all of the form:

$$\psi(\pi\text{-bond}) = \sum_{i=5}^{15} \psi_{2p_xC_i} + \psi_{2p_xO}$$

All the remaining C atoms each have four sp^3 hybrid atomic orbitals that form σ bonds with their neighboring atoms.

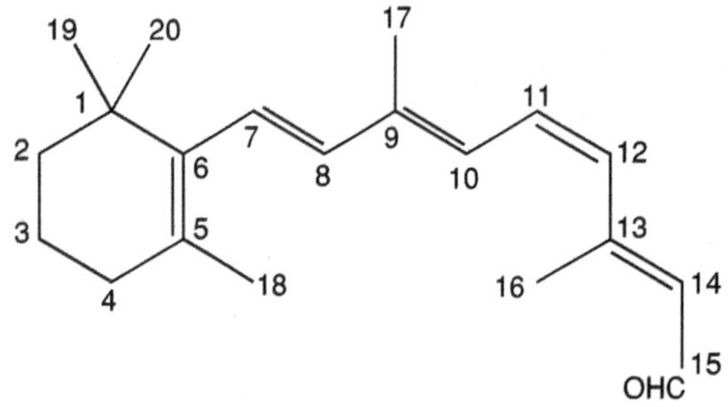

Figure 10.1

E10.14 The atomic orbital φ is normalized when $\int \varphi^2 d\tau = 1$, where the integral over τ represents an integral over all possible values of x, y, and z. Two atomic orbitals φ_1 and φ_2 are orthogonal if $\int \varphi_1\varphi_2 d\tau = 0$. Each AO of the set s, p_x, p_y, and p_z is both normalized and orthogonal to other members of the set so they are said to be an **orthonormal** set.

$$h_1 = s + p_x + p_y + p_z \qquad \text{and} \qquad h_2 = s - p_x - p_y + p_z$$

h_1 and h_2 are orthogonal providing that $\int h_1 h_2 d\tau = 0$. We check that this condition is satisfied.

$$\begin{aligned}\int h_1h_2 d\tau &= \int (s + p_x + p_y + p_z)(s - p_x - p_y + p_z)d\tau \\ &= \int (s^2 - sp_x - sp_y + sp_z)d\tau + \int (sp_x - p_x^2 - p_xp_y + p_xp_z)d\tau \\ &\quad + \int (sp_y - p_xp_y - p_y^2 + p_yp_z)d\tau + \int (sp_z - p_xp_z - p_yp_z + p_z^2)d\tau\end{aligned}$$

All of the above integrals of the type $\int (sp_x)d\tau$ and $\int (p_xp_y)d\tau$ vanish because the integrand AOs are orthogonal. Therefore,

$$\begin{aligned}\int h_1h_2\, d\tau &= \int (s^2)d\tau - \int (p_x^2)d\tau - \int (p_y^2)d\tau + \int (p_z^2)d\tau \\ &= 1 - 1 - 1 + 1 \text{ because the AO are normalized} \\ &= 0\end{aligned}$$

Thus, h_1 and h_2 are orthogonal.

E10.15 We need to demonstrate that $\int h_{sp^2}^2 d\tau = 1$, where $h_{sp^2} = \dfrac{s + \sqrt{2}p}{\sqrt{3}}$.

$$\begin{aligned}\int h_{sp^2}^2 d\tau &= \tfrac{1}{3}\int (s + \sqrt{2}p)^2 d\tau \\ &= \tfrac{1}{3}\int (s^2 + 2p^2 + 2\sqrt{2}sp)d\tau \\ &= \tfrac{1}{3}(1 + 2 + 0) \quad \text{as } \int s^2 d\tau = 1, \int p^2 d\tau = 1, \text{ and } \int sp d\tau = 0 \text{ (orthonormality)} \\ &= 1\end{aligned}$$

Thus, this hybrid orbital is normalized to 1.

E10.16 Rewrite the normalized sp^2 hybrid orbital of part (b) as $h_1 = \dfrac{s + \sqrt{2}p_x}{\sqrt{3}}$. This hybrid orbital points along the line l_1 on the x-axis as shown in Figure 10.2. There are two additional sp^2 hybrid orbitals, h_2 and h_3; they point along the lines l_2 and l_3. These three orbitals have the same size and shape. They differ only in the direction to which they point. To construct h_2, we appropriately weigh the s, p_x, and p_y AOs so that the sum points along l_2 while simultaneously being normalized and orthogonal to h_1. In order to point along l_2, the weight of p_y must be positive but the weight of p_x must be negative. Furthermore, as shown in Figure 10.2, the weight of p_y must be $3^{1/2}$ times the weight of p_x. Thus,

$h_2 = as - bp_x + 3^{1/2}bp_y$ where the weights a and b are positive numbers

The values of the weights a and b are found with the orthonormal conditions. The orthogonality of h_1 and h_2 provides a useful relation:

$$\int h_1 h_2 d\tau = \int \left(\frac{s + 2^{1/2}p_x}{3^{1/2}}\right)(as - bp_x + 3^{1/2}bp_y)d\tau$$

$$= \frac{a}{3^{1/2}}\int s^2 d\tau - \frac{2^{1/2}b}{3^{1/2}}\int p_x^2 d\tau \quad \text{because terms like } \int sp_x d\tau \text{ equal zero (orthgonality)}$$

$$= \frac{a}{3^{1/2}} - \frac{2^{1/2}b}{3^{1/2}} \quad \text{because the orbitals are normalized}$$

The above expression equals zero when $a = 2^{1/2}b$ so $h_2 = b(2^{1/2}s - p_x + 3^{1/2}p_y)$. Now, we use the normalization condition $\int h_2^2 d\tau = 1$ to determine the value of b.

$$\int h_2^2 d\tau = b^2\int(2^{1/2}s - p_x + 3^{1/2}p_y)^2 d\tau$$
$$= b^2\{2\int s^2 d\tau + \int p_x^2 d\tau + 3\int p_y^2 d\tau\} \quad \text{because terms like } \int sp_x d\tau \text{ equal zero (orthgonality)}$$
$$= 6b^2 \quad \text{because the orbitals are normalized}$$

Thus, $b = 6^{-1/2}$ and substitution gives

$$\boxed{h_2 = \sqrt{\frac{1}{3}}s - \sqrt{\frac{1}{6}}p_x + \sqrt{\frac{1}{2}}p_y}$$

The sp^2 hybrid h_3 is a reflection of h_2 through the y-axis so we need only change the weight of p_y by changing the weight sign, thereby, giving the last of the three sp^2 hybrids.

$$\boxed{h_3 = \sqrt{\frac{1}{3}}s - \sqrt{\frac{1}{6}}p_x - \sqrt{\frac{1}{2}}p_y}$$

Note that our solution here goes somewhat beyond the form of the hybrids listed in eqns 10.2b of the text in that the orbitals here have been normalized and include the normalization factor $\sqrt{\frac{1}{3}}$. Also we arbitrarily switched the x and y labels on orbitals. This is inconsequential.

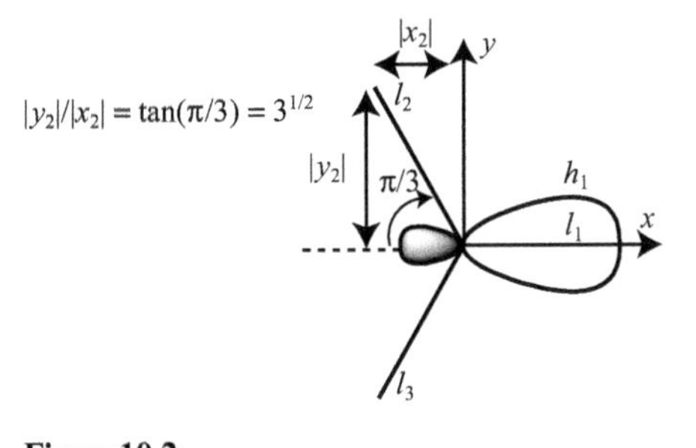

Figure 10.2

E10.17 Covalent structures are shown in Figure 10.3(a), while ionic structures are shown in Figure 10.3(b).

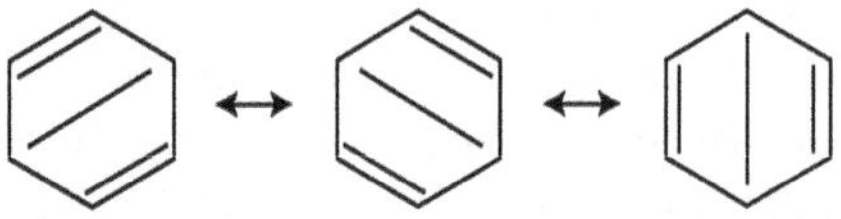

Figure 10.3(a)

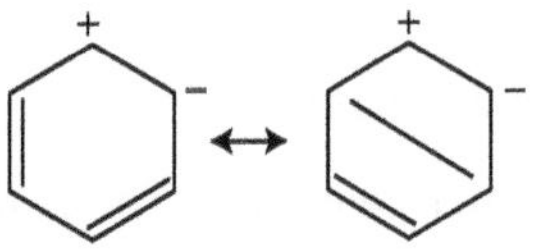

Figure 10.3(b)

In addition, there are many other possible ionic structures. These structures can be safely ignored in simple descriptions of the molecule because the coefficients of the wavefunction representing these structures in the linear combination of wavefunctions for the entire resonance hybrid are very small. Benzene is a very symmetrical molecule, and we expect that all the C atoms will be equivalent. Hence, those structures in which the C atoms are not equivalent should contribute little to the resonance hybrid.

E10.18 When the s orbital shares a common center with the p orbital, the constructive interference between the s orbital and one of the p orbital lobes is exactly balanced with the destructive interference between the s orbital and the other p orbital lobe (see Figure 10.4). Thus, when $R = 0$, $S = 0$. As the two orbitals separate the overlap increases to a maximum after which larger separation results in an ever decreasing overlap, which is sketched in Figure 10.5.

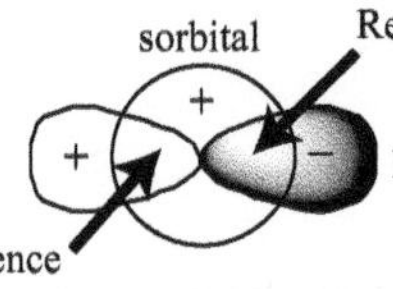

Figure 10.4

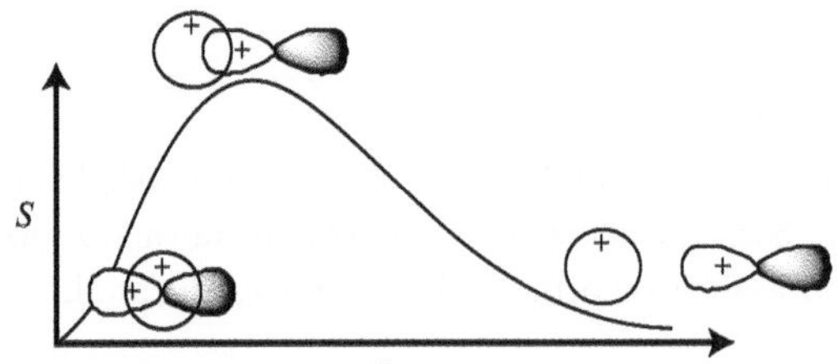

Figure 10.5

$$S = (R/a_0)\{1 + (R/a_0) + \tfrac{1}{3}(R/a_0)^2\}e^{-R/a_0}$$

Draw up the following table and plot S against R/a_0 as shown in Figure 10.6. A maximum occurs at $\boxed{R/a_0 = 2.11}$.

R/a_0	0	1	2	3	4	5	6	7	8	9	10
S	0	0.858	1.173	1.046	0.757	0.483	0.283	0.155	0.081	0.041	0.02

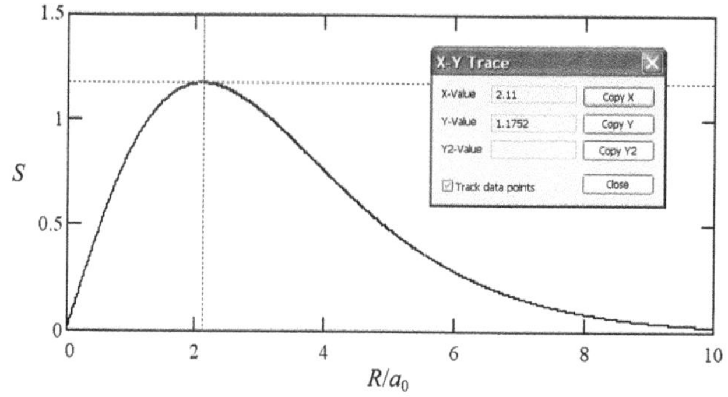

Figure 10.6

E10.19 We seek an orbital of the form $aA + bB$, where a and b are constants, which is orthogonal to the orbital $N(0.245A + 0.644B)$. Orthogonality implies

$$\int(aA + bB)N(0.245A + 0.644B)\mathrm{d}\tau = 0$$

$$N\int[0.245aA^2 + (0.245b + 0.644a)AB + 0.644bB^2]\mathrm{d}\tau = 0$$

The integrals of squares of orbitals are 1 and the integral $\int AB\mathrm{d}\tau$ is the overlap integral S, so

$$0 = (0.245 + 0.644\ S)a + (0.245\ S + 0.644)b \quad \text{so} \quad a = -\frac{0.245\ S + 0.644}{0.245 + 0.644\ S}b$$

This would make the orbitals orthogonal, but not necessarily normalized. If $S = 0$, the expression simplifies to

$$a = -\frac{0.644}{0.245}b$$

and the new orbital would be normalized if $a = 0.644N$ and $b = -0.245N$. That is

$$\boxed{N(0.644A - 0.245B)}$$

E10.20 In general, we have $\psi = c_A\,\psi_A + c_B\,\psi_B$

We need to determine the coefficients c_A and c_B.

A systematic way of finding the coefficients in the linear combinations used to build molecular orbitals is provided by the **variation principle**:

If an arbitrary wavefunction is used to calculate the energy, then the value calculated is never less than the true energy.

The arbitrary wavefunction is called the **trial wavefunction**. The principle implies that if we vary the coefficients in the trial wavefunction until we achieve the lowest energy, then those coefficients will be the best. We might get a lower energy if we use a more complicated wavefunction (for example, by taking a linear combination of several atomic orbitals on each atom), but we shall have the optimum molecular orbital that can be built from the given set of atomic orbitals.

The method can be illustrated by the trial wavefunction

$$\psi = c_A\psi_A + c_B\psi_B$$

This function is real but not normalized (because the coefficients can take arbitrary values), so in the following we cannot assume that $\int\psi^2 d\tau = 1$. The energy of the orbital is the expectation value of the energy operator

$$E = \frac{\int\psi H\psi d\tau}{\int\psi^2 d\tau}$$

We must search for values of the coefficients in the trial function that minimize the value of E. This is a standard problem in calculus, and is solved by finding the coefficients for which

$$\frac{\partial E}{\partial c_A} = 0 \quad \text{and} \quad \frac{\partial E}{\partial c_B} = 0$$

The first step is to express the two integrals in terms of the coefficients. The denominator is

$$\begin{aligned}\int\psi^2 d\tau &= \int\{c_A\psi(A) + c_B\psi(B)\}^2 d\tau \\ &= c_A^2\int\psi(A)^2 d\tau + c_B^2\int\psi(B)^2 d\tau + 2c_Ac_B\int\psi(A)\psi(B)d\tau \\ &= c_A^2 + c_B^2 + 2c_Ac_BS \quad (1)\end{aligned}$$

because the individual atomic orbitals are normalized and the third integral is the overlap integral S. The numerator is

$$\begin{aligned}\int\psi H\psi d\tau &= \int\{c_A\psi_A + c_B\psi_B\}H\{c_A\psi_A + c_B\psi_B\}d\tau \\ &= c_A^2\int\psi_A H\psi_A d\tau + c_B^2\int\psi_B H\psi_B d\tau + 2c_Ac_B\int\psi_A H\psi_B d\tau\end{aligned}$$

There are some complicated integrals in this expression, but we can denote them by the constants

$$\alpha_A = \int\psi(A)H\psi(A)d\tau \quad \alpha_B = \int\psi(B)H\psi(B)d\tau \quad \beta = \int\psi(A)H\psi(B)d\tau$$

Then

$$\int\psi H\psi d\tau = c_A^2\alpha_A + c_B^2\alpha_B + 2c_Ac_B\beta$$

α is called a **Coulomb integral**. It is negative, and can be interpreted as the energy of the electron when it occupies ψ_A (for α_A) or ψ_B (for α_B). In a homonuclear diatomic molecule, $\alpha_A = \alpha_B$. β is called a **resonance integral** (for classical reasons). It vanishes when the orbitals do not overlap, and at equilibrium bond lengths it is normally negative.

The complete expression for E is

$$E = \frac{c_A^2\alpha_A + c_B^2\alpha_B + 2c_Ac_B\beta}{c_A^2 + c_B^2 + 2c_Ac_BS}$$

Its minimum is found by differentiation with respect to the two coefficients. This involves elementary but slightly tedious work, the end result being the two **secular equations**

$$(\alpha_A - E)c_A + (\beta - ES)c_B = 0$$

$$(\beta - ES)c_A + (\alpha_B - E)c_B = 0$$

They have a solution if the determinant of the coefficients, the *secular determinant* vanishes; that is, if

$$\begin{vmatrix}\alpha_A - E & \beta - ES \\ \beta - ES & \alpha_B - E\end{vmatrix} = 0$$

This determinant expands to a quadratic equation in E, which may be solved. Its two roots give the energies of the bonding and antibonding MOs formed from the basis set and, according to the variation principle, these are the best energies for the given basis set. The corresponding values of the coefficients are then obtained by solving the secular equations using the two energies: the lower energy gives the coefficients for the bonding MO, the upper energy the coefficients for the antibonding MO. The secular equations give expressions for the *ratio* of the coefficients in each case, and so we need a further equation in order to find their individual values. This is obtained by demanding that the best wavefunction should be normalized, which means that we must also ensure (from eqn (1) above) that

$$\int \psi^2 d\tau = c_A^2 + c_B^2 + 2c_A c_B S = 1.$$

There are two cases where the roots can be written down very simply. First, when the two atoms are the same, and we can write $\alpha_A = \alpha_B = \alpha$, the solutions are

$$E_+ = \frac{\alpha+\beta}{1+S} \qquad c_A = \left\{\frac{1}{2(1+S)}\right\}^{1/2} \qquad c_B = c_A$$

$$E_- = \frac{\alpha-\beta}{1-S} \qquad c_A = \left\{\frac{1}{2(1-S)}\right\}^{1/2} \qquad c_B = -c_A$$

In this case, the best bonding function has the form

$$\psi_+ = \left\{\frac{1}{2(1+S)}\right\}^{1/2} \{\psi_A + \psi_B\}$$

and the corresponding antibonding function is

$$\psi_- = \left\{\frac{1}{2(1-S)}\right\}^{1/2} \{\psi_A - \psi_B\}$$

(a) When it is justifiable to neglect overlap, the secular determinant is

$$\begin{vmatrix} \alpha_A - E & \beta \\ \beta & \alpha_B - E \end{vmatrix} = 0$$

and its solutions can be expressed in terms of the parameter θ, with

$$\tan(2\theta) = \frac{2\beta}{\alpha_A - \alpha_B}$$

The solutions are:

$$E_- = \alpha_A - \beta\cot\theta \qquad \psi_- = -\sin\theta\,\psi_A + \cos\theta\,\psi_B$$

$$E_+ = \alpha_B + \beta\cot\theta \qquad \psi_+ = \cos\theta\,\psi_A + \sin\theta\,\psi_B$$

If $\theta = 0$, the wavefunction $\psi_+ = \psi_A$; if $\theta = \pi/2$, the wavefunction $\psi_+ = \psi_B$. So we see that this wavefunction can describe a polar covalent bond, the degree of polarity is dependent on θ. If $\theta = \pi/4$, the bond is completely covalent.

(b) We need to evaluate $\int \psi^2 d\tau$ to see if it equals 1.

$$\begin{aligned}\int(\psi_A\cos\theta + \psi_B\sin\theta)^2 d\tau &= \int(\psi_A^2\cos^2\theta + \psi_B^2\sin^2\theta + 2\psi_A\psi_B\sin\theta\cos\theta)d\tau \\ &= \cos^2\theta\int\psi_A^2 d\tau + \sin^2\theta\int\psi_B^2 d\tau + 2\sin\theta\cos\theta\int\psi_A\psi_B d\tau \\ &= \cos^2\theta + \sin^2\theta + 2\sin\theta\cos\theta\, S \\ &= 1\end{aligned}$$

We have used the facts that ψ_A and ψ_B are each normalized and that S is zero (i.e. ψ_A and ψ_B are orthogonal). Remember θ is a constant.

(c) In a homonuclear diatomic molecule we set $\alpha_A = \alpha_B$. Therefore, $\tan(2\theta) = \infty$ which solves to $\theta = \frac{1}{2}\arctan(\infty) = \frac{1}{2} \times 1.5708 = 0.7854 \text{ radian} = \boxed{45°}$

E10.21 σ-bonding with a $d_{x^2-y^2}$ orbital is shown in Figure 10.7.

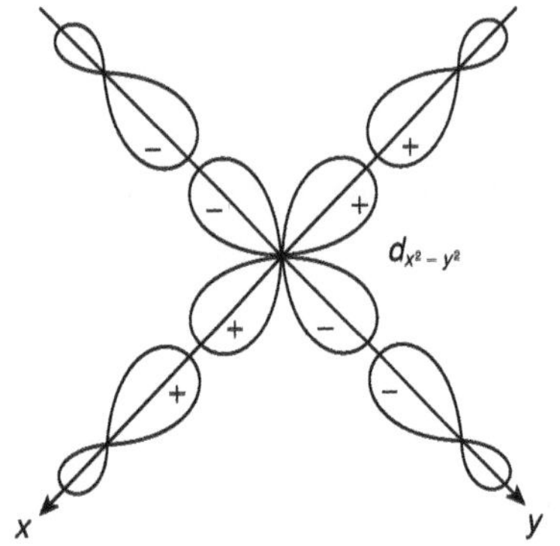

Figure 10.7

The σ-antibonding orbital looks the same but with the p-orbital lobes pointed in the opposite direction to cause destructive interference.

The σ-bonding and antibonding diagrams with the d_{xy}, d_{yz}, and d_{xz} orbitals have the same appearance as the diagram above except that the d-orbital lobes are pointed between the axes rather than along them.

σ-bonding with d_{z^2} orbital is shown in Figure 10.8.

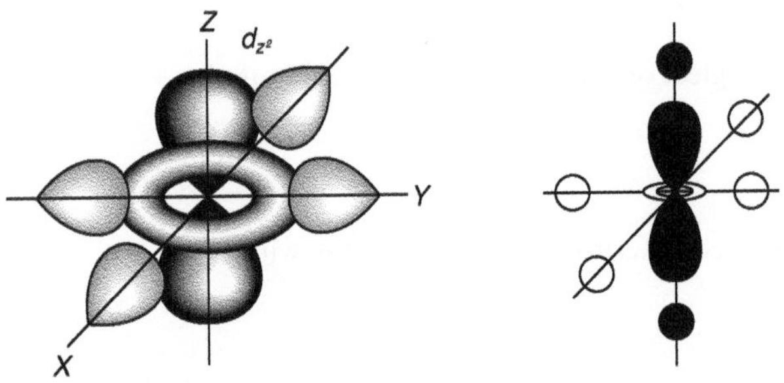

Figure 10.8

In this figure, only one of the p-orbital lobes is shown in each p-orbital. p-orbitals with positive lobes may also approach this orbital along the + and − z-direction as indicated in the smaller diagram on the right. The antibonding diagrams are similar, but with the signs of the p-orbital lobes reversed.

π-bonding (See Figure 10.9)

Only the d_{xy}, d_{yz}, and d_{xz} orbitals undergo π-bonding with p-orbitals on neighboring atoms. The bonding arrangement is pictured below with the d_{xy} orbital. The diagrams for the d_{yz} and d_{xz} orbitals are similar. The antibonding diagrams have the signs of the p-orbital lobes reversed.

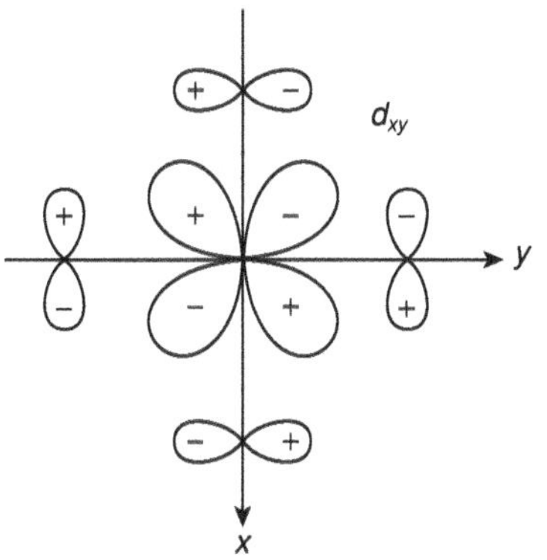

Figure 10.9

E10.22 The total number of atomic orbitals in a single set of s, p, d, and f subshells is

$$1 + 3 + 5 + 7 = 16$$

In a diatomic molecule there would then be 32 AO from which $\boxed{\text{32 molecular orbitals}}$ can be constructed.

E10.23

(a)	H_2^-	$1\sigma_g^2 1\sigma_u^1$	$b = \frac{1}{2}$
(b)	N_2	$1\sigma_g^2 1\sigma_u^2 1\pi_u^4 2\sigma_g^2$	$b = 3$
(c)	O_2	$1\sigma_g^2 1\sigma_u^2 2\sigma_g^2 1\pi_u^4 1\pi_g^2$	$b = 2$

E10.24

(a)	CO	$1\sigma^2 2\sigma^{*2} 1\pi^4 3\sigma^2$	$b = 3$
(b)	NO	$1\sigma^2 2\sigma^{*2} 1\pi^4 3\sigma^2 2\pi^{*1}$	$b = \frac{5}{2}$
(c)	CN^-	$1\sigma^2 2\sigma^{*2} 1\pi^4 3\sigma^2$	$b = 3$

E10.25 Decide whether the electron added or removed increases or decreases the bond order. The simplest procedure is to decide whether the electron occupies or is removed from a bonding or antibonding orbital. The levels for the homonuclear diatomics are shown in text Figures 10.30 and 10.31. The level for the heteronuclear diatomics is shown in text Figure 10.35.

The following table gives the orbital involved

		N_2	NO	O_2	C_2	F_2	CN
(a)	AB^-	$1\pi_g$	2π	$1\pi_g$	$2\sigma_g$	$2\sigma_u$	3σ
	Δb	$-\frac{1}{2}$	$-\frac{1}{2}$	$-\frac{1}{2}$	$+\frac{1}{2}$	$-\frac{1}{2}$	$+\frac{1}{2}$
(b)	AB^+	$2\sigma_g$	2π	$1\pi_g$	$1\pi_u$	$1\pi_g$	3σ
	Δb	$-\frac{1}{2}$	$+\frac{1}{2}$	$+\frac{1}{2}$	$-\frac{1}{2}$	$+\frac{1}{2}$	$-\frac{1}{2}$

Therefore,

$\boxed{\text{C}_2 \text{ and CN are stabilized by anion formation. NO, O}_2\text{, and F}_2 \text{ are stabilized by cation formation}}$

E10.26 The wavefunctions are

$$\psi_n(x) = \left(\frac{2}{L}\right)^{1/2} \sin\left(\frac{n\pi x}{L}\right) \text{[Chapter 9]}, \quad \text{where} \quad n = 1,2,3\ldots \quad \text{and} \quad 0 \leqslant x \leqslant L$$

Each of these wavefunctions has a center of symmetry at $L/2$ so we invert through $x = \frac{L}{2}$. For $n = 1$, when $x < \frac{L}{2}$, $\psi_1 = +$, when $x > \frac{L}{2}$, $\psi_1 = +$, therefore ψ_1 is $\boxed{g}$. In a similar fashion we determine

$n = 2$, $\boxed{\text{u}}$; $n = 3$, $\boxed{\text{g}}$; $n = 4$, $\boxed{\text{u}}$.

Figure 10.10 below illustrates the operation of inversion through the center of symmetry. First, identify the center of symmetry. Then, pick any non-nodal position and note the sign and magnitude of the wavefunction. Draw an arrow from that wavefunction point through the center of symmetry to a point that is an equal distance on the opposite side of the center of symmetry. If the wavefunction has the same magnitude and sign after the inversion, the wavefunction has gerade (g) symmetry. If the wavefunction has the same magnitude and opposite sign, the wavefunction has ungerade (u) symmetry. We quickly see that all odd-numbered ($n = 1, 3, 5, \ldots$) quantum states of the particle in a box have gerade symmetry or "parity" while all even-numbered ($n = 2, 4, 6, \ldots$) states have ungerade symmetry or "parity".

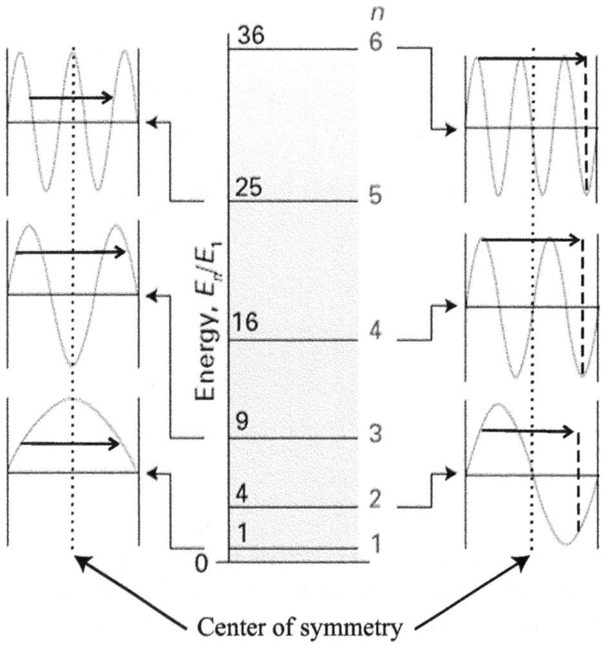

Figure 10.10

E10.27 Refer to Figure 9.39(a) of the text. Examine the inversion through the center of these functions, i.e. replace x with $-x$.

(a) $v = 0$, $\boxed{\text{g}}$

$v = 1$, $\boxed{\text{u}}$

$v = 2$, $\boxed{\text{g}}$

$v = 3$, u (Figure 9.39(a) of the text)

(b) If v is even, ψ_v is g.

If v is odd, ψ_v is u.

E10.28 The parities can be deduced from Figure 10.39 of the text:

Orbital at $\alpha + 2\beta$ has u parity.

Doubly degenerate orbitals at $\alpha + \beta$ have g parity.

Doubly degenerate orbitals at $\alpha - \beta$ have u parity.

Orbital at $\alpha - 2\beta$ has g parity.

Figure 10.11 below shows the inversion operation that indicates the ungerade symmetry of one of the MOs at energy $\alpha - \beta$.

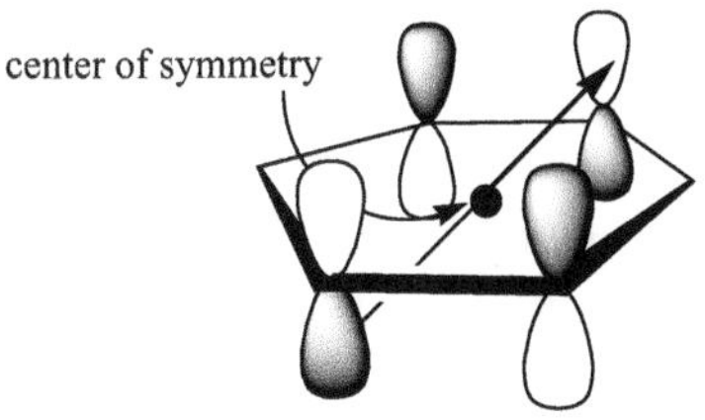

Figure 10.11

E10.29 NO $1\sigma^2 2\sigma^{*2} 1\pi^4 3\sigma^2 2\pi^{*1}$ $b = \frac{5}{2}$

N_2 $1\sigma_g^2 1\sigma_u^2 1\pi_u^4 2\sigma_g^2$ $b = 3$

Because the bond order of N_2 is greater, $\boxed{N_2}$ is likely to have the shorter bond length.

E10.30 O_2^+(11 electrons): $1\sigma_g^2 1\sigma_u^2 2\sigma_g^2 1\pi_u^4 1\pi_g^1$ $b = \frac{5}{2}$

O_2(12 electrons): $1\sigma_g^2 1\sigma_u^2 2\sigma_g^2 1\pi_u^4 1\pi_g^2$ $b = 2$

O_2^-(13 electrons): $1\sigma_g^2 1\sigma_u^2 2\sigma_g^2 1\pi_u^4 1\pi_g^3$ $b = \frac{3}{2}$

O_2^{2-}(14 electrons): $1\sigma_g^2 1\sigma_u^2 2\sigma_g^2 1\pi_u^4 1\pi_g^4$ $b = 1$

Each electron added to O_2^+ is added to an antibonding $1\pi_g$ orbital, thus increasing the length. So the sequence $\boxed{O_2^+, O_2, O_2^-, O_2^{2-}}$ has progressively longer bonds.

E10.31 (a) The molecular orbital energy level diagram of the hybridized CH_2 fragments and ethene, $CH_2{=}CH_2$, is shown in Figure 10.12(a).

COMMENT. Note that the π-bonding orbital must be lower in energy than the σ-antibonding orbital for π-bonding to exist in ethene.

(b) The molecular orbital energy level diagram of the hybridized CH fragments and ethyne, $CH{\equiv}CH$ is shown in Figure 10.12(b).

QUESTION. Would the ethyne molecule exist if the order of the energies of the π and σ* orbitals were reversed?

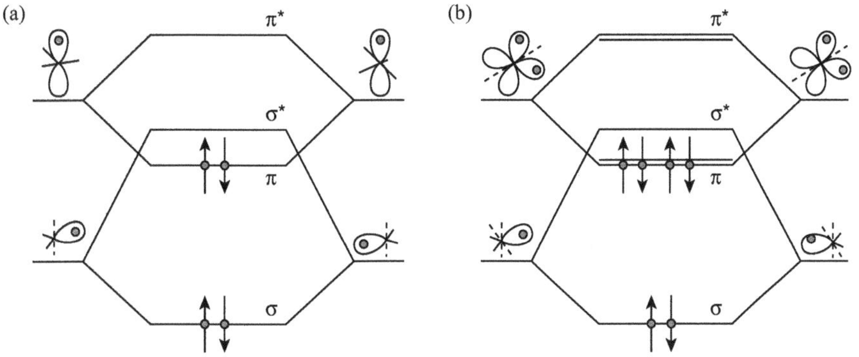

Figure 10.12

E10.32 $$\psi_n(x) = \left(\frac{2}{L}\right)^{1/2} \sin\left(\frac{n\pi x}{L}\right) \text{[Chapter 9]}, \quad \text{where} \quad n = 1,2,3,\ldots \quad \text{and} \quad 0 \leqslant x \leqslant L$$

$$E_n = \frac{n^2h^2}{8m_eL^2}$$

Two electrons occupy each level by the Pauli principle, and so butadiene, which has four π electrons, has two electrons in ψ_1 and two electrons in ψ_2. The occupied particle in a box wavefunctions are

$$\psi_1 = \left(\frac{2}{L}\right)^{1/2} \sin\left(\frac{\pi x}{L}\right) \quad \text{and} \quad \psi_2 = \left(\frac{2}{L}\right)^{1/2} \sin\left(\frac{2\pi x}{L}\right)$$

These FEMO wavefunctions are sketched and compared with the Hückel π model MOs in Figure 14.19.

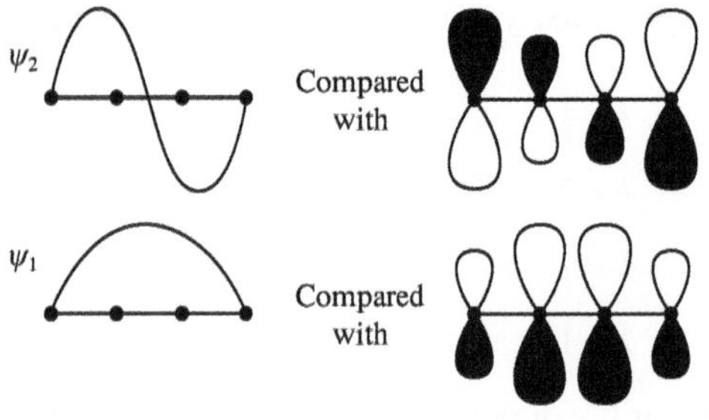

Figure 10.13

The minimum excitation energy for butadiene in the FEMO model is

$$\Delta E = E_3 - E_2 = (3^2 - 2^2)\left(\frac{h^2}{8m_eL^2}\right) = \frac{5h^2}{8m_eL^2} \quad \text{where} \quad L \simeq 4R = 4 \times 140 \text{ pm}$$
$$= \frac{5 \times (6.626 \times 10^{-34} \text{ J s})^2}{8 \times (9.109 \times 10^{-31} \text{ kg}) \times (4 \times 140 \times 10^{-12} \text{ m})^2}$$
$$= 9.61 \times 10^{-19} \text{ J} = 6.00 \text{ eV}$$

In CH_2=CH–CH=CH–CH=CH–CH=CH_2 there are eight π electrons to accommodate, so the HOMO will be ψ_4 and the LUMO ψ_5. The minimum excitation energy in the FEMO model is

$$\Delta E = E_5 - E_4 = (5^2 - 4^2)\left(\frac{h^2}{8m_eL^2}\right) = \frac{9h^2}{8m_eL^2}, \quad \text{where} \quad L \simeq 8R = 8 \times 140 \text{ pm}$$
$$= \frac{9 \times (6.626 \times 10^{-34} \text{ J s})^2}{8 \times (9.109 \times 10^{-31} \text{ kg}) \times (8 \times 140 \times 10^{-12} \text{ m})^2}$$
$$= 4.32 \times 10^{-19} \text{ J} = \boxed{2.70 \text{ eV}}$$

The FEMO model ψ_4 HOMO and ψ_5 LUMO are sketched in Figure 10.14.

COMMENT: Can you identify the positions of the carbon nuclei in Figure 10.14 and superimpose the p_z orbitals of the simple Hückel π model with their wave sign upon both the HOMO and LUMO of the figure?

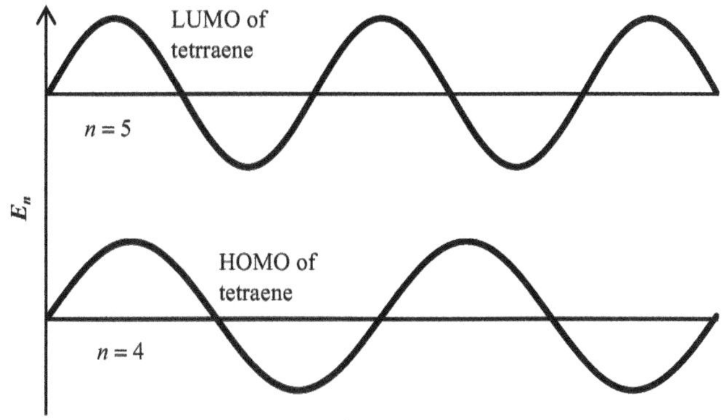

Figure 10.14

COMMENT. In the FEMO model the wavelength of the minimum excitation energy is given by

$$\lambda = \frac{hc}{\Delta E} = \frac{(6.626 \times 10^{-34}\ \text{J s}) \times (2.998 \times 10^{8}\ \text{m s}^{-1})}{4.32 \times 10^{-19}\ \text{J}} = 4.60 \times 10^{-7}\ \text{m} \quad \text{or} \quad \boxed{460\ \text{nm}}$$

The wavelength 460 nm corresponds to blue light; so on the basis of this calculation alone the molecule would appear $\boxed{\text{orange}}$ in white light (because blue is subtracted). The experimental value of λ_{max} is 304 nm, and the compound is colorless. This illustrates the very approximate nature of the calculation we have performed.

E10.33 The secular determinant of cyclobutadiene in the Hückel approximation is:

$$\begin{vmatrix} \alpha - E & \beta & 0 & \beta \\ \beta & \alpha - E & \beta & 0 \\ 0 & \beta & \alpha - E & \beta \\ \beta & 0 & \beta & \alpha - E \end{vmatrix} = 0$$

E10.34 In setting up the secular determinant we use the approximations of Section 10.7

$$\begin{vmatrix} \alpha - E & \beta & 0 \\ \beta & \alpha - E & \beta \\ 0 & \beta & \alpha - E \end{vmatrix} = 0$$

The atomic orbital basis consists of unhybridized $2p_z$ orbitals centered on each carbon atom. We ignore overlap between the terminal carbon atoms because they are not neighboring. To find the energies, expand the determinants:

$$(\alpha - E) \times \{(\alpha - E)(\alpha - E) - \beta^2\} - \beta \times \beta(\alpha - E) = 0,$$

$$0 = (\alpha - E) \times \{(\alpha - E)(\alpha - E) - 2\beta^2\} = (\alpha - E)(\alpha - E + 2^{1/2}\beta)(\alpha - E - 2^{1/2}\beta)$$

The roots (the orbital energies) are $E = \alpha, \alpha \pm 2^{1/2}\beta$. The binding energy is the sum of the orbital energies of the occupied orbitals

$$E_\pi = 2(\alpha + 2^{1/2}\beta) + \alpha = 3\boxed{\alpha + 2^{3/2}\beta}.$$

E10.35 The energy of the bonding π-molecular orbital of ethene is $\alpha + \beta$ [10.13c]. Since there are two electrons to accommodate in that π–orbital, the π-electron binding energy, E_π of ethene is:

$$E_\pi = 2(\alpha + \beta) = \boxed{2\alpha + 2\beta}$$

In the case of butadiene, there are two electrons in the π-molecular orbital with an energy of $\alpha + 1.62\beta$ and two electrons in the π-molecular orbital with an energy of $\alpha + 0.62\beta$. Thus the π-electron binding energy, E_π is:

$$E_\pi = 2(\alpha + 1.62\beta) + 2(\alpha + 0.62\beta) = \boxed{4\alpha + 4.48\beta}.$$

Therefore, the energy of the butadiene molecule lies $\boxed{\text{lower}}$ by 0.48β than the sum of two individual π bonds. Recall that β is a negative quantity.

E10.36 We use the molecular orbital energy level diagram in Figure 10.39 of the main text. As usual, we fill the orbitals starting with the lowest-energy orbital, obeying the Pauli principle and Hund's rule. We then write (where the π orbitals of benzene are labeled sequentially from lowest to highest energy)

(a) $C_6H_6^-$ (7 electrons): $1\pi^2 2\pi^2 3\pi^2 4\pi^1$

$$E_\pi = 2(\alpha + 2\beta) + 4(\alpha + \beta) + (\alpha - \beta) = \boxed{7\alpha + 7\beta}.$$

(b) $C_6H_6^+$ (5 electrons): $1\pi^2 2\pi^2 3\pi^1$

$$E_\pi = 2(\alpha + 2\beta) + 3(\alpha + \beta) = \boxed{5\alpha + 7\beta}.$$

The delocalization energy is the difference between E_π and the energy of isolated π bonds:

$$E_{\text{deloc}} = E_\pi - N_{\text{db}}\,(2\alpha + 2\beta).$$

(a) $E_\pi = 7\alpha + 7\beta$,
and $E_{\text{delocal}} = 7\alpha + 7\beta - 3.5(2\alpha + 2\beta) = \boxed{0}$.

COMMENT. With an odd number of π electrons, we do not have a whole number of π bonds in the formula for delocalization energy. In effect, we compare the π-electron binding energy to the energy of 3.5 isolated π bonds—whatever that means. The result is that the benzene anion has none of the "extra" stabilization we associate with aromaticity.

(b) $E_\pi = 5\alpha + 7\beta$,
and $E_{\text{deloc}} = 5\alpha + 7\beta - 2.5(2\alpha + 2\beta) = \boxed{2\beta}$.

E10.37 (a) The transitions occur for photons whose energies are equal to the difference in energy between the highest occupied and lowest unoccupied orbital energies:

$$E_{\text{photon}} = E_{\text{LUMO}} - E_{\text{HOMO}}$$

If N_c is the number of carbon atoms in these species, then the number of π electrons is also N_c. These N_c electrons occupy the first $N_c/2$ orbitals, so orbital number $N_c/2$ is the HOMO and orbital number $1 + N_c/2$ is the LUMO. Writing the photon energy in terms of the wavenumber, and substituting the given energy expressions with this identification of the HOMO and LUMO gives

$$hc\tilde{\nu} = \left(\alpha + 2\beta\cos\frac{(\frac{1}{2}N_c + 1)\pi}{N+1}\right) - \left(\alpha + 2\beta\cos\frac{\frac{1}{2}N_c\pi}{N+1}\right)$$

$$= 2\beta\left(\cos\frac{(\frac{1}{2}N_c + 1)\pi}{N_c+1} - \cos\frac{\frac{1}{2}N_c\pi}{N_c+1}\right).$$

Solving for β yields

$$\beta = \frac{hc\tilde{\nu}}{2\left(\cos\frac{(\frac{1}{2}N_c + 1)\pi}{N_c+1} - \cos\frac{\frac{1}{2}N_c\pi}{N_c+1}\right)}.$$

Draw up the following table

Species	N_c	$\tilde{\nu}/cm^{-1}$	Estimated β/eV
C_2H_4	2	61 500	−3.813
C_4H_6	4	46 080	−4.623
C_6H_8	6	39 750	−5.538
C_8H_{10}	8	32 900	−5.873

(b) The total energy of the π electron system is the sum of the energies of occupied orbitals weighted by the number of electrons that occupy them. In C_8H_{10}, each of the first four orbitals is doubly occupied, so

$$E_\pi = 2\sum_{k=1}^{4} E_k = 2\sum_{k=1}^{4}\left(\alpha + 2\beta\cos\frac{k\pi}{9}\right) = 8\alpha + 4\beta\sum_{k=1}^{4}\cos\frac{k\pi}{9} = 8\alpha + 9.518\beta$$

The delocalization energy is the difference between this quantity and that of four isolated double bonds:

$$E_{\text{deloc}} = E_\pi - 8(\alpha+\beta) = 8\alpha + 9.518\beta - 8(\alpha+\beta) = \boxed{1.518\beta}.$$

Using the estimate of β from part (a) yields $E_{\text{deloc}} = \boxed{8.913\ \text{eV}}$.

(c) Draw up the following table, in which the orbital energy decreases as we go down. For the purpose of comparison, we express orbital energies as $(E_k - \alpha)/\beta$. Recall that β is negative (as is α for that matter), so the orbital with the greatest value of $(E_k - \alpha)/\beta$ has the lowest energy.

Orbital	Energy $(E_k-\alpha)/\beta$	Coefficients 1	2	3	4	5	6
6	−1.8019	0.2319	−0.4179	0.5211	−0.5211	0.4179	−0.2319
5	−1.2470	0.4179	−0.5211	0.2319	0.2319	−0.5211	0.4179
4	−0.4450	0.5211	−0.2319	−0.4179	0.4179	0.2319	−0.5211
3	0.4450	0.5211	0.2319	−0.4179	−0.4179	0.2319	0.5211
2	1.2470	0.4179	0.5211	0.2319	−0.2319	−0.5211	−0.4179
1	1.8019	0.2319	0.4179	0.5211	0.5211	0.4179	0.2319

The orbitals are shown schematically in Figure 10.15, with each vertical pair of lobes representing a p orbital on one of the carbons in hexatriene. Shaded lobes represent one sign of the wavefunction (say positive) and unshaded lobes the other sign. Where adjacent atoms have atomic orbitals of the same sign, the resulting molecular orbital is bonding with respect to those atoms; where adjacent atoms have different sign, there is a node between the atoms and the resulting molecular orbital is antibonding with respect to them. The lowest energy orbital is totally bonding (no nodes between atoms) and the highest-energy orbital is totally antibonding (nodes between each adjacent pair). Note that the orbitals have increasing antibonding character as their energy increases. The size of each atomic p orbital is proportional to the magnitude of the coefficient of that orbital in the molecular orbital. So, for example, in orbitals 1 and 6, the largest lobes are in the center of the molecule, so electrons that occupy those orbitals are more likely to be found near the center of the molecule than on the ends. In the ground state of the molecule, there are two electrons in each of orbitals 1, 2, and 3, with the result that the probability of finding a π electron in hexatriene is uniform over the entire molecule.

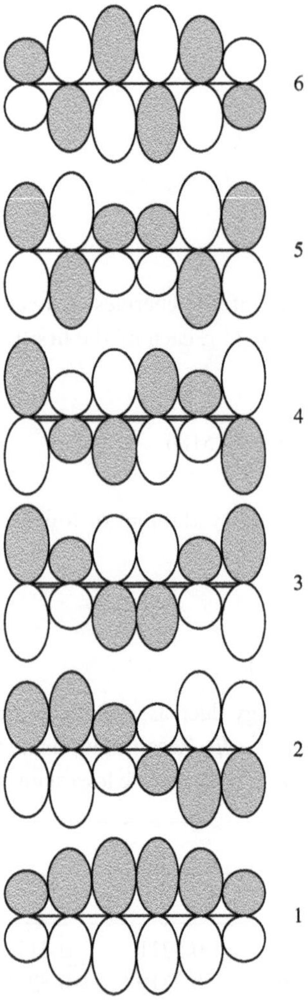

Figure 10.15

E10.38 (a) In the absence of numerical values for α and β, we express orbital energies as $(E_k - \alpha)/\beta$ for the purpose of comparison. Recall that β is negative (as is α for that matter), so the orbital with the greatest value of $(E_k - \alpha)/\beta$ has the lowest energy. Draw up the following table, evaluating the Hückel theory expression

$$\frac{E_k - \alpha}{\beta} = 2\cos\frac{2k\pi}{N}$$

$$k = 0, \pm1, \pm2, \ldots \pm N/2 \text{ for even } N \quad \text{or} \quad k = 0, \pm1, \pm2, \ldots \pm(N-1)/2 \text{ for odd } N$$

where N is the number of carbon atoms contributing an electron in a $2p_z$ orbital to the π system of a monocyclic conjugated polyene.

	Energy $(E_k - \alpha)/\beta$	
Orbital, k	benzene, C_6H_6 $N = 6$	cyclooctatetraene, C_8H_8 $N = 8$
±4		−2.000
±3	−2.000	−1.414
±2	−1.000	0
±1	1.000	1.414
0	2.000	2.000

In each case, the lowest and highest energy levels are non-degenerate, while the other energy levels are doubly degenerate. The degeneracy is clear for all energy levels except, perhaps, the highest: each value of the quantum number k corresponds to a separate MO, and positive and negative values of k therefore give rise to a pair of MOs of the same energy. This is not the case for the highest energy level, though, because there are only as many MOs as there were AOs input to the calculation, which is the same as the number of carbon atoms; having a doubly degenerate top energy level would yield one extra MO.

(b) The total energy of the π electron system is the sum of the energies of occupied orbitals weighted by the number of electrons that occupy them.

In C_6H_6, each of the first three orbitals are doubly occupied, but the second level ($k = \pm 1$) is doubly degenerate, so

$$E_\pi = 2E_0 + 4E_1 = 2(\alpha + 2\beta\cos 0) + 4\left(\alpha + 2\beta\cos\frac{2\pi}{6}\right) = 6\alpha + 8\beta$$

The delocalization energy is the difference between this quantity and that of three isolated double bonds:

$$E_{\text{deloc}} = E_\pi - 6(\alpha + \beta) = 6\alpha + 8\beta - 6(\alpha + \beta) = \boxed{2\beta}$$

For linear hexatriene refer to E10.37(c) for its energy levels. The six electrons occupy the three lowest energy orbitals resulting in $E_{\text{deloc}} = 0.988\beta$, so benzene has considerably more delocalization energy (assuming that β is similar in the two molecules). This extra stabilization is an example of the special stability of $\boxed{\text{aromatic}}$ compounds.

(c) In C_8H_8, each of the first three orbitals is doubly occupied, but the second level ($k = \pm 1$) is doubly degenerate. The next level is also doubly degenerate, with a single electron occupying each orbital. So the energy is

$$\begin{aligned}E_\pi &= 2E_0 + 4E_1 + 2E_2\\ &= 2(\alpha + 2\beta\cos 0) + 4\left(\alpha + 2\beta\cos\frac{2\pi}{8}\right) + 2\left(\alpha + 2\beta\cos\frac{4\pi}{8}\right)\\ &= 8\alpha + 9.657\beta\end{aligned}$$

The delocalization energy is the difference between this quantity and that of four isolated double bonds:

$$E_{\text{deloc}} = E_\pi - 8(\alpha + \beta) = 8\alpha + 9.657\beta - 8(\alpha + \beta) = \boxed{1.657\beta}$$

This delocalization energy is not much different from that of linear octatetraene (1.518β). See the solution to E10.37(b) where this value is calculated. Thus cyclooctatetraene does not have much additional stabilization over the linear structure. Once again, though, we do see that the delocalization energy stabilizes the π orbitals of the closed-ring conjugated system to a greater extent than what is observed in the open-chain conjugated system. However, the benzene/hexatriene comparison shows a much greater stabilization than does the cyclooctatetraene/octatetraene system. This is a demonstration of the Hückel $4n+2$ rule, which states that any planar, cyclic, conjugated system exhibits unusual aromatic stabilization if it contains $(4n+2)\pi$ electrons where "n" is an integer. Benzene with its 6π electrons has this aromatic stabilization, whereas cyclooctatetraene with 8π electrons does not have this unusual stabilization. We can say that it is $\boxed{\text{not aromatic}}$, consistent with indicators of aromaticity such as the Hückel $4n+2$ rule.

E10.39 (a) In the spectrochemical series of ligands, water, H_2O, is usually considered a weak field or small Δ_O ligand. Thus, 3 of the 5 unpaired d-electrons of Fe^{3+} are expected to go unpaired into the lower-lying t_{2g} orbitals shown in text Figure 10.41 and the remaining 2 electrons will go unpaired into the higher e_g orbitals. On the other hand, CN^- is a strong field ligand with large Δ_O, so all electrons are expected to go into the three lower-lying t_{2g} orbitals.

(b) The number of unpaired electrons in $[Fe(H_2O)_6]^{3+}$ is thus $\boxed{5}$. Due to the Pauli principle, 4 of the 5 d-electrons in $[Fe(CN)_6]^{3-}$ must be paired, so the number of unpaired electrons is $\boxed{1}$.

E10.40 The number of d-electrons in metal ions can range from 1 to 10. When the number of d-electrons is $\boxed{4, 5, 6, \text{ or } 7}$ different electron configurations are possible in weak and strong field ligand environments. This is easily seen by placing the electrons in parts (a) and (b) of text Figure 10.42 in turn.

E10.41 Both octahedral and tetrahedral complexes of a Ni^{2+} ion would be expected to have 2 unpaired electrons as can be seen from text Figure 10.41 for the octahedral arrangement and from text Figure 10.41 turned upside down for the tetrahedral arrangement. On the other hand for a $\boxed{\text{square planar arrangement}}$ all 8 electrons go paired into the lower-lying 4 orbitals if the separation between the d_{xy} and $d_{x^2-y^2}$ orbitals is sufficiently large, as is expected here.

E10.42 (a) Ligands such as Cl^- supply electrons and are Lewis π bases; they have π-donor orbitals. In the case of Cl^-, the ligand π orbitals are full and lie below the metal orbitals in energy, Figure 10.45(a). The effect of the bonding is then to decrease the ligand field splitting parameter, Δ_O. On the other hand, Lewis π acids such as CO with initially empty π-antibonding orbitals, which lie close to the d-electron energies Figure 10.45(b), are π-acceptors. The effect of π overlap is then to increase Δ_O.

(b) O_2 has the electron configuration $1\sigma_g^2 1\sigma_u^{*2} 2\sigma_g^2 1\pi_u^4 1\pi_g^{*2}$ and has a half-filled π-antibonding orbital and is thus a Lewis acid. As such, it is expected to be somewhat of a moderate-field ligand.

(c) Based upon the *Case study* mentioned, it can be seen that O_2 bonds reversibly to the Fe(II) atom of haemoglobin, which implies that the bonding is only moderately strong. On the other hand, CO is a strong field ligand and therefore binds very strongly to the Fe(II) atom. This bonding is for all practical purposes irreversible, and CO forms a very stable complex with haemoglobin that does not allow for the transport of O_2 by haemoglobin.

Solutions to projects

P10.43 (a) The orbitals are sketched in Figure 10.16(a). ψ_1 is a bonding orbital, showing no nodes between adjacent atoms, and ψ_3 is antibonding with respect to all three atoms. ψ_2 is non-bonding, with neither constructive nor destructive interaction of the atomic orbitals of adjacent atoms.

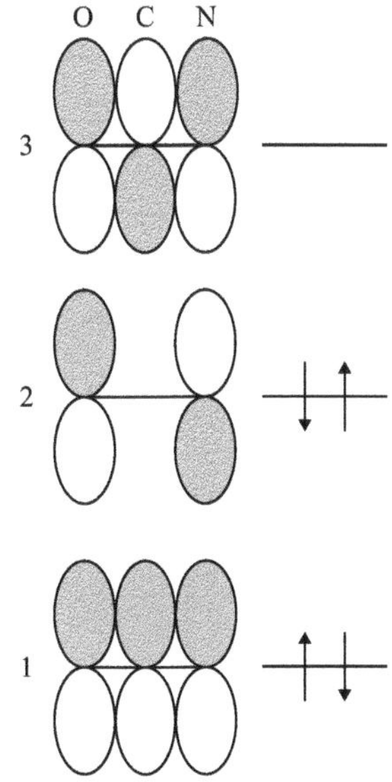

Figure 10.16(a)

(b) This arrangement only works if the entire peptide link is coplanar. Let us call the plane defined by the O, C, and N atoms the xy plane; therefore, the p orbitals used to make the three MOs sketched above are p_z orbitals. If the p_z orbital of N is used in the π system, then the σ bonds it makes must be in the xy plane. Hence, the H atom and the atom labeled $C_{\alpha 2}$ must also be in the xy plane. Likewise, if the p_z orbital of the C atom in the peptide link is used in the π system, then its σ bonds must also lie in the xy plane, putting the atom labeled $C_{\alpha 1}$ in that plane as well.

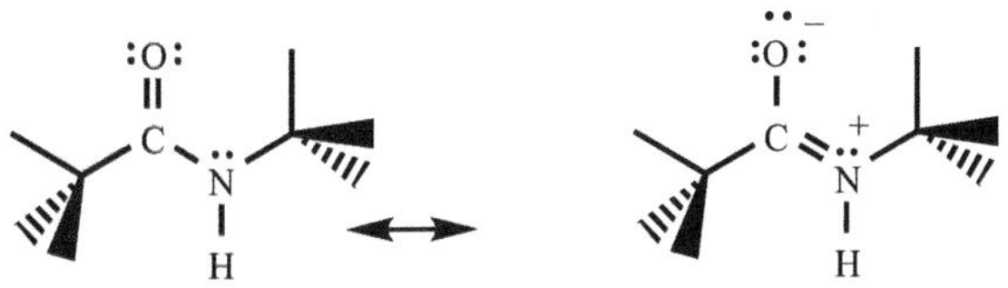

(c) The relative energies of the orbitals and their occupancy are shown in Figure 10.16(a). There are four electrons to be distributed. If we look at the conventional representation of the peptide link, the two electrons represented by the C=O π bond are obviously part of the π system, leaving the two lone pairs on O, the C–O σ bond, and the two other σ bonds of C as part of the σ system. Turning now to the Lewis octet of electrons around the N atom, we must assign two electrons to each of the σ bonds involving N; clearly they cannot be part of the π system. That leaves the lone pair on N, which must occupy the other orbital that N contributes to the molecule, namely the p_z orbital that is part of the π system.

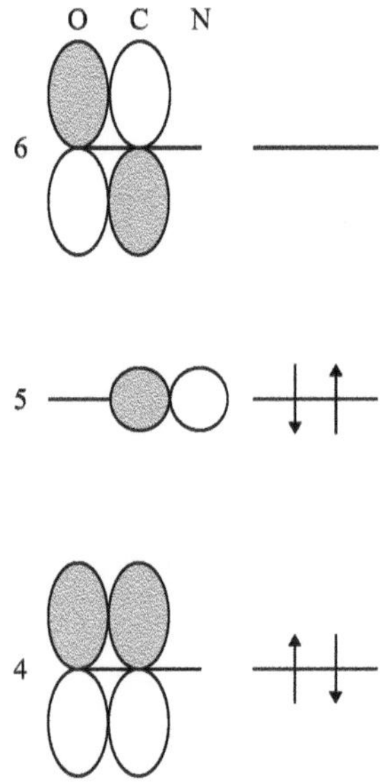

Figure 10.16(b)

(d) The orbitals are sketched in Figure 10.16(b). ψ_4 is a bonding orbital with respect to C and O, and ψ_6 is antibonding with respect to C and O. ψ_5 is non-bonding, involving only the N atom. There are four electrons to be placed in this system, as before, two each in a bonding and non-bonding orbital.

(e) This system cannot be planar. As before, the atom labeled $C_{\alpha 1}$ must be in the xy plane. As before, the atoms bound to N must be in a plane perpendicular to the orbital that N contributes to this system, which is itself in the xy plane; the bonding partners of N are therefore forced out of the xy plane.

(f) The bonding MO ψ_1 must have a lower energy than the bonding MO ψ_4, for ψ_1 is bonding (stabilizing) with respect to all three atoms, while ψ_4 is bonding with respect to only two of them. Likewise, the antibonding MO ψ_3 must have a higher energy than the antibonding MO ψ_6, for ψ_3 is antibonding (destabilizing) with respect to all three atoms pairwise, while ψ_6 is antibonding only with respect to two of them. The non-bonding MOs ψ_2 and ψ_5 must have similar energies, not much different than the parameter α, for there is no significant constructive or destructive interference between adjacent atoms in either one.

(g) Because bonding orbital ψ_1 has a lower energy than ψ_4, the planar arrangement has a lower energy than the non-planar one. The total energy of the planar arrangement is

$$E_{\text{planar}} = 2E_1 + 2E_2.$$

Compare this to the energy of the non-planar arrangement:

$$E_{\text{non-planar}} = 2E_4 + 2E_5 > 2E_1 + 2E_2 = E_{\text{planar}}.$$

The fact that $E_3 > E_6$ is immaterial, for neither of those orbitals is occupied.

P10.44 (a) The table displays computed orbital energies and experimental $\pi^* \leftarrow \pi$ wavenumbers of ethene and the first few conjugated linear polyenes as given in Table 10.3. The computed values in Table 10.3 are taken from the results of the Hartree–Fock computation with an STO-3G basis set in part (b) of the solution below. They clearly do not match well the experimental spectroscopic data. The computed energy separations are plotted against the experimental frequencies (in wavenumbers) in Figure 10.17(a) below.

Species	ΔE/eV*(calc)	$\tilde{\nu}$/cm^{-1}(calc)	$\tilde{\nu}$/cm^{-1}(exp)
C_2H_4	18.1	1.46×10^5	6.13×10^4
C_4H_6	14.5	1.17×10^5	4.61×10^4
C_6H_8	12.7	1.02×10^5	3.97×10^4
C_8H_{10}	11.6	9.36×10^4	3.29×10^4

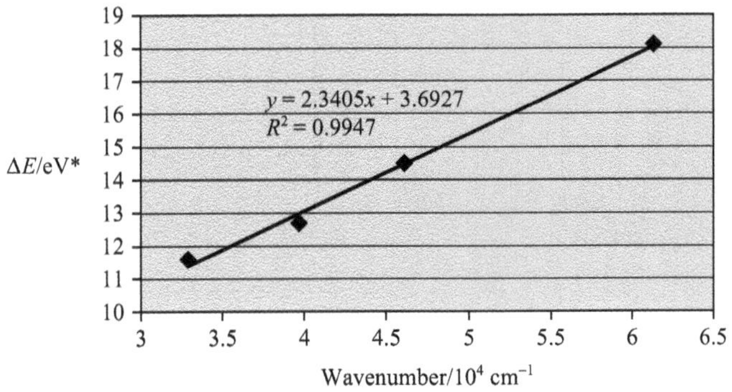

Figure 10.17(a)

The data points clearly fit a straight line. In view of the limited number of data there is no point in exploring whether another polynomial would fit the data better. The linear best fit is:

$$\boxed{\Delta E/\text{eV} = 3.6927 + 2.3405 \times 10^{-4}\ \tilde{\nu}/\text{cm}^{-1}}$$

(b) A Hartree–Fock computation with an STO–3G basis set is shown in the table below.

Species	E_{LUMO}/eV*	E_{HOMO}/eV*	ΔE/eV*	$\tilde{\nu}$/cm^{-1}(calc)	$\tilde{\nu}$/cm^{-1}(exp)
C_2H_4	8.9335	−9.1288	18.0623	1.46×10^5	6.13×10^4
C_4H_6	6.9667	−7.5167	14.4834	1.17×10^5	4.61×10^4
C_6H_8	6.0041	−6.6783	12.6824	1.02×10^5	3.97×10^4
C_8H_{10}	5.4488	−6.1811	11.6299	9.36×10^4	3.29×10^4
$C_{10}H_{12}$	5.0975	−5.8621	10.9596		

**Ab initio*, STO–3G, PC Spartan Pro™

This calculation is clearly the source of the *ab initio* values given in Table 10.3. As stated above the calculated and experimental values do not match well. A semiempirical, PM3 level calculation shown in the table below gives a somewhat better, though not great, fit to the data.

Species	E_{LUMO}/eV*	E_{HOMO}/eV*	ΔE/eV*	$\tilde{\nu}$/cm^{-1} (calc)	$\tilde{\nu}$/cm^{-1} (exp)
C_2H_4	1.2282	−10.6411	11.8693	9.57×10^4	6.13×10^4
C_4H_6	0.2634	−9.4671	9.7305	7.85×10^4	4.61×10^4
C_6H_8	−0.2494	−8.8993	8.6499	6.98×10^4	3.97×10^4
C_8H_{10}	−0.5568	−8.5767	8.0199	6.47×10^4	3.29×10^4
$C_{10}H_{12}$	−0.7556	−8.3755	7.6199		

*Semiempirical, PM3 level, PC Spartan Pro™

A plot of the computed energy difference in the table above versus experimental wavenumbers appears in Figure 10.17(b). The computed points fall on a rather good straight line. Of course a better fit can be obtained to a quadratic and a perfect fit to a cubic polynomial; however, the improvement would be slight and the justification even more slight. The linear least-squares best fit is:

$$\boxed{\Delta E/\text{eV} = 3.3534 + 1.3791 \times 10^{-4}\ \tilde{\nu}/\text{cm}^{-1}} \quad (r^2 = 0.994)$$

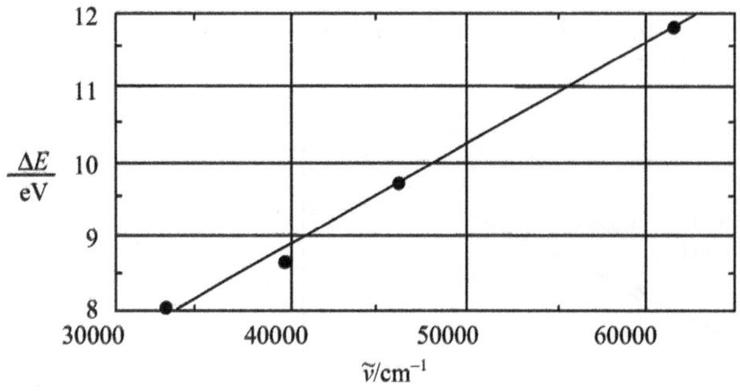

Figure 10.17(b)

(c) The STO–3G data calculated in part (b) and plotted in Figure 10.17(a) fit a straight line. That fit can also be used to estimate the transition in $C_{10}H_{12}$:

$$\tilde{\nu}/\text{cm}^{-1} = \frac{\Delta E/\text{eV} - 3.6927}{2.3045 \times 10^{-4}}.$$

so, for $C_{10}H_{12}$ we expect a transition at:

$$\tilde{\nu}/\text{cm}^{-1} = \frac{10.9596 - 3.6927}{2.3045 \times 10^{-4}} = 31534.$$

$$\tilde{\nu} = \boxed{31534\ \text{cm}^{-1}}.$$

(d) The fitting procedure is necessary because the orbital energies are only approximate. Remember that an orbital wavefunction is itself an approximation. A semiempirical computation is a further approximation. If the orbitals were exact, then we would expect the energy difference to be directly proportional to the spectroscopic wavenumbers with the following proportionality:

$$\Delta E = hc\tilde{\nu} = \frac{(6.26 \times 10^{-34}\ \text{J s})(2.998 \times 10^{10}\ \text{cm s}^{-1})\tilde{\nu}}{1.602 \times 10^{-19}\ \text{J/eV}},$$

so $\Delta E/\text{eV} = 1.240 \times 10^{-4}\tilde{\nu}/\text{cm}^{-1}$.

Clearly this is different from the fits reported above. A further illustration of why the fitting procedure is necessary can be discerned by comparing the two tables in part (b) based on the two different computational methods. Obviously these two calculated energy differences are not the same. Nor are they the energy differences that correspond to the experimental frequencies.

(e) In all of the molecules considered, the HOMO is bonding with respect to the carbon atoms connected by double bonds, but antibonding with respect to the carbon atoms connected by single bonds. (The bond lengths returned by the modeling software suggest that it makes sense to talk about double bonds and single bonds. Despite the electron delocalization, the nominal double bonds are consistently shorter than the nominal single bonds.) The LUMO had just the opposite character, tending to weaken the C=C bonds but strengthen the C–C bonds. To arrive at this conclusion, examine the nodal surfaces of the orbitals. An orbital has an antibonding effect on atoms between which nodes occur, and it has a binding effect on atoms that lie within regions in which the orbital does not change sign. The $\pi^* \leftarrow \pi$ transition, then, would lengthen and weaken the double bonds and shorten and strengthen the single bonds, bringing the different kinds of polyene bonds closer to each other in length and strength. Since each molecule has more double bonds than single bonds, there is an overall weakening of bonds. (See Figure 10.18.)

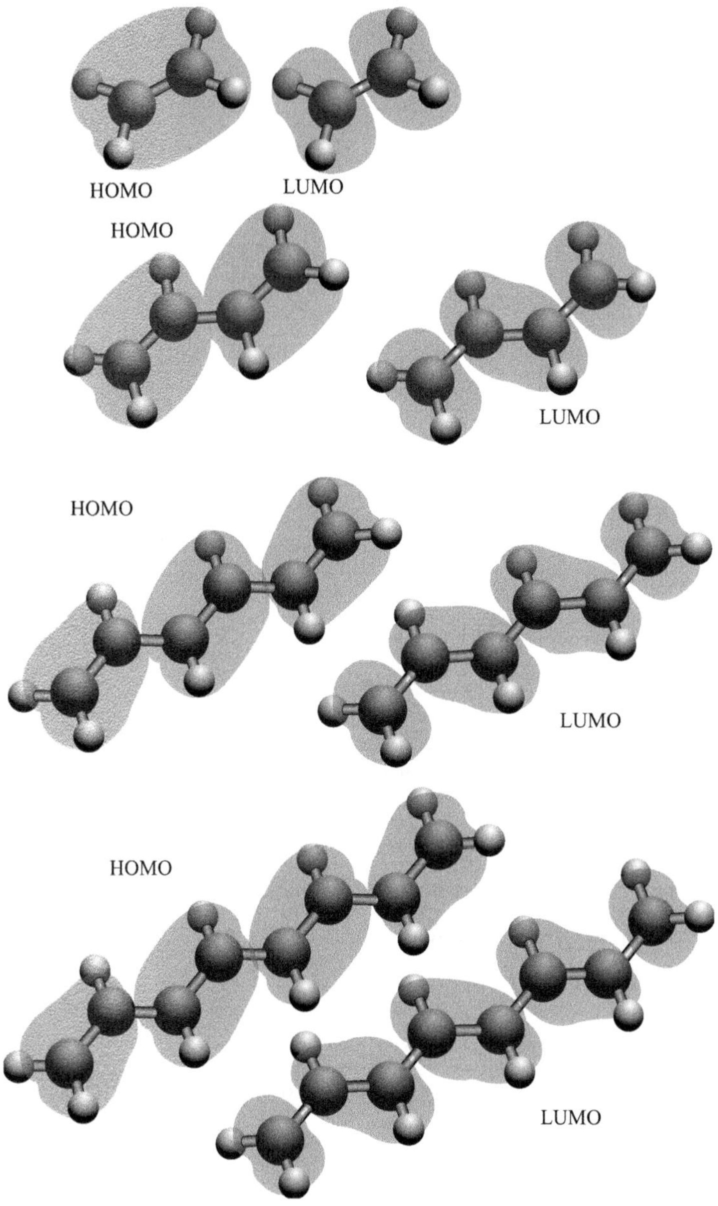

Figure 10.18

P10.45 This question refers to six 1,4-benzoquinones: the unsubstituted, four methyl-substituted, and a dimethyldimethoxy species. The table below defines the molecules and displays reduction potentials and computed LUMO energies.

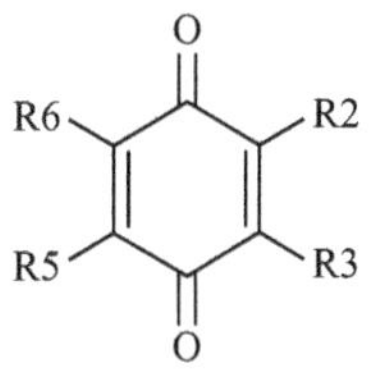

Species	R2	R3	R5	R6	E°/V	E_{LUMO}/eV*
1	H	H	H	H	−0.078	−1.706
2	CH_3	H	H	H	−0.023	−1.651
3	CH_3	H	CH_3	H	−0.067	−1.583
4	CH_3	CH_3	CH_3	H	−0.165	−1.371
5	CH_3	CH_3	CH_3	CH_3	−0.260	−1.233
6	CH_3	CH_3	CH_3O	CH_3O		−1.446

*Semiempirical, PM3 level, PC Spartan Pro™

(a) The LUMO energies for species 1–5 are plotted against standard potentials in Figure 10.19. The figure shows that a linear relationship between the plotted quantities is consistent with these calculations.

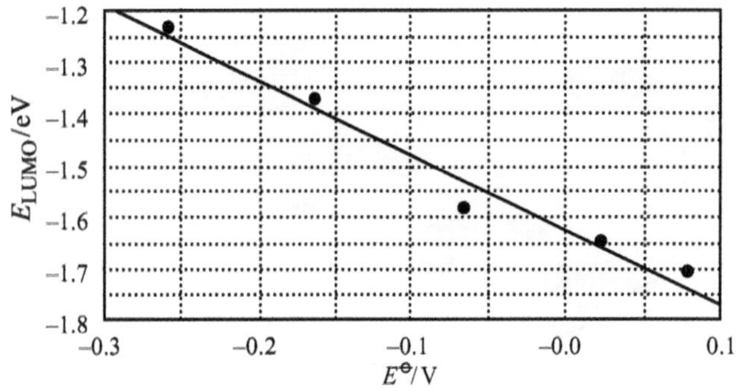

Figure 10.19

(b) The linear least-squares fit from the plot of E_{LUMO} versus E° is:

$$E_{LUMO}/\text{eV} = -1.621 - 1.435E^\circ/\text{V} \quad (r^3 = 0.927)$$

Solving for E° yields:

$$E^\circ/\text{V} = -(E_{LUMO}/\text{eV} + 1.621)/1.435$$

Substituting the computed LUMO energy for compound 6 that is a model for coenzyme Q (another name for ubiquinone) yields

$$E^\circ = [-(-1.446 + 1.621)/1.435] = \boxed{-0.122\text{ V}}.$$

(c) The model of plastoquinone defined in the problem is compound 4 in the table above. Its experimental reduction potential is known; however, a comparison to the ubiquinone analog based on E_{LUMO} ought to use a computed reduction potential:

$$E^\circ = [-(-1.371 + 1.621)/1.435] = \boxed{-0.174\text{ V}}.$$

The better oxidizing agent is the one that is more easily reduced, the one with the less negative reduction potential. Thus, we would expect compound 6 to be a better oxidizing agent than compound 4, and ubiquinone, i.e. coenzyme Q, is a better oxidizing agent than plastiquinone.

(d) Respiration oxidizes organic compounds like glucose to carbon dioxide and water; hence oxidizing agents are required. Photosynthesis reduces carbon dioxide and water to glucose; hence reducing agents are required. It stands to reason, then, that the better oxidizing agent, ubiquinone, is employed in oxidizing glucose (i.e. in respiration), while the better reducing agent (that is, the poorer oxidizing agent) is used in reduction, i.e. in photosynthesis. (Note, however, that both species are recycled to their original forms: reduced ubiquinone is oxidized by iron (III) and oxidized plastoquinone is reduced by water.)

11 Macromolecules and self-assembly

Answers to discussion questions

D11.1 The X-ray diffraction pattern of fibrous B-DNA, shown in both text Figure 11.6 and Figure 11.1 below, provides a seminal example of a diffraction pattern of a helical conformation of a macromolecule. Features are discussed in Case Study 11.1. *The structure of DNA from X-ray diffraction studies*. This particular pattern was obtained with a fiber consisting of many DNA molecules orientated with their axes parallel to the axis of the fiber, with X-rays incident from a perpendicular direction. All the molecules in the fiber are parallel (or nearly so) but are randomly distributed in the perpendicular directions; as a result, the diffraction pattern exhibits the periodic structure parallel to the fiber axis superimposed on a general background of scattering from the distribution of molecules in the perpendicular directions.

Each turn of a helix defines two planes, shown in text Figure 11.22, one orientated at an angle α to the horizontal and the other at $-\alpha$. As a result, to a first approximation, a helix can be thought of as consisting of an array of planes at an angle α to the horizontal together with an array of planes at an angle $-\alpha$ with a separation within each set determined by the pitch p, which is the vertical rise per turn of the helix. Thus, a DNA molecule is like two arrays of planes, each set corresponding to those treated in the derivation of the Bragg law, with a perpendicular separation $d = p\cos\alpha$. The diffraction spots from one set of planes therefore occur at an angle α to the vertical, giving one leg of the X character of the diffraction pattern, and those of the other set occur at an angle $-\alpha$, giving rise to the other leg of the pattern. The experimental arrangement has up–down symmetry, so the diffraction pattern repeats to produce the lower half of the X. The sequence of spots outward along a leg corresponds to first-, second-,...order diffraction ($n = 1, 2, \ldots$ in eqn 11.11b). Figure 11.2 serves to illustrate the reflections off the two sets of planes that form the angle 2α; it also shows the definitions of d and p.

As an aid to understanding the B-DNA X-ray diffraction pattern we represent the nucleotide bases by points as shown in text Figure 11.23 and see that there is an additional periodicity of separation h, forming planes that are perpendicular to the axis to the molecule (and the fiber). These planes give rise to the strong meridional diffraction at the top and bottom of the pattern with an angle that allows us to determine the layer spacing from Bragg's law.

The characteristic X-shape of the B-DNA diffraction pattern shown in Figure 11.1 is that of a helix with incident radiation (Cu K_α 0.1542 nm) perpendicular to the cylindrical axis. An angle $\theta = 2.6°$ between the line of the incident radiation and the line from the sample to the first spot on the X gives $p = \lambda/\sin\theta = 0.1542\ \text{nm}/\sin(2.6°) = 3.4\ \text{nm}$. 10 spots (counting two "missing fourth" spots) along the X diagonal indicate that there are 10 base-planes per turn of the helix with each accounting for a turn of 40°. The very large spot is at a distance $(1/h)$ that is 10 times the distance $1/p$ shown in the diagram. Consequently, $h = 0.34$ nm. The missing fourth spots on the X diagonals indicate two coaxial sugar–phosphate backbones that are separated by $3p/8$ along the axis. The periodic h spacing of the large, very electron-dense phosphorous atoms in the sugar–phosphate backbone of the macromolecule causes the $1/h$ spots to be very intense. The fact that the fibrous X-ray sample was saturated with water suggests that the phosphates are to the outside.

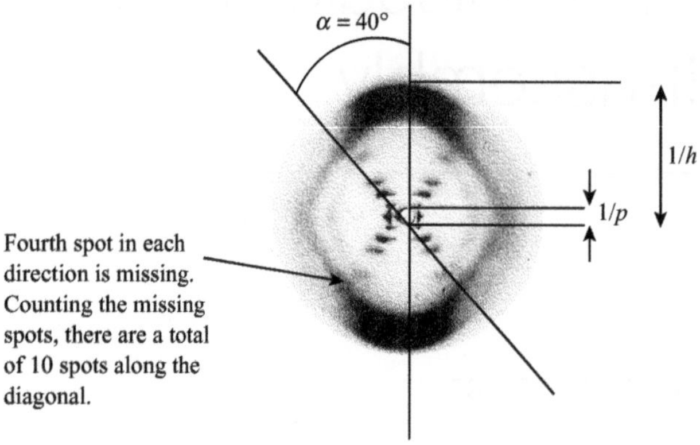

Figure 11.1

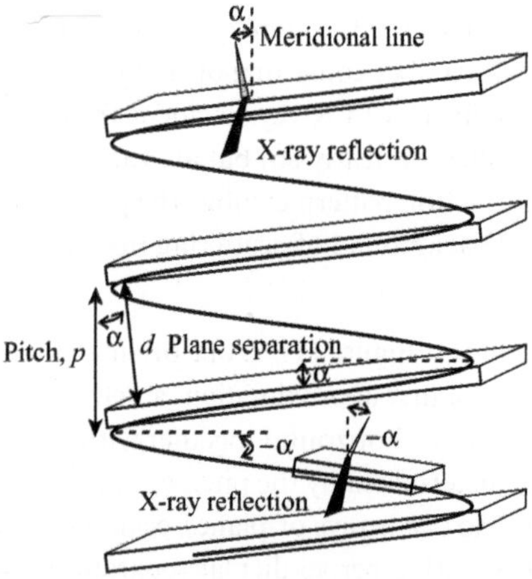

Figure 11.2

Figure 11.3 shows the two-dimensional zig-zag projection of the helical sugar–phosphate backbone onto a plane along the central axis. It serves to show the definitions of the projection length l, perpendicular distance d between backbone planes, and the helix radius r. Examination of the right triangle that shows the definition of α yields:

$$\tan(\alpha) = \frac{p}{4r} \quad \text{or} \quad r = \frac{p}{4\tan(\alpha)} = \frac{3.4\text{ nm}}{4\tan(40°)} = 1.0\text{ nm}$$

Examination of the right triangle containing the angle α also shows that $l\sin(\alpha) = p/2$, while the right triangle containing the angle 2α shows that $l\sin(2\alpha) = d$. Dividing these two equations yields:

$$\frac{\sin(2\alpha)}{\sin(\alpha)} = \frac{2d}{p} \quad \text{or} \quad \frac{2\sin(\alpha)\cos(\alpha)}{\sin(\alpha)} = \frac{2d}{p} \quad \text{or} \quad \cos(\alpha) = \frac{d}{p}$$

$$d = p\cos(\alpha) = (3.4\text{ nm})\cos(40°) = 2.6\overline{0}\text{ nm}$$

Finishing,

$$l = \frac{p}{2\sin(\alpha)} = \frac{3.4\text{ nm}}{2\sin(40°)} = 2.6\overline{5}\text{ nm}$$

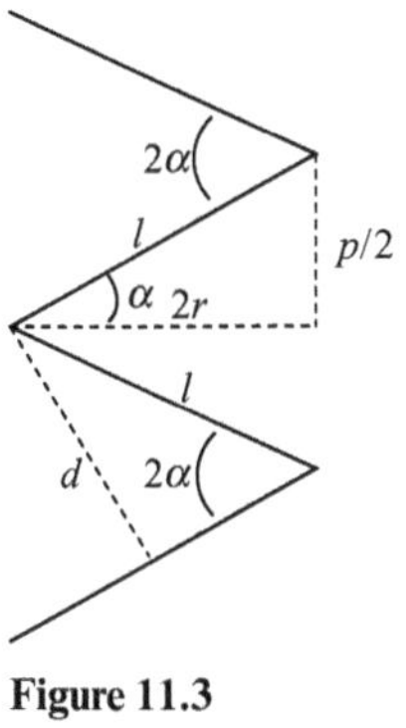

Figure 11.3

D11.2 The **structure factor** F_{hkl} is the sum over all j atoms of terms each of which has a scattering factor f_j:

$$F_{hkl} = \sum_j f_j e^{i\phi_{hkl}(j)}, \quad \text{where} \quad \phi_{hkl}(j) = 2\pi(hx_j + ky_j + lz_j)$$

The importance of the structure factor to the X-ray crystallographic method of structure determination is its **Fourier synthesis** relationship to the electron density distribution, $\rho(r)$, within the crystal:

$$\rho(r) = \frac{1}{V}\sum_{hkl} F_{hkl} e^{-2\pi i(hx+ky+lz)} \quad \text{[generalization of eqn 11.12].}$$

The Fourier synthesis reveals that, if the structure factors for the lattice planes can be measured, the electron density distribution can be calculated by performing the indicated sum. Therein lies the **phase problem**. Measurement detectors yield only the intensity of scattered radiation, which is proportional to $|F_{hkl}|^2$, and give no direct information about F_{hkl}. To see this, consider the structure factor form $F_{hkl} = |F_{hkl}|\, e^{i\alpha_{hkl}}$, where α_{hkl} is the phase of the hkl reflection plane. Then,

$$|F_{hkl}|^2 = \{|F_{hkl}|e^{i\alpha_{hkl}}\}^*\{|F_{hkl}|e^{i\alpha_{hkl}}\} = |F_{hkl}| \times |F_{hkl}|\, e^{-i\alpha_{hkl}}e^{i\alpha_{hkl}} = |F_{hkl}| \times |F_{hkl}|$$

and we see that all information about the phase is lost in an intensity measurement. It seems impossible to perform the sum of the Fourier synthesis since we do not have the important factor $e^{i\alpha_{hkl}}$. Crystallographers have developed numerous methods to resolve the phase problem. In the **Patterson synthesis**, X-ray diffraction spot intensities are used to acquire separation and relative orientations of atom pairs. Another method uses the dominance of heavy-atom scattering to deduce phase. **Heavy-atom replacement** may be necessary for this type of application. **Direct methods** dominate modern X-ray diffraction analysis. These methods use statistical techniques, and the considerable computational capacity of the modern computer, to compute the probabilities that the phases have a particular value.

D11.3 Molecules with a permanent separation of electric charge have a **permanent dipole moment** $\boldsymbol{\mu}$. In molecules containing atoms of differing electronegativity, the bonding electrons may be displaced in such a way as to produce a net separation of charge in the molecule. Separation of charge may also arise from a difference in the atomic radii of the bonded atoms. The separation of charges in the bonds is usually, though not always, in the direction of the more electronegative atom but depends on the precise bonding situation in the molecule as described in Section 11.6. A heteronuclear diatomic molecule necessarily has a dipole moment if there is a difference in electronegativity between the atoms, but the situation in polyatomic molecules is more complex. A polyatomic molecule has a permanent dipole moment only if at least one of its μ_x, μ_y, μ_z components is nonzero (see eqn 11.15a). Thus, the tetrahedral CCl_4 molecule has polar bonds but the sums of the polar components balance so as to cancel and give $\mu = 0$. Molecular symmetry is reduced in $CHCl_3$, a molecule that has a permanent dipole moment because the C–H bond does not balance the polarity of the C–Cl bonds. Similarly, 1,4-dichlorobenzene is non-polar, while 1,2-dichlorobenzene is a polar molecule.

QUESTION: Why does a molecule of trans-1e,4e-dichlorocyclohexane have no permanent dipole while a molecule of cis-1a,4e-dichlorocyclohexane does have a permanent dipole?

See the discussion of ozone and carbon dioxide in Section 11.6 for further examples of the importance of molecular symmetry and electronegativities in deciding whether a polyatomic molecule is polar or not. The discussion of carbon monoxide is a very important example of a molecule that has a dipole moment in the opposite direction to that expected from electronegativity considerations alone because the polarity of the HOMO antibonding orbital, which is reversed from the electronegativity expectation, provides a large contribution to the observed polarity.

Both non-polar and polar molecules may acquire a temporary **induced dipole moment** μ^* as a result of the influence of an **electric field** $\mathcal{E}$ generated by a nearby ion or polar molecule. The field distorts the electron distribution of the molecule, and gives rise to an electric dipole. The induced dipole moment is proportional to the field [11.19] and the constant of proportionality is called the **polarizability** α. Molecular structure features that affect polarizability include molecular size, nuclear control, ionization energy, and the relative orientation of the molecule with the external electric field.

D11.4 A **hydrogen bond** ($\cdots$) is an attractive interaction between two species that arises from a link of the form A–H$\cdots$B, where A and B are highly electronegative elements (usually nitrogen, oxygen, or fluorine) and B possesses a lone pair of electrons. It is a contact-like attraction that requires AH to touch B. Experimental evidence supports a linear or near-linear structural arrangement and a bond strength of about 20 kJ mol^{-1}. The hydrogen bond strength is considerably weaker than a covalent bond but it is larger than, and dominates, other intermolecular attractions such as dipole–dipole attractions. Its formation can be understood in terms of either the (a) electrostatic interaction model or with (b) molecular orbital calculations.

(a) A and B, being highly electronegative, are viewed as having partial negative charges (δ^-) in the electrostatic interaction model of the hydrogen bond. Hydrogen, being less electronegative than A, is viewed as having a partial positive (δ^+). The linear structure maximizes the electrostatic attraction between H and B:

$$\overset{\delta^-}{\mathrm{A}}\text{——}\overset{\delta^+}{\mathrm{H}}\text{-------------------}\overset{\delta^-}{:\mathrm{B}}$$

This model is conceptually very useful. However, it is impossible to exactly calculate the interaction strength with this model because the partial atomic charges cannot be precisely defined. There is no way to define which fraction of the electrons of the AB covalent bond should be assigned to one or the other nucleus.

(b) *Ab initio* MO quantum calculations are needed in order to explore questions about the linear structure, the role of the lone pair, the shape of the potential energy surface, and the extent to which the hydrogen bond has covalent sigma bond character. Yes, the hydrogen bond appears to have some sigma bond character. This was initially suggested by Linus Pauling in the 1930s and more recent experiments with Compton scattering of X-rays and NMR techniques indicate that the covalent character may provide as much as 20% of the hydrogen bond strength. A three-center molecular orbital model provides a degree of insight. A linear combination of an appropriate sigma orbital on A, the 1s hydrogen orbital, and an appropriate orbital for the lone pair on B yields a total of three molecular orbitals of the form:

$$\psi = c_1\psi_A + c_2\psi_H + c_3\psi_B$$

One of the MOs is bonding, one is almost non-bonding, and the third is antibonding (see text Figure 11.29). Both the bonding MO and the almost non-bonding orbital are occupied by two electrons (the sigma bonding electrons of A–H and the lone pair of B). The antibonding MO is empty. Thus, depending on the precise location of the almost non-bonding orbital, the non-bonding orbital may lower the total energy and account for the hydrogen bond.

D11.5 **Contour length, R_c**: the length of the macromolecule measured along its backbone, the length of all its monomer units placed end to end. This is the stretched-out length of the macromolecule with bond angles maintained within the monomer units and 180° angles at unit links. It is proportional to the number of monomer units, N, and to the length of each unit [11.29].

Root mean square separation, R_{rms}: one measure of the average separation of the ends of a random coil. It is the square root of the mean value of R^2, where R is the separation of the two ends of the coil. R_{rms} is proportional to $N^{1/2}$ and the length of each unit [11.28].

Radius of gyration, R_g: the radius of a thin hollow spherical shell of the same mass and moment of inertia as the macromolecule. In general, it is not easy to visualize this distance geometrically. However, for the simple case of a molecule consisting of a chain of identical atoms this quantity can be visualized as the root mean square distance of the atoms from the center of mass. It also depends on $N^{1/2}$, but is smaller than the root mean square separation by a factor of $(1/6)^{1/2}$ [11.30].

D11.6 (a) $V = -\dfrac{Q_2\mu_1}{4\pi\varepsilon_0 r^2}$ [11.16a]

V is the potential energy of interaction between a point dipole μ_1 and the point charge Q_2 at the separation r. The point charge lies on the axis of the dipole and the separation r is much larger than the separation of charge within the dipole so that the partial charges of the dipole seem to merge and cancel to create the so-called **point dipole**.

(b) $V = -\dfrac{Q_2\mu_1\cos\theta}{4\pi\varepsilon_0 r^2}$ [11.16b]

V is the potential energy of interaction between a point dipole μ_1 and the point charge Q_2 at the separation r. The point charge lies at an angle θ to the axis of the dipole and the separation r is much larger than the separation of charge within the dipole so that the partial charges of the dipole seem to merge and cancel.

(c) $V = \dfrac{\mu_1\mu_2 f(\theta)}{4\pi\varepsilon_0 r^3}$, where $f(\theta) = 1 - 3\cos^2\theta$ [11.17]

V is the potential energy of interaction between the two point dipoles μ_1 and μ_2 at the separation r. The dipoles are parallel and the separation distance is at angle θ to the dipoles. The separation r is much larger than the separation of charge within the dipoles so that the partial charges of the dipoles seem to merge and cancel.

(d) $R_{rms} = N^{1/2}l$

Each of N monomers occupies length l along the polymer chain. R_{rms} is the root mean square separation between the ends of the polymer chain when the polymer is modeled as a freely jointed chain in which any bond between two monomer residues is free to make any angle with respect to the previous one. This random-coil model ignores the volume occupied by each residue, electronic restrictions on bond angles, and solvent effects.

(e) $R_g = (N/6)^{1/2}l$

Each of N monomers occupies length l along the polymer chain, which is modeled as a freely jointed chain. R_g is the radius of gyration of the randomly coiled polymer. That is, R_g is the radius of a thin shell that has the same mass and the same moment of inertia as the coiled polymer.

D11.7 The α helix and β sheets exhibited as secondary structure in polypeptide macromolecules involve hydrogen bonding between widely separated peptides. We begin by recognizing that the peptide group is planar due to delocalization of carbonyl π electrons, which is illustrated with the resonance structures of Figure 11.4.

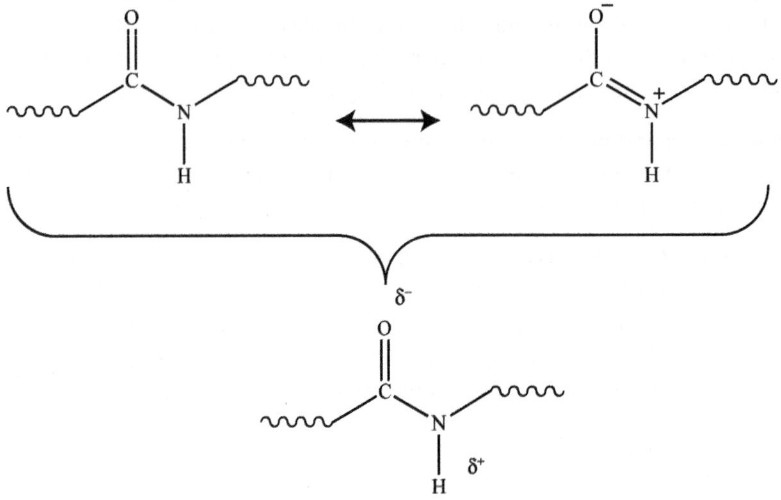

Figure 11.4

The nitrogen atom of the peptide group is incapable of hydrogen bonding but both the oxygen and hydrogen atoms are hydrogen bonded. Text Figure 11.40 and the sketch of Figure 11.5 illustrate the α helix secondary structure, which is stabilized by the hydrogen bond between each peptide of the helix and one within an amino acid residue 4 units along the chain. The hydrogen atom is always hydrogen bonded to the oxygen atom.

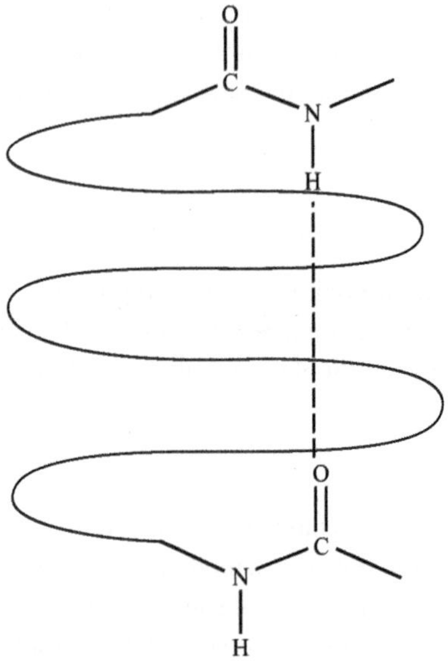

Figure 11.5

The antiparallel and parallel β sheet of hydrogen-bonded secondary structure are illustrated in text Figures 11.43a and 11.43b, respectively. Two separated segments of the peptide chain are hydrogen bonded in sequence with the N-terminal-end to C-terminal-end of each segment running together in the parallel β shell but in the antiparallel β sheet the N-terminal-end to C-terminal-end of each peptide segment run in opposite directions. The **Corey–Pauling rules**, discussed in text

Section 11.13(a), summarize the observed characteristics of the peptide-to-peptide hydrogen-bonding pattern in protein.

D11.8 A polypeptide residue side chain R that has polar bonds, thereby, has the capacity to participate in Coulombic interactions. If the R group contains an alcohol, an amine, a carboxylate group, it can participate in hydrogen bonding. Should the R group be non-polar, the hydrophobic interaction predominates (see *Resource section 1: Atlas A, amino acids*)

(a) Residues capable of the Coulombic interaction at pH 7: Arg, Asn, Asp, Gln, Glu, His
(b) Capable of hydrogen bonding at pH 7: Arg, Asn, Asp, Gln, Glu, His, Hys, Ser, Thr
(c) Hydrophobic interactions: Ala, Iso, Leu, Met, Phe, Trp, Tyr, Val

D11.9 Because there are three hydrogen bonds per C−G base pair, a DNA sequence that is rich in C−G base pairs is more stable than sequences that are rich in A−T base pairs for which only two hydrogen bonds form per pair.

D11.10 The glycosidic link between α and β cyclic forms of glucose (α and β pyranose rings) determines the bent, linear, and branched character of polysaccharides. The "bent" structure of amylose originates with α-1,4-glycosidic linkages between the glucose monomer unit as shown in text structure **16**. The "branched" structure of amylopectin, shown in text structure **18**, may be viewed as having α-1,6-glycosidic linkages off a bent parent of α-1,4-glycosidically linked glucose units. Glycogen has linkages like those of amylopectin but is more heavily branched. The "linear" structure of cellulose originates with β-1,4-glycosidic linkages between the glucose units as shown in text structure **17**. Figure 11.6 uses Haworth diagrams to illustrate the α-1,4-glycosidic link of maltose and the β-1,4-glycosidic link of lactose. Can you add an additional glucose unit with an α-1,6-glycosidic link to the maltose figure?

Starch is 10–20% amylose and 80–90% amylopectin. Glycogen ("animal starch") is stored in liver and muscle cells.

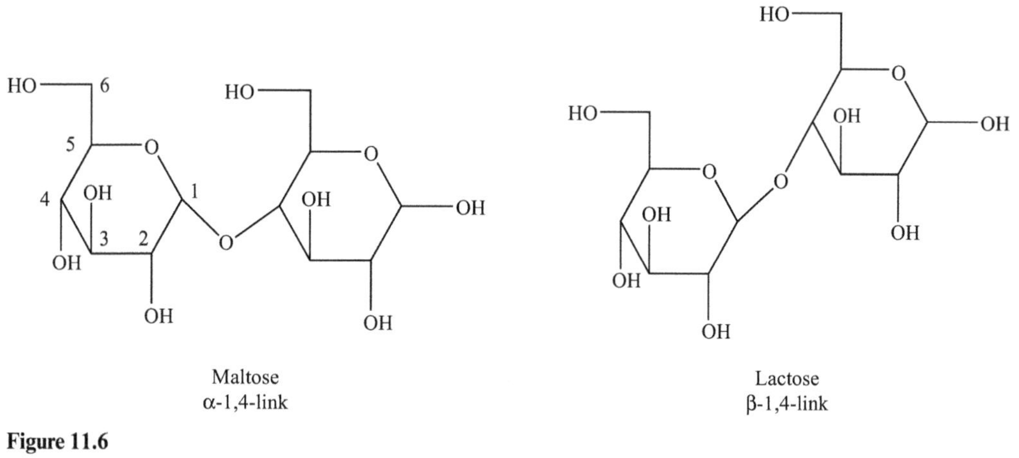

Figure 11.6

D11.11 (a) In aqueous environments amphipathic molecules can group together as **micelles**, in which hydrophobic tails congregate, leaving hydrophilic heads exposed to the solvent (text Figure 11.51). **Liposomes** form with an inward-pointing inner surface of molecules surrounded by an outward-pointing outer layer (text Figure 11.52).

Planar bilayers consist of extended parallel sheets two molecules thick. The individual molecules lie perpendicular to the sheets, with hydrophilic groups on the outside in aqueous solution and on the inside in non-polar media. When segments of planar bilayers fold back on themselves, **unilamellar vesicles** may form where the spherical hydrophobic bilayer shell separates an inner aqueous compartment from the external aqueous environment.

Membranes have a complex bilayer structure consisting of phospholipids, cholesterol, and an assortment of other lipids. **Peripheral proteins** are proteins attached to the membrane bilayer. **Integral proteins** are proteins embedded in the mobile but viscous bilayer. Examples include complexes I–IV of oxidative phosphorylation (Section 5.10), ion channels, and ion pumps (Section 5.3). Integral proteins may span the depth of the bilayer and consist of tightly packed α helices or, in some cases, β sheets containing hydrophobic residues that sit within the hydrocarbon region of the bilayer. The hydrophobicity of a protein residue can be assessed by measuring the Gibbs energy of transfer of the corresponding amino acid from an aqueous solution to the interior of a membrane (Table 11.9). Amino acids with negative values of the Gibbs energy of transfer are likely to be found in the membrane-spanning regions of integral proteins.

(b) The **surfactant parameter** is defined by $N_s = V/Al$ [11.36] where V is the volume of the hydrophobic tail, A is the area of the hydrophilic head group, and l is the maximum length of the tail. The parameter is a predictor of the shape of the micelle as illustrated by the Table 11.8 correlations of parameter values and ranges with micelle shape. A detailed discussion of this parameter is presented by R. Nagarajan in *Langmuir*, 2002, **18(1)**, pp 31–38.

D11.12 The formation of micelles is favored by the interaction between hydrocarbon tails and is opposed by charge repulsion of the polar groups that are placed close together at the micelle surface. As salt concentration is increased, the repulsion of head groups is reduced because their charges are partly shielded by the ions of the salt. This favors micelle formation causing the micelles to be larger and the critical micelle concentration to be smaller.

D11.13 In the **fluid mosaic model**, shown in text Figure 11.53, the proteins are mobile, but their diffusion coefficients are much smaller than those of the lipids. In the **lipid raft model**, a number of lipid and cholesterol molecules form ordered structures, or "rafts," that envelope proteins and help carry them to specific parts of the cell.

D11.14 For a **molecular mechanics** calculation, potential-energy functions are chosen for all the interactions between the atoms in the molecule; the calculation itself is a mathematical procedure that locates the conformational energy minima (local and global) as a function of bond distances and bond angles. Terms are included to account for potential-energy variations with bond stretching, bond bending, torsional forces, dispersive, electrostatic, and hydrogen-bonding interactions. Because only the potential energy is included in the calculation, contributions to the total energy from the kinetic energy are excluded in the result. Thus, the global energy minimum is a snapshot of the molecular structure at $T = 0$. No equations of motion are solved in a molecular mechanics calculation.

In a **molecular dynamics** calculation, equations of motion are numerically integrated to determine the trajectories of all atoms in the molecule. The equations of motion can, in principle, be either classical (Newton's laws of motion) or quantum mechanical. But, in practice, due to the very large number of atoms in a macromolecule, Newton's equations of motion are used. The history of an initial arrangement is followed by calculating the trajectories of all the particles under the influence of the intermolecular potentials. Equations of motion predict where each particle will be after a short time interval (about 1 fs, which is shorter than the average time between collisions), and then the calculation is repeated for tens of thousands of such steps. The time-consuming part of the calculation is the evaluation of the net force on the molecule arising from all the other molecules present in the system. The method has been used to provide insight into the molecular motions in simple liquids and to give a series of snapshots of the wiggling, vibrational motion of macromolecule atoms during unaided folding.

In the **Monte Carlo method**, the atoms of the molecule are moved through small but otherwise random distances, and the change in conformational energy, ΔV_C, is calculated. Whether this new configuration is accepted is then judged from the following rules:

1. If the potential energy is not greater than before the change, then the configuration is accepted.

2. If the potential energy is greater than before the change, the Boltzmann factor $e^{-\Delta V_C/kT}$ is compared with a random number between 0 and 1; if the factor is larger than the random number, the configuration is accepted; if the factor is not larger, the configuration is rejected. This procedure ensures that at equilibrium the probability of occurrence of any configuration is proportional to the Boltzmann factor.

The structure of a macromolecule (or any molecule, for that matter) can, in principle, be determined by solving the time-independent **Schrödinger equation of quantum mechanics** by computational methods similar to those described in Chapter 10. But due to the very large size of macromolecules, these methods may be impractical and, due to approximations to make them tractable, inaccurate. Consequently, quantum-mechanical methods are too time consuming, complicated, and at this stage too inaccurate to be popular in the field of biopolymer chemistry.

Solutions to exercises

E11.15 Angular frequency, $\omega = (\text{rotation rate}) \times (2\pi \text{ radians cycle}^{-1})$

$$\omega = (50\,000 \text{ cycles min}^{-1}) \times (2\pi \text{ radians cycle}^{-1}) \times \left(\frac{1 \text{ min}}{60 \text{ s}}\right) = 5236 \text{ radians s}^{-1}$$

$$M = \frac{2RT}{b\omega^2} \times (\text{slope} \times \text{cm}^{-2}) \text{ [Example 11.1]}$$

In the molar mass expression $b = 1 - \rho v_s$, ρ is the solvent density, and v_s is the specific volume of the solute. The "slope" is that of the linear $\ln c$ against r^2 plot. Here, slope $\times$ cm^{-2} = 729 cm^{-2} = 729×10^4 m^{-2} so

$$M = \frac{2 \times (8.3145 \text{ J K}^{-1} \text{ mol}^{-1}) \times (300 \text{ K})}{\{1 - (1.00 \text{ g cm}^{-3}) \times (0.61 \text{ cm}^3 \text{ g}^{-1})\} \times (5236 \text{ s}^{-1})^2} \times (729 \times 10^4 \text{ m}^{-2}) \text{ [1 J = 1 kg m}^2 \text{ s}^{-2}]$$
$$= \boxed{3.40 \times 10^3 \text{ kg mol}^{-1}}$$

E11.16 The net force acting upon the settling particle equals zero because of the balance between the gravitation pull, $m_{\text{eff}}g$, and the frictional force, fs, where s is the drift speed and the frictional coefficient f is given by the Stokes–Einstein relation $f = kT/D = 6\pi a\eta$ [8.14] for a particle of radius a. Thus,

$$fs = m_{\text{eff}}g$$

$$s = \frac{m_{\text{eff}}g}{f} = \frac{bmg}{f} = \frac{(1 - \rho v_s) \times \{(\tfrac{4}{3}\pi a^3) \times \rho_s\}g}{6\pi a\eta} \quad [v_s = \rho_s^{-1}]$$

$$= \frac{2(\rho_s - \rho)a^2 g}{9\eta}$$

$$= \frac{2 \times (1750 \text{ kg m}^{-3} - 1000 \text{ kg m}^{-3}) \times (20 \times 10^{-6} \text{ m})^2 \times (9.8067 \text{ m s}^{-2})}{9 \times (8.9 \times 10^{-4} \text{ kg m}^{-1} \text{ s}^{-1})}$$

$$= \boxed{0.73 \text{ mm s}^{-1}}$$

E11.17

$$M = \frac{SRT}{(1-\rho v_s)D} \quad [11.2]$$

$$= \frac{(3.2\ \text{Sv}) \times (10^{-13}\ \text{s Sv}^{-1}) \times (8.3145\ \text{J mol}^{-1}\ \text{K}^{-1}) \times (293.15\ \text{K})}{\{1-(1.06\ \text{g cm}^{-3}) \times (0.656\ \text{cm}^3\ \text{g}^{-1})\} \times (8.3 \times 10^{-11}\ \text{m}^2\ \text{s}^{-1})}$$

$$= \boxed{31\ \text{kg mol}^{-1}}$$

E11.18

$$M = \frac{2RT}{(r_2^2 - r_1^2)b\omega^2} \ln\frac{c_2}{c_1} [11.3a] = \frac{2RT}{(r_2^2 - r_1^2)b(2\pi v)^2} \ln\frac{c_2}{c_1} [\omega = 2\pi v]$$

Solving for the squared rotational frequency v and substitution of $b = 1 - \rho v_s$ gives

$$v^2 = \frac{RT\ln(c_2/c_1)}{2\pi^2 M \times (1-\rho v_s) \times (r_2^2 - r_1^2)}$$

$$= \frac{(8.3145\ \text{J K}^{-1}\ \text{mol}^{-1}) \times (298\ \text{K}) \times \ln 5}{2\pi^2 \times (1 \times 10^2\ \text{kg mol}^{-1}) \times (1-0.75) \times (7.0^2 - 5.0^2) \times 10^{-4}\ \text{m}^2}$$

$$= 33\overline{67}\ \text{s}^{-2}$$

$$v = (58\ \text{s}^{-1}) \times (60\ \text{s min}^{-1}) = \boxed{3500\ \text{r.p.m.}}$$

E11.19 (a) The plots of n_{bp} against t and against t^2 are shown in Figures 11.7 and 11.8, respectively. The plot of n_{bp} against t is non-linear but a second-order polynomial regression fit is an excellent description of the variation. The plot of $\boxed{n_{bp} \text{ against } t^2 \text{ is linear}}$ and the linear regression fit is shown as a box insert within the figure. The linearity of the latter plot is in agreement with eqn 11.5, which indicates that the molecule mass is proportional to t^2 in a TOF spectrometer. Since the molecule mass is proportional to the number of DNA base pairs, n_{bp}, the number of base pairs must also be proportional to t^2 in a TOF spectrometer.

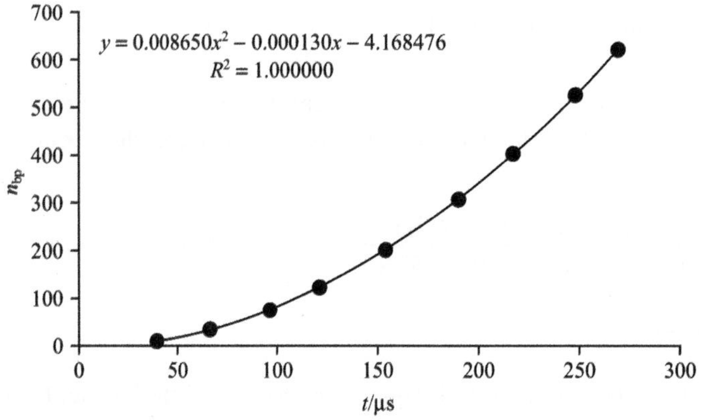

Figure 11.7

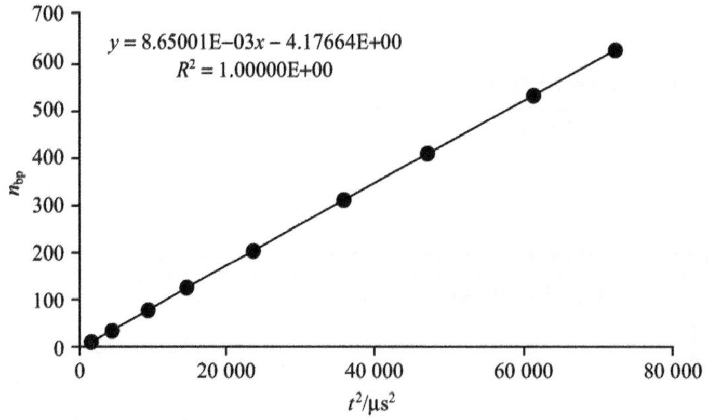

Figure 11.8

(b) $n_{bp} = 8.650 \times 10^{-3} \times (t/\mu s)^2 - 4.177$

$$t = \left(\frac{n_{bp} + 4.177}{8.650 \times 10^{-3}}\right)^{1/2} \mu s$$

$$= \left(\frac{238 + 4.177}{8.650 \times 10^{-3}}\right)^{1/2} \mu s$$

$$= \boxed{167\ \mu s}$$

E11.20 The points and planes are shown in Figure 11.9.

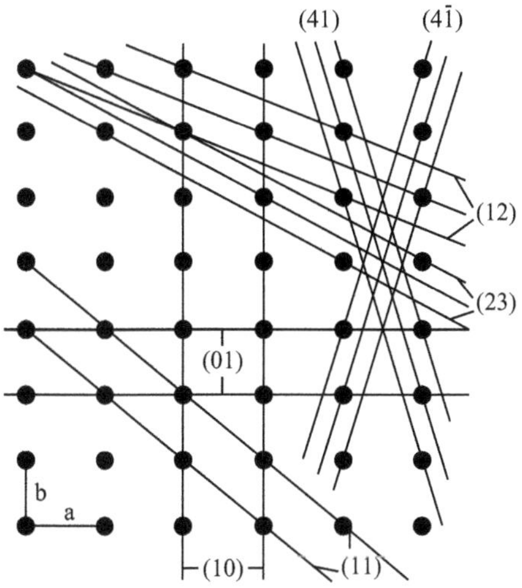

Figure 11.9

E11.21 The points and planes are shown in Figure 11.10.

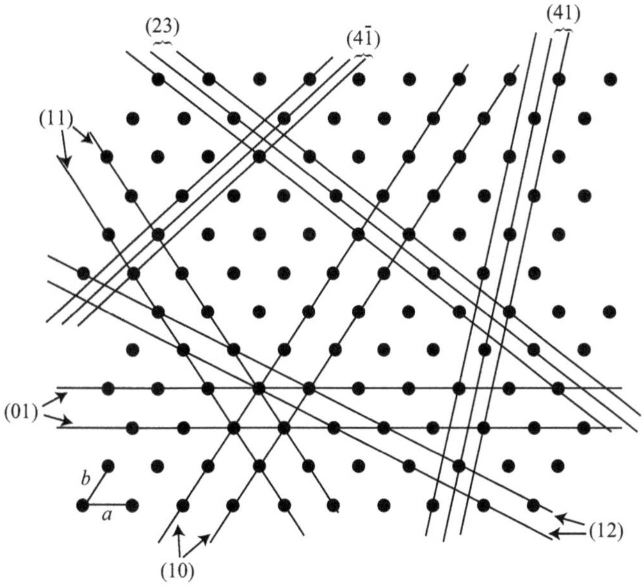

Figure 11.10

E11.22 Draw up the following table, using the procedure set out in Section 11.4(c).

Original	Reciprocal	Clear fractions	Miller indices
$(2a, 3b, c)$ or $(2, 3, 1)$	$(\frac{1}{2}, \frac{1}{3}, 1)$	$(3, 2, 6)$	(326)
(a, b, c) or $(1, 1, 1)$	$(1, 1, 1)$	$(1, 1, 1)$	(111)
$(6a, 3b, 3c)$ or $(6, 3, 3)$	$(\frac{1}{6}, \frac{1}{3}, \frac{1}{3})$	$(1, 2, 2)$	(122)
$(2a, -3b, -3c)$ or $(2, -3, -3)$	$(\frac{1}{2}, -\frac{1}{3}, -\frac{1}{3})$	$(3, -2, -2)$	$(3\bar{2}\bar{2})$

E11.23 See Figure 11.11(a) for the (100), (010), (001), (011), (101), and (111) planes. See Figure 11.11(b) for the $(10\bar{1})$ plane.

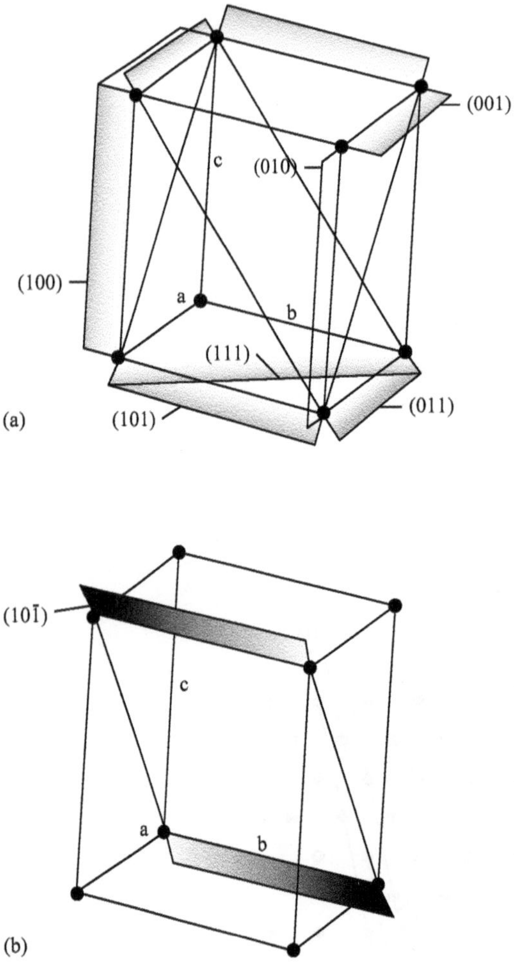

Figure 11.11

E11.24
$$\frac{1}{d_{hkl}^2} = \frac{h^2}{a^2} + \frac{k^2}{b^2} + \frac{l^2}{c^2} \quad [11.10]$$

(a) For a cubic unit cell in which $a = b = c$

$$d_{hkl} = \frac{a}{(h^2 + k^2 + l^2)^{1/2}}$$

Therefore,

$$d_{111} = \frac{a}{3^{1/2}} = \frac{532\ \text{pm}}{3^{1/2}} = \boxed{307\ \text{pm}}$$

$$d_{211} = \frac{a}{6^{1/2}} = \frac{532\ \text{pm}}{6^{1/2}} = \boxed{217\ \text{pm}}$$

$$d_{100} = a = \boxed{532\ \text{pm}}$$

(b) For an orthorhombic crystal of sides $a = 0.754$ nm, $b = 0.623$ nm, $c = 0.433$ nm

$$d_{hkl} = \left(\frac{h^2}{a^2} + \frac{k^2}{b^2} + \frac{l^2}{c^2}\right)^{-1/2}$$

$$d_{123} = \left(\frac{1^2}{0.754^2} + \frac{2^2}{0.623^2} + \frac{3^2}{0.433^2}\right)^{-1/2} \text{nm} = \boxed{0.129\ \text{nm}}$$

$$d_{236} = \left(\frac{2^2}{0.754^2} + \frac{3^2}{0.623^2} + \frac{6^2}{0.433^2}\right)^{-1/2} \text{nm} = \boxed{0.0671\ \text{nm}}$$

E11.25 $\lambda = 2d\sin\theta$ [11.11a] $= 2 \times (97.3\ \text{pm}) \times (\sin 19.85°) = \boxed{66.1\ \text{pm}}$

E11.26 $V\rho(x) = F_0 + 2\sum_{h=1}^{\infty} F_h \cos(2h\pi x)$ [11.12]

$$= 30 + 16.4\cos(2\pi x) + 13.0\cos(4\pi x) + 8.2\cos(6\pi x) + 11\cos(8\pi x) - 4.8\cos(10\pi x) + 10.8\cos(12\pi x) + 6.4\cos(14\pi x) - 2.0\cos(16\pi x) + 2.2\cos(18\pi x) + 13.0\cos(20\pi x) + 10.4\cos(22\pi x) - 8.6\cos(24\pi x) - 2.4\cos(26\pi x) + 0.2\cos(28\pi x) + 4.2\cos(30\pi x)$$

A plot of $V\rho(x)$ is shown in Figure 11.12.

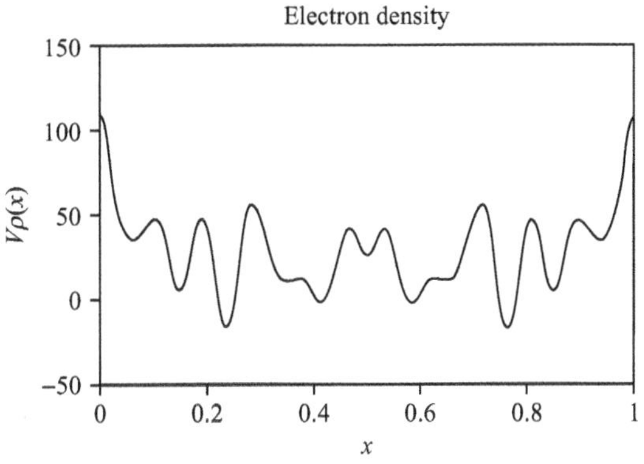

Figure 11.12

E11.27 We assume a linear alignment of the hydrogen bond C=O···H−N between the two peptide groups. The distance between the C and N atoms is given as $r_{CN} = 290$ pm and text structure **5** (the peptide group properties) is used to deduce the bond lengths $r_{CO} = 124$ pm and $r_{NH} = 100$ pm. From these values we deduce the distances within the hydrogen bond to be $r_{CH} = 190$ pm, $r_{OH} = 66$ pm, and $r_{ON} = 166$ pm. Text structure **5** also provides values for the partial charges on each atom within the hydrogen bond. They are $Q_H = +0.18e$, $Q_N = -0.36e$, $Q_C = +0.45e$, and $Q_O = -0.38e$, where $e = 1.6022 \times 10^{-19}$ C. When estimating the hydrogen bond strength with the electrostatic model, we need only consider the interactions between the two peptides, which means that the hydrogen bond strength is given by $-V = V_{CH} + V_{CN} + V_{OH} + V_{ON}$.

(a) The relative permittivity, ε_r, equals 1 in a vacuum so the hydrogen bond strength is given by

$$-V=-\left\{\frac{Q_CQ_H}{4\pi\varepsilon_0 r_{CH}}+\frac{Q_CQ_N}{4\pi\varepsilon_0 r_{CN}}+\frac{Q_OQ_H}{4\pi\varepsilon_0 r_{OH}}+\frac{Q_OQ_N}{4\pi\varepsilon_0 r_{ON}}\right\}\quad[11.13a]$$

$$=-\frac{e^2}{4\pi\varepsilon_0\times(1\times10^{-12}\,\text{m})}\left\{\frac{Q_CQ_H/e^2}{r_{CH}/\text{pm}}+\frac{Q_CQ_N/e^2}{r_{CN}/\text{pm}}+\frac{Q_OQ_H/e^2}{r_{OH}/\text{pm}}+\frac{Q_OQ_N/e^2}{r_{ON}/\text{pm}}\right\}$$

$$=-\frac{(1.6022\times10^{-19}\,\text{C})^2}{(1.1127\times10^{-10}\,\text{J}^{-1}\,\text{C}^2\,\text{m}^{-1})\times(1\times10^{-12}\,\text{m})}\left\{\frac{Q_CQ_H/e^2}{r_{CH}/\text{pm}}+\frac{Q_CQ_N/e^2}{r_{CN}/\text{pm}}+\frac{Q_OQ_H/e^2}{r_{OH}/\text{pm}}+\frac{Q_OQ_N/e^2}{r_{ON}/\text{pm}}\right\}$$

$$=-(2.307\times10^{-16}\,\text{J})\times\left\{\frac{Q_CQ_H/e^2}{r_{CH}/\text{pm}}+\frac{Q_CQ_N/e^2}{r_{CN}/\text{pm}}+\frac{Q_OQ_H/e^2}{r_{OH}/\text{pm}}+\frac{Q_OQ_N/e^2}{r_{ON}/\text{pm}}\right\}$$

$$=-(2.307\times10^{-16}\,\text{J})\times\left\{\frac{(+0.45)(+0.18)}{190}+\frac{(+0.45)(-0.36)}{290}+\frac{(-0.38)(+0.18)}{66}+\frac{(-0.38)(-0.36)}{166}\right\}$$

$$=7.9\bar{5}\times10^{-20}\,\text{J}$$

$$-VN_A=(7.9\bar{5}\times10^{-20}\,\text{J})\times(6.022\times10^{23}\,\text{mol}^{-1})=\boxed{47.\bar{9}\,\text{kJ mol}^{-1}}$$

(b) The relative permittivity, ε_r, equals 2.0 in a membrane so the hydrogen bond strength is reduced from vacuum by the factor 1/2.0, thereby, giving $\boxed{24\,\text{kJ mol}^{-1}}$.

(c) The relative permittivity, ε_r, equals 80.0 in water so the hydrogen bond strength is reduced from vacuum by the factor 1/80.0, thereby, giving $\boxed{0.60\,\text{kJ mol}^{-1}}$.

E11.28 From Table 10.2: $\chi(\text{H})=2.1$, $\chi(\text{Cl})=3.0$, and $\Delta\chi_{HCl}=0.9$. Thus,

$$\mu_{HCl}\approx\Delta\chi_{HCl}\,\text{D}\,[11.14]=\boxed{0.9\,\text{D}}$$

$$\mu_{HCl}\approx(0.9\,\text{D})\times(3.33564\times10^{-30}\,\text{C m D}^{-1})=\boxed{3\times10^{-30}\,\text{C m}}$$

The chlorine atom has the greater electronegativity and a partial negative charge. Consequently, the dipole moment points toward the hydrogen atom. The experimental dipole moment of HCl(g) is 1.08 D so the estimate based on electronegativity differences is low by 17%.

E11.29 (a) $\mu_{res}=(\mu_1^2+\mu_2^2+2\mu_1\mu_2\cos\theta)^{1/2}$

$$=\{(1.50)^2+(0.80)^2+2\times1.50\times0.80\times\cos109.5°\}^{1/2}\,\text{D}=\boxed{1.45\,\text{D}}$$

(b) The *ortho* disubstituted benzene orients two identical dipoles μ at 60°.

$$\mu_{ortho}=\{1+1+2\times\cos60°\}^{1/2}\mu=\sqrt{3}\,\mu$$

The *meta* disubstituted benzene orients two identical dipoles μ at 120°.

$$\mu_{meta}=\{1+1+2\times\cos120°\}^{1/2}\mu=\mu$$

Thus, $\boxed{\mu_{ortho}/\mu_{meta}=\sqrt{3}}$

E11.30 This computation is most conveniently performed with spreadsheet software. Equation 11.15b is used for the computation of the x, y, and z components of the dipole moment after which eqn 11.15a is used to calculate the dipole moment. We find that $\boxed{\mu = 2.72\ \text{D}}$.

Atom	H	H	N	H	H	C	C	O	O	H
Charge/e	0.18	0.18	−0.36	0.02	0.02	0.06	0.45	−0.38	−0.38	0.42
x/pm	34	−199	−101	−86	34	−195	82	199	49	129
y/pm	146	−1	−11	118	146	70	−15	16	−107	−146
z/pm	−98	−100	−126	37	−98	−38	34	−38	88	126
Qx/e pm	6.12	−35.82	36.36	−1.72	0.68	−11.7	36.9	−75.62	−18.62	54.18
Qy/e pm	26.28	−0.18	3.96	2.36	2.92	4.2	−6.75	−6.08	40.66	−61.32
Qz/e pm	−17.64	−18	45.36	0.74	−1.96	−2.28	15.3	14.44	−33.44	52.92

Sum of all Qx: $\mu_x = -9.24\ e$ pm	-1.48043E-30 C m	-0.4438227 D
Sum of all Qy: $\mu_y = 6.05\ e$ pm	9.69331E-31 C m	0.2905982 D
Sum of all Qz: $\mu_z = 55.44\ e$ pm	8.8826E-30 C m	2.6629363 D
Dipole moment, μ =		2.7152634 D

E11.31 First, determine the dipole moment of the OH fragment, $\mu_{\text{O-H}}$, of the H_2O molecule by considering the total dipole moment of the molecule, $\mu_{\text{H-O-H}} = 1.85$ D, to be the resultant of the dipoles of two identical OH fragments at an angle θ equal to 104.5° with respect to each other. This is the bond angle in water. Use the equation provided in Exercise 11.29 for the resultant of two dipole moments that make an angle θ to each other:

$$\mu_{\text{res}} = (\mu_1^2 + \mu_2^2 + 2\mu_1\mu_2\cos\theta)^{1/2}$$

We substitute $\mu_1 = \mu_2 = \mu_{\text{O-H}}$ and solve for $\mu_{\text{O-H}}$.

$$\mu_{\text{H-O-H}} = (2\mu_{\text{O-H}}^2 + 2\mu_{\text{O-H}}^2\cos\theta)^{1/2} = \sqrt{2}\,\mu_{\text{O-H}}(1+\cos\theta)^{1/2} = 2\mu_{\text{O-H}}\cos(\theta/2)$$

$$\mu_{\text{O-H}} = \frac{\mu_{\text{H-O-H}}}{2\cos(\theta/2)} = \frac{1.85\ \text{D}}{2\cos(52.25°)} = 1.51\ \text{D}$$

Then, using ϕ to represent the angle between the two O–H bond dipoles in H_2O_2, we have the estimate

$$\mu_{\text{H-O-O-H}} = \boxed{2\mu_{\text{O-H}}\cos(\phi/2)}, \quad \text{where } \mu_{\text{O-H}} = 1.51\ \text{D}$$

(a) $\mu_{\text{H-O-O-H}}$ is plotted as a function of ϕ in Figure 11.13. At 90°, $\mu_{\text{H-O-O-H}}$ is $\boxed{2.13\ \text{D}}$, which is the experimental value.

(b) The angle can be related to the dipole moment as follows

$$\phi = 2\arccos(\mu_{\text{H-O-O-H}}/2\mu_{\text{O-H}}) = \boxed{2\arccos(\mu_{\text{H-O-O-H}}/3.02\ \text{D})}$$

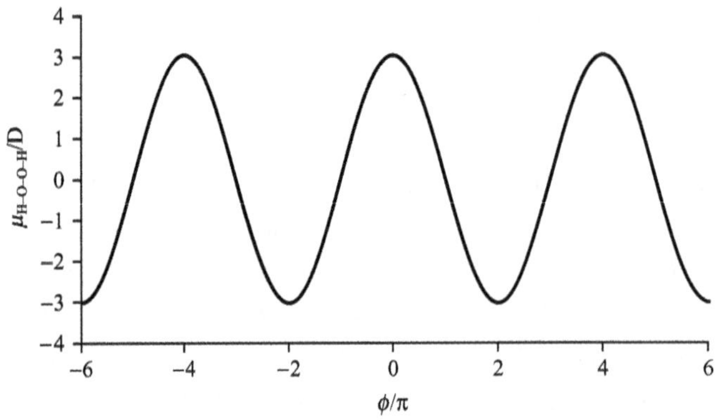

Figure 11.13

E11.32 We assume that the dipole of the water molecule and the Li^+ ion are collinear and that the separation of charges in the dipole is smaller than the distance to the ion. With these assumptions we can use eqn 11.16a of the text. To flip the water molecule over requires twice the energy of interaction given by eqn 11.16a.

$$E=\left(-\frac{Q_2(-\mu_1)}{4\pi\varepsilon_0 r^2}\right)-\left(-\frac{Q_2(+\mu_1)}{4\pi\varepsilon_0 r^2}\right)=\frac{2Q_2\mu_1}{4\pi\varepsilon_0 r^2} \quad Q_2=e=1.602\times10^{-19}\text{ C}$$

$$=\frac{2\times(1.602\times10^{-19}\text{ C})\times(1.85\text{ D})\times(3.336\times10^{-30}\text{ C m D}^{-1})}{(4\pi\times8.854\times10^{-12}\text{ J}^{-1}\text{ C}^2\text{ m}^{-1})\times r^2}$$

$$=\frac{1.777\times10^{-38}\text{ J m}^2}{r^2}$$

(a) $$E=\frac{1.777\times10^{-38}\text{ J m}^2}{(100\times10^{-12}\text{ m})^2}=1.78\times10^{-18}\text{ J}$$

molar energy $= N_A\times E=\boxed{1.07\times10^3\text{ kJ mol}^{-1}}$

(b) $$E=\frac{1.777\times10^{-38}\text{ J m}^2}{(300\times10^{-12}\text{ m})^2}=1.97\times10^{-19}\text{ J}$$

molar energy $= N_A\times E=\boxed{119\text{ kJ mol}^{-1}}$

E11.33 Figure 11.14 summarizes the definitions and basic relations needed to analyze the dipole–dipole interactions between molecule 1–2 and molecule 3–4. We begin by finding simplified expressions for $1/r_{14}$ and $1/r_{23}$ under the condition that $x=l/r\ll 1$. The Taylor series

$$(1+z)^{-1/2}=1-\tfrac{1}{2}z+\tfrac{3}{8}z^2-\cdots \quad -1<z\le 1$$

is applied to the Law of Cosines relation given in Figure 11.14.

$(r/r_{14})=(1+x^2+2x\cos\theta)^{-1/2}$ [Apply Taylor series and truncate after second-order term because $x\ll 1$.]

$$=1-\tfrac{1}{2}(x^2+2x\cos\theta)+\tfrac{3}{8}(x^2+2x\cos\theta)^2$$

$$=1-\tfrac{1}{2}x^2-x\cos\theta+\tfrac{3}{2}x^2\cos^2\theta$$

Likewise,

$$(r/r_{23}) = (1 + x^2 - 2x\cos\theta)^{-1/2} \quad \text{[Apply Taylor series and truncate after second-order term because } x \ll 1.]$$
$$= 1 - \tfrac{1}{2}(x^2 - 2x\cos\theta) + \tfrac{3}{8}(x^2 - 2x\cos\theta)^2$$
$$= 1 - \tfrac{1}{2}x^2 + x\cos\theta + \tfrac{3}{2}x^2\cos^2\theta$$

The dipole–dipole interaction is

$$V = \frac{Q_1Q_3}{4\pi\varepsilon_0}\left\{\frac{1}{r_{13}} - \frac{1}{r_{14}} - \frac{1}{r_{23}} + \frac{1}{r_{24}}\right\} \quad \text{[Coulomb potential, } r = r_{13} = r_{24}]$$
$$= \frac{Q_1Q_3}{4\pi\varepsilon_0 r}\{2 - (1 - \tfrac{1}{2}x^2 - x\cos\theta + \tfrac{3}{2}x^2\cos^2\theta) - (1 - \tfrac{1}{2}x^2 + x\cos\theta + \tfrac{3}{2}x^2\cos^2\theta)\}$$
$$= \frac{Q_1Q_3}{4\pi\varepsilon_0 r}\{x^2 - 3x^2\cos^2\theta\} \; [x = l/r]$$
$$= \frac{Q_1Q_3l^2}{4\pi\varepsilon_0 r^3}\{1 - 3\cos^2\theta\} = \frac{\mu_{12}\mu_{34}}{4\pi\varepsilon_0 r^3}\{1 - 3\cos^2\theta\}$$

This is identical to eqn 11.17 and, thus, provides justification for the relation.

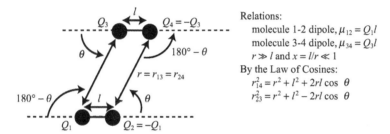

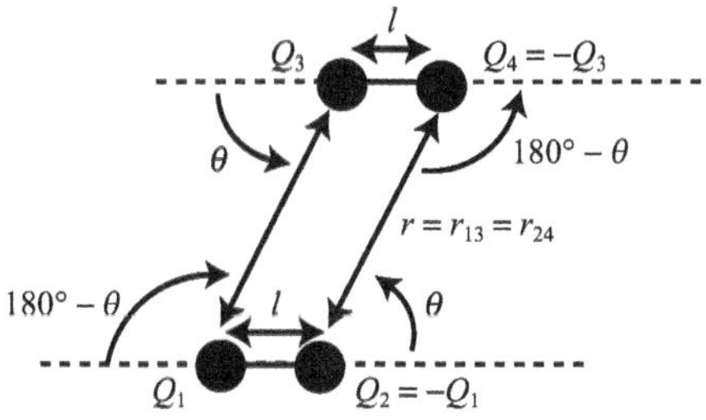

Relations:
- molecule 1-2 dipole, $\mu_{12} = Q_1 l$
- molecule 3-4 dipole, $\mu_{34} = Q_3 l$
- $r \gg l$ and $x = l/r \ll 1$

By the Law of Cosines:
- $r_{14}^2 = r^2 + l^2 + 2rl\cos\theta$
- $r_{23}^2 = r^2 + l^2 - 2rl\cos\theta$

Figure 11.14

E11.34 (a) Polarizability, $\alpha = \mu^*/\mathcal{E}$ [11.19], where μ^*, the induced dipole moment has the SI unit "C m". The electric field strength, $\mathcal{E}$, is the force per unit charge experienced by a charge. It has the unit "N C^{-1}" or "J C^{-1} m^{-1}". Consequently, polarizability has the SI unit

$$(\text{C m})/(\text{J C}^{-1}\,\text{m}^{-1}) = \text{C}^2\,\text{m}^2\,\text{J}^{-1}.$$

(b) Polarizability volume, $\alpha' = \alpha/4\pi\varepsilon_0$ [11.20], where α has the SI unit "C^2 m^2 J^{-1}" and ε_0 has the "C^2 J^{-1} m^{-1}" unit. Consequently, the polarizability volume has the SI unit

$$(\text{C}^2\,\text{m}^2\,\text{J}^{-1})/(\text{C}^2\,\text{J}^{-1}\,\text{m}^{-1}) = \text{m}^3.$$

E11.35

$$\mu^* = \alpha\mathcal{E}\,[11.19] = 4\pi\varepsilon_0\alpha'\mathcal{E}\,[11.20] = 4\pi\varepsilon_0\alpha'\left(\frac{e}{4\pi\varepsilon_0 r^2}\right) = 1.85\text{ D} \quad [Q = e \text{ for a proton}]$$

$$\frac{\alpha' e}{r^2} = 1.85\text{ D}$$

Solve for r,

$$r = \left(\frac{\alpha' e}{1.85\text{ D}}\right)^{1/2} = \left(\frac{(1.48\times10^{-30}\,\text{m}^3)\times(1.602\times10^{-19}\,\text{C})}{(1.85\text{ D})\times(3.336\times10^{-30}\,\text{C m D}^{-1})}\right)^{1/2}$$
$$= 1.96\times10^{-10}\,\text{m} = \boxed{196\text{ pm}}$$

E11.36 The interaction is a dipole–induced-dipole interaction. The energy is given by eqn 11.21:

$$V = -\frac{\mu_1^2\alpha_2'}{4\pi\varepsilon_0 r^6}\ [11.21]$$

$$= -\frac{\{(2.7\ \text{D})\times(3.336\times10^{-30}\ \text{C m D}^{-1})\}^2\times(1.04\times10^{-29}\ \text{m}^3)}{4\pi(8.854\times10^{-12}\ \text{J}^{-1}\ \text{C}^2\ \text{m}^{-1})\times(4.0\times10^{-9}\ \text{m})^6}$$

$$V = \boxed{-1.8\times10^{-27}\ \text{J} \quad \text{or} \quad 1.1\times10^{-3}\ \text{J mol}^{-1}}$$

E11.37

$$V = -\tfrac{3}{2}\times\frac{\alpha_1'\alpha_2'}{r^6}\times\frac{I_1I_2}{I_1+I_2}\ [11.22] = -\tfrac{3}{4}\frac{(\alpha')^2I}{r^6}$$

$$= -\tfrac{3}{4}\times\frac{(10.4\times10^{-30}\ \text{m}^3)^2\times(5.0\ \text{eV})}{(4.0\times10^{-9}\ \text{m})^6}$$

$$= -9.9\times10^{-8}\ \text{eV} = 1.6\times10^{-27}\ \text{J} \quad \text{or} \quad \boxed{-9.6\ \text{mJ mol}^{-1}}$$

E11.38 The geometry at the hydrogen bond is linear (see Figure 11.15). The partial charges are as given in Table 11.2. Distances in structure (**21**) are $r_{\text{O–H}} = 97.5$ pm, $r_{\text{H}\cdots\text{N}} = 104.3$ pm, and $r_{\text{O}\cdots\text{N}} = 201.8$ pm.

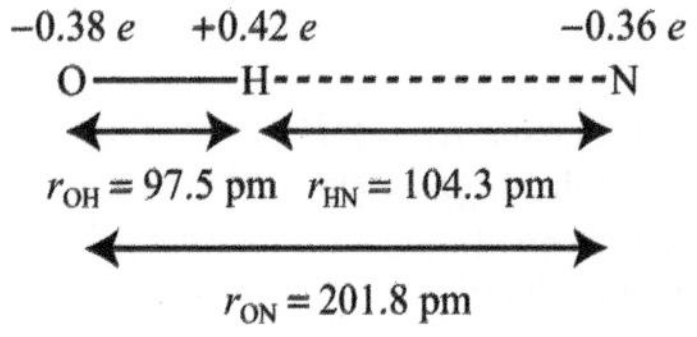

Figure 11.15

$$V_{\text{hydrogen bond}} = N_A\{V_{\text{ON}} + V_{\text{HN}}\}$$

$$= N_A\left\{\frac{Q_OQ_N}{4\pi\varepsilon_0 r_{\text{ON}}} + \frac{Q_HQ_N}{4\pi\varepsilon_0 r_{\text{HN}}}\right\}$$

$$= \frac{N_Ae^2}{4\pi\varepsilon_0}\left\{\frac{(Q_O/e)(Q_N/e)}{r_{\text{ON}}} + \frac{(Q_H/e)(Q_N/e)}{r_{\text{HN}}}\right\}$$

$$= \frac{(6.022\times10^{23}\ \text{mol}^{-1})\times(1.6022\times10^{-19}\ \text{C})^2}{1.1127\times10^{-10}\ \text{J}^{-1}\ \text{C}^2\ \text{m}^{-1}}\left\{\frac{(-0.38)\times(-0.36)}{201.8\times10^{-12}\ \text{m}} + \frac{(+0.42)\times(-0.36)}{104.3\times10^{-12}\ \text{m}}\right\}$$

$$= \boxed{-1.1\times10^5\ \text{J mol}^{-1}}$$

E11.39 We assume that the Lennard-Jones potential, eqn [11.25], adequately represents the potential energy in this case.

$$V(R) = 4\varepsilon\left\{\left(\frac{\sigma}{R}\right)^{12} - \left(\frac{\sigma}{R}\right)^{6}\right\} = \frac{4\varepsilon\sigma^6}{R^6}\left\{\left(\frac{\sigma}{R}\right)^{6} - 1\right\}$$

$$V(R+\delta R) = 4\varepsilon\left\{\left(\frac{\sigma}{R+\delta R}\right)^{12} - \left(\frac{\sigma}{R+\delta R}\right)^{6}\right\} = \frac{4\varepsilon\sigma^6}{R^6}\left\{\frac{\sigma^6}{R^6}\left(\frac{1}{1+\frac{\delta R}{R}}\right)^{12} - \left(\frac{1}{1+\frac{\delta R}{R}}\right)^{6}\right\}$$

Since $(1+x)^{-12} \approx 1 - 12x$ and $(1+x)^{-6} \approx 1 - 6x$ when $x \ll 1$, and since $\delta R/R \ll 1$,

$$V(R+\delta R) = \frac{4\varepsilon\sigma^6}{R^6}\left\{\frac{\sigma^6}{R^6}\left(1 - 12\frac{\delta R}{R}\right) - \left(1 - 6\frac{\delta R}{R}\right)\right\}$$

$$\Delta V = V(R+\delta R) - V(R) = \frac{4\varepsilon\sigma^6}{R^6}\left\{\frac{\sigma^6}{R^6}\left(1-12\frac{\delta R}{R}\right)-\left(1-6\frac{\delta R}{R}\right)\right\} - \frac{4\varepsilon\sigma^6}{R^6}\left\{\left(\frac{\sigma}{R}\right)^6 - 1\right\}$$

$$= \frac{24\varepsilon\sigma^6}{R^6}\left(\frac{\delta R}{R}\right)\left\{1-2\left(\frac{\sigma}{R}\right)^6\right\}$$

$$F = -\frac{\Delta V}{\Delta r} = -\left[\frac{\frac{24\varepsilon\sigma^6}{R^6}\left(\frac{\delta R}{R}\right)\left\{1-2\left(\frac{\sigma}{R}\right)^6\right\}}{(R+\delta R)-R}\right] = -\frac{24\varepsilon\sigma^6}{R^7}\left\{1-2\left(\frac{\sigma}{R}\right)^6\right\}$$

Thus, F equals zero when the factor $1-2\left(\frac{\sigma}{R}\right)^6$ equals zero or $\boxed{R = 2^{1/6}\sigma}$.

E11.40 $$F = -\frac{\mathrm{d}V}{\mathrm{d}r} = -4\varepsilon\frac{\mathrm{d}}{\mathrm{d}r}\left\{\left(\frac{\sigma}{r}\right)^{12} - \left(\frac{\sigma}{r}\right)^6\right\} = -4\varepsilon\left\{\frac{12\sigma^{12}}{r^{13}} - \frac{6\sigma^6}{r^7}\right\} = -\frac{24\varepsilon\sigma^6}{r^7}\left\{\frac{2\sigma^6}{r^6} - 1\right\}$$

The minimum occurs when

$\frac{2\sigma^6}{r^6} - 1 = 0$, which solves to $\boxed{r = 2^{1/6}\sigma}$.

E11.41 Individual acetic acid molecules have a non-zero dipole moment but the two dipoles of the dimer are exactly opposed and cancel. At low temperature a significant fraction of molecules are a part of a dimer, which means that their individual dipole moments will not be observed. However, as temperature is increased, hydrogen bonds of the dimer are broken, thereby, releasing individual molecules and the apparent dipole moment increases.

E11.42 By the law of cosines $r_{\mathrm{O\cdots H}}^2 = r^2 + R^2 - 2rR\cos\theta$. Therefore, $r_{\mathrm{O\cdots H}} = (r^2 + R^2 - 2rR\cos\theta)^{1/2}$

With $\delta_{\mathrm{O}} = -0.83$, $\delta_{\mathrm{H}} = +0.45$, $r = 95.7$ pm, and $R = 200$ pm the molar hydrogen bond potential is given by

$$V_{\mathrm{m,hydrogen\ bond}} = \frac{N_{\mathrm{A}}e^2}{4\pi\varepsilon_0}\left\{\frac{\delta_{\mathrm{O}}\delta_{\mathrm{H}}}{r_{\mathrm{O\cdots H}}} + \frac{\delta_{\mathrm{O}}^2}{R}\right\}$$

$$= \frac{N_{\mathrm{A}}e^2}{4\pi\varepsilon_0\times(10^{-12}\,\mathrm{m})}\left\{\frac{\delta_{\mathrm{O}}\delta_{\mathrm{H}}}{\{(r^2+R^2-2rR\cos\theta)^{1/2}\}/\mathrm{pm}} + \frac{\delta_{\mathrm{O}}^2}{R/\mathrm{pm}}\right\}$$

$$= (138.94\ \mathrm{MJ\ mol^{-1}})\times\left\{\frac{(-0.83)\times(+0.45)}{(95.7^2+200^2-2\times(95.7)\times(200)\cos\theta)^{1/2}} + \frac{(-0.83)^2}{200}\right\}$$

$$= (138.94\ \mathrm{MJ\ mol^{-1}})\times\left\{\frac{-0.37\overline{35}}{(95.7^2+200^2-2\times(95.7)\times(200)\cos\theta)^{1/2}} + \frac{0.68\overline{89}}{200}\right\}$$

A plot of this function of θ is presented in Figure 11.16. As expected, the potential is a minimum (-19.0 kJ mol^{-1}) when $\theta = 0$ because at that angle the hydrogen lies directly between the two oxygen atoms, which repel. The plot shows that as the absolute value of θ increases from zero the hydrogen bond strength decreases and becomes unstable ($V \geq 0$) when $|\theta| \geq 12°$. The employed electrostatic model has created this effect by minimizing the $r_{\mathrm{O\cdots H}}$ distance when $\theta = 0$. This causes the O$\cdots$H attraction to become stronger than the oxygen-to-oxygen repulsion, which is independent of θ. At larger angles, the $r_{O\cdots H}$ distance is larger and the oxygen-to-oxygen repulsion predominates.

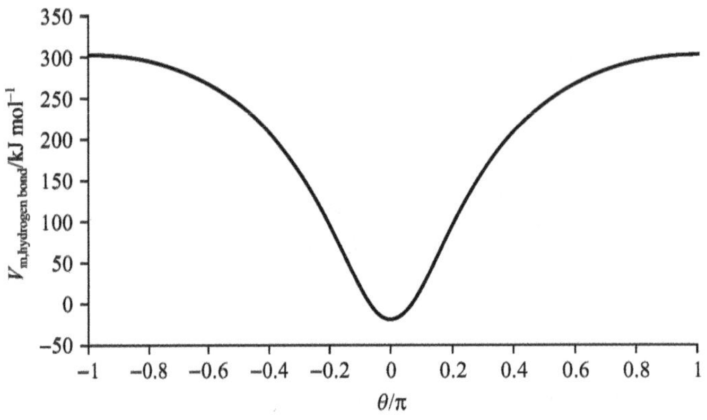

Figure 11.16

E11.43 In a **parallel β sheet** (text Figure 11.43b), $\phi = -119°$ and $\psi = +113°$, and the N–H–O atoms of the hydrogen bonds are not perfectly aligned. This arrangement is a result of the parallel arrangement of the chains: each N–H bond on one chain is aligned with a N–H bond of another chain and, as a result, each C–O bond of one chain is aligned with a C–O bond of another chain. These structures are not common in proteins because the non-linear alignment of the hydrogen bond reduces the hydrogen bond strength by increasing $r_{O\cdots H}$ distance and, thereby, decreasing the magnitude of the O---H attraction. The resulting increase in the importance of the nitrogen-to-oxygen repulsion relative to the attraction destabilizes the structure.

E11.44 $V = \frac{1}{2}V_0(1 + \cos 3\phi)$ where ϕ is the azimuthal angle and $V_0 = 11.6$ kJ mol^{-1}

(a) $\phi = 0$ in the maximum potential eclipsed conformation. In the minimum potential staggered conformation, $\phi = (2n + 1)\pi/3$ where $n = 0, 1, 2, 3, \ldots$. For convenience, use the potential minimum that occurs at $\phi = \pi/3$.

$$\begin{aligned}\Delta V = V_{\text{eclipsed}} - V_{\text{staggered}} &= V(0) - V(\pi/3)\\ &= \tfrac{1}{2}V_0\{1 + \cos(3\times 0)\} - \tfrac{1}{2}V_0\{1 + \cos(3\times \pi/3)\}\\ &= \tfrac{1}{2}V_0\{1 + 1\} - \tfrac{1}{2}V_0\{1 + (-1)\}\\ &= \boxed{V_0}\end{aligned}$$

(b) The harmonic potential has a potential minimum at a displacement of $x = 0$ and an increasing potential that is given by $V_{\text{harmonic}} = \frac{1}{2}kx^2$ [12.12] for both very small negative and positive values of displacement x. The frequency of the harmonic oscillator is given by $\nu = (2\pi)^{-1}(k/\mu)^{1/2}$ [12.13b], where μ is the effective particle mass. To aid the comparison with the harmonic oscillator, let $\varphi = (\chi + \pi)/3$. ϕ is the displacement angle from the eclipsed conformation; χ is the displacement angle from the equilibrium, staggered conformation. The potential energy of a CH_3 group in ethane is a minimum when $\chi = 0$ and by writing the potential in terms of χ we can more readily compare it to the harmonic oscillator potential.

$$\begin{aligned}V &= \tfrac{1}{2}V_0\{1 + \cos(3\phi)\} = \tfrac{1}{2}V_0\{1 + \cos(\chi + \pi)\} = \tfrac{1}{2}V_0\{1 + \cos(\chi)\cos(\pi) - \sin(\chi)\sin(\pi)\}\\ &= \tfrac{1}{2}V_0\{1 + \cos(\chi)(-1) - \sin(\chi)\times 0\} = \tfrac{1}{2}V_0\{1 - \cos(\chi)\}\end{aligned}$$

For small values of χ the function $\cos(\chi)$ may be expanded in a Taylor series around $\chi = 0$, and the series may be truncated at fourth-order terms because the higher-order terms are negligibly small when $\chi \ll 1$.

$$V = \tfrac{1}{2}V_0\{1 - [1 - \chi^2/2! + \chi^4/4! - \chi^6/6! + \cdots]\} = \tfrac{1}{2}V_0\{1 - [1 - \chi^2/2]\}$$
$$= \tfrac{1}{4}V_0\chi^2$$

The square dependence on χ clearly shows that the torsional motion around the C–C bond of ethane has the harmonic oscillator behavior around the equilibrium azimuthal angle.

(c) We can estimate the frequency of the torsional oscillation by comparing a reasonable total energy for this motion with the total energy and frequency of the harmonic oscillator. Since the rotational motion around the C–C bond involves the kinetic energy of six hydrogen atoms traveling in circles of radius r perpendicular to the axis of the C–C bond, we estimate that each hydrogen atom has kinetic energy provided by its rotational energy (text Section 9.5). Adding the potential energy gives:

$$E_{\text{rotation}} = 6 \times \left(\frac{J^2_{\text{C–C}}}{2m_{\text{H}}r^2}\right) + \tfrac{1}{4}V_0\chi^2 \text{ [9.17]}$$

where $r = R_{\text{C–H}}\sin(180° - \theta_{\text{H–C–C}}) = (110.7\text{ pm})\sin(180° - 109.5°) = 104.4\text{ pm}$

The total energy of the harmonic oscillator is

$$E_{\text{vibration}} = \frac{p_x}{2\mu} + \tfrac{1}{2}kx^2$$

Comparison of the two relationships shows that the harmonic oscillator energy transforms into the rotational energy using $\mu \to m_{\text{H}}r^2/6$ and $k \to V_0/2$. Using these transformations in the eqn. 12.13b expression for frequency, yields an estimate of the frequency of the torsional oscillation.

$$\nu = \frac{1}{2\pi}\left(\frac{V_0/2}{m_{\text{H}}r^2/6}\right)^{1/2} = \frac{1}{2\pi}\left(\frac{3V_0}{m_{\text{H}}r^2}\right)^{1/2}$$
$$= \frac{1}{2\pi}\left(\frac{3(11.6 \times 10^3\text{ J mol}^{-1})}{(0.001\,00\text{ kg mol}^{-1}) \times (104.4 \times 10^{-12}\text{ m})^2}\right)^{1/2}$$
$$= \boxed{8.99 \times 10^{12}\text{ Hz}}$$

Expressing this as a wavenumber gives

$$\tilde{\nu} = \frac{\nu}{c} = \frac{8.99 \times 10^{12}\text{ s}^{-1}}{3.00 \times 10^8\text{ m s}^{-1}} = \boxed{300.0\text{ cm}^{-1}}$$

E11.45 $R_{\text{rms}} = N^{1/2}l\text{ [11.28]} = (700)^{1/2} \times (0.90\text{ nm}) = \boxed{24\text{ nm}}$

E11.46 The repeating unit (monomer) of polyethylene is ($-CH_2-CH_2-$), which has a molar mass of 28 g mol^{-1}. The number of repeating units, N, is therefore

$$N = \frac{280\,000\text{ g mol}^{-1}}{28\text{ g mol}^{-1}} = 1.0 \times 10^4$$

and $l = 2R(\text{C–C})$ [Add half a bond length on either side of monomer.]

Apply the formulas for the contour length (11.29) and rms separation (11.28):

$$R_c = Nl = 2 \times (1.0 \times 10^4) \times (154\text{ pm}) = 3.1 \times 10^6\text{ pm} = \boxed{3.1\ \mu\text{m}}$$

$$R_{\text{rms}} = N^{1/2}l = 2 \times (1.0 \times 10^4)^{1/2} \times (154\text{ pm}) = 3.1 \times 10^4\text{ pm} = \boxed{31\text{ nm}}$$

E11.47 For a random coil, the radius of gyration is (11.30)

$$R_g=\left(\frac{N}{6}\right)^{1/2} l, \quad \text{so} \quad N=6\left(\frac{R_g}{l}\right)^2=6\times\left(\frac{7.3\text{ nm}}{0.154\text{ nm}}\right)^2=\boxed{1.3\times 10^4}$$

E11.48 (a) In this exercise we discuss two different ways to find the desired relations. Our first method involves a detailed analysis based upon the application of the integral calculus. Our second method suggests how to pursue the solution as a library exercise. Here's the method that uses fundamental definitions and the integral calculus:

R_g is the radius of rotation of a point mass that has the same mass m and moment of inertia I as the object of interest. For an object that has a continuum of mass within its macroscopic boundaries:

$I=mR_g^2$ where $I=\int \rho R^2 \mathrm{d}\tau$ and R is the distance from the axis of rotation.

For a homogeneous object $\rho = m/V$ so

$R_g^2 = V^{-1}\int R^2 \mathrm{d}\tau$ (The integrand is the square of the distance from the axis of rotation.)

For a solid sphere of radius a: $V=\frac{4}{3}\pi a^3$, $\mathrm{d}\tau = r^2\sin\theta\, \mathrm{d}r\, \mathrm{d}\theta\, \mathrm{d}\phi$, and $R^2=x^2+y^2$.

$$\begin{aligned}
R_g^2 V &= \int_{\phi=0}^{2\pi}\int_{\theta=0}^{\pi}\int_{r=0}^{a}(x^2+y^2)r^2\sin\theta\, \mathrm{d}r\, \mathrm{d}\theta\, \mathrm{d}\phi \\
&= \int_{\phi=0}^{2\pi}\int_{\theta=0}^{\pi}\int_{r=0}^{a}(r^2\sin^2\theta\cos^2\phi + r^2\sin^2\theta\sin^2\phi)r^2\sin\theta\, \mathrm{d}r\, \mathrm{d}\theta\, \mathrm{d}\phi \\
&= \int_{\phi=0}^{2\pi}\int_{\theta=0}^{\pi}\int_{r=0}^{a} r^4\sin^3\theta\, \mathrm{d}r\, \mathrm{d}\theta\, \mathrm{d}\phi \\
&= \left\{\int_0^{2\pi}\mathrm{d}\phi\right\}\left\{\int_0^{\pi}\sin^3\theta \mathrm{d}\theta\right\}\left\{\int_0^{a} r^4 \mathrm{d}r\right\} \\
&= 2\pi\times\{-\cos\theta+\tfrac{1}{3}\cos^3\theta\}\Big|_{\theta=0}^{\theta=\pi}\times\left[\frac{r^5}{5}\right]_{r=0}^{r=a} \\
&= \frac{8\pi a^5}{15}
\end{aligned}$$

$$R_g^2=\frac{8\pi a^5}{15V}=\frac{8\pi a^5}{15\times(\frac{4}{3}\pi a^3)}=\tfrac{2}{5}a^2$$

$$R_g=\boxed{\sqrt{\tfrac{2}{5}}\,a}$$

COMMENT: A common error involves using r^2 in place of the squared distance from the axis of rotation, which is actually equal to x^2+y^2 not to $x^2+y^2+z^2=r^2$. The common error gives the result $\sqrt{\frac{3}{5}}a$.

For a spherical macromolecule, the specific volume is:

$$v_s=\frac{V}{m}=\frac{4\pi a^3}{3}\times\frac{N_A}{M} \quad \text{so} \quad a=\left(\frac{3v_s M}{4\pi N_A}\right)^{1/3}$$

Therefore,

$$R_g = \left(\frac{2}{5}\right)^{1/2} \times \left(\frac{3v_s M}{4\pi N_A}\right)^{1/3}$$

$$= \left(\frac{2}{5}\right)^{1/2} \times \left(\frac{(3v_s/\text{cm}^3\,\text{g}^{-1}) \times \text{cm}^3\,\text{g}^{-1} \times (M/\text{g mol}^{-1}) \times \text{g mol}^{-1}}{4\pi \times (6.022 \times 10^{23}\,\text{mol}^{-1})}\right)^{1/3}$$

$$= (4.6460 \times 10^{-9}\,\text{cm}) \times \{(v_s/\text{cm}^3\,\text{g}^{-1}) \times (M/\text{g mol}^{-1})\}^{1/3}$$

$$R_g/\text{nm} = \boxed{0.046460 \times \{(v_s/\text{cm}^3\,\text{g}^{-1}) \times (M/\text{g mol}^{-1})\}^{1/3}}$$

COMMENT: The common error, discussed above, gives 0.056902 × $\{(v_s/\text{cm}^3\,\text{g}^{-1}) \times (M/\text{g mol}^{-1})\}^{1/3}$.

For a solid rod of radius a and length l there are two moments of inertia. One about the axis of its length ($R_{g,\parallel}$, $R = r$) and another about the axis that is perpendicular to its length ($R_{g,\perp}$, $R = z$).

$$R_{g,\parallel}^2 = (\pi a^2 l)^{-1} \int r^2 \text{d}\tau = (\pi a^2 l)^{-1} \int_{r=0}^{a} \int_{z=0}^{l} \int_{\theta=0}^{2\pi} r^3 \text{d}\theta \text{d}z \text{d}r \quad \text{[cylindrical coordinates]}$$

$$= (\pi a^2 l)^{-1}\, 2\pi l \int_0^a r^3 \text{d}r$$

$$= \frac{2}{a^2}\left[\frac{r^4}{4}\right]_{r=0}^{r=a} = \tfrac{1}{2}a^2$$

$$R_{g,\perp} = \boxed{\sqrt{\tfrac{1}{2}}a}$$

$$R_{g,\perp}^2 = (\pi a^2 l)^{-1} \int z^2 \text{d}\tau = (\pi a^2 l)^{-1} \int_{r=0}^{a} \int_{z=-l/2}^{l/2} \int_{\theta=0}^{2\pi} z^2 r \text{d}\theta \text{d}z \text{d}r \quad \text{[cylindrical coordinates]}$$

$$= \tfrac{1}{12} l^2$$

$$R_{g,\perp} = \boxed{\sqrt{\tfrac{1}{12}}l}$$

For a rod-like macromolecule, the specific volume is:

$$v_s = \frac{V}{m} = \pi a^2 l \times \frac{N_A}{M} \quad \text{so} \quad l = \frac{v_s M}{\pi a^2 N_A}$$

Here's the type of thinking that we might use when approaching the problem as a library exercise: Classical rotation of a solid object of uniform density is an important aspect of physics. One application involves using its mathematical formulas to model protein shapes in solution so as to estimate whether the shape is spherical or ellipsoidal. In this problem we have an interest in estimating whether the shape is spherical-like or rod-like, or disk-like so we make that lovely trip to the physics stacks within the library and find that the mathematics of rotation is simplified by expressing the motion in terms of a point particle that has circular motion, the same mass and moment of inertia as the object of interest. The radius of the circular motion is called the radius of gyration, R_g and it is related to the moment of inertia of the object by the expression

$$R_g = \left(\frac{I}{m}\right)^{1/2} \quad \text{where } I \text{ and } m \text{ are the moment of inertia and mass of the object.}$$

So far, so good. But, what are the formulas for the moment of inertia of the different objects? Well, this is a purely mathematical problem so we expect that the formulas can be found in any basic handbook of mathematics and sure enough we find the formulas of the following table.

Object	I	Motion
solid sphere, radius r	$2mr^2/5$	about axis through center
thin rod, length l	$ml^2/12$	about axis perpendicular to rod and through center of mass
thin disk, radius r	$mr^2/2$	about axis perpendicular to disk and through center
thin ring, radius r	mr^2	about axis perpendicular to ring and through center

We need only examine the expression for the solid sphere. Substitution gives

$$R_{\mathrm{g,sphere}} = \left(\frac{2mr^2}{5m}\right)^{1/2} = \left(\frac{2}{5}\right)^{1/2} r$$

Two relations are used to find an expression for r in terms of data provided in the exercise. They are

$V_{\text{sphere}} = \frac{4}{3}\pi r^3$ and $V_{\text{sphere}} = mv_s = Mv_s/N_A$ where v_s is the specific volume of the object.

Solving them for r gives

$$r = \left(\frac{3Mv_s}{4N_A}\right)^{1/3}$$

and the working equation for the radius of gyrations becomes

$$R_{\mathrm{g,sphere}} = \left(\frac{2}{5}\right)^{1/2}\left(\frac{3Mv_s}{4\pi N_A}\right)^{1/3} = \boxed{0.046460 \times \{(v_s/\mathrm{cm^3\,g^{-1}}) \times (M/\mathrm{g\,mol^{-1}})\}^{1/3}}$$

(b) Using the data for serum albumin, we find

$$R_{\mathrm{g,sphere}} = 0.046460 \times \{(0.752) \times (66 \times 10^3)\}^{1/3}\ \mathrm{nm} = 1.7\ \mathrm{nm}$$

This radius of gyration value and those calculated for bushy stunt virus and DNA are summarized along with the experimental data in the following table. The values calculated for serum albumin and bushy stunt virus are in reasonable agreement with the experiment but that of DNA does not agree. Therefore, serum albumin and bushy stunt virus resemble solid spheres, but DNA does not. We have found that for a rod: $R_{\mathrm{g},\perp} = \sqrt{1/12} \times$ (rod length). This relation along with the observation of the extremely large experimental R_g value for DNA (compared to the value predicted for a sphere) leads us to speculate that DNA has a rod-like shape with a very large length-to-radius ratio.

	$M/(\mathrm{g\,mol^{-1}})$	$v_s/(\mathrm{cm^3\,g^{-1}})$	$(R_g/\mathrm{nm})_{\mathrm{expt}}$	$(R_g/\mathrm{nm})_{\mathrm{sphere}}$
Serum albumin	66×10^3	0.752	2.98	1.7
Bushy stunt virus	10.6×10^6	0.741	12.0	9.2
DNA	4×10^6	0.556	117.0	6.0

E11.49 The intensity of scattered light is measured as a function of the angle θ that the direction of the laser beam makes with the direction of the detector from the sample at a distance r. The intensity, $I(\theta)$, of light scattered is written as the **Rayleigh ratio**:

$$R(\theta)=\frac{I(\theta)}{I_0}\times r^2\,[11.6]=KP(\theta)c_{\mathrm{M}}M\,[11.7]$$

where the structure factor $P(\theta)$ for small molecules is given by

$$P(\theta)\approx 1-\frac{16\pi^2R_g^2\sin^2(\frac{1}{2}\theta)}{3\lambda^2}\,[11.8]=\frac{3\lambda^2-16\pi^2R_g^2\sin^2(\frac{1}{2}\theta)}{3\lambda^2}$$

(a) We seek an expression for a ratio of scattering intensities of a macromolecule in two different conformations, a rigid rod or a closed circle (cc). For any given scattering angle, the ratio of scattered intensity of two conformations is the same as the ratio of their Rayleigh ratios:

$$\frac{I_{\mathrm{rod}}}{I_{\mathrm{cc}}}=\frac{R_{\mathrm{rod}}}{R_{\mathrm{cc}}}=\frac{P_{\mathrm{rod}}}{P_{\mathrm{cc}}}.$$

The radius of gyration of a rod of length l is

$$R_{g,\perp}=\sqrt{\tfrac{1}{12}}\,l\quad\text{[See Exercise 11.48]}$$

For a closed circle, the radius of gyration is simply the radius of a circle whose circumference is l:

$$l=2\pi R_{\mathrm{cc}},\quad\text{so}\quad R_{\mathrm{cc}}=\frac{l}{2\pi}.$$

The intensity ratio is:

$$\frac{I_{\mathrm{rod}}}{I_{\mathrm{cc}}}=\frac{3\lambda^2-\frac{4}{3}\pi^2l^2\sin^2(\frac{1}{2}\theta)}{3\lambda^2-4l^2\sin^2(\frac{1}{2}\theta)}.$$

Substituting $\lambda=488$ nm and $l=250$ nm at the scattering angles 20°, 45°, and 90° yields:

$\theta/°$	20	45	90
$I_{\mathrm{rod}}/I_{\mathrm{cc}}$	0.976	0.876	0.514

(b) It is best to work at a detection angle at which the ratio $I_{\mathrm{rod}}/I_{\mathrm{cc}}$ shows the greatest difference between I_{rod} and I_{cc}. The table of part (a) indicates that this occurs at $\boxed{90°}$.

E11.50 When a coil containing N_{A} bonds of length l is stretched or compressed by nl so that $\nu=n/N_{\mathrm{A}}=0.1$, the conformational entropy change is given by

$$\Delta S_{\mathrm{m}}=-\tfrac{1}{2}kN_{\mathrm{A}}\ln\{(1+\nu)^{1+\nu}(1-\nu)^{1-\nu}\}\,[11.31]=-\tfrac{1}{2}R\ln\{(1+\nu)^{1+\nu}(1-\nu)^{1-\nu}\}$$
$$=-\tfrac{1}{2}\times(8.3145\ \mathrm{J\,K^{-1}\,mol^{-1}})\ln\{(1.1)^{1.1}\times(0.90)^{0.9}\}$$
$$=\boxed{-0.042\ \mathrm{J\,K^{-1}\,mol^{-1}}}$$

E11.51 (a) We want to construct the exponential-6 potential so that $V(r_0)=0$ and the depth of the potential well is $-\varepsilon$ when $r=r_{\mathrm{e}}$. Consequently, we write

$$V(r)=A\varepsilon\left[Be^{-r/r_0}-\left(\frac{r_0}{r}\right)^6\right]$$

where the constants A and B are chosen to satisfy the conditions. For example, examine $V(r_0)$ to find an expression for B.

$$V(r_0) = A\varepsilon\left[Be^{-r_0/r_0} - \left(\frac{r_0}{r_0}\right)^6\right] = A\varepsilon[Be^{-1} - 1] = 0$$

So, $B = \mathrm{e}$ and the potential becomes $V(r) = A\varepsilon\left[\mathrm{e}^{1-r/r_0} - \left(\frac{r_0}{r}\right)^6\right]$.

The constant A is evaluated at the equilibrium distance r_e.

$$-\varepsilon = A\varepsilon\left[\mathrm{e}^{1-r_e/r_0} - \left(\frac{r_0}{r_e}\right)^6\right]$$

So, $A = -\left[\mathrm{e}^{1-r_e/r_0} - \left(\frac{r_0}{r_e}\right)^6\right]^{-1}$, where $r_e > r_0$ and $A < 0$

The repulsive exponential term, the attractive term, and the potential (in the unit $|A\varepsilon|$) are sketched in Figure 11.17. The point at which the potential is a minimum is labeled as $x_e = r_e/r_0$.

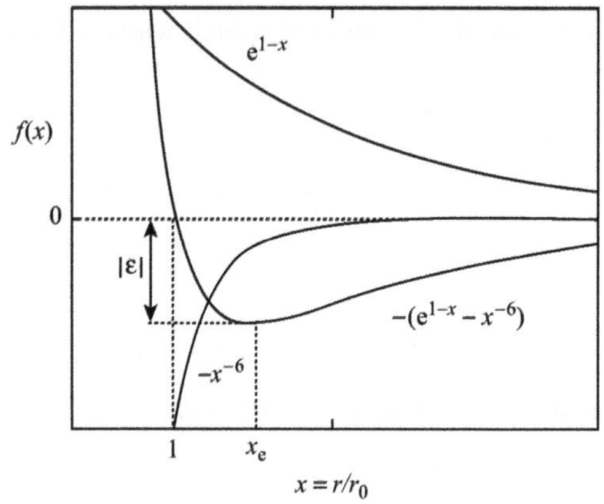

Figure 11.17

(b) Let $x = r/r_0$, then $V(x) = A\varepsilon\left[\mathrm{e}^{1-x} - \frac{1}{x^6}\right]$ and the potential minimum occurs when $\left.\frac{\mathrm{d}V(x)}{\mathrm{d}x}\right|_{x=x_e} = 0$

$$\frac{\mathrm{d}V(x)}{\mathrm{d}x} = A\varepsilon\left[-\mathrm{e}^{1-x} + \frac{6}{x^7}\right]$$

$$\left.\frac{\mathrm{d}V(x)}{\mathrm{d}x}\right|_{x=x_e} = A\varepsilon\left[-\mathrm{e}^{1-x_e} + \frac{6}{x_e^7}\right] = 0$$

Thus, the solution of the transcendental equation

$$\mathrm{e}^{1-x_e} = \frac{6}{x_e^7}$$

gives the value x_e at which V is a minimum.

x_e may be found as the intersection of the curves e^{1-x} and $6/x^7$ or it may be found using the numeric solver of a scientific calculator. Here is a short Mathcad worksheet solution for x_e:

$x_e := 1$ Estimate of x_e for the Given/Find solve block. ($x_e = r_e/r_0$)

Given $e^{1-x_e} = 6x_e^{-7}$ $x_e := \text{Find}(x_e)$

$x_e = 1.35985$

Thus, $\boxed{r_e = 1.3598 r_0}$ and, in addition to $B = e$,

$$A = -\left[e^{1-r_e/r_0} - \left(\frac{r_0}{r_e}\right)^6\right]^{-1} = -\left[e^{-0.3598} - \left(\frac{1}{1.3598}\right)^6\right]^{-1} = -1.8531$$

E11.52 $\log A = b_0 + b_1 S + b_2 W$, where S is a parameter related to the solubility of TIBO derivatives in water and W is a parameter related to the width of the first atom in a substituent X.

(a) The following Mathcad worksheet determines the coefficients b_0, b_1, and b_2 using a multiple linear regression analysis. The row of the array called Data are log A, S, and W; the column are values for X = H, Cl, SCH_3, OCH_3, CN, CHO, Br, CH_3, and CCH.

$$\text{Data} := \begin{pmatrix} 7.36 & 8.37 & 8.3 & 7.47 & 7.25 & 6.73 & 8.52 & 7.87 & 7.53 \\ 3.53 & 4.24 & 4.09 & 3.45 & 2.96 & 2.89 & 4.39 & 4.03 & 3.80 \\ 1.00 & 1.80 & 1.70 & 1.35 & 1.60 & 1.60 & 1.95 & 1.60 & 1.60 \end{pmatrix}$$

$\log A := (\text{Data}^T)^{\langle 0 \rangle}$ $S := (\text{Data}^T)^{\langle 1 \rangle}$ $W := (\text{Data}^T)^{\langle 2 \rangle}$ $\text{Mxy} := \text{augment}(S, W)$

info := regress(Mxy, log A, 1) b := submatrix(info, 3, 5, 0, 0) $\boxed{b = \begin{pmatrix} 0.957 \\ 0.362 \\ 3.59 \end{pmatrix} \begin{matrix} b_0 \\ b_1 \\ b_2 \end{matrix}}$

(b) Here, a simple Mathcad worksheet for the computation of W for a drug with $S = 4.84$ and log $A = 7.60$. The method used avoids algebraic manipulations. The expressions b_0, b_1, and b_2 within the worksheet are elements of the matrix b found in part (a).

$W := 1.5$ Estimate for Given/Find Solve Block

$S := 4.84$ $\log A := 7.60$

Given $\log A = b_0 + b_1 \cdot S + b_2 \cdot W$ $W := \text{Find}(W)$ $\boxed{W = 1.362}$

Solutions to projects

P11.53 (a) The ground equilibrium geometry total dipole moment of trans-*N*-methylacetamide as computed with the density functional B3LYP method with 6-31G* basis set (Spartan '06 software) is

$$\mu = \boxed{3.788 \text{ D}}$$
$$= (3.788 \text{ D}) \times (3.33564 \times 10^{-30} \text{ C m D}^{-1}) = 1.264 \times 10^{-29} \text{ C m}$$

The dipole moment, shown in Figure 11.18, is roughly along the line from the positive hydrogen to the negative oxygen of the amide group.

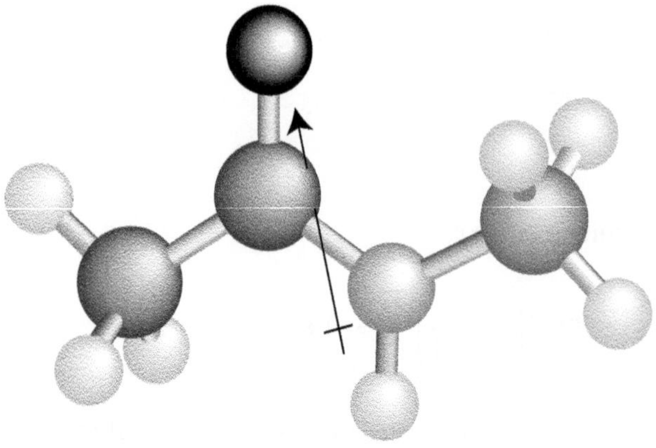

Figure 11.18

(b) The interaction potential of two parallel dipoles is given by eqn 11.17:

$$V = \frac{\mu_1\mu_2 f(\theta)}{4\pi\varepsilon_0 r^3}, \text{ where } f(\theta) = 1 - 3\cos^2\theta$$

where r is the distance between two parallel dipoles and θ the angle between the direction of the dipoles and the line that joins them. The angular dependence of the interaction is shown in Figure 11.19. $V(\theta)$ is at a minimum for $\theta = 0$ and 180° while it is at a maximum for 90°.

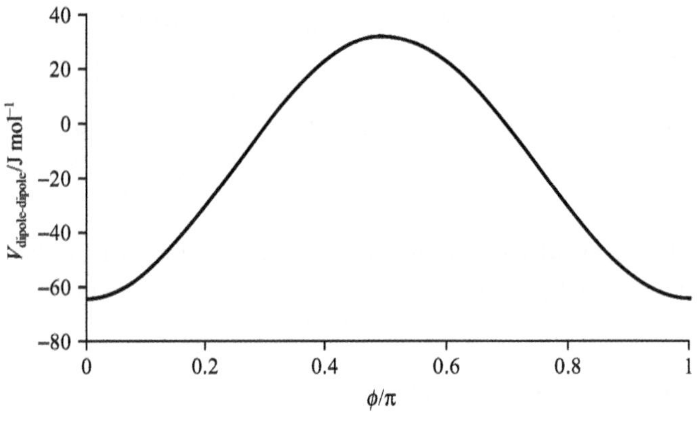

Figure 11.19

(c) If two dipoles are parallel and separated by 3.0 nm, then the maximum attractive interaction occurs when $\theta = 0$.

$$V_{\text{m,max}} = \frac{N_A\mu_1\mu_2(1 - 3\cos^2\theta)}{4\pi\varepsilon_0 r^3} = -\frac{N_A 2\mu_1\mu_2}{4\pi\varepsilon_0 r^3}$$

$$= -\frac{2 \times (6.02214 \times 10^{23}\ \text{mol}^{-1}) \times (1.264 \times 10^{-29}\ \text{C m})^2}{(1.11265 \times 10^{-10}\ \text{J}^{-1}\ \text{C}^2\ \text{m}^{-1}) \times (3.0 \times 10^{-9}\ \text{m})^3}$$

$= -64\ \text{J mol}^{-1}$ [The negative indicates an attractive interaction between the two dipoles.]

Thus, the dipole–dipole interactions at 3.0 nm are $\boxed{\text{dwarfed by hydrogen-bonding interactions}}$, which occur at much shorter distances and have a strength of about 20 kJ mol^{-1}.

P11.54 (a) The left-side molecule in Figure 11.20 is methyladenine. Please note that we have taken the liberty of placing the methyl group in the position that would be occupied by a sugar in RNA and DNA. The wavefunction, structure, and atomic electrostatic charges (shown in the figure) calculation was performed with Spartan '06™ using a Hartree–Fock procedure with a 6-31G* basis set. The atomic electrostatic charge (ESP) numerical method generates charges that reproduce the electrostatic field from the entire wavefunction. The right-side molecule in Figure 11.20 is methylthymine.

(b) The two molecules will hydrogen bond into a stable dimer in an orientation for which hydrogen bonding is linear, maximized, and steric hindrance is avoided. We expect hydrogen bonds of the type N–H···O and N–H···N with the N and O atoms having large negative electrostatic charges and the H atoms having large positive charges. These atoms are evident in Figure 11.20.

(c) Figure 11.20 shows one of three arrangements of hydrogen bonding between the two molecules. Another can be drawn by rotating methylthymine over the top of methyladenine and a third involves rotation to the bottom. The dashed lines show the alignments of two strong hydrogen bonds between the molecules.

(d) The A-to-T base pairing shown in Figure 11.20 has the largest charges in the most favorable positions for strong hydrogen bonding. Also, the N-to-O distance of one hydrogen bond equals the N-to-N distance of the other, a favorable feature in RNA and DNA polymers where this pairing and alignment is observed naturally.

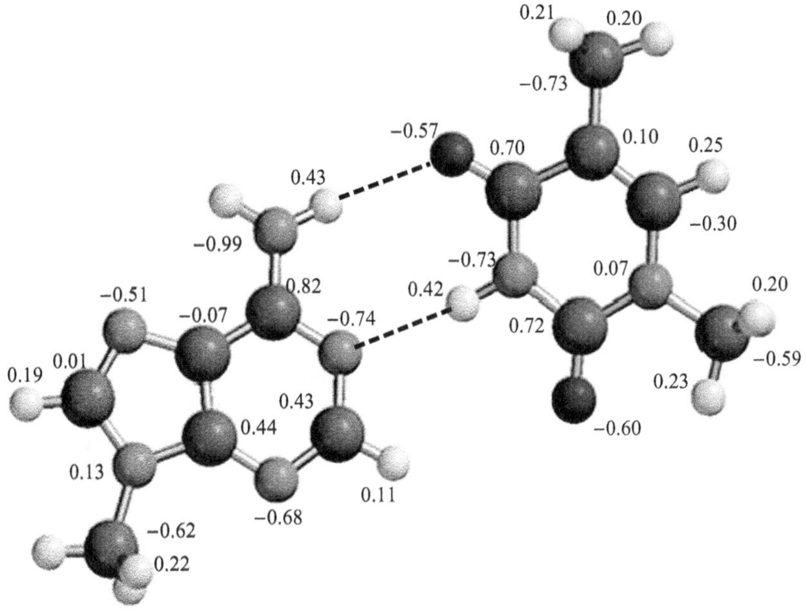

Figure 11.20

(e) The favorable orientation for hydrogen bonding between methylguanine (left-side molecule) and methylcytosine (right-side molecule) is shown in Figure 11.21. Large counter charges align within distances that result in three strong hydrogen bonds.

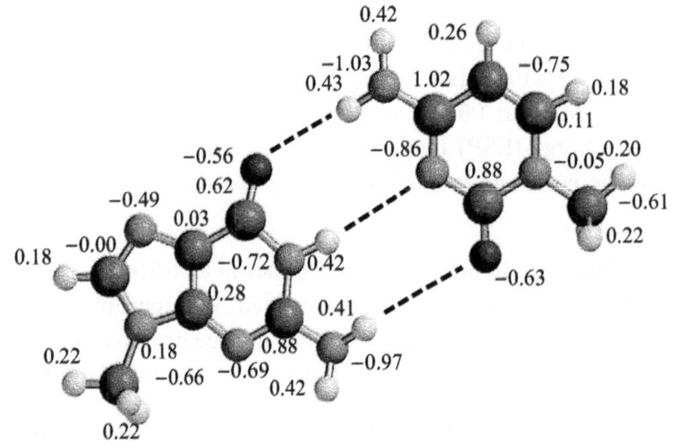

Figure 11.21

P11.55 The exercise explicitly indicates exploration of a molecular mechanic software package for the analysis of ground-state geometry of the molecule shown in Figure 11.22. We have selected the MMFF94 method as implemented by Spartan '06™, a method developed by Merck Pharmaceuticals for the prediction of geometries of ordinary organic molecules and biopolymers. The "steric energy" values obtained with this method have no validity in the absolute sense; however, the values do reasonably account for conformational energy differences (i.e. differences in agreement with high-level quantum chemical calculations). The method allows small deviations from planar for the peptide group.

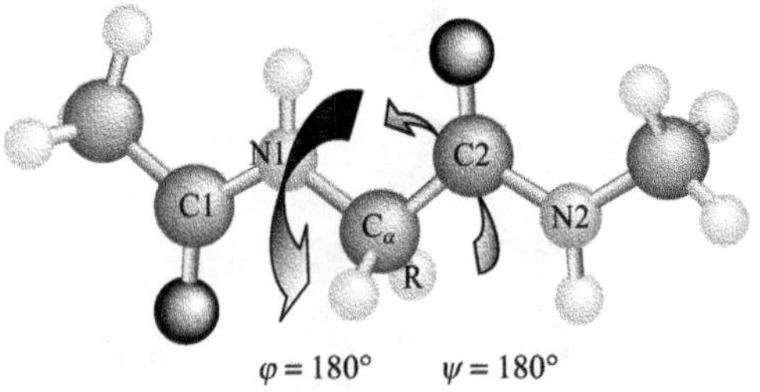

Figure 11.22

The search for the ground-state geometry begins by building the molecule as shown in Figure 11.22 with the torsional (dihedral) angles $\varphi = 180°$ and $\psi = 180°$. MMFF94 is set to search for the equilibrium geometry after which the software is used to measure φ, ψ, the total steric energy, and the dipole moment. The process is repeated for each set of initial (φ,ψ) values. When setting the initial (φ,ψ) values, we use the torsional angle of the C1–N1–C_α–C2 sequence to set φ and the torsional angle of the N1–C_α–C2–N2 sequence to set ψ. In each case we use a Newman projection, as shown in Figure 11.23, to visualize φ and ψ.

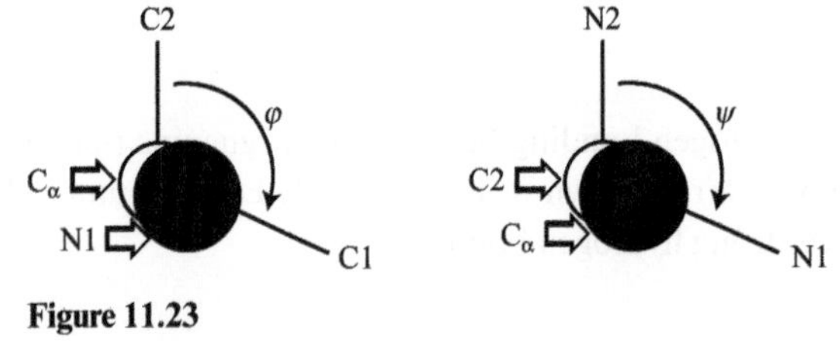

Figure 11.23

(a) Calculations for R = H. The initial conformers are shown in Figures 11.22, 11.24, and 11.25. The following table summarizes results.

Initial		Optimized			
$\varphi/°$	$\psi/°$	$\varphi/°$	$\psi/°$	V_{total}/kJ mol^{-1}	μ/D
180	180	−178.37	177.19	−77.5223	3.049
75	−65	82.71	−48.62	−83.0073	3.750
65	35	82.71	−48.62	−83.0073	3.750

There appears to be a local minimum in the model potential near the point $(\varphi,\psi) = (-178°,177°)$ and a global minimum at about $(\varphi,\psi) = (83°,-49°)$ that has a greater dipole moment. The global minimum is about 5.5 kJ mol^{-1} below the minimum at $(\varphi,\psi) = (-178°,177°)$.

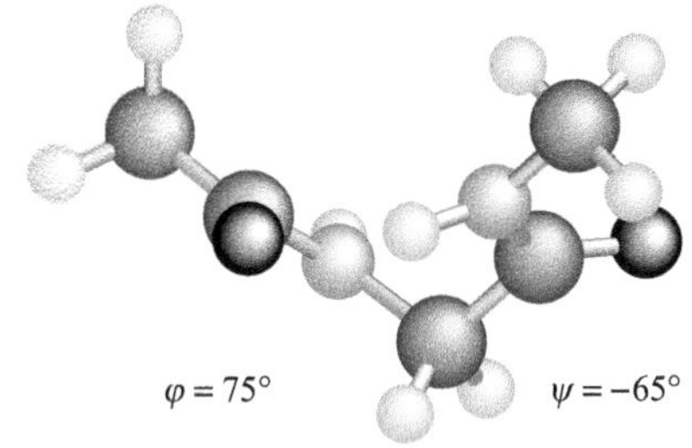

Figure 11.24

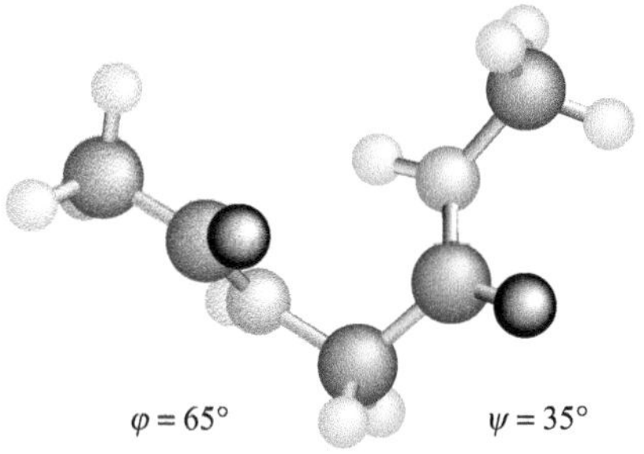

Figure 11.25

(b) Calculations for R = CH_3. The following table summarizes results.

Initial		Optimized			
$\varphi/°$	$\psi/°$	$\varphi/°$	$\psi/°$	V_{total}/kJ mol^{-1}	μ/D
180	180	157.19	−169.30	−66.1664	3.330
75	−65	82.82	−56.49	−73.5552	3.281
65	35	82.82	−56.49	−73.5552	3.281
-90	90	−71.55	75.62	−64.7178	3.162

There appears to be a local minimum in the model potential near the point $(\varphi,\psi) = (157°,-169°)$ and a global minimum at about $(\varphi,\psi) = (83°,-56°)$ that has a slightly smaller dipole moment. The global minimum is about 7.4 kJ mol^{-1} below the minimum at $(\varphi,\psi) = (157°,-169°)$. Furthermore, it is only slightly displaced from the global minimum found with R = H in part (a). An additional calculation with the initial torsional angles $(\varphi,\psi) = (-90°,90°)$ found a second, very shallow local minimum at $(\varphi,\psi) = (-72°,76°)$, which suggests that computation of a complete **Ramachandran plot** (a contour plot of potential energy against the angles φ and ψ) is needed to explore what may be a complicated potential-energy surface.

PART 4 Biochemical spectroscopy

12 Optical spectroscopy and photobiology

Answers to discussion questions

D12.1 The absorption or emission band of an electronic transition is usually very broad because of a multitude of accompanying vibrational transitions that add a progressive series of energies to that of the electronic transition as shown in text Figure 12.12. For example, an electronic spectral transition at 500 nm may have a bandwidth anywhere between about 50 nm and 200 nm. Each of the vibrational transitions of an electronic spectral band has a width as shown in text Figure 12.12 and the overlap of these makes the electronic band a continuum of energies rather than a series of discrete lines. The width of the vibrational component may be due to interactions of the solvent with the vibrational modes of the absorbing (or emitting) molecule. The range of interaction strengths that arise as solvent molecules enter and leave the immediate vicinity of the chemical bond slightly alter the value of the bond force constant of eqn 12.12, thereby smearing each of the vibrational absorption (or emission) lines over a range of values.

Lifetime broadening, which occurs in all states of matter, is also an important source of broadening. This kind of broadening is a quantum-mechanical effect related to the uncertainty principle in the form $\delta E \approx \hbar/\tau$ [12.11a] and is due to the finite lifetime τ of the state involved in the transition. δE is the width, or uncertainty, in the state energy. When τ is finite, the energy of the states is smeared out and hence the transition frequency is broadened. The rate of spontaneous emission cannot be changed; hence lifetime broadening is a **natural linewidth** limit to the breadth of a spectral line. Collisions shorten the lifetime of an excited state and, therefore, produce **collisional broadening** through **collisional deactivation**.

D12.2 (a) In the case of infrared vibrational spectroscopy, the physical basis of the gross selection rule is that the molecule must have a structure that allows for the existence of an oscillating dipole moment when the molecule vibrates. Polar molecules necessarily satisfy this requirement, but non-polar molecules may also have a fluctuating dipole moment upon vibration. See Figure 12.17 of the text. Homonuclear diatomic molecules like O_2 and N_2 are infrared inactive because their dipole moment does not change during vibration.

The gross selection rule for vibrational Raman spectroscopy is that the polarizability of the molecule must change as the molecule vibrates. All diatomic molecules satisfy this condition as the molecules swell and contract during a vibration, the control of the nuclei over the electrons varies, and the molecular polarizability changes. Hence, both homonuclear and heteronuclear diatomics are vibrationally Raman active. In polyatomic molecules it is usually quite difficult to judge by inspection whether or not the molecule is anisotropically polarizable; hence group theoretical methods are relied on for judging the Raman activity of the various normal modes of vibration. The procedure is an advanced topic but the **exclusion rule** can be useful. It states that, if the molecule has a center of inversion, then no modes can be both infrared and Raman active.

(b) The exclusion rule applies to the benzene molecule because it has a center of inversion. Consequently, none of the normal modes of vibration of benzene can be both infrared and Raman active. If we wish to characterize all the normal modes, we must obtain both kinds of spectra.

D12.3 Color can arise by emission, absorption, or scattering of electromagnetic radiation by an object. Many molecules have electronic transitions that have wavelengths in the visible portion of the electromagnetic spectrum. When a substance emits radiation the perceived color of the object will be that of the emitted radiation and it may be an additive color resulting from the emission of more than one wavelength of radiation. When a substance absorbs radiation its color is determined by the subtraction of those wavelengths from white light. For example, absorption of red light results in the object being perceived as green. Scattering, including the diffraction that occurs when light falls on a material with a grid of variation in texture or refractive index having dimensions comparable to the wavelength of light, for example, a bird's plumage, may also form color.

As an addition to general interest in color, recent submicrometer investigations of fossilized anon-avian dinosaurs, *Sinosauropteryx* and *Sinornithosaurus*, have found evidence of color-bearing organelles called melanosomes. (See Zhang, F. *et al.*, *Nature*, doi:101038/nature08740 (Jan. 27, 2010); Fossilized melanosomes and the color of Cretaceous dinosaurs and birds.) From the study of living organisms the melanosomes, eumelanosomes, and phaeomelanosomes are associated with dark colors and lighter yellow-to-red colors, respectively. Their placement within fossils allows for the reconstruction of dinosaur color patterns.

D12.4 The Franck–Condon principle states that because electrons are so much lighter than nuclei an electronic transition occurs so rapidly compared to vibrational motions that the internuclear distance is relatively unchanged as a result of the transition. This implies that the most probable transitions $\upsilon_f \leftarrow \upsilon_i$ are **vertical transitions** when depicted between the ground and excited electronic curves of potential energy against internuclear distance as in text Figure 12.33. This vertical line will, however, intersect any number of vibrational levels υ_f in the upper electronic state. Hence, transitions to many vibrational levels of the excited state will occur with transition probabilities proportional to the Frank–Condon factors that are in turn proportional to the overlap integral of the wavefunctions of the initial and final vibrational states. A vibrational transition progression is observed in the absorption spectrum, the shape of which is determined by the relative horizontal positions of the two electronic potential-energy curves. The most probable transitions are those to excited vibrational states with wavefunctions having large amplitude at the internuclear position R_e. The equilibrium separation of the nuclei in the initial electronic state therefore becomes a **turning point**, one of the end points of a nuclear vibration, in the final electronic state.

D12.5 Groups with characteristic ultraviolet and visible absorptions are called **chromophores** (from the Greek for "color bringer"), and their presence accounts for the colors of many substances. Transition-metal complexes, such as iron porphyrins, often exhibit d–d transitions in the visible. They may also show **charge–transfer transitions** in which a ligand electron jumps into a metal d-orbital creating an intense absorption due to the large transition dipole moment of the jump; an example is provided by the complex of copper in the bacterial protein azurin, which absorbs in the 500–700 nm range and accounts for its intense blue color.

Carbonyl groups (C=O) and unconjugated alkene groups (C=C) have absorptions in the ultraviolet that correspond to n-to-π* and π-to-π* transitions, respectively. Peptide bonds also absorb in the UV as do the aromatic side chains of amino acids and the aromatic bases of nucleic acids. The conjugated π systems of carotenes, the flavins, biological pigments called melanins, and retinal within rhodopsin absorb in the visible.

D12.6 The overall process associated with fluorescence involves the following steps. The molecule is first promoted from the vibrational ground state of a lower electronic level to a higher vibrational-electronic energy level by absorption of energy from a radiation field. Because of the requirements of the Franck–Condon principle, the transition is to excited vibrational levels of the upper electronic state. See Figures 12.33 and 12.43 of the text. Therefore, the absorption spectrum shows a vibrational structure characteristic of the upper state. The excited-state molecule can now lose energy to the surroundings through radiationless transitions and decay to the lowest vibrational level of the upper state. A spontaneous radiative transition now occurs to the lower electronic level and this fluorescence spectrum has a vibrational structure characteristic of the lower state. The fluorescence spectrum is not the mirror image of the absorption spectrum because the vibrational frequencies of the upper and lower states are different due to the difference in their potential-energy curves.

The first steps of phosphorescence are the same as in fluorescence: a singlet state absorbs energy in a transition to an excited singlet state and non-radiative vibrational relaxation occurs within the excited singlet state. The presence of a heavy atom, which can provide angular momentum via strong spin–orbit coupling, and a triplet state of energy similar to the excited singlet, provide the critical step needed for phosphorescence. It is the **intersystem crossing** step in which an electron undergoes a spin-flip from the excited singlet to the triplet state. The electron may remain in the triplet state for an unusually long period because an emission transition from a triplet state to a single state is spin-forbidden.

D12.7 (a) The Förster theory of resonance energy transfer examines the interaction between an induced oscillating dipole moment in chromophore S, the energy donor, with a second chromophore Q, the energy acceptor. The oscillating dipole moment of S is induced by incident electromagnetic radiation and the chromophores are separated by distance R. S transfers the excitation energy of the radiation to Q via a mechanism in which its oscillating dipole moment induces an oscillating dipole moment in Q. Resonance energy transfer can be efficient when R is short (typically less than about 9 nm) and when the absorption spectrum of the acceptor overlaps with the emission spectrum of the donor.

(b) Fluorescence resonance energy transfer (FRET) experiments commonly use the fluorescent spectrum and relaxation times of the Förster donor and acceptor chromophores to find the distances between fluorescent dyes at labelled sites in protein, DNA, RNA, etc. FRET is a type of spectroscopic "ruler". The computation uses either experimental quantum yields or relaxation lifetimes to calculate the efficiency of resonance energy transfer η_T.

$$\eta_T = 1 - \frac{\phi_F}{\phi_{F,0}} = 1 - \frac{\tau}{\tau_0} \quad [12.26 \text{ and } 12.23]$$

η_T is used to calculate R.

$$\eta_T = \frac{R_0^6}{R_0^6 + R^6} \quad \text{or} \quad R = R_0\left(\frac{1-\eta_T}{\eta_T}\right)^{1/6} \quad [12.27]$$

Solutions to exercises

E12.8 (a) $$\nu = \frac{c}{\lambda} = \frac{2.998\times10^{8}\ \text{m s}^{-1}}{670\times10^{-9}\ \text{m}} = 4.47\times10^{14}\ \text{s}^{-1} = \boxed{447\ \text{THz}}$$

(b) $$\tilde{\nu} = \frac{1}{\lambda} = \frac{1}{670\times10^{-9}\ \text{m}} = 1.49\times10^{6}\ \text{m}^{-1} = \boxed{1.49\times10^{4}\ \text{cm}^{-1}}$$

E12.9 (a) $\tilde{\nu}=\frac{\nu}{c}=\frac{92.0\times10^{6}\ \mathrm{s^{-1}}}{2.998\times10^{8}\ \mathrm{m\ s^{-1}}}=\boxed{0.307\ \mathrm{m^{-1}}}$

(b) $\lambda=\frac{1}{\tilde{\nu}}=\frac{1}{0.307\ \mathrm{m^{-1}}}=\boxed{3.26\ \mathrm{m}}$

E12.10 $\varepsilon=\frac{A}{[\mathrm{J}]L}\,[12.5]=-\frac{\log T}{[\mathrm{J}]L}\quad[12.4\mathrm{a\ and\ b}]$

$$\varepsilon_{410\ \mathrm{nm}}=-\frac{\log(0.715)}{(0.433\times10^{-3}\ \mathrm{mol\ dm^{-3}})\times(2.5\ \mathrm{mm})}=\boxed{13\overline{5}\ \mathrm{dm^{3}\ mol^{-1}\ mm^{-1}}}$$

Expressing all lengths in cm yields: $\boxed{1.3\overline{5}\times10^{6}\ \mathrm{cm^{2}\ mol^{-1}}}$

E12.11 $[\mathrm{J}]=\frac{n}{V}=\frac{0.0302\ \mathrm{g}/(602\ \mathrm{g\ mol^{-1}})}{0.500\ \mathrm{dm^{3}}}=1.00\times10^{-4}\ \mathrm{mol\ dm^{-3}}$

(a) $\varepsilon=\frac{A}{[\mathrm{J}]L}\quad[12.5]$

$$\varepsilon=\frac{1.011}{(1.00\times10^{-4}\ \mathrm{mol\ dm^{-3}})\times(1.00\ \mathrm{cm})}=\boxed{1.01\times10^{4}\ \mathrm{dm^{3}\ mol^{-1}\ cm^{-1}}}$$

(b) Since the solution is twice as concentrated, the absorbance must be twice as large. Thus,

$A=2.022$ and $T=10^{-A}\,[12.5]=10^{-2.022}=0.00951$ or $\boxed{0.951\%}$

E12.12 (a) Since $I=\frac{1}{2}I_0$, we calculate that $T=I/I_0=\frac{1}{2}$ and $A=-\log T=0.301$

For water, $[\mathrm{H_2O}]\approx\frac{1.00\ \mathrm{kg/dm^{3}}}{18.02\ \mathrm{g\ mol^{-1}}}=55.5\ \mathrm{mol\ dm^{-3}}$

$$\text{Depth, } L=\frac{A}{\varepsilon[\mathrm{H_2O}]}$$
$$=\frac{0.301}{(6.2\times10^{-5}\ \mathrm{dm^{3}\ mol^{-1}\ cm^{-1}})\times(55.5\ \mathrm{mol\ dm^{-3}})}$$
$$=\boxed{87\ \mathrm{cm}}$$

(b) Since $I=\frac{1}{10}I_0$, we calculate that $T=I/I_0=\frac{1}{10}$ and $A=-\log T=1.00$

$$\text{Depth, } L=\frac{A}{\varepsilon[\mathrm{H_2O}]}$$
$$=\frac{1.00}{(6.2\times10^{-5}\ \mathrm{dm^{3}\ mol^{-1}\ cm^{-1}})\times(55.5\ \mathrm{mol\ dm^{-3}})}$$
$$=\boxed{2.9\ \mathrm{m}}$$

E12.13 The absorbances A_1 and A_2 at wavelengths λ_1 and λ_2 are the sum of the individual absorbances in the mixture of A and B. Let λ_1 be the isosbestic wavelength for which $\varepsilon_{\mathrm{A1}}=\varepsilon_{\mathrm{B1}}\equiv\varepsilon^{\circ}$. Then, at each wavelength we have

$$A_1=\varepsilon_{\mathrm{A1}}L[\mathrm{A}]+\varepsilon_{\mathrm{B1}}L[\mathrm{B}]=\varepsilon^{\circ}L\times([\mathrm{A}]+[\mathrm{B}])\quad\text{(i)}$$

$$A_2=\varepsilon_{\mathrm{A2}}L[\mathrm{A}]+\varepsilon_{\mathrm{B2}}L[\mathrm{B}]\quad\text{(ii)}$$

Solving (i) for [A] gives

$$[\mathrm{A}] = \frac{A_1}{\varepsilon^\circ L} - [\mathrm{B}] \quad \text{(iii)}$$

Substitution of (iii) into (ii) and solving for [B] gives

$$A_2 = \varepsilon_{\mathrm{A2}} L\left(\frac{A_1}{\varepsilon^\circ L} - [\mathrm{B}]\right) + \varepsilon_{\mathrm{B2}} L[\mathrm{B}]$$

$$\varepsilon^\circ A_2 = \varepsilon_{\mathrm{A2}} A_1 - \varepsilon_{\mathrm{A2}}\varepsilon^\circ L[\mathrm{B}] + \varepsilon^\circ \varepsilon_{\mathrm{B2}} L[\mathrm{B}]$$

$$[\mathrm{B}] = \frac{\varepsilon^\circ A_2 - \varepsilon_{\mathrm{A2}} A_1}{\varepsilon^\circ(\varepsilon_{\mathrm{B2}} - \varepsilon_{\mathrm{A2}})L}$$

$$= \frac{A_2 - \frac{\varepsilon_{\mathrm{A2}}}{\varepsilon^\circ} A_1}{(\varepsilon_{\mathrm{B2}} - \varepsilon_{\mathrm{A2}})L}$$

$$\boxed{[\mathrm{B}] = \frac{A_2 - r_{\mathrm{A}} A_1}{(\Delta\varepsilon_2)L}} \quad \text{(iv), where } r_{\mathrm{A}} \equiv \tfrac{\varepsilon_{\mathrm{A2}}}{\varepsilon^\circ} \text{ and } \Delta\varepsilon_2 \equiv \varepsilon_{\mathrm{B2}} - \varepsilon_{\mathrm{A2}}$$

Substitution of (iv) into (iii) and simplifying gives

$$\boxed{[\mathrm{A}] = \frac{r_{\mathrm{B}} A_1 - A_2}{(\Delta\varepsilon_2)L}}, \quad \text{where } r_{\mathrm{B}} \equiv \tfrac{\varepsilon_{\mathrm{B2}}}{\varepsilon^\circ} \text{ and } \Delta\varepsilon_2 \equiv \varepsilon_{\mathrm{B2}} - \varepsilon_{\mathrm{A2}} = (r_{\mathrm{B}} - r_{\mathrm{A}})\varepsilon^\circ$$

E12.14 Let [tryptophan] = [A] and [tyrosine] = [B] in the following analysis. The data using a cell of thickness 1.00 cm is:

At 240 nm, $\varepsilon_{\mathrm{A},240} = 2.00 \times 10^3\ \mathrm{dm^3\ mol^{-1}\ cm^{-1}}$, $\varepsilon_{\mathrm{B},240} = 1.12 \times 10^4\ \mathrm{dm^3\ mol^{-1}\ cm^{-1}}$, $A_{240} = 0.660$

At 280 nm, $\varepsilon_{\mathrm{A},280} = 5.40 \times 10^3\ \mathrm{dm^3\ mol^{-1}\ cm^{-1}}$, $\varepsilon_{\mathrm{B},280} = 1.50 \times 10^3\ \mathrm{dm^3\ mol^{-1}\ cm^{-1}}$, $A_{280} = 0.221$

Substitution of these values into eqn 12.7 yields the concentrations of A and B.

$$[\text{tryptophan}] = [\mathrm{A}] = \frac{\varepsilon_{\mathrm{B},280} A_{240} - \varepsilon_{\mathrm{B},240} A_{280}}{(\varepsilon_{\mathrm{A},240}\varepsilon_{\mathrm{B},280} - \varepsilon_{\mathrm{A},280}\varepsilon_{\mathrm{B},240})L}$$

$$= \frac{\{(1.50 \times 10^3) \times (0.660) - (1.12 \times 10^4) \times (0.221)\}\ \mathrm{mol\ dm^{-3}\ cm}}{\{(2.00 \times 10^3) \times (1.50 \times 10^3) - (5.40 \times 10^3) \times (1.12 \times 10^4)\} \times (1.00\ \mathrm{cm})}$$

$$= \boxed{25.8\ \mu\mathrm{mol\ dm^{-3}}}$$

$$[\text{tyrosine}] = [\mathrm{B}] = \frac{\varepsilon_{\mathrm{A},240} A_{280} - \varepsilon_{\mathrm{A},280} A_{240}}{(\varepsilon_{\mathrm{A},240}\varepsilon_{\mathrm{B},280} - \varepsilon_{\mathrm{A},280}\varepsilon_{\mathrm{B},240})L}$$

$$= \frac{\{(2.00 \times 10^3) \times (0.221) - (5.40 \times 10^3) \times (0.660)\}\ \mathrm{mol\ dm^{-3}\ cm}}{\{(2.00 \times 10^3) \times (1.50 \times 10^3) - (5.40 \times 10^3) \times (1.12 \times 10^4)\} \times (1.00\ \mathrm{cm})}$$

$$= \boxed{54.3\ \mu\mathrm{mol\ dm^{-3}}}$$

E12.15 We will use the definitions and concentration equations that are derived in Exercise E12.13. Let [tryptophan] = [A] and [tyrosine] = [B] in the following analysis.

$$r_{\mathrm{A}} = \tfrac{\varepsilon_{\mathrm{A2}}}{\varepsilon^\circ} = (5.23 \times 10^3)/(2.38 \times 10^3) = 2.20$$

$$r_{\mathrm{B}} = \tfrac{\varepsilon_{\mathrm{B2}}}{\varepsilon^\circ} = (1.58 \times 10^3)/(2.38 \times 10^3) = 0.664$$

$$\Delta\varepsilon_2 = (r_B - r_A)\varepsilon° = (0.664 - 0.220) \times 2.38 \times 10^3\ \text{dm}^3\ \text{mol}^{-1}\ \text{cm}^{-1}$$
$$= -3.67 \times 10^3\ \text{dm}^3\ \text{mol}^{-1}\ \text{cm}^{-1}$$

$$[\text{tryptophan}] = [\text{A}] = \frac{r_B A_1 - A_2}{(\Delta\varepsilon_2)L}$$
$$= \frac{(0.664) \times (0.468) - 0.676}{(-3.67 \times 10^3\ \text{dm}^3\ \text{mol}^{-1}\ \text{cm}^{-1}) \times (1.00\ \text{cm})}$$
$$= \boxed{99.5\ \mu\text{mol dm}^{-3}}$$

$$[\text{tyrosine}] = [\text{B}] = \frac{A_2 - r_A A_1}{(\Delta\varepsilon_2)L}$$
$$= \frac{0.676 - (2.20) \times (0.468)}{(-3.67 \times 10^3\ \text{dm}^3\ \text{mol}^{-1}\ \text{cm}^{-1}) \times (1.00\ \text{cm})}$$
$$= \boxed{96.3\ \mu\text{mol dm}^{-3}}$$

E12.16 (a) The absorption band of text Figure 12.8 illustrates the concept of the **integrated absorption coefficient, $\mathcal{A}$.** It can often be estimated with a normal Gaussian lineshape having the form

$\varepsilon = \varepsilon_{\max} e^{-(\tilde{\nu}-\tilde{\nu}_{\text{peak}})^2/a^2}$, where a is a constant related to the half-width $\Delta\tilde{\nu}_{1/2}$

$$\mathcal{A} = \int_{\text{band}} \varepsilon(\tilde{\nu})\,d\tilde{\nu}\ [\text{Text Fig. 12.8}] = \varepsilon_{\max}\int_{-\infty}^{\infty} e^{-(\tilde{\nu}-\tilde{\nu}_{\text{peak}})^2/a^2}\,d\tilde{\nu}$$
$$= \varepsilon_{\max} a\sqrt{\pi} \quad \text{(standard integral)}$$

The relationship between the half-width and a is found by evaluation of the lineshape at $\varepsilon(\tilde{\nu}_{1/2}) = \varepsilon_{\max}/2$.

$$\varepsilon_{\max}/2 = \varepsilon_{\max} e^{-(\tilde{\nu}_{1/2}-\tilde{\nu}_{\text{peak}})^2/a^2}$$

$$\ln(1/2) = -(\tilde{\nu}_{1/2} - \tilde{\nu}_{\text{peak}})^2/a^2$$

$$a^2 = \frac{(\tilde{\nu}_{1/2} - \tilde{\nu}_{\text{peak}})^2}{\ln(2)} = \frac{(\Delta\tilde{\nu}_{1/2}/2)^2}{\ln(2)}$$

$$a = \frac{\Delta\tilde{\nu}_{1/2}}{2\sqrt{\ln 2}}$$

Thus,

$$\mathcal{A} = \boxed{\tfrac{1}{2}\Delta\tilde{\nu}_{1/2}\varepsilon_{\max}\sqrt{\pi/\ln(2)}} = 1.0645\Delta\tilde{\nu}_{1/2}\varepsilon_{\max}$$

(b) (i) $\mathcal{A} = \frac{1}{2}(5000\ \text{cm}^{-1}) \times (1.00 \times 10^4\ \text{dm}^3\ \text{mol}^{-1}\ \text{cm}^{-1})\sqrt{\pi/\ln(2)}$

$$= \boxed{5.32 \times 10^7\ \text{dm}^3\ \text{mol}^{-1}\ \text{cm}^{-2}}$$

(ii) $\mathcal{A} = \frac{1}{2}(5000\ \text{cm}^{-1}) \times (5.00 \times 10^2\ \text{dm}^3\ \text{mol}^{-1}\ \text{cm}^{-1})\sqrt{\pi/\ln(2)}$

$$= \boxed{2.66 \times 10^6\ \text{dm}^3\ \text{mol}^{-1}\ \text{cm}^{-2}}$$

E12.17 (a) The Dobson unit (DU) is the height of a pure ozone column defined by the factor 10^{-3} cm DU^{-1} with the ozone at 1 atm and 0 °C. The ozone concentration under the conditions of the Dobson unit is

$$[O_3] = \frac{n}{V} = \frac{p}{RT} = \frac{1\ \text{atm}}{(0.0820574\ \text{dm}^3\ \text{atm mol}^{-1}\ \text{K}^{-1}) \times (273.15\ \text{K})} = 4.4615 \times 10^{-2}\ \text{mol dm}^{-3}$$

The absorbance of the ozone column is given by

$$A = \varepsilon[O_3]h_{\text{column}} \quad [12.5]$$
$$= (476\ \text{dm}^3\ \text{mol}^{-1}\ \text{cm}^{-1}) \times (4.4615 \times 10^{-2}\ \text{mol dm}^{-3}) \times h_{\text{column}} = (21.2\overline{4}\ \text{cm}^{-1}) \times h_{\text{column}}$$

For 300 DU: $A = (21.2\overline{4}\ \text{cm}^{-1}) \times (300\ \text{DU}) \times (10^{-3}\ \text{cm DU}^{-1}) = \boxed{6.37}$

For 100 DU: $A = (21.2\overline{4}\ \text{cm}^{-1}) \times (100\ \text{DU}) \times (10^{-3}\ \text{cm DU}^{-1}) = \boxed{2.12}$

(b) The integrated absorption coefficient is $\mathcal{A} = \int_{\text{band}} \varepsilon(\tilde{\nu})\text{d}\tilde{\nu}$ [Text Fig. 12.8].

If we can express ε as an analytical function of $\tilde{\nu}$, we can carry out the integration analytically. Following the hint in the problem, we seek to fit ε to an exponential function of $\tilde{\nu}$, which means that a plot of $\ln\varepsilon$ versus $\tilde{\nu}$ ought to be a straight line. Denoting the slope as m and the intercept at $\tilde{\nu} = 0$ as b the linear and corresponding exponential expressions are

$$\ln(\varepsilon/\text{dm}^3\ \text{mol}^{-1}\ \text{cm}^{-1}) = m(\tilde{\nu}/\text{cm}^{-1}) + b \quad \text{and} \quad \varepsilon/\text{dm}^3\ \text{mol}^{-1}\ \text{cm}^{-1} = \text{e}^{b}\text{e}^{m(\tilde{\nu}/\text{cm}^{-1})}$$

We draw up the following table of data, ln ε, and $\tilde{\nu}$ values and we prepare the ln ε against $\tilde{\nu}$ plot, which is shown in Figure 12.1. Visual inspection of the plot indicates that it is reasonably linear so the plot is fitted with a linear regression, which is shown as an insert within Figure 12.1.

λ/nm	ε/dm^3 mol^{-1} cm^{-1}	$\tilde{\nu}$/cm^{-1}	ln(ε/dm^3 mol^{-1} cm^{-1})
292.0	1512	34247	7.321
296.3	865	33750	6.762
300.8	477	33245	6.167
305.4	257	32744	5.549
310.1	135.9	32248	4.912
315.0	69.5	31746	4.241
320.0	34.5	31250	3.541

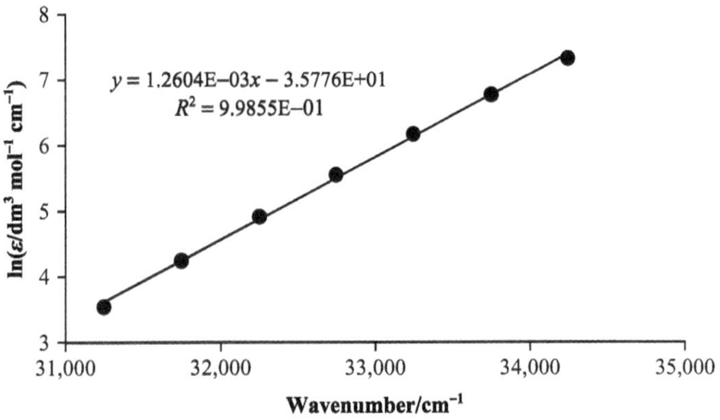

Figure 12.1

The linear regression fit of the ln ε against $\tilde{\nu}$ plot is valid over the range of the data wavenumbers only (i.e. the UVB range of 290–320 nm or 31 250–34 500 cm^{-1}). So with the slope and intercept of Figure 12.1, the integrated absorption coefficient is found to be

$$\mathcal{A}_{\text{UVB}} = \int_{\text{UVB band}} \varepsilon(\tilde{\nu})\text{d}\tilde{\nu}$$

$$= \text{e}^b \int_{31\,250\ \text{cm}^{-1}}^{34\,500\ \text{cm}^{-1}} \text{e}^{m(\tilde{\nu}/\text{cm}^{-1})}\text{d}\tilde{\nu} \times \text{dm}^3\ \text{mol}^{-1}\ \text{cm}^{-1}$$

$$= \frac{\text{e}^b}{m/\text{cm}^{-1}} \left\{ \text{e}^{m(\tilde{\nu}/\text{cm}^{-1})} \Big|_{\tilde{\nu}=31\,250\ \text{cm}^{-1}}^{\tilde{\nu}=34\,500\ \text{cm}^{-1}} \right\} \times \text{dm}^3\ \text{mol}^{-1}\ \text{cm}^{-1} = \frac{\text{e}^b}{m}\{\text{e}^{m\times(34\,500)} - \text{e}^{m\times(31\,250)}\} \times \text{dm}^3\ \text{mol}^{-1}\ \text{cm}^{-2}$$

$$= \frac{\text{e}^{-35.776}}{1.2604\times 10^{-3}}\{\text{e}^{(1.2604\times 10^{-3})\times(34\,500)} - \text{e}^{(1.2604\times 10^{-3})\times(31\,250)}\} \times \text{dm}^3\ \text{mol}^{-1}\ \text{cm}^{-2}$$

$$= \boxed{1.74\times 10^6\ \text{dm}^3\ \text{mol}^{-1}\ \text{cm}^{-2}}$$

E12.18 Using the definitions and thought processes illustrated in text *Justification* 12.1, we write

$$\frac{\text{d}I}{I} = -\kappa[\text{J}]\,\text{d}x$$

To obtain the intensity that emerges from a sample of thickness L when the intensity incident on one face of the sample is I_0, we sum all the successive changes. Because a sum over infinitesimally small increments is an integral, we write

$$\int_{I_0}^{I} \frac{\text{d}I}{I} = -\kappa \int_0^L [\text{J}]\,\text{d}x$$

We make the substitution $[\text{J}] = [\text{J}]_0\text{e}^{-x/\lambda}$ and analytically perform the integrals.

$$\int_{I_0}^{I} \frac{\text{d}I}{I} = -\kappa[\text{J}]_0 \int_0^L \text{e}^{-x/\lambda}\text{d}x$$

$$\ln\frac{I}{I_0} = \frac{-\kappa[\text{J}]_0}{-1/\lambda}\text{e}^{-x/\lambda}\Big|_{x=0}^{x=L}$$

$$= \kappa\lambda[\text{J}]_0\{\text{e}^{-L/\lambda} - 1\}$$

But $\text{e}^{-L/\lambda} \approx 0$ when $L \gg \lambda$, so the expression simplifies to

$$-\ln\frac{I}{I_0} = \kappa\lambda[\text{J}]_0$$

Because the relation between natural and common logarithms is $\ln x = \ln 10 \times \log x$, we can write $\varepsilon = \kappa/\ln 10$ and substitute $A = -\log\frac{I}{I_0}$ to obtain

$$\boxed{A = \varepsilon\lambda[\text{J}]_0}$$

E12.19 We need to establish whether the transition dipole moments

$$\mu_{\text{fi}} = \int \psi_\text{f}^* \mu \psi_\text{i}\,\text{d}\tau\ [12.9]$$

connecting the states 1 and 2 and the states 1 and 3 are zero or non-zero. The particle in a box wavefunctions are $\psi_n = (2/L)^{1/2}\sin(n\pi x/L)$ [9.7]

Thus, $\mu_{2,1} \propto \int \sin\left(\frac{2\pi x}{L}\right) x \sin\left(\frac{\pi x}{L}\right)\text{d}x \propto \int x\left[\cos\left(\frac{\pi x}{L}\right) - \cos\left(\frac{3\pi x}{L}\right)\right]\text{d}x$

and $\mu_{3,1} \propto \int \sin\left(\frac{3\pi x}{L}\right) x \sin\left(\frac{\pi x}{L}\right) dx \propto \int x\left[\cos\left(\frac{2\pi x}{L}\right) - \cos\left(\frac{4\pi x}{L}\right)\right] dx$

having used $\sin\alpha\sin\beta = \frac{1}{2}\cos(\alpha-\beta) - \frac{1}{2}\cos(\alpha+\beta)$. Both of these integrals can be evaluated using the standard form

$$\int x(\cos ax)dx = \frac{1}{a^2}\cos ax + \frac{x}{a}\sin ax$$

$$\int_0^L x\cos\left(\frac{\pi x}{L}\right)dx = \frac{1}{(\pi/L)^2}\cos\left(\frac{\pi x}{L}\right)\bigg|_0^L + \frac{x}{(\pi/L)}\sin\left(\frac{\pi x}{L}\right)\bigg|_0^L = -2\left(\frac{L}{\pi}\right)^2 \neq 0$$

$$\int_0^L x\cos\left(\frac{3\pi x}{L}\right)dx = \frac{1}{(3\pi/L)^2}\cos\left(\frac{3\pi x}{L}\right)\bigg|_0^L + \frac{x}{(3\pi/L)}\sin\left(\frac{3\pi x}{L}\right)\bigg|_0^L = -2\left(\frac{L}{3\pi}\right)^2 \neq 0$$

Thus, $\mu_{2,1} \neq 0$

In a similar manner, $\mu_{3,1} = 0$.

COMMENT: A general formula for μ_{fi} applicable to all possible particle in a box transitions may be derived. The result is (n = f, m = i)

$$\mu_{nm} = -\frac{eL}{\pi^2}\left[\frac{\cos(n-m)\pi - 1}{(n-m)^2} - \frac{\cos(n+m)\pi - 1}{(n+m)^2}\right]$$

For m and n both even or both odd numbers, $\mu_{nm} = 0$; if one is even and the other odd, $\mu_{nm} \neq 0$. Can you establish the general relation for μ_{nm} above?

E12.20 $\delta\tilde{\nu} = \frac{5.3\ \text{cm}^{-1}}{\tau/\text{ps}}$ [12.11b] implying that $\tau = \frac{5.3\ \text{ps}}{\delta\tilde{\nu}/\text{cm}^{-1}}$

(a) $\tau = \frac{5.3\ \text{ps}}{0.1} = \boxed{53\ \text{ps}}$

(b) $\tau = \frac{5.3\ \text{ps}}{1} = \boxed{5\ \text{ps}}$

(c) $\lambda = \frac{2.998\times10^8\ \text{m s}^{-1}}{1.0\times10^9\ \text{s}^{-1}} = 0.300\ \text{m} = 30.0\ \text{cm}$

$\tilde{\nu} = \frac{1}{\lambda} = 0.0333\ \text{cm}^{-1}$

$\tau = \frac{5.3\ \text{ps}}{0.0333} = \boxed{1.6\times10^2\ \text{ps}}$

E12.21 $\delta\tilde{\nu} = \frac{5.3\ \text{cm}^{-1}}{\tau/\text{ps}}$ [12.11b]

(a) $\tau \approx 1\times10^{-13}\ \text{s} = 0.1\ \text{ps}$, implying that $\boxed{\delta\tilde{\nu} = 53\ \text{cm}^{-1}}$.

(b) $\tau \approx 200\times(1\times10^{-13}\ \text{s}) = 20\ \text{ps}$, implying that $\boxed{\delta\tilde{\nu} = 0.27\ \text{cm}^{-1}}$.

E12.22 $\nu = \frac{1}{2\pi}\left(\frac{k_f}{\mu}\right)^{1/2}$ [12.13b; isotopic masses are found in the *CRC Handbook of Chemistry and Physics*]

(a) $\mu = \frac{m_{^{12}C}m_{^{16}O}}{m_{^{12}C}+m_{^{16}O}} = \frac{(12.0000\ m_u)\times(15.9949\ m_u)}{(12.0000+15.9949)\ m_u}\times(1.66054\times10^{-27}\ \text{kg}\ m_u^{-1}) = 1.139\times10^{-26}\ \text{kg}$

$$\nu = \frac{1}{2\pi}\left(\frac{908\ \text{N m}^{-1}}{1.139\times10^{-26}\ \text{kg}}\right)^{1/2} = 4.49\times10^{13}\ \text{s}^{-1} = \boxed{4.49\times10^{13}\ \text{Hz}}$$

(b) $\mu = \frac{m_{^{13}C}m_{^{16}O}}{m_{^{13}C}+m_{^{16}O}} = \frac{(13.0034\ m_u)\times(15.9949\ m_u)}{(13.0034+15.9949)\ m_u}\times(1.66054\times10^{-27}\ \text{kg}\ m_u^{-1}) = 1.191\times10^{-26}\ \text{kg}$

$$\nu = \frac{1}{2\pi}\left(\frac{908\ \text{N m}^{-1}}{1.191\times10^{-26}\ \text{kg}}\right)^{1/2} = 4.39\times10^{13}\ \text{s}^{-1} = \boxed{4.39\times10^{13}\ \text{Hz}}$$

E12.23 (a) We begin by finding a relation between the wavenumber and the bond force constant.

$\nu = \frac{1}{2\pi}\left(\frac{k_f}{\mu}\right)^{1/2}$ [12.13b] implies that $k_f = (2\pi\nu)^2\mu = (2\pi c\tilde{\nu})^2\mu$ [12.2b]

Thus,

$$\begin{aligned} k_f &= \{2\pi\times(2.9979\times10^{10}\ \text{cm s}^{-1})\times\text{cm}^{-1}\}^2(\tilde{\nu}/\text{cm}^{-1})^2(\mu)\times(1.66054\times10^{-27}\ \text{kg}/m_u) \\ &= (5.89174\times10^{-5}\ \text{kg s}^{-2})\times(\tilde{\nu}/\text{cm}^{-1})^2\ (\mu/m_u) \\ &= (5.89174\times10^{-5}\ \text{N m}^{-1})\times(\tilde{\nu}/\text{cm}^{-1})^2\ (\mu/m_u) \end{aligned}$$

The fundamental wavenumber of each molecule is provided in the exercise. Isotopic masses, found in the *CRC Handbook of Chemistry and Physics*, are used to calculate the effective masses after which we calculate the bond force constant. The calculations for $^1H^{19}F$ are

$$\mu_{HF} = \frac{1.0078\times18.9984}{1.0078+18.9984}m_u = 0.9570\ m_u$$

$$k_{HF} = (5.89174\times10^{-5}\ \text{N m}^{-1})\times(4141.3)^2(0.9570) = 967.0\ \text{N m}^{-1}$$

We summarize the computations for all molecules in the following table.

	$^1H^{19}F$	$^1H^{35}Cl$	$^1H^{80}Br$	$^1H^{127}I$
$\tilde{\nu}/\text{cm}^{-1}$	4141.3	2988.9	2649.7	2309.5
μ/m_u	0.9570	0.9796	0.9954	0.9999
$k_f/\text{N m}^{-1}$	967.0	515.6	411.8	314.2

(b) The bonding force constant is an electronic property that does not depend upon the effective mass so we use the force constant values above for the deuterium halides as well as the hydrogen halides. This suggests that by taking the wavenumber ratio of two molecules that are identical except for the substitution of the deuterium for hydrogen we can efficiently calculate the wavenumber of the deuterium halides. Since

$$\tilde{\nu} = \frac{\nu}{c} = \frac{1}{2\pi c}\left(\frac{k_f}{\mu}\right)^{1/2}$$

the ratio for $^2D^{19}F$ is

$$\frac{\tilde{\nu}_{DF}}{\tilde{\nu}_{HF}} = \left(\frac{\mu_{HF}}{\mu_{DF}}\right)^{1/2} \quad \text{and} \quad \tilde{\nu}_{DF} = \left(\frac{\mu_{HF}}{\mu_{DF}}\right)^{1/2} \tilde{\nu}_{HF}$$

So, we calculate μ_{DF} and substitute values from the above table in the calculation of $\tilde{\nu}_{DF}$.

$$\mu_{DF} = \frac{2.0140 \times 18.9984}{2.0140 + 18.9984} m_u = 1.8210\, m_u$$

$$\tilde{\nu}_{DF} = \left(\frac{0.9570}{1.8210}\right)^{1/2} \times (4141.3\ \text{cm}^{-1}) = 3002.2\ \text{cm}^{-1}$$

We summarize the computations for all molecules in the following table.

	$^2D^{19}F$	$^2D^{35}Cl$	$^2D^{80}Br$	$^2D^{127}I$
μ/m_u	1.8210	1.9043	1.9651	1.9825
$\tilde{\nu}/\text{cm}^{-1}$	3002.2	2143.7	1885.8	1640.2

E12.24 Select those molecules in which a vibration gives rise to a change in dipole moment. It is helpful to write down the structural formulas of the compounds. The molecules that show infrared absorption are:

(b) HCl, (c) CO_2, (d) H_2O, (e) CH_3CH_3, (f) CH_4, and (g) CH_3Cl

E12.25 For non-linear molecules the number of normal modes is given by $3N - 6$ where N is the number of atoms in the molecules; for linear molecules the number of normal modes is $3N - 5$. Thus, we need to establish the linearity of the molecules listed. Molecules (c) and (d) are clearly non-linear. From the Lewis structures of molecules (a) and (b) and VSEPR we decide that they are non-linear and linear, respectively.

(a) NO_2, non-linear, $3N - 6 = 9 - 6 = \boxed{3}$

(b) N_2O, linear, $3N - 5 = 9 - 5 = \boxed{4}$

(c) C_6H_{12}, non-linear, $3N - 6 = 3 \times 18 - 6 = \boxed{48}$

(d) C_6H_{14}, non-linear, $3N - 6 = 3 \times 20 - 6 = \boxed{54}$

E12.26 The uniform expansion mode of planar, centrosymmetric benzene is depicted in Figure 12.2. It is apparent that this "breathing" motion leaves the molecular dipole moment unchanged so we conclude that the mode is infrared inactive. The polarizability of the mode does change with the motion so it is Raman active. This confirms the **exclusion rule** that specifies that a vibrational mode of a molecule that has a center of inversion cannot be both infrared and Raman active. See text Section 12.5(c),

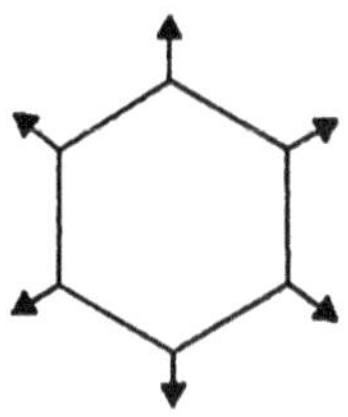

Figure 12.2

E12.27 Tabulate the six observed absorptions.

$\tilde{\nu}_{\text{Infrared}}/\text{cm}^{-1}$	870	1370	2869	3417
$\tilde{\nu}_{\text{Raman}}/\text{cm}^{-1}$	877	1408	1435	3407

(a) If H_2O_2 were linear, it would have $3N - 5 = \boxed{7}$ vibrational modes.

(b) The exclusion rule applies to structure **8** because it has a center of inversion: no vibrational modes can be both IR and Raman active. But the absorptions at about 872 cm^{-1} and 3410 cm^{-1} indicate modes that are both infrared and Raman active so we conclude that $\boxed{\text{structure } \mathbf{8} \text{ is inconsistent with these absorptions}}$.

E12.28 A typical n-to-π* transition at about 4 eV (32000 cm^{-1}) characterizes an absorption in which a carbonyl lone pair electron on oxygen is excited to a π* orbital, while an unconjugated alkene exhibits a π-to-π* electron transition at about 7 eV (56000 cm^{-1}). Thus, the 30000 cm^{-1} absorption of $CH_3CH{=}CHCHO$ is a n-to-π* transition and the 46950 cm^{-1} absorption is a π-to-π* transition, which lies lower than the norm because of conjugation between the alkene and carbonyl π bonds.

E12.29 Tryptophan (Trp) and tyrosine (Tyr) show the characteristic absorption of a phenyl group at about 280 nm. Cysteine (Cys) and glycine (Gly) lack the phenyl group, as is evident from their spectra.

E12.30 The transition dipole moment, μ_{fi}, between states i and f determines the intensity of a transition. If it has a high absolute value, the transition probability and intensity are high (see text Section 12.2). We need to determine how it depends on the length of the chain. We assume that wavefunctions of the conjugated electrons in the linear polyene can be approximated by the wavefunctions of a particle in a one-dimensional box. Then, for a transition between the states n' and n

$$\mu_x = -e\int_0^L \psi_{n'}(x)x\psi_n(x)\mathrm{d}x\ [12.9], \quad \text{where} \quad \psi_n = \left(\frac{2}{L}\right)^{1/2}\sin\left(\frac{n\pi x}{L}\right)\ [9.7]$$

$$= -\frac{2e}{L}\int_0^L x\sin\left(\frac{n'\pi x}{L}\right)\sin\left(\frac{n\pi x}{L}\right)\mathrm{d}x$$

$$= \begin{cases} 0 & \text{if } n' = n+2 \\ \left(\dfrac{8eL}{\pi^2}\right)\dfrac{n(n+1)}{(2n+1)^2} & \text{if } n' = n+1 \end{cases}$$

Thus, the selection rule for radiation absorption is $\Delta n = +1$ and the transition integral is proportional to L. We conclude that $\boxed{\text{longer lengths of the dye's conjugated electronic } \pi \text{ system are expected to yield greater absorption intensity}}$. To examine the effect that changing the length has on the apparent color, consider the energy of the absorption.

$$h\nu = E_{n+1} - E_n = \frac{(n+1)^2h^2}{8m_eL^2} - \frac{n^2h^2}{8m_eL^2} = (2n+1)\frac{h^2}{8m_eL^2}$$

$\boxed{\text{Therefore, since } L \text{ appears in the denominator, increasing the length } L \text{ of the polyene shifts the absorption to lower energy. This is a red shift}}$.

E12.31 (a) The fluorescence intensity is proportional to the concentration of excited singlet state S* so at time t: $I_F/I_0 = [S^*]/[S^*]_0$. Thus, by eqn 12.20

$$I_F/I_0 = e^{-t/\tau_0} \quad \text{and} \quad \ln(I_F/I_0) = -t/\tau_0 \quad \text{and} \quad \tau_0 = -t/\ln(I_F/I_0)$$

We simply calculate $-t/\ln(I_F/I_0)$ at each time and, if the results are constant within experimental uncertainties, average the results.

t/ns	5.0	10.0	15.0	20.0
I_F/I_0	0.45	0.21	0.11	0.05
$-t/\ln(I_F/I_0)$/nm	6.26	6.41	6.80	6.68

The average gives $\tau_0 = \boxed{6.5\bar{4}\ \text{ns}}$.

(b) $k_F = \phi_f/\tau_0 = 0.70/(6.5\bar{4}\ \text{ns}) = \boxed{0.11\ \text{ns}^{-1}}$ [eqn 12.23]

E12.32 (a) The molar concentration corresponding to 1 molecule per μm^3 is:

$$\frac{n}{V} = \left(\frac{1\ \text{molecule}}{1.0\times10^{-18}\ \text{m}^3}\right)\times\left(\frac{1\ \text{mol}}{6.022\times10^{23}\ \text{molecules}}\right)\times\left(\frac{1\ \text{m}^3}{10^3\ \text{dm}^3}\right) = 1.7\times10^{-9}\ \text{mol dm}^{-3}$$

$$= \boxed{1.7\ \text{nmol dm}^{-3}}$$

(b) An impurity of a compound of molar mass 100 g mol^{-1} present at a concentration of 0.10 mg per 1.0 kg water (i.e. 1.0 μmol kg^{-1}). So, the impurity number density is

$$\frac{N}{V} = \left(\frac{1.0\times10^{-6}\ \text{mol impurity}}{1.0\ \text{kg water}}\right)\times\left(\frac{6.022\times10^{23}\ \text{molecules}}{\text{mol}}\right)\times\left(\frac{1.0\ \text{kg water}}{1\ \text{dm}^3}\right)\times\left(\frac{10^3\ \text{dm}^3}{10^{18}\ \mu\text{m}^3}\right)$$

$$= \boxed{6.0\times10^2\ \text{molecules}\ \mu\text{m}^{-3}}$$

Pure as it seems, the solvent is much too contaminated for single-molecule spectroscopy.

E12.33 The laser is delivering photons of energy

$$E = h\nu = \frac{hc}{\lambda} = \frac{(6.626\times10^{-34}\ \text{J s})\times(2.998\times10^8\ \text{m s}^{-1})}{488\times10^{-9}\ \text{m}} = 4.07\times10^{-19}\ \text{J}$$

Since the laser is putting out 1.0 mJ of these photons every second, the rate of photon emission is:

$$v = \frac{1.0\times10^{-3}\ \text{J s}^{-1}}{4.07\times10^{-19}\ \text{J}} = 2.5\times10^{15}\ \text{s}^{-1}$$

The time it takes the laser to deliver 10^6 photons (and therefore the time the dye remains fluorescent) is

$$t = \frac{10^6}{2.5\times10^{15}\ \text{s}^{-1}} = 4\times10^{-10}\ \text{s} = \boxed{0.4\ \text{ns}}$$

E12.34 The lifetime of the unimolecular photochemical reaction is $\tau = \dfrac{1}{k_r} = \dfrac{1}{1.7\times10^4\ \text{s}^{-1}} = 5.9\times10^{-5}$ s. This is a much longer lifetime than that of fluorescence (1.0×10^{-9} s) so we conclude that the excited singlet state decays too rapidly to be the precursor of this photochemical reaction. The lifetime of phosphorescence (1.0×10^{-3} s) is longer than the reaction lifetime, making the $\boxed{\text{triplet state}}$ the likely reaction precursor.

E12.35 $A \rightarrow 2\,B + C$

$$\text{Number of photons absorbed} = \frac{\text{number of events}}{\phi} \quad [12.17]$$

$$= \frac{\frac{1}{2}n_B}{\phi}$$

$$= \left(\frac{\frac{1}{2}(2.28 \times 10^{-3}\ \text{mol})}{2.1 \times 10^{2}\ \text{mol einstein}^{-1}}\right) \times (6.022 \times 10^{23}\ \text{einstein}^{-1})$$

$$= \boxed{3.3 \times 10^{18}}$$

E12.36 For a source of power P and wavelength λ, the amount of photons (n_λ) generated in a time t is

$$n_\lambda = \frac{Pt}{h\nu N_A} = \frac{Pt\lambda}{hcN_A}$$

$$= \frac{(100\ \text{W}) \times (45\ \text{min}) \times (60\ \text{s min}^{-1}) \times (490 \times 10^{-9}\ \text{m})}{(6.626 \times 10^{-34}\ \text{J s}) \times (2.998 \times 10^{8}\ \text{m s}^{-1}) \times (6.022 \times 10^{23}\ \text{mol}^{-1})}$$

$$= 1.11\ \text{mol}$$

The amount of photons absorbed is 60 per cent of this incident flux, or 0.664 mol. Therefore,

$$\phi = \frac{0.344\ \text{mol}}{0.664\ \text{mol}} = \boxed{0.518}$$

E12.37 $M + h\nu_i \rightarrow M^*$ $\quad I_{abs}$ [M = benzophenone]

$M^* + Q \rightarrow M + Q,$ $\quad k_Q$ [Q = triethylamine]

$M^* \rightarrow M + h\nu_{phos},$ $\quad k_{phos}$

$$\frac{d[M^*]}{dt} = I_{abs} - k_{phos}[M^*] - k_Q[Q][M^*] \approx 0 \text{ [steady state]}$$

Hence, $[M^*] = \dfrac{I_{abs}}{k_{phos} + k_Q[Q]}$

$$I_{phos} = k_{phos}[M^*] = \frac{k_{phos}I_{abs}}{k_{phos} + k_Q[Q]} \quad \text{and so} \quad \boxed{\frac{1}{I_{phos}} = \frac{1}{I_{abs}} + \frac{k_Q[Q]}{k_{phos}I_{abs}}}$$

This analysis indicates that a plot of $1/I_{phos}$ against [Q] should be linear with an intercept equal to $1/I_{abs}$ and a slope equal to $k_Q/k_{phos}I_{abs}$.

Thus, $k_Q = \dfrac{\textit{slope} \times k_{phos}}{\textit{intercept}}$, where $k_{phos} = \dfrac{\ln 2}{t_{1/2,phos}} = \dfrac{\ln 2}{29 \times 10^{-6}\ \text{s}} = 2.3\overline{9} \times 10^{4}\ \text{s}^{-1}$.

We draw up the following table and prepare the plot shown in Figure 12.3 with the linear regression fit shown as an insert.

$[Q]/\text{mol dm}^{-3}$	0.0010	0.0050	0.0100
$1/I_{phos}$	$2.4\overline{4}$	$4.0\overline{0}$	$6.2\overline{5}$

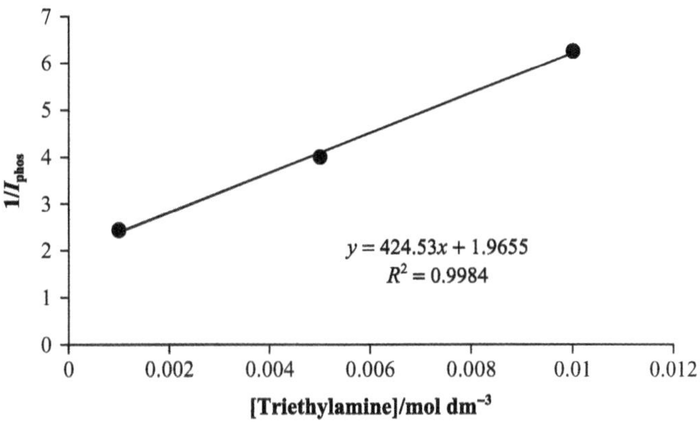

Figure 12.3

$$k_Q = \frac{slope \times k_{phos}}{intercept} = \frac{(425\ \text{dm}^3\ \text{mol}^{-1}) \times (2.3\overline{9} \times 10^4\ \text{s}^{-1})}{1.97}$$

$$= \boxed{5.2 \times 10^6\ \text{dm}^3\ \text{mol}^{-1}\ \text{s}^{-1}}$$

E12.38 $\frac{1}{\tau} = \frac{1}{\tau_0} + k_Q[\text{Q}]$ [Stern–Volmer equation, 12.25]

A plot of $1/\tau$ against [Q] should give a straight line with a slope equal to k_Q and an intercept at $[\text{Q}] = 0$ equal to $1/\tau_0$, where τ_0 is the fluorescence lifetime in the absence of a quencher. The half-life of the fluorescence in the absence of a quencher is determined from $t_{1/2,0} = \tau_0 \ln 2$ [6.13 and 6.14]. We draw up the following table of data and prepare the plot shown in Figure 12.4. The plot appears to be linear so the linear regression fit is calculated and presented as an insert within the figure.

[Q]/mmol dm^{-3}	1.0	2.0	3.0	4.0	5.0
I_{abs}/I_F	3.2$\overline{3}$	5.5$\overline{6}$	7.6$\overline{9}$	10.$\overline{0}$	12.$\overline{3}$
$(\tau/\text{ns})^{-1}$	0.013$\overline{2}$	0.022$\overline{2}$	0.031$\overline{3}$	0.040$\overline{0}$	0.050$\overline{0}$

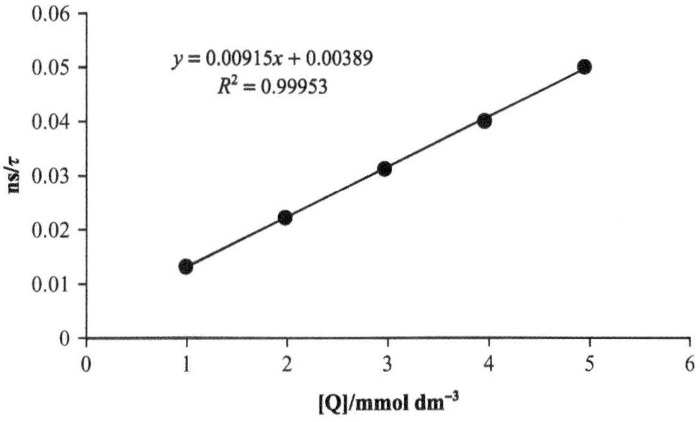

Figure 12.4

Thus, $k_Q = slope = 9.2 \times 10^{-3}\ \mathrm{dm^3\ mmol^{-1}\ ns^{-1}} = \boxed{9.2 \times 10^9\ \mathrm{dm^3\ mol^{-1}\ s^{-1}}}$

and $\tau_0 = 1/intercept = \mathrm{ns}/3.8\bar{9} \times 10^{-3} = \boxed{25\bar{7}\ \mathrm{ns}}$

and $t_{1/2,0} = \tau_0 \ln 2 = (25\bar{7}\ \mathrm{ns}) \times \ln 2 = \boxed{17\bar{8}\ \mathrm{ns}}$

E12.39 $\eta_T = \dfrac{R_0^6}{R_0^6 + R^6}$ or $\dfrac{1}{\eta_T} = 1 + (R/R_0)^6$ [12.27]

The data plot of η_T^{-1} values against R^6 (Figure 12.5) appears to be linear with an intercept equal to 1. Consequently, we perform the calculation of a linear regression fit of the plot, shown as a figure insert, with the intercept constrained to equal 1. The slope of the regression fit equals $1/R_0^6$ so

$$R_0 = \left(\frac{1}{5.4\bar{6} \times 10^{-4}\ \mathrm{nm^{-6}}}\right)^{1/6} = \boxed{3.5\ \mathrm{nm}}$$

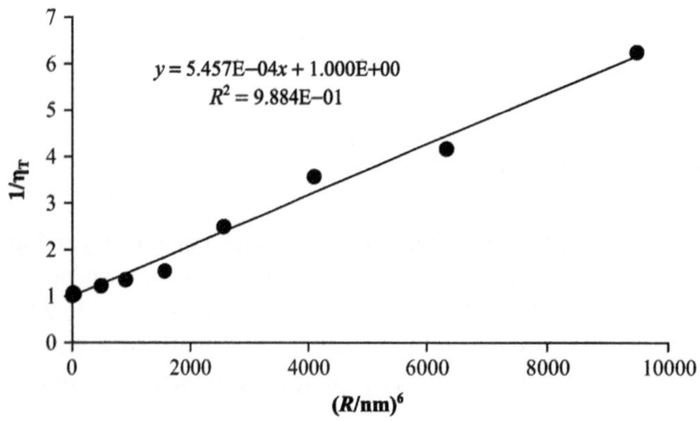

Figure 12.5

E12.40 The efficiency of resonance energy transfer is given by [12.26]:

$$\eta_T = 1 - \frac{\phi_F}{\phi_{F,0}} = 1 - \frac{0.9\phi_{F,0}}{\phi_{F,0}} = 0.10$$

Förster theory relates this quantity to the distance R between donor–acceptor pairs by

$$\eta_T = \frac{R_0^6}{R_0^6 + R^6}\ [12.27]$$

where $R_0 = 4.9$ nm is the empirical parameter listed in Table 12.5 for the 1.5-I-AEDANS-FITC donor–acceptor pair. Solving for the distance between the amino acids yields

$$R = R_0\left(\frac{1}{\eta_T} - 1\right)^{1/6} = (4.9\ \mathrm{nm}) \times \left(\frac{1}{0.10} - 1\right)^{1/6} = \boxed{7.1\ \mathrm{nm}}$$

E12.41 The percentage of visible photons transmitted to the retina from the North Star, or any star, is

$$\begin{aligned} &70\% - 0.25 \times 70\% - 0.09 \times 0.75 \times 70\% - 0.43 \times (1 - 0.25 - 0.09 \times 0.75) \times 70\% \\ &= 70\% - (0.25 + 0.0675 + 0.390) \times 70\% \\ &= 20.5\% \end{aligned}$$

Number of North Star photons focused on the retina in 0.1 s is

$$(0.205) \times (40\ \text{mm}^2) \times (0.1\ \text{s}) \times (4 \times 10^3\ \text{mm}^{-2}\ \text{s}^{-1}) = \boxed{3 \times 10^3}$$

This is perhaps more than what one might have guessed.

E12.42 $\eta_T = 1 - \dfrac{\phi_F}{\phi_{F,0}} = 1 - \dfrac{\tau}{\tau_0}$ [12.26 and 12.23], where τ is the lifetime in the presence of the quencher.

Also, according to Förster theory: $\eta_T = \dfrac{R_0^6}{R_0^6 + R^6}$ [12.27]

Equating these two expressions for η_T and solving for R gives:

$$\frac{R_0^6}{R_0^6 + R^6} = 1 - \frac{\tau}{\tau_0}$$

$$\frac{R_0^6 + R^6}{R_0^6} = \frac{1}{1 - \dfrac{\tau}{\tau_0}}$$

$$\left(\frac{R}{R_0}\right)^6 = \frac{1}{1 - \dfrac{\tau}{\tau_0}} - 1 = \frac{\dfrac{\tau}{\tau_0}}{1 - \dfrac{\tau}{\tau_0}} \quad \text{or} \quad R = R_0\left(\frac{\dfrac{\tau}{\tau_0}}{1 - \dfrac{\tau}{\tau_0}}\right)^{1/6}$$

$$\tau/\tau_0 = 10\ \text{ps}/10^3\ \text{ps} = 0.010 \quad \text{and} \quad R = 5.6\ \text{nm}\left(\frac{0.010}{1 - 0.010}\right)^{1/6} = \boxed{2.6\ \text{nm}}$$

E12.43

$$\underset{\text{Chlorophyll}}{\text{C}} + \underset{\text{Quinone}}{\text{Q}} \xrightarrow{h\nu} \text{C}^* + \text{Q} \xrightarrow{\text{electron transfer}} \text{C}^+ + \text{Q}^-$$

Direct electron transfer from the ground state of C is not spontaneous. It is spontaneous from the excited state. The difference between the ΔGs of the two processes is given by the expression:

$$\Delta(\Delta G) = \Delta G_{C*} - \Delta G_C \approx U_C - U_{C*} \approx -(U_{LUMO} - U_{HOMO}) < 0$$

where U_{LUMO} and U_{HOMO} are energies of the lowest unoccupied and highest occupied molecular orbitals of chlorophyll, respectively, and by definition $U_{LUMO} > U_{HOMO}$. Since $\Delta(\Delta G) < 0$, the electron transfer is exergonic and, therefore, spontaneous when the electron is transferred from the excited state of chlorophyll.

E12.44 *Hypothesis*: The 1270 nm emission band is the emission of the first excited state of O_2 as it returns to the O_2 ground state according to the following porphyrin photosensitization and O_2 emission scheme.

$$\underset{\text{porphyrin}}{\text{P}} + h\nu_1 \xrightarrow{\text{absorption}} \text{P}^*$$

$$\text{P}^* + \text{O}_2\ (\text{triplet ground state}) \xrightarrow{\text{energy transfer}} \text{P} + \text{O}_2\ (\text{singlet excited state})$$

$$\text{O}_2\ (\text{singlet excited state}) \xrightarrow{\text{emission}} \text{O}_2\ (\text{triplet ground state}) + h\nu_2$$

Test of hypothesis: It is well known that the first dioxygen excited state is 0.977 eV (1270 nm) above the ground state. If the hypothesis is correct the emission intensity should be proportional to both the concentration of dissolved oxygen and to the intensity of the porphyrin absorption.

Solutions to projects

P12.45 (a) This report is left for the student.

(b) The following excerpts are from P. Atkins and J. de Paula, *Elements of Physical Chemistry*, 5th edn, W.H. Freeman and Company (2009). They serve as background preparation for your literature search into spectroscopic studies of protein denaturation. See the text section *In The Laboratory 1.1(c)* for a discussion of differential scanning calorimetry (DSC) as a technique for the study of **thermal denaturation**.

Proteins are polymers that attain well-defined three dimensional structures both in solution and in biological cells. They are polypeptides formed from different amino acids strung together by the peptide link, —CONH—. Hydrogen bonds between amino acids of a polypeptide give rise to stable helical or sheet structures, which may collapse into a random coil when certain conditions are changed. For example, the synthetic polypeptide poly-γ-benzyl-glutamate is helical in a nonhydrogen-bonding solvent, but in a hydrogen-bonding solvent it forms a random coil. The unwinding, or **denaturation**, of a helix into a random coil is a cooperative transition, in which the polymer becomes increasingly more susceptible to structural changes once the process has begun. Denaturation can be brought about by changes in temperature or pH, and by reaction with certain compounds, such as urea or guanidinium hydrochloride, known as denaturants.

Earlier work on folding and unfolding of small polypeptides and large proteins relied primarily on rapid mixing and stopped-flow techniques. In a typical stopped-flow experiment, a sample of the protein with a high concentration of a chemical denaturant is mixed with a solution containing a much lower concentration of the same denaturant. Upon entering the mixing chamber, the denaturant is diluted and the protein refolds. Unfolding is observed by mixing a sample of folded protein with a solution containing a high concentration of denaturant. These experiments are ideal for sorting out events in the millisecond time scale, such as the formation of contacts between helical segments in a large protein. However, the available data also indicate that, in a number of proteins, a significant portion of the folding process occurs in less than 1 ms, a time range not accessible by the stopped-flow technique. More recent temperature-jump and flash-photolysis experiments have uncovered faster events. For example, at room temperature the formation of a loop between helical or sheet segments may be as fast as 1 μs and the formation of tightly packed cores with significant tertiary structure occurs in 10–100 μs. Among the fastest events are the formation of helices and sheets from fully unfolded peptide chains and we examine how the laser-induced temperature-jump technique has been used in the study of the helix–coil transition.

The laser-induced temperature-jump technique takes advantage of the fact that proteins unfold, or "melt", at high temperatures and each protein has a characteristic melting temperature. Proteins also lose their native structures at very low temperatures, a process known as **cold denaturation**, and refold when the temperature in increased but kept significantly below the melting temperature. Hence, a temperature-jump experiment can be configured to monitor either folding or unfolding of a polypeptide, depending on the initial and final temperatures of the sample. The challenge of using melting or cold denaturation as the basis of kinetic measurements lies in increasing the temperature of the sample very quickly so that fast relaxation processes can be monitored. A number of clever strategies have been employed. In one, a pulsed laser excites dissolved dye molecules that discard the extra energy largely by heat transfer to the solution. Another variation makes use of direct excitation of H_2O or D_2O with a pulsed infrared laser. The latter strategy leads to temperature jumps in a small irradiated volume of about 20 K in less than 100 ps. Relaxation of the sample can then be probed by a variety of spectroscopic techniques.

P12.46 (a) There are three isosbestic wavelengths (or wavenumbers). The presence of two or more isosbestic points is good evidence that $\boxed{\text{only two solutes in equilibrium with each other are present}}$. The solutes here being Her(CNS)$_8$ and Her(OH)$_8$.

(b) Resonance Raman spectroscopy is preferable to vibrational spectroscopy for studying the O–O stretching mode because such a mode would be $\boxed{\text{infrared inactive}}$, or at best only weakly active. The mode is sure to be inactive in free O_2, because it would not change the molecule's dipole moment. In a complex in which O_2 is bound, the O–O stretch may change the dipole moment, but it is not certain to do so at all, let alone strongly enough to provide a good signal.

(c) The vibrational wavenumber is proportional to the frequency, and it depends on the effective mass as follows:

$$\tilde{\nu} \propto \left(\frac{k}{\mu}\right)^{1/2}, \quad \text{where } \mu = \frac{m_A m_A}{m_A + m_A} = \tfrac{1}{2}m_A = \tfrac{1}{2}M_A/N_A \text{ for a homonuclear diatomic}$$

$$\text{so} \quad \tilde{\nu} \propto \left(\frac{k}{M_A}\right)^{1/2}$$

By taking the ratio of the vibrational wavenumbers of the two isotopic oxygen molecules the constant of proportionality and the force constant cancel, giving

$$\frac{\tilde{\nu}(^{18}O_2)}{\tilde{\nu}(^{16}O_2)} = \left(\frac{M(^{16}O)}{M(^{18}O)}\right)^{1/2} = \left(\frac{16.0\,m_u}{18.0\,m_u}\right)^{1/2} = 0.943$$

$$\tilde{\nu}(^{18}O_2) = (0.943) \times (844\ \text{cm}^{-1}) = \boxed{796\ \text{cm}^{-1}}$$

Note the assumption that the effective masses are proportional to the isotopic masses. This assumption is valid in the free molecule, where the effective mass of O_2 is equal to half the mass of the O atom; it is also valid if the O_2 is strongly bound at one end, such that one atom is free and the other is essentially fixed to a very massive unit.

(d) Text Figure 10.30 is the appropriate MO energy level diagram for O_2. There are 12 valence electrons in the configuration $1\sigma_g^2 1\sigma_u^2 2\sigma_g^2 1\pi_u^4 1\pi_g^2$. Because $1\sigma_g$, $2\sigma_g$, and $1\pi_u$ are regarded as bonding and $1\sigma_u$ and $1\pi_g$ as antibonding, the bond order is $b = \tfrac{1}{2}(8-4) = 2$, a value that is consistent with the classical view that O_2 has a double bond. The $1\pi_g$ antibonding HOMO of O_2 has two vacancies, which accept electrons of anions. Thus, O_2^- has the configuration $1\sigma_g^2 1\sigma_u^2 2\sigma_g^2 1\pi_u^4 1\pi_g^3$ and bond order $b = 1\tfrac{1}{2}$, while O_2^{2-} has the configuration $1\sigma_g^2 1\sigma_u^2 2\sigma_g^2 1\pi_u^4 1\pi_g^4$ and bond order $b = 1$. In summary, molecular orbital analysis of O_2, O_2^-, and O_2^{2-} results in bond orders of 2, 1.5, and 1, respectively.

The vibrational wavenumber is proportional to the square root of the force constant. The force constant is itself a measure of the strength of the bond (technically of its stiffness, which correlates with strength), which in turn is characterized by bond order. Given decreasing bond order with increasing negative charge on the O_2^{2-} species, one would expect decreasing force constants, and decreasing vibrational wavenumbers. Weaker bonding results in lower vibrational wavenumbers (and frequencies).

(e) The 844 cm^{-1} wavenumber of the her(O–O) stretch is very similar to that of the peroxide anion (878 cm^{-1}), suggesting that the oxygenated haemerythrin species involves $\boxed{Fe_2^{3+}O_2^{2-}}$.

(f) The detection of two bands due to $^{16}O^{18}O$ implies that the two O atoms occupy non-equivalent positions in the complex. Structures **14** and **15** are consistent with this observation, but structures **12** and **13** are not.

P12.47 (a) $\text{A (solution)} + \text{B (surface)} \underset{k_{\text{off}}}{\overset{k_{\text{on}}}{\rightleftharpoons}} \text{AB (suface)} \quad K = k_{\text{on}}/k_{\text{off}}$

$$\frac{\text{d[AB]}}{\text{d}t} = k_{\text{on}}[\text{A}][\text{B}] - k_{\text{off}}[\text{AB}]$$

In a typical experiment, the flow rate of A is sufficiently high that $[\text{A}] = a_0$ is essentially constant. We can also write $[\text{B}] = b_0 - [\text{AB}]$ from mass-balance considerations, where b_0 is the total concentration of B. Finally, the SPR signal, denoted by R, is often observed to be proportional to [AB]. The maximum value that R can have is $R_{\text{max}} \propto b_0$, which would be measured if all B molecules were ligated to A. We may then write

$$\begin{aligned}\frac{\text{d}R}{\text{d}t} &= k_{\text{on}}a_0(R_{\text{max}} - R) - k_{\text{off}}R \\ &= k_{\text{on}}a_0R_{\text{max}} - (k_{\text{on}}a_0 + k_{\text{off}})R \\ &= k_{\text{off}}\{a_0KR_{\text{max}} - (a_0K + 1)R\}\end{aligned}$$

At equilibrium $R = R_{\text{eq}}$, $\text{d}R/\text{d}t = 0$, and the above expression becomes

$$\boxed{\frac{R_{\text{eq}}}{R_{\text{max}}} = \frac{a_0K}{a_0K+1}}$$

(b) Taking the inverse of the above relation, we find a useful expression for the analysis of experimental data.

$$\frac{a_0}{R_{\text{eq}}} = \frac{a_0K+1}{KR_{\text{max}}} = \frac{1}{KR_{\text{max}}} + \left(\frac{1}{R_{\text{max}}}\right)a_0$$

This suggests measuring values of R_{eq} over a range of a_0 values and preparing a plot a_0/R_{eq} against a_0. If this association model is applicable, the plot will be linear with a slope and intercept at $a_0 = 0$ equal to $1/R_{\text{max}}$ and $1/KR_{\text{max}}$, respectively. Thus, by computing the linear regression fit of the plot we use the resulting intercept and slope to compute R_{max} and K with the relations

$$\boxed{R_{\text{max}} = 1/\textit{slope} \quad \text{and} \quad K = \textit{slope}/\textit{intercept}}$$

(c) Let $A = k_{\text{on}}a_0R_{\text{max}}$ and $B = -(k_{\text{on}}a_0 + k_{\text{off}})$ and rewrite the differential expression of part (a) as

$$\frac{\text{d}R}{\text{d}t} = A + BR$$

$$\frac{\text{d}R}{A+BR} = \text{d}t$$

Now integrate between $R = 0$ at $t = 0$ and $R = R(t)$ at $t = t$.

$$\int_{R=0}^{R=R(t)} \frac{\text{d}R}{A+BR} = \int_{t=0}^{t=t} \text{d}t \quad \text{[Standard integrals]}$$

$$\frac{1}{B}\ln(BR+A)\Big|_{R=0}^{R=R(t)} = t$$

$$\frac{1}{B}\ln\frac{BR(t)+A}{A} = t$$

$$\frac{BR(t)+A}{A} = \text{e}^{Bt} \quad \text{[Solve for } R(t)\text{]}$$

$$R(t) = \frac{A}{B}(e^{Bt} - 1)$$

We now find an expression for the ratio A/B in terms of the association constants.

$$\frac{A}{B} = -\frac{k_{on}a_0}{k_{on}a_0 + k_{off}}R_{max} = -\frac{a_0K}{a_0K+1}R_{max} = -R_{eq} \quad \text{[See part (a)]}$$

Substitution of the expressions for B and A/B into the $R(t)$ expression yields the desired equation:

$$\boxed{R(t) = R_{eq}(1 - e^{-k_{obs}t}), \quad \text{where} \quad k_{obs} = k_{on}a_0 + k_{off}}$$

(d) The decay shown in text Figure 12.56 occurs after solution containing no A is directed to flow over the surface. Any A that dissociates from B is swept away so that the dissociation process alone occurs. The differential expression for R is

$$\frac{dR}{dt} = -k_{off}R \quad \text{or} \quad \frac{dR}{R} = -k_{off}dt$$

Integrate between $R = R_{max}$ at $t = 0$ and $R = R(t)$ at $t = t$.

$$\int_{R=R_{max}}^{R=R(t)} \frac{dR}{R} = -k_{off}\int_{t=0}^{t=t} dt$$

$$\ln\frac{R(t)}{R_{max}} = -k_{off}t$$

$$\boxed{R(t) = R_{max}e^{-k_{obs}t}, \text{ where } k_{obs} = k_{off}}$$

P12.48 In an effort to simplify the look of the equations, we will indicate that a property is a function of wavenumber with a subscript ν for the absorption spectrum and ν_F for the fluorescence spectrum. For example, A_ν and ε_ν indicate that the absorbance and molar absorption coefficient are functions of incident radiation wavenumber, while I_{F,ν_F} and ϕ_{F,ν_F} indicate that the fluorescence emission intensity and fluorescence quantum yield are functions of the fluorescence spectrum wavenumber. We assume the intensity of the incident radiation, I_0, is independent of wavenumber (i.e. identical at all wavenumbers).

(a) The Beer–Lambert Law is: $A_\nu = \log(I_0/I_\nu) = \varepsilon_\nu[\text{J}]L$

The absorbed intensity is: $I_{abs,\nu} = I_0 - I_\nu$ so $I_\nu = I_0 - I_{abs,\nu}$

Substitute this expression into the Beer–Lambert law and solve for $I_{abs,\nu}$:

$$\log\frac{I_0}{I_0 - I_{abs,\nu}} = \varepsilon_\nu[\text{J}]L \quad \text{so} \quad I_0 - I_{abs,\nu} = I_0 \times 10^{-\varepsilon_\nu[\text{J}]L}$$

and $I_{abs,\nu} = \boxed{I_0 \times (1 - 10^{-\varepsilon_\nu[\text{J}]L})}$

(b) The problem states that I_{F,ν_F} is proportional to ϕ_{F,ν_F} and to $I_{abs,\nu}$, so:

$$I_{F,\nu_F} \propto \phi_{F,\nu_F} I_0 \times (1 - 10^{-\varepsilon_\nu[\text{J}]L})$$

If the exponent is small, we can expand $1 - 10^{-\varepsilon_\nu[\text{J}]L}$ in a power series:

$$\begin{aligned} 1 - 10^{-\varepsilon_\nu[\text{J}]L} &= 1 - e^{-\varepsilon_\nu[\text{J}]L\ln 10} \approx 1 - \{1 - \varepsilon_\nu[\text{J}]L\ln 10\} \\ &\approx \varepsilon_\nu[\text{J}]L\ln 10 \end{aligned}$$

Thus,

$$I_{F,\nu_F} \propto \boxed{\phi_{F,\nu_F} I_0 \varepsilon_\nu [J] L}$$

(c) When the incident radiation consists of a single wavenumber in resonance with a permitted vibrational absorption transition as shown in Figure 12.6, an adaptation of text Figure 12.43, many vibrational transitions are observed in the fluorescence spectrum. The transitions are from the $v = 0$ vibrational level of the excited singlet to any of a number of vibrational levels of the ground singlet state. The absorption spectrum of this hypothetical experiment consists of a single, narrow line, shown in Figure 12.6, while the fluorescence spectrum is a wide band of many overlapping transitions, which are also shown in Figure 12.6. The fluorescence spectrum contains no information about the vibrational levels of the excited state. It does contain information about the separation of vibrational levels in the ground state. The equation of part (b) agrees with this pictorial analysis because ε_ν has but a single possible value (it is a constant) and we find that $I_{F,\nu_F} \propto \phi_{F,\nu_F}$ for the experiment.

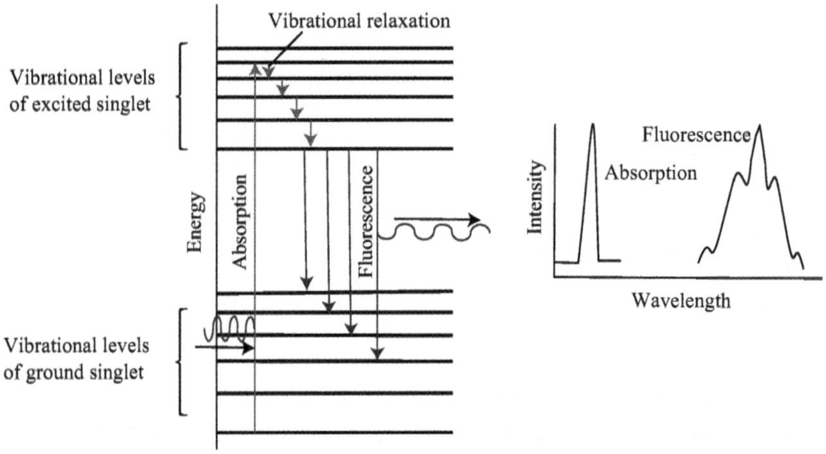

Figure 12.6

When the fluorescence is observed at a single wavenumber in resonance with a permitted transition as shown in Figure 12.7, the intensity of the fluorescence line varies as the incident radiation is scanned over the energy range of the permitted transitions. This happens because the absorption depends upon the molar absorption coefficient, which varies as the wavenumbers are scanned. Therefore, there is variation in the number of radiationless decays from the upper vibrational levels to the $v = 0$ vibrational level of the excited singlet and the fluorescence intensity must vary. As revealed by Figure 12.7, a plot of the fluorescence intensity against incident radiation contains no information about the ground vibrational levels but necessarily contains information about the excited vibrational levels. That is, this fluorescence excitation experiment yields a spectrum and vibrational information that is identical to the absorption spectrum. The equation of part (b) agrees with this pictorial analysis because ϕ_{F,ν_F} has but a single possible value (it is a constant) and we find that $I_{F,\nu_F} \propto \varepsilon_\nu$ for the experiment.

Absorption and fluorescence spectra are often displayed together as shown in Figure 12.8. Again, the absorption spectrum shows the vibrational structure characteristic of the excited state while the fluorescence spectrum shows the structure characteristic of the ground state. The fluorescence spectrum is always displaced to lower wavenumbers (but the 0–0 transitions are coincident) and resembles a mirror image of the absorption.

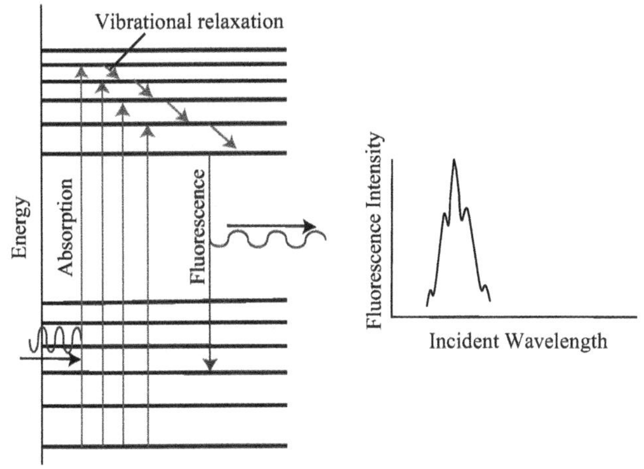

Figure 12.7

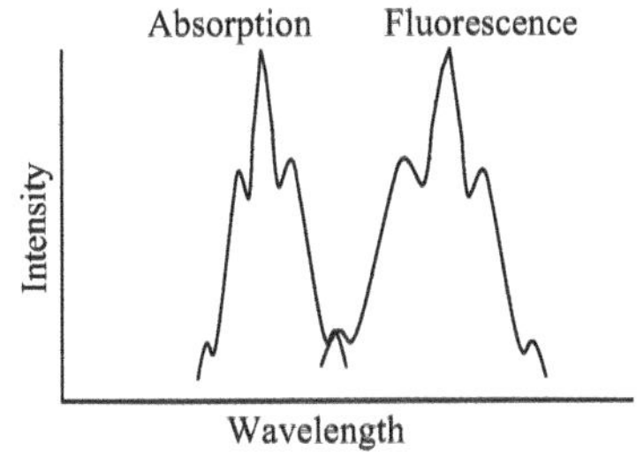

Figure 12.8

(d) $S^* + Q \rightarrow S^+ + Q^-$

Diminished line intensities in the fluorescence excitation spectrum in the presence of a quencher, overlap of the absorption spectrum of Q with the emission spectrum of S*, and a decrease in the fluorescence yield of S with an appearance of lines attributable to the fluorescence of Q^- provide evidence for resonance energy transfer between S and Q.

P12.49 (a) These calculations, which use either the Møller–Plesset (MP2) model with two different basis sets or density functional theory (DFT) technique, were performed with Spartan '06™ running in Windows on a PC. The star indicates that the basis set adds d-type polarization functions for each atom other than hydrogen. Other basis sets give comparable results. 1 au = 27.2114 eV. In addition to reporting the vibrational wavenumbers of the normal modes, calculation of bond lengths, angles, ground-state energies, and dipole moments have been included.

Water. The DFT calculation method is DFT(B3LYP).

	H_2O ground state				
	MP2/6-31G*	MP2/6-311G*	DFT/6-31G*	DFT/6-311G*	Exp.
Basis fns	19	24	19	24	
R/pm	96.9	95.7	96.8	96.3	95.8
E_0/eV	−2073.4	−2074.5	−2074.6	−2079.9	
Angle/°	104.00	106.58	103.72	105.91	104.45
$\tilde{\nu}_1$/cm^{-1}	3774.25	3858.00	3731.72	3764.70	3652
$\tilde{\nu}_2$/cm^{-1}	1735.35	1739.88	1709.79	1705.47	1595
$\tilde{\nu}_3$/cm^{-1}	3915.76	3994.30	3853.53	3877.60	3756
μ/D	n.s.	n.s.	2.0950	2.2621	1.854

Normal modes and IR spectrum:

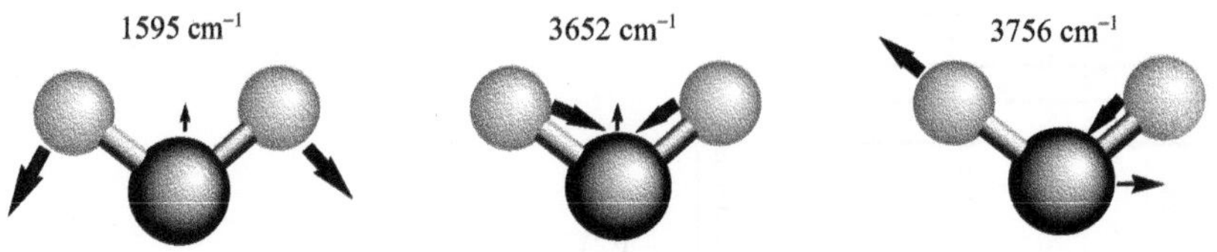

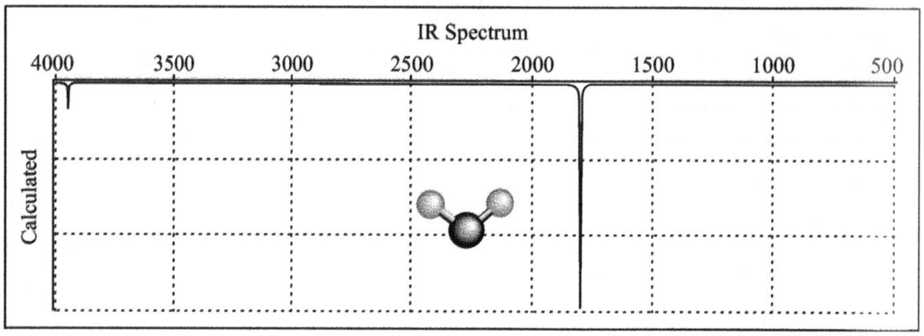

Carbon dioxide. The DFT calculation method is DFT(B3LYP).

CO_2 ground state					
	MP2/6-31G*	MP2/6-311G*	DFT/6-31G*	DFT/6-311G*	Exp.
Basis fns	45	54	45	54	
R/pm	118.0	116.9	116.9	116.0	116.3
E_0/eV	−5118.7	−5121.2	−5131.6	−5133.2	
Angle/°	180.00	180.00	180.00	180.00	180
$\tilde{\nu}_1$/cm^{-1}	1332.82	1341.46	1373.05	1376.55	1388
$\tilde{\nu}_2$/cm^{-1}	636.22	657.60	641.47	666.39	667
$\tilde{\nu}_3$/cm^{-1}	2446.78	2456.16	2438.17	2437.85	2349
μ/D	n.s.	n.s.	0.0000	0.0000	0

Normal modes and IR spectrum:

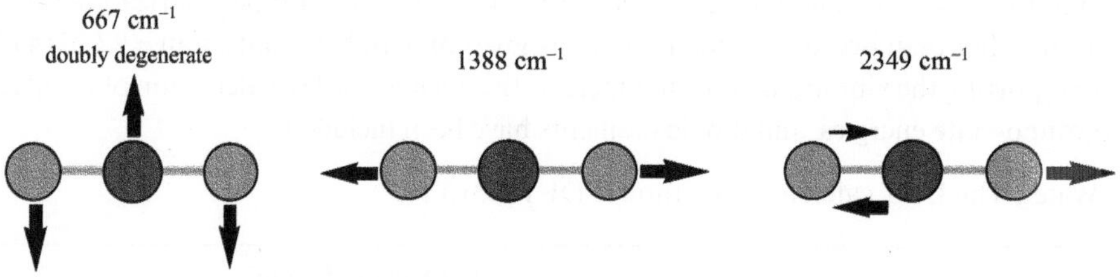

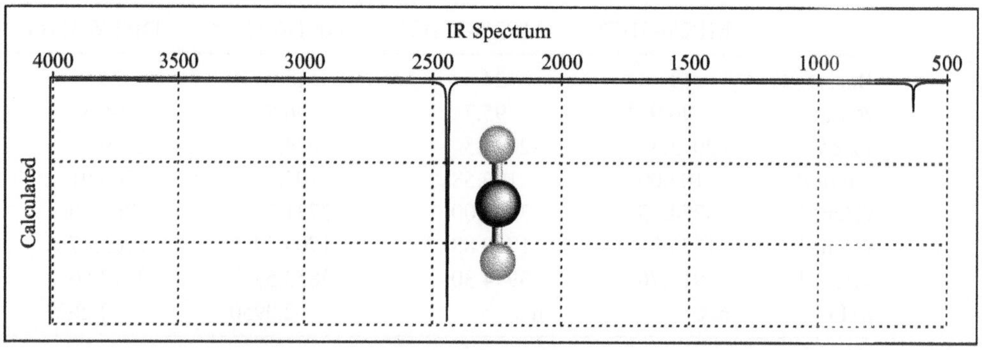

Methane, CH_4. This tetrahedral molecule has nine normal modes that you can view with your software. The following IR spectrum is calculated with the PM3 semiempirical method.

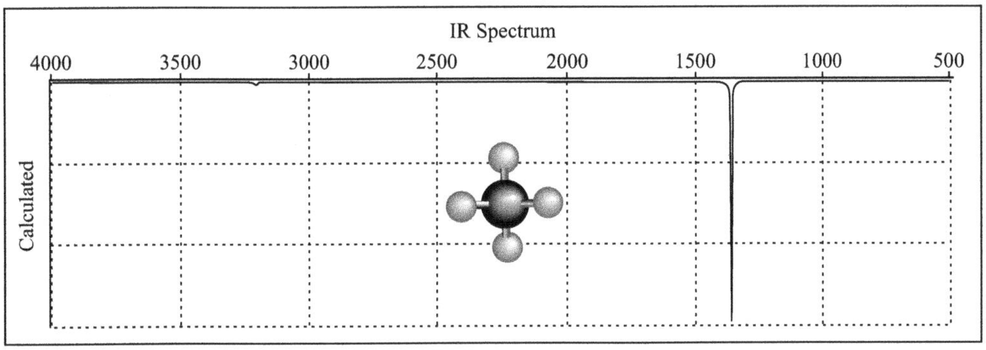

Except for the dipole moment, all calculations are typically within a reasonable 1–3% of the experimental value. The dipole moment is very sensitive to the distribution of charge density. The significant difference between the dipole moment calculations and the experimental dipole moment may indicate that the computation methods do not adequately account for charge distribution in the very polar water molecule.

(b) Water: all modes of water are infrared active. However, as seen in the above spectrum, not all have a strong intensity.

Carbon dioxide: the symmetric stretch is infrared inactive; the bend and asymmetric stretch are infrared active.

Methane: the vibrational modes that cause the molecular dipole to change as the molecule vibrates are infrared active.

P12.50 (a) and (b) The following table summarizes AM1 calculations (an extended Hückel method) of the LUMO–HOMO separation in the 11-*cis* (Figure 12.9) and 11-*trans* molecule **(16)** model of retinal. The –46.0° torsional angle between the first two alternate double bonds indicates that they are not coplanar. In contrast, the C11C12C13C14 torsion angle shows that the C11C12 double bond is close to coplanar with neighboring double bonds. The aromatic character of the alternating π-bond system is evidenced by contrasting the computed bond lengths at a single bond away from the π-system (C1–C2), a double bond (C11–C12), and a single bond between doubles (C12–C13) within the Lewis structure. We see a typical single bond length, a slightly elongated double bond length, and a bond length that is intermediate between a single and a double, respectively. The latter lengths are characteristic of aromaticity. The HOMO and LUMO of 11-*cis* are shown in Figure 12.9.

Conformation	11-*trans*	11-*cis*
$\Delta_f H^{\circ}/\mathrm{kJ\ mol^{-1}}$	725.07	738.1
$E_{\mathrm{LUMO}}/\mathrm{eV}$	−5.142	−5.138
$E_{\mathrm{HOMO}}/\mathrm{eV}$	−10.770	−10.888
$\Delta E/\mathrm{eV}$	(a) 5.628	(b) 5.750
λ/nm	(a) 220.3	(b) 215.6
C5C6C7C8 torsion angle/°	−44.5	−46.0
C11C12C13C14 torsion angle/°	179.7	−165.5
C1–C2/pm	153.2	153.2
C11–C12/pm	137.3	136.7
C12–C13/pm	142.0	142.1

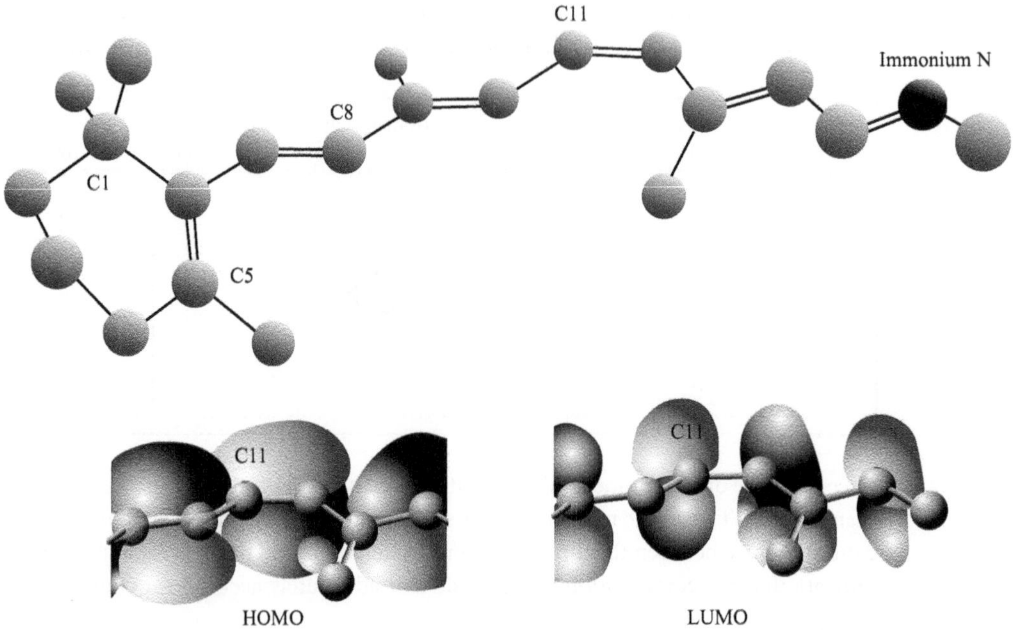

Figure 12.9

(c) The lowest $\pi^* \leftarrow \pi$ transition occurs in the ultraviolet with the 11-*cis* transition at higher energy (higher frequency, lower wavelength). It is apparent that important interactions between retinal and a surrounding opsin molecule are responsible for reducing the transition energy to the observed strong absorption in the 400 to 600 nm visible range.

13 Magnetic resonance

Answers to discussion questions

D13.1 Resonance is a mechanical phenomenon in which energy is exchanged between two systems that have the same natural frequency. An example of classical resonance is the exchange of energy between two similar pendulums connected by a light spring. When one pendulum is set in motion the other pendulum starts to oscillate. Energy is transferred from the first pendulum to the second. The transfer of energy is most efficient when the two pendulums have the same natural frequency. The pendulums are then said to be in resonance. Quantum-mechanical resonance has some of the same characteristics. An important case is when one system is the electromagnetic field and the other is an atom or molecule. In spectroscopy, resonance denotes the strong coupling and efficient energy transfer that occurs when the frequency of the electromagnetic field just matches the transition frequency associated with the energy separation between states of the atomic or molecular system. In principle, all spectroscopic techniques are resonance techniques, but the term is most commonly applied to NMR and EPR. In NMR and EPR either the energy levels of the system can be adjusted to achieve resonance with the frequency of the electromagnetic field or the frequency of the field can be adjusted to match the energy separation of the states of the atom or molecule.

D13.2 Discussions of the origins of the local, neighboring group, and solvent contributions to the shielding constant can be found in text Section 13.3(b). The local contribution is essentially the contribution of the electrons in the atom that contains the nucleus being observed. It can be expressed as the sum of a diamagnetic and paramagnetic parts, that is $\sigma(\text{local}) = \sigma_d + \sigma_p$. The diamagnetic part arises because the applied field generates a circulation of charge in the ground state of the atom. In turn, the circulating charge generates a magnetic field. The direction of this field can be found through Lenz's law that states that the induced magnetic field must be opposite in direction to the field producing it. Thus, it shields the nucleus. The diamagnetic contribution is roughly proportional to the electron density on the atom and it is the only contribution for closed-shell free atoms and for distributions of charge that have spherical or cylindrical symmetry. The local paramagnetic contribution is somewhat harder to visualize since there is no simple and basic principle analogous to Lenz's law that can be used to explain the effect. The applied field adds a term to the hamiltonian of the atom that mixes in excited electronic states into the ground state and any theoretical calculation of the effect requires detailed knowledge of the excited-state wavefunctions. It is to be noted that the paramagnetic contribution does not require that the atom or molecule be paramagnetic. It is paramagnetic only in the sense that it results in an induced field in the same direction as the applied field.

The neighboring group contributions arise in a manner similar to the local contributions. Both diamagnetic and paramagnetic currents are induced in the neighboring atoms and these currents result in shielding contributions to the nucleus of the atom being observed. However, there are some differences: The magnitude of the effect is much smaller because the induced currents in neighboring atoms are much farther away. It also depends on the anisotropy of the magnetic susceptibility of the neighboring group. Only anisotropic susceptibilities result in a contribution.

Solvents can influence the local field in many different ways. Detailed theoretical calculations of the effect are difficult due to the complex nature of the solute–solvent interaction. Polar solvent–polar solute interactions are an electric-field effect that usually causes deshielding of the solute protons. Solvent magnetic anisotropy can cause shielding or deshielding, for example, for solutes in benzene solution. In addition, there are a variety of specific chemical interactions between solvent and solute that can affect the chemical shift.

D13.3 Examination of eqn 13.16 implies that in an NMR spectrometer operating at a fixed frequency, the external magnetic field required to fulfill the resonance condition is given by

$$\mathcal{B}_0 = \frac{2\pi\nu_L}{\gamma(1-\sigma)}.$$

Thus, a positive value of the shielding constant, σ, shifts the resonance field to "high field", a negative value of σ shifts the resonance field to "low field". Conversely, in a spectrometer at fixed external magnetic field, positive values of σ shift the resonant frequency to lower values and negative values of σ shift it to higher values. Chemical shifts are also reported in terms of the parameter, δ, called the "chemical shift" and defined by eqn 13.17. σ and δ are independent of the external magnetic field, but the chemical shift in frequency units, $\nu - \nu_0$, is not. See the Brief Illustration in text Section 13.3(a).

D13.4 Spin–spin couplings in NMR are due to a polarization mechanism, which is transmitted through bonds. The following description applies to the coupling between the protons in the H_X–C–H_Y group as is typically found in organic compounds. See Figure 13.16 of the text. On H_X, the Fermi contact interaction causes the spins of its proton and electron to be aligned antiparallel. The spin of the electron from C in the H_X–C bond is then aligned antiparallel to the electron from H_X due to the Pauli exclusion principle. The spin of the C electron in the bond H_Y is then aligned parallel to the electron from H_X because of Hund's rule. Finally, the alignment is transmitted through the second bond in the same manner as the first. This progression of alignments (antiparallel × antiparallel × parallel × antiparallel × antiparallel) yields an overall energetically favorable parallel alignment of the two proton nuclear spins. Therefore, in this case the coupling constant, ${}^2J_{HH}$, is negative in sign.

D13.5 Both spin–lattice and spin–spin relaxation are caused by fluctuating magnetic and electric fields at the nucleus in question and these fields result from the random thermal motions present in the solution or other form of matter. These random motions can be the result of a number of processes and it is hard to summarize all that could be important. In theory, every known nuclear interaction coupled with every type of motion can contribute to relaxation and detailed treatments can be exceedingly complex. However, they all depend on the magnetogyric ratio of the atom in question and the magnetogyric ratio of the proton is much larger than that of ^{13}C. Hence, the interaction of the proton with fluctuating local magnetic fields caused by the presence of neighboring magnetic nuclei will be greater, and the relaxation will be quicker, corresponding to a shorter relaxation time for protons. Another consideration is the structure of compounds containing carbon and hydrogen. Typically, the C atoms are in the interior of the molecule bonded to other C atoms, 99% of which are non-magnetic, so the primary relaxation effects are due to bonded protons. Protons are on the outside of the molecule and are subject to many more interactions and hence faster relaxation.

D13.6 At, say, room temperature, the tumbling rate of benzene, the small molecule, in a mobile solvent, may be close to the Larmor frequency, and hence its spin–lattice relaxation time will be short. As the temperature increases, the tumbling rate may increase well beyond the Larmor frequency, resulting in an increased spin–lattice relaxation time.

For the larger oligopeptide at room temperature, the tumbling rate may be well below the Larmor frequency, but with increasing temperature, it will approach the Larmor frequency due to the increased thermal motion of the molecule combined with the decreased viscosity of the solvent. Therefore, the spin-lattice relaxation time may decrease.

D13.7 In the nuclear Overhauser effect (NOE) in NMR, spin relaxation processes are used to transfer the population difference typical of one species of nucleus X to another nucleus A, thereby enhancing the intensity of the signal produced by A. Text eqn 13.22 shows that the signal enhancement is given by

$$\frac{I}{I_0} = 1 + \eta$$

where η is the enhancement parameter. In the case of maximal enhancement it is possible to show that

$$\eta = \frac{\gamma_X}{2\gamma_A}.$$

NOE can be used to determine interproton distances in biopolymers. This application makes use of the fact that when the dipole-dipole mechanism is not the only relaxation mechanism, the NOE is given by

$$\frac{I}{I_0} = 1 + \eta = 1 + \frac{\gamma_X}{2\gamma_A} \times \frac{T_1}{T_{1,\text{dip-dip}}}$$

where T_1 is the total relaxation time and $T_{1,\text{dip-dip}}$ is the relaxation time due to the dipole-dipole mechanism. Here, A and X are both protons. The enhancement depends strongly on the separation, r, of the two spins, for the strength of the dipole–dipole interaction is proportional to $1/r^3$, and its effect depends on the square of that strength and therefore on $1/r^6$. This sharp dependence on separation is used to build up a picture of the conformation of the biopolymer by using NOE to identify which protons can be regarded as neighbors.

D13.8 The basic COSY experiment uses the simplest of all two-dimensional pulse sequences: a single 90° pulse to excite the spins at the end of the preparation period and a mixing period containing just a second 90° pulse.

The key to the COSY technique is the effect of the second 90° pulse, which can be illustrated by consideration of the four energy levels of an AX system (as shown in Figures 13.30 and 13.31 of the text). At thermal equilibrium, the population of the $\alpha_A\alpha_X$ level is the greatest, and that of the $\beta_A\beta_X$ level is the smallest; the other two levels have the same energy and an intermediate population. After the first 90° pulse, the spins are no longer at thermal equilibrium. If a second 90° pulse is applied at a time t_1 that is short compared to the spin–lattice relaxation time T_1 the extra input of energy causes further changes in the populations of the four states. The changes in populations will depend on how far the individual magnetizations have precessed during the evolution period.

For simplicity, let us consider a COSY experiment in which the second 90° pulse is split into two selective pulses, one applied to X and one to A. Depending on the evolution time t_1, the 90° pulse that excites X may leave the population differences across each of the two X transitions unchanged, inverted, or somewhere in between. Consider the extreme case in which one population difference is inverted and the other unchanged. The 90° pulse that excites A will now generate an FID in which one of the two A transitions has increased in intensity, and the other has decreased. The overall effect is that precession of the X spins during the evolution period determines the amplitudes of the signals from the A spins obtained during the detection period. As the evolution time t_1 is increased, the intensities of the signals from A spins oscillate at rates determined by the frequencies of the two X transitions.

This transfer of information between spins is at the heart of two-dimensional NMR spectroscopy and leads to the correlation of different signals in a spectrum. In this case, information transfer tells us that there is a scalar coupling between A and X. If we conduct a series of experiments in which t_1 is incremented, Fourier transformation of the FIDs on t_2 yields a set of spectra $I(\nu_1, \nu_2)$ in which the A signal amplitudes oscillate as a function of t_1. A second Fourier transformation, this time on t_1, converts these oscillations into a two-dimensional spectrum $I(\nu_1, \nu_2)$. The signals are spread out in ν_1 according to their precession frequencies during the detection period. Thus, if we apply the COSY pulse sequence to our AX spin system, the result is a two-dimensional spectrum that contains four groups of signals centered on the two chemical shifts in ν_1 and ν_2. Each group will show fine structure, consisting of a block of four signals separated by J_{AX}. The diagonal peaks are signals centered on $(\delta_A\ \delta_A)$ and $(\delta_X\ \delta_X)$ and lie along the diagonal $\nu_1 = \nu_2$. They arise from signals that did not change chemical shift between t_1 and t_2. The cross-peaks (or *off-diagonal peaks*) are signals centered on $(\delta_A\ \delta_X)$ and $(\delta_X\ \delta_A)$ and owe their existence to the coupling between A and X. Consequently, cross-peaks in COSY spectra allow us to map the couplings between spins and to trace out the bonding network in complex molecules. Text Figure 13.37 shows a simple example of a proton COSY spectrum of isoleucine.

D13.9 The molecular orbital occupied by the unpaired electron in an organic radical can be identified through the observation of hyperfine splitting in the EPR spectrum of the radical. The magnitude of this splitting is related to the spin density of the unpaired electron at those positions in the radical having atoms with nuclear moments. In addition, the spin density on carbon atoms adjacent to the magnetic nuclei can be determined indirectly through the McConnell relation, text eqn 13.26. Thus, for example, in the benzene negative ion, unpaired spin densities on both the carbon atoms and hydrogen atoms can be determined from the EPR hyperfine splittings. The next step then is to construct a molecular orbital that will theoretically reproduce these experimentally determined spin densities. A good match indicates that we have found a good molecular orbital for the radical. For nuclei other than protons in aromatic radicals similar, although more complicated equations arise; but in all cases the spin densities can be related to the coefficients of the basis functions used to describe the molecular orbital of the unpaired electron.

D13.10 The ESR spectrum of a spin probe, such as the di-*tert*-butyl nitroxide radical, broadens with restricted motion of the probe. This suggests that the width of spectral lines may correlate with the depth to which a probe may enter into a biopolymer crevice. Deep crevices are expected to severely restrict probe motion and broaden the spectral lines. Additionally, the splitting and center of ESR spectra of an oriented sample can provide information about the shape of the biopolymer–probe environment because the probe ESR signal is anisotropic and depends upon the orientation of the probe with respect to the external magnetic field. Oriented biopolymers occur in lipid membranes and in muscle fibers.

Solutions to exercises

E13.11 $E_{m_s} = g_e\mu_B\mathcal{B}_0 m_s$ [13.4], where $m_s = +\frac{1}{2}$ or $-\frac{1}{2}$ (α and β spin states, respectively)

$$\Delta E = E_\alpha - E_\beta = E_{m_s=1/2} - E_{m_s=-1/2} = g_e\mu_B\mathcal{B}_0(\tfrac{1}{2} - (-\tfrac{1}{2})) = g_e\mu_B\mathcal{B}_0 \quad [13.8]$$

$$= (2.0023) \times (9.274 \times 10^{-24}\ \mathrm{J\ T^{-1}}) \times (0.300\ \mathrm{T}) = \boxed{5.57 \times 10^{-24}\ \mathrm{J}}$$

E13.12 $E_{m_I} = -\gamma_N\hbar\mathcal{B}_0 m_I = -g_I\mu_N\mathcal{B}_0 m_I$ [13.5, 13.7] $m_I = 3/2, 1/2, -1/2, -3/2$

$$E_{m_I} = -0.4289 \times (5.051 \times 10^{-27}\ \mathrm{J\ T^{-1}}) \times (7.500\ \mathrm{T}) \times m_I$$

$$= \boxed{-1.625 \times 10^{-26}\ \mathrm{J} \times m_I}$$

E13.13 (a) Unit of γ_N = unit of $\left(\dfrac{g_I\mu_N}{\hbar}\right) = \dfrac{\text{J T}^{-1}}{\text{J s}} = \text{T}^{-1}\,\text{s}^{-1} = \boxed{\text{T}^{-1}\,\text{Hz}}$

(b) $1\,\text{T} = 1\,\text{kg s}^{-2}\,\text{A}^{-1}$

Unit of $\gamma_N = \text{kg}^{-1}\,\text{s}^2\,\text{A} \times \text{s}^{-1} = \boxed{\text{A s kg}^{-1}}$

E13.14 $\gamma_N\hbar = g_I\mu_N$ [13.5, 13.7]

Therefore,

$$g_I = \frac{\gamma_N\hbar}{\mu_N} = \frac{1.0840\times10^8\,\text{T}^{-1}\,\text{s}^{-1}\times1.05457\times10^{-34}\,\text{J s}}{5.051\times10^{-27}\,\text{J T}^{-1}}$$
$$= \boxed{2.263}$$

E13.15 We assume a temperature of 300 K.

$$\frac{N_\beta - N_\alpha}{N} = \frac{N_\beta - N_\alpha}{N_\beta + N_\alpha} \approx \frac{g_e\mu_B\mathcal{B}_0}{2kT}$$ (See the derivation of eqn 13.14, noting that $g_e\mu_B$ substitutes for $-\gamma_N\hbar$ for the case of electrons)

(a) $$\frac{N_\beta - N_\alpha}{N} = \frac{(2.0023)\times(9.274\times10^{-24}\,\text{J T}^{-1})\times(0.300\,\text{T})}{2(1.381\times10^{-23}\,\text{J K}^{-1})\times(300\,\text{K})} = \boxed{6.72\times10^{-4}}$$

(b) $$\frac{N_\beta - N_\alpha}{N} = \frac{(2.0023)\times(9.274\times10^{-24}\,\text{J T}^{-1})\times(1.1\,\text{T})}{2(1.381\times10^{-23}\,\text{J K}^{-1})\times(300\,\text{K})} = \boxed{2.47\times10^{-3}}$$

E13.16 $\nu = \dfrac{g_e\mu_B\mathcal{B}_0}{h}$ [13.10], where $g_e = 2.0023$ and μ_B is the Bohr magneton

$$\nu = \frac{2.0023\times9.274\times10^{-24}\,\text{J T}^{-1}\times0.330\,\text{T}}{6.626\times10^{-34}\,\text{J s}}$$
$$= 9.248\times10^9\,\text{s}^{-1} = \boxed{9.248\,\text{GHz}}$$

$$\lambda = \frac{c}{\nu} = \frac{2.998\times10^8\,\text{m s}^{-1}}{9.248\times10^9\,\text{s}^{-1}} = \boxed{0.0324\,\text{m}}$$

EPR employs microwave radiation, rather than the radio frequency radiation of NMR.

E13.17 We assume a temperature of 300 K.

$$\frac{N_\alpha - N_\beta}{N} = \frac{N_\alpha - N_\beta}{N_\alpha + N_\beta} \approx \frac{\gamma_N\hbar\mathcal{B}}{2kT}$$ [13.14] for spin-$\tfrac{1}{2}$ nuclei like ^{1}H and ^{13}C

Nuclear magnetogyric ratios are found in Table 13.2.

(a) ^{1}H: $$\frac{N_\alpha - N_\beta}{N} = \frac{(26.752\times10^7\,\text{T}^{-1}\,\text{s}^{-1})\times(1.055\times10^{-34}\,\text{J s})\times(10\,\text{T})}{2(1.381\times10^{-23}\,\text{J K}^{-1})\times(300\,\text{K})} = \boxed{3.4\times10^{-5}}$$

(b) ^{13}C: $$\frac{N_\alpha - N_\beta}{N} = \frac{(6.7272\times10^7\,\text{T}^{-1}\,\text{s}^{-1})\times(1.055\times10^{-34}\,\text{J s})\times(10\,\text{T})}{2(1.381\times10^{-23}\,\text{J K}^{-1})\times(300\,\text{K})} = \boxed{8.6\times10^{-6}}$$

E13.18 $\delta N \approx \dfrac{Ng_I\mu_N\mathcal{B}_0}{2kT}$ [Exercise 13.15] $= \dfrac{Nh\nu}{2kT}$

Thus, $\delta N \propto \nu$

$$\frac{\delta N(800\text{ MHz})}{\delta N(60\text{ MHz})} = \frac{800\text{ MHz}}{60\text{ MHz}} = \boxed{13}$$

This ratio is not dependent on the nuclide as long as the approximation $\Delta E \ll kT$ holds [Exercise 13.15].

E13.19 $\nu = \dfrac{\gamma_N\mathcal{B}_0}{2\pi}$ [31.12] $= \dfrac{2.5177\times10^8\text{ T}^{-1}\text{ s}^{-1}\times 8.200\text{ T}}{2\pi}$

$= 3.285\times10^8\text{ s}^{-1} = \boxed{328.5\text{ MHz}}$

E13.20 $\nu = \dfrac{\gamma_N\mathcal{B}_0}{2\pi}$ [31.12] $= \dfrac{g_I\mu_N}{h}\mathcal{B}_0$

$$\nu = \frac{0.4036\times5.051\times10^{-27}\text{ J T}^{-1}\times15.00\text{ T}}{6.626\times10^{-34}\text{ J s}}$$

$= 4.615\times10^7\text{ s}^{-1} = \boxed{46.15\text{ MHz}}$

E13.21 $\mathcal{B}_0 = \dfrac{h\nu}{\gamma_N\hbar}$ [31.12] $= \dfrac{2\pi\nu}{\gamma_N}$, where $\gamma_N = 26.752\times10^7\text{ T}^{-1}\text{ s}^{-1}$ for ^{1}H (Table 13.2)

$$= \frac{2\pi(500.0\times10^6\text{ Hz})}{26.752\times10^7\text{ T}^{-1}\text{ s}^{-1}} = \boxed{11.74\text{ T}}$$

E13.22 $\delta = \dfrac{\nu-\nu^\circ}{\nu^\circ}\times10^6$ [13.17]

$$\text{shift} = \nu-\nu^\circ = \frac{\delta\times\nu^\circ}{10^6} = \frac{6.33\times420\times10^6\text{ Hz}}{10^6}$$

$= 2.66\times10^3\text{ Hz} = \boxed{2.66\text{ kHz}}$

E13.23 (a) $\delta = \dfrac{\nu-\nu^\circ}{\nu^\circ}\times10^6$ [13.17]

Since both ν and ν° depend upon the magnetic field in the same manner, namely

$$\nu = \frac{g_I\mu_N\mathcal{B}}{h} \quad\text{and}\quad \nu^\circ = \frac{g_I\mu_N\mathcal{B}_0}{h} \quad [13.5, 13.7]$$

δ is $\boxed{\text{independent}}$ of both B and ν. Note: Equation 13.10 in the text relates specifically to the case of electrons, but a similar equation holds for nuclei. We can also see from eqn 13.12 that ν and ν° depend similarly on the magnetic field, so δ is therefore independent of both B_0 and ν.

(b) Rearranging [13.17] we see $\nu-\nu^\circ = \nu^\circ\delta\times10^{-6}$

and we see that the relative chemical shift is

$$\frac{\nu-\nu^\circ(800\text{ MHz})}{\nu-\nu^\circ(60\text{ MHz})} = \frac{(800\text{ MHz})}{(60\text{ MHz})} = \boxed{13}$$

E13.24 The direct proportionality between $\nu - \nu°$ and $\nu°$ demonstrated in Exercise 13.23 is one of the major reasons for operating an NMR spectrometer at the highest frequencies possible. Exercise 13.18 also demonstrates that the population difference and hence the signal intensity depends upon ν. The splittings observed in an NMR spectrum are proportional to the spectrometer frequency.

E13.25 $\mathcal{B}_{\text{loc}} = (1-\sigma)\mathcal{B}_0$ [equation above eqn 13.16]

$|\Delta\mathcal{B}_{\text{loc}}| = |(\Delta\sigma)|\mathcal{B}_0 \approx |[\delta(\text{CH}_3) - \delta(\text{CHO})]|\mathcal{B}_0$ [$|\Delta\sigma| = |\Delta\delta \times 10^{-6}|$ which follows from eqns 13.16 & 13.17]
$= |(2.20 - 9.80)| \times 10^{-6}\mathcal{B}_0 = 7.60 \times 10^{-6}\mathcal{B}_0$

(a) $\mathcal{B}_0 = 1.5\text{ T}$, $|\Delta\mathcal{B}_{\text{loc}}| = 7.60 \times 10^{-6} \times 1.5\text{ T} = \boxed{9.5\ \mu\text{T}}$

(b) $\mathcal{B}_0 = 6.0\text{ T}$, $|\Delta\mathcal{B}_{\text{loc}}| = 7.60 \times 10^{-6} \times 6.0\text{ T} = \boxed{46\ \mu\text{T}}$

E13.26 $$\nu - \nu° = \frac{\nu°\delta}{10^6} \quad [13.17]$$

$$\Delta(\nu - \nu°) = \frac{\nu°\Delta\delta}{10^6}$$

(a) $\Delta(\nu - \nu°) = 300\text{ MHz} \times 10^{-6} \times (9.5 - 1.5) \approx \boxed{2.4\text{ kHz}}$

(b) $\Delta(\nu - \nu°) = 500\text{ MHz} \times 10^{-6} \times (9.5 - 1.5) \approx \boxed{4.0\text{ kHz}}$

E13.27 For identical nuclei with spin $\frac{1}{2}$, there will be $N + 1$ lines from the splitting. In this case 8 lines. The lines will have relative intensities of $\boxed{1:7:21:35:35:21:7:1}$.

These relative intensities can be determined by extending Pascal's triangle shown in (**1**) of the text three more rows to $N + 1 = 8$. Alternatively the intensities can also be determined from the coefficients in the expansion of $(1 + x)^N$.

E13.28 Because each resonance is split into three lines by a single N nucleus, the result will be:

(a) $\boxed{\text{quintet } 1:2:3:2:1}$ and (b) $\boxed{\text{septet } 1:3:6:7:6:3:1}$

Also, see the solution to Exercise 13.32 which discusses the procedure for determining the splitting pattern.

E13.29 $$E = -\gamma_N\hbar(1-\sigma_A)\mathcal{B}m_A - \gamma_N\hbar(1-\sigma_X)\mathcal{B}m_{X_1} - \gamma_N\hbar(1-\sigma_X)\mathcal{B}m_{X_2}$$

As m_A, m_{X_1} and m_{X_2} can each be $\pm\frac{1}{2}$, there are a total of 6 energy levels, two of which are two-fold degenerate, for a total of eight levels. These are shown on the left of Figure 13.1. The allowed transitions are indicated by arrows. There are 7 transitions, but only 2 transition frequencies. This follows from the selection rule for magnetic resonance transitions, which is $\Delta(m_1 + m_2) = \pm 1$. The shorter arrows represent the X transitions, the larger arrows the A transitions. Spin–spin splitting perturbs these levels as follows:

$$E_{\text{spin-spin}} = hJm_Am_{X_1} + hJm_Am_{X_2}$$

	$\alpha\alpha_1\alpha_2$	$\alpha\alpha_1\beta_2$	$\alpha\beta_1\alpha_2$	$\alpha\beta_1\beta_2$
$E_{\text{spin-spin}}$	$\frac{1}{2}hJ$	0	0	$-\frac{1}{2}hJ$
	$\beta\beta_1\beta_2$	$\beta\alpha_1\beta_2$	$\beta\beta_1\alpha_2$	$\beta\alpha_1\alpha_2$

There are again a total of 6 energy levels (two of which are two-fold degenerate), but they are perturbed by the amounts in the above chart. The perturbed levels are shown on the right in the figure below.

The frequencies of the X transitions are changed by $\pm\frac{1}{2}J$, the frequencies of the A transitions by $-J$, 0, $+J$. A stick diagram representing the spectrum is shown in Figure 13.2.

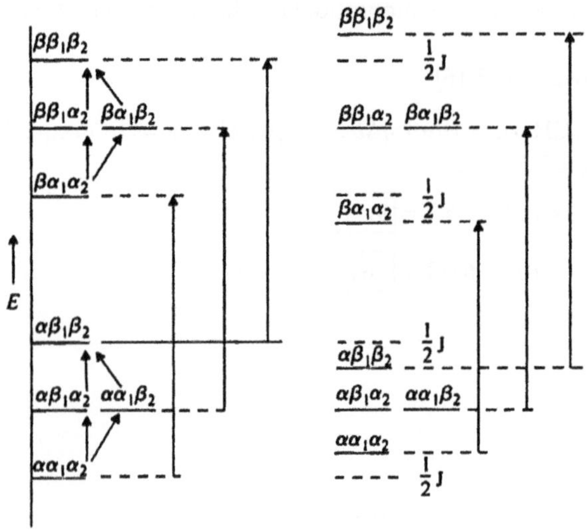

Figure 13.1

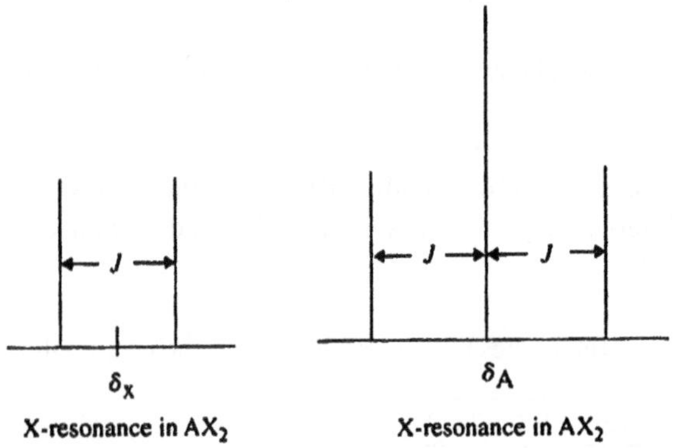

Figure 13.2

E13.30

$$\nu - \nu^\circ = \nu^\circ\delta \times 10^{-6}$$

$$|\Delta\nu| \equiv (\nu - \nu^\circ)(\text{CHO}) - (\nu - \nu^\circ)(\text{CH}_3)$$
$$= \nu(\text{CHO}) - \nu(\text{CH}_3)$$
$$= \nu^\circ\,[\delta(\text{CHO}) - \delta(\text{CH}_3)] \times 10^{-6}$$
$$= (9.80 - 2.20) \times 10^{-6}\nu^\circ = 7.60 \times 10^{-6}\,\nu^\circ \text{ [chemical shift values from Exercise 13.25]}$$

(a) $\nu^\circ = 300\text{ MHz}\ |\Delta\nu| = 7.60 \times 10^{-6} \times 300\text{ MHz} = \boxed{2.28\text{ kHz}}$

(b) $\nu^\circ = 500\text{ MHz}\ |\Delta\nu| = 7.60 \times 10^{-6} \times 500\text{ MHz} = \boxed{3.80\text{ kHz}}$

(a) The spectrum is shown in Figure 13.3.

(b) When the frequency is changed to 500 MHz, the separation of the CH_3 and CHO resonance increases (3.80 kHz), the fine structure remains unchanged, and the intensity increases.

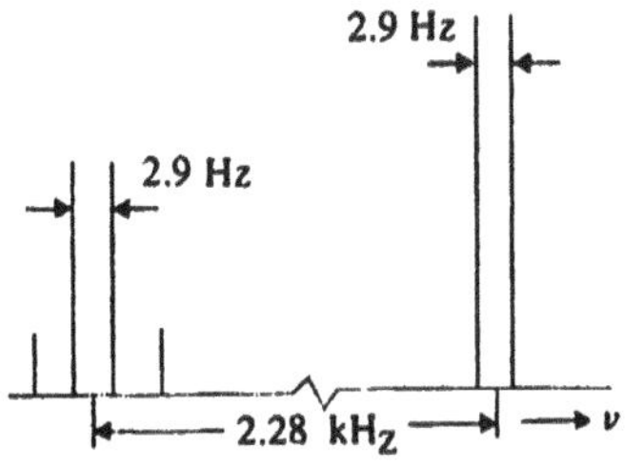

Figure 13.3

E13.31 There is a typographical error in the statement of the problem. The second inequality should be the greater than symbol, >. Thus $J_{AM} > J_{AX} > J_{MX}$. In order to draw the figure the relative values of all splitting constants have to be known. Hence, the A, M, and X resonances lie in distinctly different groups. The A, M, and X resonances lie in distinctly different groups. The A resonance is split into a 1:2:1 triplet by the M nuclei, and each line of that triplet is split into a 1:4:6:4:1 quintet by the X nuclei, (with $J_{AM} > J_{AX}$). The M resonance is split into a 1:3:3:1 quartet by the A nuclei and each line is split into a quintet by the X nuclei (with $J_{AM} > J_{MX}$). The X resonance is split into a quartet by the A nuclei and then each line is split into a triplet by the M nuclei (with $J_{AX} > J_{MX}$). The spectrum is sketched in Figure 13.4.

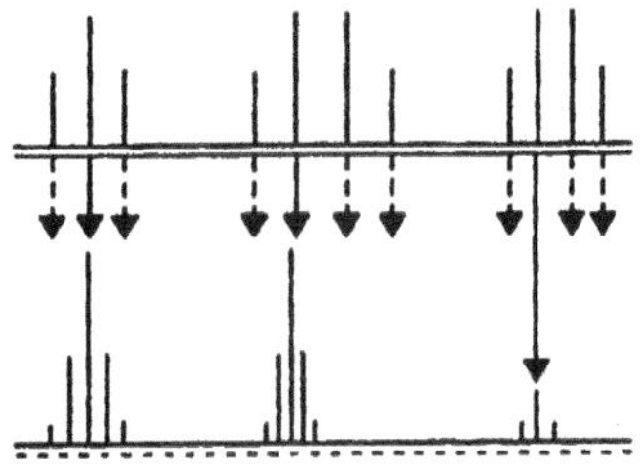

Figure 13.4

E13.32 See Self-test 13.3 and the two paragraphs above the Self-test as well as Pascal's triangle (**1**) for the approach to the solution to this exercise. That Self-test and that figure are followed in the process of determining the intensity pattern in the fine structure of the A resonance in an AX_N NMR spectrum. See the table below for the version of Pascal's triangle for up to 5 spin-1 nuclei. Each number in the table is the sum of the three ($I = 1, 2I + 1 = 3$) numbers in the three adjacent cells above it (one to the right, one in the middle, and one to the left).

					1					
				1	1	1				
			1	2	3	2	1			
		1	3	6	7	6	3	1		
	1	4	10	16	19	16	10	4	1	
1	5	15	30	45	51	45	30	15	5	1

E13.33
$$^3J_{HH} = A + B\cos\phi + C\cos 2\phi \quad [13.20]$$

$$\frac{d}{d\phi}(^3J_{HH}) = -B\sin\phi - 2C\sin 2\phi = 0$$

This equation has a number of solutions:

$$\phi = 0,\ \phi = n\pi,\ \phi = \pi - \arccos\left(\frac{B}{4C}\right) = \arccos\left(\frac{B}{4C}\right)$$

The first two are trivial solutions.

$$\text{If } \phi = \arccos\left(\frac{B}{4C}\right) \text{ then, } \sin\phi = \sqrt{1 - \frac{B^2}{16C^2}}$$

$$\sin 2\phi = 2\sin\phi\cos\phi = 2\sqrt{1 - \frac{B^2}{16C^2}}\left(\frac{B}{4C}\right)$$

$$B\sin\phi + 2C\sin 2\phi = B\sqrt{1 - \frac{B^2}{16C^2}} + 4C\sqrt{1 - \frac{B^2}{16C^2}}\left(\frac{B}{4C}\right) = 0$$

So, $\dfrac{B}{4C} = \cos\phi$ clearly satisfies the condition for an extremum.

The second derivative is

$$\frac{d^2}{d\phi^2}(^3J_{HH}) = -B\cos\phi - 4C\cos 2\phi = -B\cos\phi - 4C(2\cos^2\phi - 1)$$

$$= -B\left(\frac{B}{4C}\right) - 4C\left(2\frac{B^2}{16C^2} - 1\right) = -\frac{B^2}{4C} - \frac{2B^2}{4C} + 4C$$

This quantity is positive if

$$16C^2 > 3B^2$$

This is certainly true for typical values of B and C, namely $B = -1$ Hz and C = 5 Hz. Therefore, the condition for a minimum is as stated, namely, $\boxed{\cos\phi = B/4C}$.

E13.34 The frequency difference between the two signals is

$$\Delta\nu = [\nu^\circ + \nu^\circ \times 10^{-6}\,\delta'] - [\nu^\circ + \nu^\circ \times 10^{-6}\,\delta] = \nu^\circ \times 10^{-6}\,(\delta' - \delta) \quad [13.18]$$

$$\tau = \frac{2^{1/2}}{\pi\Delta\nu}[13.21] = \frac{2^{1/2}}{\pi\nu^\circ \times 10^{-6}(\delta' - \delta)}$$

$$= \frac{2^{1/2}}{\pi(500 \times 10^{-6}\ \text{s}^{-1}) \times 10^{-6}(4.8 - 2.7)} = 4.3 \times 10^{-4}\ \text{s}$$

Therefore, the signals merge when the conversion rate is greater than about $1/\tau = \boxed{2.6 \times 10^3\ \text{s}^{-1}}$

E13.35 The desired result is the linear equation:

$$[\text{I}]_0 = \frac{[E]_0\Delta\nu}{\delta\nu} - K_\text{I},$$

so the first task is to express quantities in terms of $[\text{I}]_0$, $[\text{E}]_0$, $\Delta\nu$, $\delta\nu$, and K_I, eliminating terms such as [I], [EI], [E], ν_I, ν_EI, and ν. (Note: symbolic mathematical software is helpful here.) Begin with ν:

$$\nu = \frac{[\text{I}]}{[\text{I}] + [\text{EI}]}\nu_\text{I} + \frac{[\text{EI}]}{[\text{I}] + [\text{EI}]}\nu_\text{EI} = \frac{[\text{I}]_0 - [\text{EI}]}{[\text{I}]_0}\nu_\text{I} + \frac{[\text{EI}]}{[\text{I}]_0}\nu_\text{EI},$$

where we have used the fact that total I (i.e. free I plus bound I)is the same as initial I. Solve this subject to the condition that it must also be much greater than [EI]:

$$[\mathrm{EI}] = \frac{[\mathrm{I}]_0(\nu - \nu_\mathrm{I})}{\nu_\mathrm{EI} - \nu_\mathrm{I}} = \frac{[\mathrm{I}]_0\delta\nu}{\Delta\nu},$$

where in the second equality we notice that the frequency differences that appear are the ones defined in the problem. Now, take the equilibrium constant:

$$K_\mathrm{I} = \frac{[\mathrm{E}][\mathrm{I}]}{[\mathrm{EI}]} = \frac{([\mathrm{E}]_0 - [\mathrm{EI}])([\mathrm{I}]_0 - [\mathrm{EI}])}{[\mathrm{EI}]}$$
$$\approx \frac{([\mathrm{E}]_0 - [\mathrm{EI}])[\mathrm{I}]_0}{[\mathrm{EI}]}$$

We have used the fact that total I is much greater than total E (from the condition that $[\mathrm{I}]_0 \gg [\mathrm{E}]_0$), so it must also be much greater than [EI], even if all E binds I. Now solve this for $[\mathrm{E}]_0$:

$$[\mathrm{E}]_0 = \frac{K_\mathrm{I} + [\mathrm{I}]_0}{[\mathrm{I}]_0}[\mathrm{EI}] = \left(\frac{K_\mathrm{I} + [\mathrm{I}]_0}{[\mathrm{I}]_0}\right)\left(\frac{[\mathrm{I}]_0\delta\nu}{\Delta\nu}\right)$$
$$= \frac{(K_\mathrm{I} + [\mathrm{I}]_0)\delta\nu}{\Delta\nu}.$$

The expression contains the desired terms and only those terms. Solving for $[\mathrm{I}]_0$ yields:

$$\boxed{[\mathrm{I}]_0 = \frac{[\mathrm{E}]_0\Delta\nu}{\delta\nu} - K_\mathrm{I}}$$

which would result in a straight line with slope $[\mathrm{E}]_0\Delta\nu$ and y-intercept K_I if one plots $[\mathrm{I}]_0$ against $1/\delta\nu$.

E13.36 Analogous to precession of the magnetization vector in the laboratory frame due to the presence of $\mathcal{B}_0$ that is

$$\nu_\mathrm{L} = \frac{\gamma\mathcal{B}_0}{2\pi} \quad \text{[Section 13.6]},$$

there is a precession in the rotating frame, due to the presence of $\mathcal{B}_1$, namely

$$\nu = \frac{\gamma_\mathrm{N}\mathcal{B}_1}{2\pi} \quad \text{[Section 13.6]} \quad \text{or} \quad \omega_1 = \gamma\mathcal{B}_1 \quad [\omega = 2\pi\nu]$$

Since ω is an angular frequency, the angle through which the magnetization vector rotates is

$$\theta = \gamma_\mathrm{N}\mathcal{B}_1 t = \frac{g_I\mu_\mathrm{N}}{\hbar}\mathcal{B}_1 t$$

and $$\mathcal{B}_1 = \frac{\theta\hbar}{g_I\mu_\mathrm{N}t} = \frac{(\frac{1}{2})\times(1.055\times10^{-34}\ \mathrm{J\ s})}{(5.586)\times(5.051\times10^{-27}\ \mathrm{J\ T^{-1}})\times(1.0\times10^{-5}\ \mathrm{s})}$$
$$= \boxed{5.9\times10^{-4}\ \mathrm{T}}$$

(A proton has been assumed in the calculation above.)

E13.37 Methionine-105 is in the vicinity of both typtophan-28 and tyrosine-23 but the latter two residues are not in the vicinity of each other. The methionine residue may lie between them as represented in Figure 13.5.

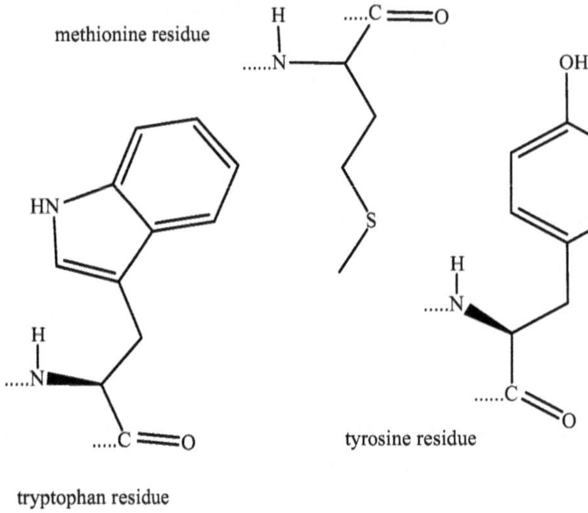

Figure 13.5

E13.38 We use $\nu = \frac{\gamma_N \mathcal{B}_{loc}}{2\pi} = \frac{\gamma_N}{2\pi}(1-\sigma)\mathcal{B}$ [13.16]

where $\mathcal{B}$ is the applied field.

Because shielding constants are quite small (a few parts per million) compared to 1, we may write for the purposes of this calculation

$$\nu = \frac{\gamma_N \mathcal{B}}{2\pi}$$

$$\nu_L - \nu_R = 100\ \text{Hz} = \frac{\gamma_N}{2\pi}(\mathcal{B}_L - \mathcal{B}_R)$$

$$\mathcal{B}_L - \mathcal{B}_R = \frac{2\pi \times 100\ \text{s}^{-1}}{\gamma_N}$$

$$= \frac{2\pi \times 100\ \text{s}^{-1}}{26.752 \times 10^7\ \text{T}^{-1}\ \text{s}^{-1}} = 2.35 \times 10^{-6}\ \text{T}$$

$$= 2.35\ \mu\text{T}$$

The field gradient required is then

$$\frac{2.35\ \mu\text{T}}{0.08\ \text{m}} = \boxed{29\ \mu\text{T m}^{-1}}$$

Note that knowledge of the spectrometer frequency, applied field, and the numerical value of he chemical shift (because constant) is not required.

E13.39 (Assume that the radius of the disk is 1 unit. The volume of each slice is proportional to the length of slice multiplied by δx. See Figure 13.6(a).

Length of slice at $x = 2 \sin\theta$
$x = \cos\theta$
$\theta = \arccos x$
x ranges from -1 to $+1$
Length of slice at $x = 2 \sin(\arccos x)$

Plot $f(x) = 2 \sin(\arccos x)$ against x between the limits -1 and $+1$. The plot is shown in Figure 13.6(b).

The volume at each value of x is proportional to $f(x)$ and the intensity of the MRI signal is proportional to the volume, so the figure represents the absorption intensity for the MRI image of the disk.

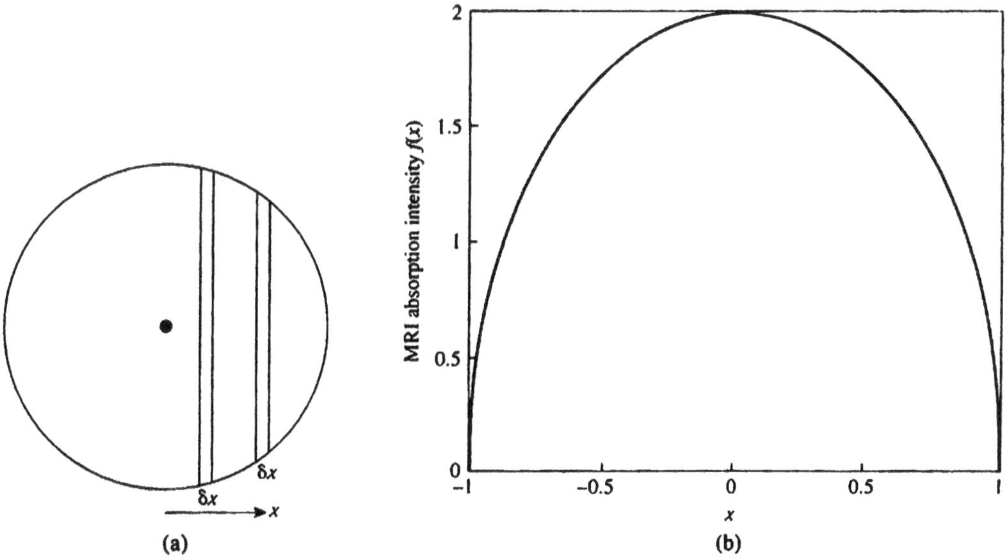

Figures 13.6(a) & (b)

E13.40 The proton COSY spectrum of 1-nitropropane shows that (a) the C_a–H resonance with $\delta = 4.3$ shares a crosspeak with the C_b–H resonance at $\delta = 2.1$ and (b) the C_b–H resonance with $\delta = 2.1$ shares a crosspeak with the C_c–H resonance at $\delta = 1.1$. Off-diagonal peaks indicate coupling between the Hs on various carbons. Thus peaks at (4,2) and (2,4) indicate that the Hs on the adjacent CH_2 units are coupled. The peaks at (1,2) and (2,1) indicate that the Hs on CH_3 and central CH_2 units are coupled.

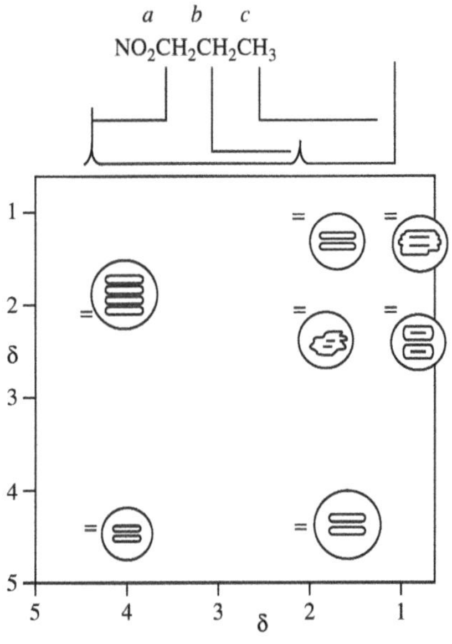

Figure 13.7

E13.41 See Figure 13.8. Only the H(N) and H(C_α) protons and the H(C_α) and H(C_β) protons are expected to show coupling. This results in a simple COSY spectrum with only two off-diagonals, one at (8.25 ppm, 4.35 ppm) and the other at (4.35 ppm, 1.39 ppm).

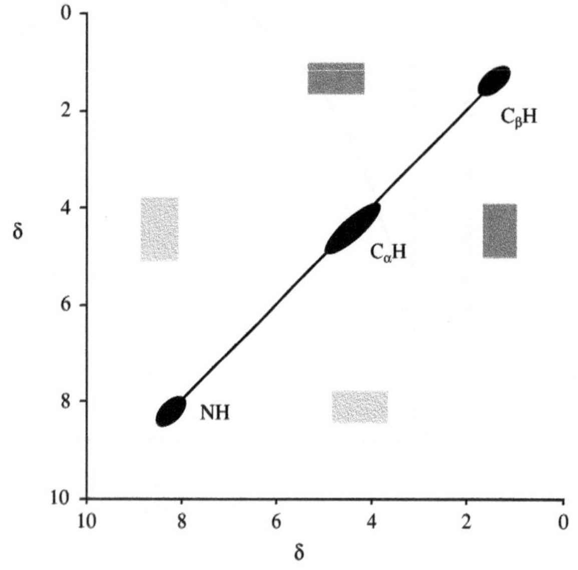

Figure 13.8

E13.42

$$g = \frac{h\nu}{\mu_B \mathcal{B}_0} \quad [13.23]$$

We shall often need the value

$$\frac{h}{\mu_B} = \frac{6.62608 \times 10^{-34}\ \text{J Hz}^{-1}}{9.27402 \times 10^{-24}\ \text{J T}^{-1}} = 7.14478 \times 10^{-11}\ \text{T Hz}^{-1}$$

Then, in this case

$$g = \frac{(7.14478 \times 10^{-11}\ \text{T Hz}^{-1}) \times (9.2231 \times 10^{9}\ \text{Hz})}{329.12 \times 10^{-3}\ \text{T}} = \boxed{2.0022}$$

E13.43

$$a = \mathcal{B}(\text{line 3}) - \mathcal{B}(\text{line 2}) = \mathcal{B}(\text{line 2}) - \mathcal{B}(\text{line 1})$$

$$\left.\begin{array}{l} \mathcal{B}_3 - \mathcal{B}_2 = (334.8 - 332.5)\ \text{mT} = 2.3\ \text{mT} \\ \mathcal{B}_2 - \mathcal{B}_1 = (332.5 - 330.2)\ \text{mT} = 2.3\ \text{mT} \end{array}\right\} a = \boxed{2.3\ \text{mT}}$$

Use the center line to calculate g

$$g = \frac{h\nu}{\mu_B \mathcal{B}_0}\ [13.23] = (7.14478 \times 10^{-11}\ \text{T Hz}^{-1}) \times \frac{9.319 \times 10^{9}\ \text{Hz}}{332.5 \times 10^{-3}\ \text{T}} = \boxed{2.002\overline{5}}$$

E13.44 We construct Figure 13.9(a) for $\cdot CH_3$ and Figure 13.9(b) for $\cdot CD_3$. The predicted intensity distribution is determined by counting the number of overlapping lines of equal intensity from which the hyperfine line is constructed.

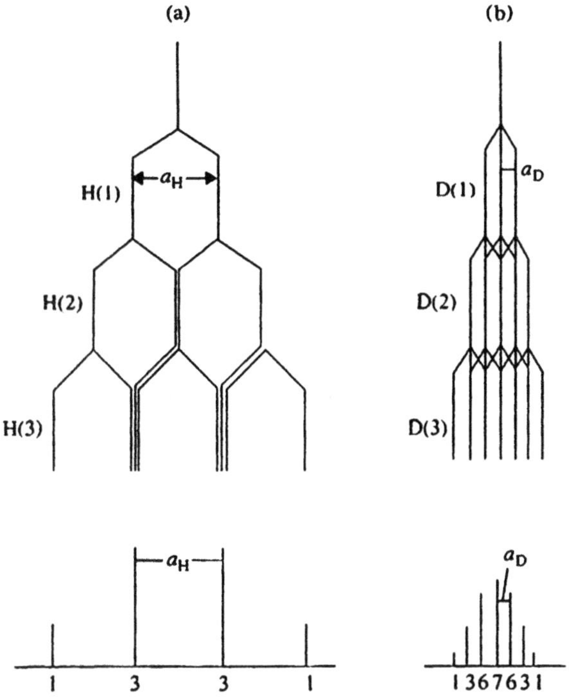

Figures 13.9(a) & (b)

E13.45 $\mathcal{B}_0 = \dfrac{h\nu}{g\mu_B}[13.23] = \dfrac{7.14478 \times 10^{-11}}{2.0025}\ \text{T Hz}^{-1} \times \nu$ [Exercise 13.42] $= 35.68\ \text{mT} \times (\nu/\text{GHz})$

(a) $\nu = 9.302$ GHz, $\mathcal{B}_0 = \boxed{331.9\ \text{mT}}$

(b) $\nu = 33.67$ GHz, $\mathcal{B}_0 = 1201\ \text{mT} = \boxed{1.201\ \text{T}}$

E13.46 If a radical contains N equivalent nuclei with spin quantum number I, then there are $2NI + 1$ hyperfine lines in the EPR spectrum. For $N = 2$ and five hyperfine lines, $\boxed{I = 1}$. The intensity ratio for the lines is 1:2:3:2:1 as demonstrated in Figure 13.10.

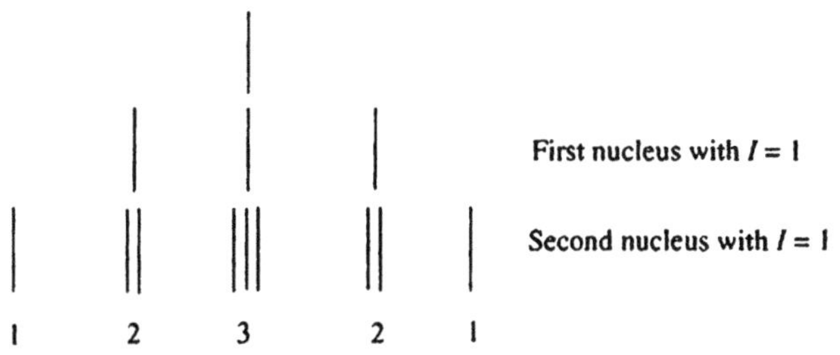

Figure 13.10

E13.47 See Example 13.3 and text Figure 13.43 for the approach to the solution to this exercise. See the table below for the version of Pascal's triangle for up to 5 spin-3/2 nuclei. Each number in the table is the sum of the four ($I = 3/2$, $2I + 1 = 4$) numbers above it (2 to the right and 2 to the left).

															1															
												1		1		1		1												
									1		2		3		4		3		2		1									
						1		3		6		10		12		12		10		6		3		1						
			1		4		10		20		31		40		44		40		31		20		10		4		1			
1		5		15		35		65		101		135		155		155		135		101		65		35		15		5		1

E13.48 When spin label molecules approach to within 800 pm, orbital overlap of the unpaired electrons and dipolar interactions between magnetic moments cause an exchange coupling interaction between the spins. The electron exchange process occurs at a rate that increases as concentration increases. Thus, the process has a lifetime that is too long at low concentrations to affect the "pure" ESR signal. As the concentration increases, the linewidths increase until the triplet coalesces into a broad singlet. Further increase of the concentration decreases the exchange lifetime and therefore the linewidth of the singlet.

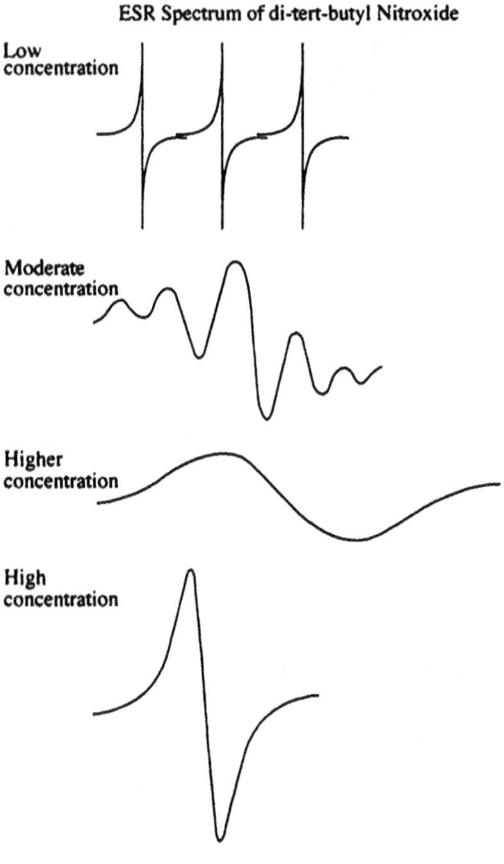

Figure 13.11

When spin labels within biological membranes are highly mobile, they may approach closely and the exchange interaction may provide the ESR spectra with information that mimics the moderate and high concentration signals above.

Solutions to projects

P13.49 This is an essay left for the student to write.

P13.50 (a) In Section 13.6 we find the key concepts and statement: "If the $\mathcal{B}_1$ field is applied in a pulse of duration $\pi/2\gamma_N\mathcal{B}_1$, the magnetization tips through 90° in the rotating frame and we say that we have applied a 90° pulse (or a "$\pi/2$ pulse")." Thus, since a 180° pulse takes twice as long,

$$\mathcal{B}_1 = \frac{\pi}{\gamma_N t} = \frac{\pi}{(26.752\times10^7\ \mathrm{T^{-1}\ s^{-1}})\times(12.5\times10^{-6}\ \mathrm{s})} = \boxed{0.934\ \mathrm{mT}}$$

Another way to arrive at this result is to recognize that a rotation through the angle θ in radians (2π radians is equivalent to 360°) is related to the angular frequency ω by the definition $\omega = \theta/t$, while the angular frequency and frequency are related by the expression $\omega = 2\pi\nu$. Furthermore, the precession frequency is $\nu = \gamma_N\mathcal{B}_1/2\pi$ [13.12] at resonance. Putting these together gives the desired relation between the duration of the pulse, t, and the angle of rotation: $t = \theta/\gamma_N\mathcal{B}_1$.

(b) Figure 13.12 is a sketch of the effects of the pulse sequence on the magnetization vector. For supplemental reading see K.R. Willams and R.W. King, *Journal of Chemical Education*, **67** (4), A93, 1990; The Fourier Transform in Chemistry–NMR: Part 3 Multiple-Pulse Experiments.

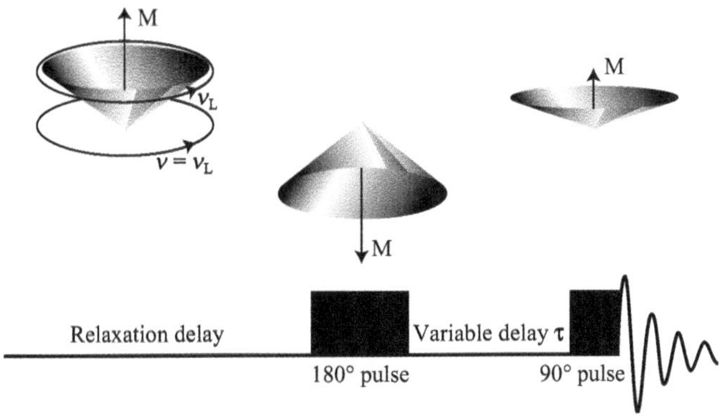

Figure 13.12

(c) The FID signal is generated after application of the 90° pulse because of the location of the detector coil, which is shown in text Figure 13.22(b). The detector coil is in the plane that is perpendicular to both the magnetic field B_0 and the magnetization vector M (prior to the 90° pulse in the inversion recovery technique) so M must be rotated into the plane of the detector in order to register a signal.

(d) After the 180° pulse and before the 90° pulse, the difference between magnetization in the direction of the B_0 field (M_z, where z is the direction of B_0) and the initial magnetization (M_0) decays exponentially with the time constant T_1, which is called the **longitudinal relaxation time**. This relaxation process involves giving up energy to the surroundings (the "lattice") as β spins revert to α spins so the time constant T_1 is also called the **spin-lattice relaxation time**:

$$M_z(t) - M_0 \propto e^{-t/T_1}$$

The constant of proportionality is chosen so that $M_z(0) = -M_0$, giving

$$M_z(t) = M_0(1 - 2e^{-t/T_1})$$

After application of the 90° pulse at time $t = \tau$, the magnetization is rotated into the xy-plane and the detector registers the y-component, which varies as the cosine of the precession angle θ. Thus,

$$\text{FID signal} \propto M_z(\tau)\cos\theta = M_0(1 - 2e^{-\tau/T_1})\cos(2\pi\nu t) \quad t \ge \tau$$

(e) The relation between the FID signal and the ratio τ/T_1 suggests measuring the FID signal over a range of τ values with all measurements at the same value of the precession angle. A plot of the signal against τ will show the exponential signal decay to equilibrium in the field $\mathcal{B}_0$ and a regression fit of the plot yields the T_1 value. Figure 13.13 illustrates a typical plot for a kidney tissue sample for which $T_1 = 0.65$ s with $B_0 = 1.5$ T.

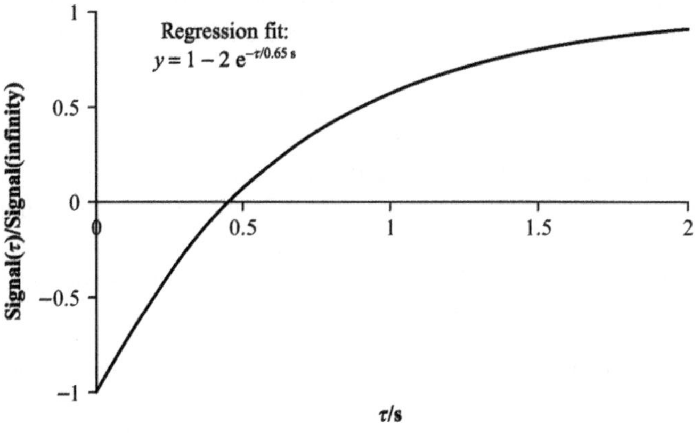

Figure 13.13

P13.51 (a) Our phenoxy radical calculations are performed with Spartan '06™ software using the following density functional method with a doublet spin multiplicity:

Job type: Geometry optimization.
Method: UB3LYP
Basis set: 6-311G*
Number of shells: 61
Number of basis functions: 165
SCF model: An unrestricted hybrid HF-DFT SCF calculation
using Pulay DIIS + Geometric Direct Minimization

The Mulliken population analysis is used to find the atomic charges and spin densities, which are summarized in Figures 13.14 and 13.15, respectively. The spin density, the electron fraction per a_0^3, is the difference between the number of α and β electrons at a point and, therefore, indicates the location of the unpaired electron in the radical. The distribution of atomic charges with the resultant dipole moment of 4.6430 D is useful for identifying sites of electrostatic interaction between residues.

(b) Examination of Figure 13.15 indicates that there are two equivalent hydrogen atoms attached to carbon atoms that are *ortho* to the carbon of C−O. These *ortho* carbon atoms each have a spin density of 0.300 and, according to the McConnell equation [13.26], the *ortho* hydrogens split the aromatic radical with the EPR hyperfine constant $a(\mathrm{H}_{ortho}) = \rho Q = (0.300) \times (2.25\ \mathrm{mT}) = 0.675\ \mathrm{mT}$.

There are also two equivalent hydrogen atoms attached to carbon atoms that are meta to the carbon of C−O. These *meta* carbon atoms each have a spin density of −0.152, a magnitude that is effectively half the magnitude of the spin density of the *ortho* carbons. Consequently, $a(\mathrm{H}_{meta}) = -\frac{1}{2}a(\mathrm{H}_{ortho})$. We now construct the expected EPR spectrum of the phenoxy radical but neglect the interaction between the methyl hydrogens and the large aromatic spin density on the para carbon.

The two equivalent *ortho* hydrogens split the radical spin flip resonance into three lines of relative intensity 1:2:1 that are separated by 0.675 mT. Each of these lines is split by the two equivalent meta hydrogens into three lines of relative intensity 1:2:1 that are separated by $\frac{1}{2} \times (0.675\ \mathrm{mT})$. As shown in Figure 13.16, this yields an EPR spectrum consisting of $\boxed{\text{seven lines separated by } \frac{1}{2} \times (0.675\ \text{mT})}$ with the relative intensities $\boxed{1{:}2{:}3{:}4{:}3{:}2{:}1}$.

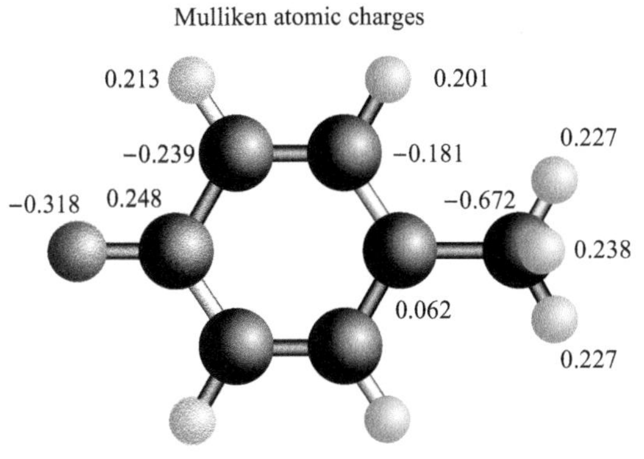

Figure 13.14

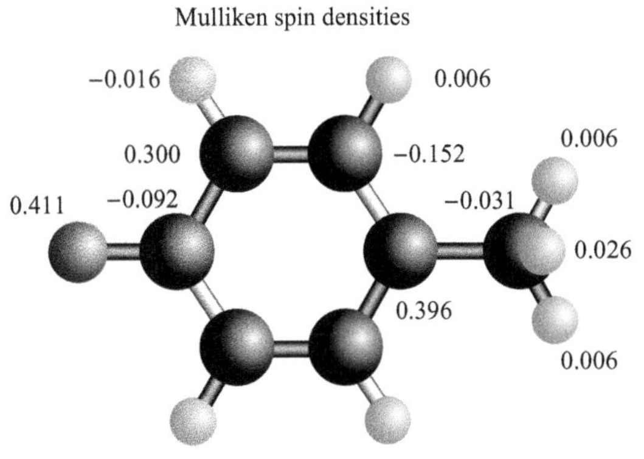

Figure 13.15

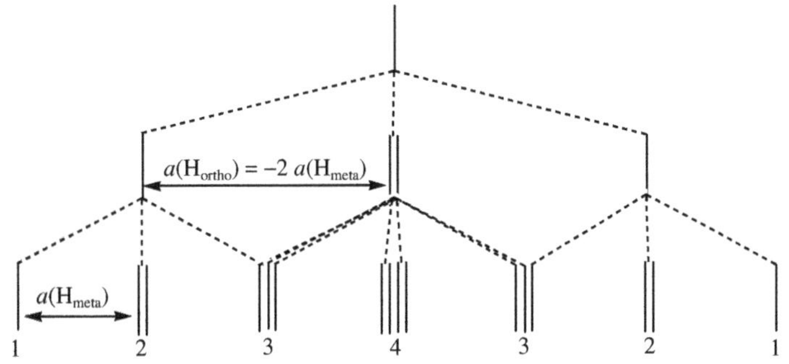

Figure 13.16